*This book is dedicated to the memory
of Wallace Waterfall (1900–1974)*

Benchmark Papers
in Acoustics

Series Editor: R. Bruce Lindsay
Brown University

**Benchmark Papers
in Acoustics / 10**

A BENCHMARK® Book Series

ARCHITECTURAL
ACOUSTICS

Edited by
THOMAS D. NORTHWOOD
National Research Council of Canada

Dowden, Hutchinson
& Ross, Inc.

STROUDSBURG, PENNSYLVANIA

79 78 77   1 2 3 4 5
Manufactured in the United States of America.

LIBRARY OF CONGRESS CATALOGING IN PUBLICATION DATA
Main entry under title:
Architectural acoustics.
   (Benchmark papers in acoustics ; 10)
   Articles chiefly in English, some in French or German, with English
summaries.
   Includes indexes.
   1. Architectural acoustics—Addresses, essays, lectures. I. Northwood,
Thomas D.
NA2800.A68   729'.29   76-54182
ISBN 0-87933-257-3

Exclusive Distributor : **Halsted Press**
A Division of John Wiley & Sons, Inc.
ISBN: 0-470-99051-1

SERIES EDITOR'S FOREWORD

The "Benchmark Papers in Acoustics" constitute a series of volumes that make available to the reader in carefully organized form important papers in all branches of acoustics. The literature of acoustics is vast in extent and much of it, particularly the earlier part, is inaccessible to the average acoustical scientist and engineer. These volumes aim to provide a practical introduction to this literature, since each volume offers an expert's selection of the seminal papers in a given branch of the subject, that is, those papers that have significantly influenced the development of that branch in a certain direction and introduced concepts and methods that possess basic utility in modern acoustics as a whole. Each volume provides a convenient and economical summary of results as well as a foundation for further study for both the person familiar with the field and the person who wishes to become acquainted with it.

Each volume has been organized and edited by an authority in the area to which it pertains. In each volume there is provided an editorial introduction summarizing the technical significance of the field being covered. Each article is accompanied by editorial commentary, with necessary explanatory notes, and an adequate index is provided for ready reference. Articles in languages other than English are either translated or abstracted in English. It is the hope of the publisher and editor that these volumes will constitute a working library of the most important technical literature in acoustics of value to students and research workers.

The present volume, *Architectural Acoustics*, has been edited by Thomas D. Northwood, Head of the Noise and Vibration Section of the Division of Building Research of the National Research Council of Canada. In its carefully chosen 30 articles it covers the principal features of the acoustics of buildings, from the standpoint of both techniques and theories as well as the subjective effects on people. Much emphasis is laid on the important topic of sound insulation in buildings. Most of the material is concerned with modern developments of the subject, though the historical background is not neglected. Adequate editorial commentaries serve to place all papers in proper perspective, and there is an extensive bibliography of associated literature. The five foreign-language articles are accompanied by explanatory abstracts in English. The student of architectural acoustics will find this volume a mine of useful information on both the practical and theoretical sides of the discipline.

R. BRUCE LINDSAY

PREFACE

The aim of this book is to present a collection of original writings that, in sum, characterize the development and current status of architectural acoustics. It is an impossible task to squeeze into the confines of one reasonable-sized book all the contributions that seem essential to such a compilation. Several of the papers selected have been abridged, to some extent, to permit inclusion of as many viewpoints as possible. It is hoped that the reader will fill the gaps by seeking out the additional references mentioned in the commentaries and in the papers themselves.

Most of the papers included here were, at the time of writing, important new contributions to architectural acoustics. The exceptions are major review papers that served to sum up certain cardinal topics. It is believed that all selections have something useful and stimulating to say to the modern student and practitioner of architectural acoustics.

In a book addressed primarily to English-speaking readers, one problem for which there is no fully satisfactory solution is the handling of material published originally in another language. To retain the flavor of the original papers, these have been reproduced here in the original language, but supplemented by English summaries. A related problem is the fact that many developments in acoustics proceeded more or less concurrently in at least two places and in two languages. In most such circumstances, the editor has chosen the English-speaking version but with, it is hoped, adequate mention of the parallel developments elsewhere.

The editor is indebted to many confreres, including authors, for advice on the selection of papers and on their preparation for the book. Special appreciation must be expressed to his colleagues, J. H. Rainer for assistance in preparing summaries of German papers, Verna Morrell for assistance in preparing the manuscript, and to the Division of Building Research of the National Research Council of Canada for various supporting services.

THOMAS D. NORTHWOOD

CONTENTS

Contents

PART IV: SOUND INSULATION—TECHNIQUE AND THEORY

CONTENTS BY AUTHOR

ARCHITECTURAL ACOUSTICS

INTRODUCTION

Architectural acoustics is concerned with the sounds people hear in buildings. Included are the physical aspects of sound propagation in enclosed spaces (room acoustics) and through separating structures (sound insulation), but always with attention to the central question of how people perceive and react to the sounds that reach them.

The single unifying concept in architectural acoustics is the perception of a signal (wanted or unwanted) over a level of background noise. Most of what may be called "signals" fluctuate in level and in frequency content, and generally the information they carry is implicit in the fluctuations. The full understanding of speech, for instance, requires perception of sounds varying over about 30 dB in level and over the frequency range of 200 to 5000 Hz. Thus, if speech is a wanted sound, the background noise should be below this 30 dB range of speech levels. Conversely, if speech is an unwanted sound, its peak levels should be below the background noise level.

Background noise may arise from extraneous sources—e.g., road traffic or ventilation noise—or from locally produced sounds. The same sound—speech, for example—may play the role of either signal or noise. In an auditorium, the concept of noise might extend to include reflected sounds that reach the listener so late that they blur subsequent direct sounds. Controlling the *reverberation* resulting from the ensemble of such sounds constitutes a major element in auditorium design.

The fourfold subdivision, physical/subjective aspects of room-acoustics/sound-insulation, constitutes the main structure of this book. In many instances, the assignment of a particular paper to a particular subdivision may seem arbitrary, since many deal with relationships between the physical and subjective processes. Nevertheless it permits a reasonably coherent analysis of each of the four topics.

1

The book is limited, perforce, to only the most basic phenomena relating to acoustics in buildings. It might be appropriate to mention some of the fringe topics that had to be omitted. In room acoustics, as presented here, attention has been on the conservation of *wanted* sounds, as for instance the transmission of speech or music from a source to an audience. On the other hand, in some rooms—such as waiting rooms, restaurants, open-plan offices—the objective is the reverse one of reducing the perception of *unwanted* sounds from other parts of the same room. This aspect of room acoustics, not covered by the papers assembled in this volume, requires tactics more or less opposite to those for conserving sounds. For example, instead of trying to keep background noise low, one may deliberately add noise to mask distant speech signals. A third category of room acoustics is that of the industrial plant or similarly noisy space where the object is to improve the working environment by reducing the level of reverberant sound.

A matter of considerable current importance is insulation from external noise sources such as aircraft and road traffic. This problem differs from the indoor sound-insulation case in three respects: the source is in a specified direction, as compared to the randomly reverberant sound usually assumed for indoors; the most important elements are usually windows and doors; and the noise spectra are usually dominated by lower frequencies.

Only briefly touched on (Paper 16) is the extensive literature relating to the special rooms used in acoustical testing. These are of interest not only because of the test methods for which they were designed but because they exemplify the limiting cases of ordinary room acoustics. One extreme is the reverberation chamber, in which one attempts to produce a perfectly diffuse reverberant sound field over as wide a frequency range as possible. Such a room provides a definitive environment in which to measure the sound-insulating properties of partitions and the sound-absorbing properties of acoustical materials. In a finite room, however, it is difficult to achieve such an ideal sound field; and it is equally difficult to devise quantitative ways of testing it. At the other extreme is the anechoic room, in which, by absorbing all sound reaching the boundaries, one can simulate free space and concentrate attention on the properties of sound sources.

Finally, the interfaces with other Benchmark books might be noted. Historically, this volume begins essentially with W. C. Sabine, although Sabine's most famous paper forms the last one in R. Bruce Lindsay's volume *Acoustics: Historical and Philosophical Development.** On the

---

*Lindsay, R. Bruce. *Acoustics: Historical and Philosophical Development*. Benchmark Papers in Acoustics Series, vol. 2. Stroudsburg, Pa.: Dowden, Hutchinson & Ross, Inc., 1973.

other hand, a few stray quotations from earlier writers are included here. Similarly, consideration of the problem of controlling or confining the vibration and noise of machinery in a building is left to be included in a Benchmark volume dealing generally with noise.

Part I

ROOM ACOUSTICS—SUBJECTIVE ASPECTS

Editor's Comments
on Papers 1 Through 10

In subjective terms, what are the acoustical attributes desired in a concert hall, a theater, or a lecture hall? Such questions have been pondered presumably since building began, and qualitative answers have existed at least since the days of Vitruvius (ca 25 BC), whose comments are quoted in Paper 11. These questions were addressed in the nineteenth century by Joseph Henry, T. Roger Smith, and W. C. Sabine.

> It must be apparent, also, that the continuance of a single sound, and the tendency to confusion in distinct perception, will depend on several conditions; ... first, on the size of the apartment; secondly, on the strength of the sound or the intensity of the impulse; thirdly, on the position of the reflecting surfaces; and fourthly, on the nature of the material of the reflecting surfaces. (Henry[1])
>
> . . . . .
>
> Music depends entirely upon the relations that subsist in pitch between different sustained sounds or notes, and upon the momentary impression of one note or chord resting in the memory of the ear when the next is sounded. It is not isolated notes, of however fine a quality, but the succession of such sounds or the blending them together that forms music; hence in all music there must be some relation between each note or chord, except the first of a passage, and those that have gone before, and some recollection of them. In articulate speech, on the other hand, each syllable is a distinct concatenation of two or more sounds only, and though it may be combined with other syllables to make a word, yet it in no way depends upon them for its own completeness.
>
> Hence upon no music, however rapid, can the consequences of prolonging the impression of sounds beyond the time that they themselves actually last, exercise so injurious an influence as upon spoken words, where each syllable ought to be heard distinctly and separately, and where only the combinations of the letters that go to form single syllables could at all bear to be run together. (T. Roger Smith[2])
>
> . . . . .
>
> In order that hearing may be good in any auditorium, it is necessary that the sound should be sufficiently loud; that the simultaneous components of a complex sound should maintain their proper relative intensities; and that the successive sounds in rapidly moving articulation, either of speech or music, should be clear and distinct, free from each other and from extraneous noises. These three are the necessary, as they are the entirely sufficient conditions for good hearing. The architectural problem is, correspondingly, threefold. Within the three fields thus defined is comprised without exception the whole of architectural acoustics. (Sabine[3])

The latter statement, by Sabine, might be said to have initiated the systematic study of architectural acoustics. Sabine, a junior member of the physics faculty at Harvard University, was assigned the task of rectifying the acoustical faults of a new lecture hall. This led first to the discovery of his reverberation-time formula, characterizing the reverberant properties of halls. The next step was to determine suitable values of reverberation time, and to this end he organized musical

7

performances in a series of halls with different reverberation times. An account of these first subjective studies forms Paper 1 of the Benchmark papers in this volume.

Following Sabine's early work, the bearing of reverberation time on musical performance became a topic of much discussion and conjecture among the practitioners of acoustics. The conjectural approach is exemplified in Paper 2, in which W. A. MacNair, an engineer at Bell Telephone Laboratories, propounded not only an optimum reverberation time, but also the optimal variation with frequency. The latter, despite its speculative origin, remains comfortably imbedded in current wisdom. Today, surveying the various criteria that have been advanced, for example by F. R. Watson,[4] H. Bagenal and A. Wood,[5] and V. O. Knudsen and C. M. Harris,[6] one finds a sort of consensus regarding reverberation time for music that seems as precise as variations in musical style and individual preferences would warrant.

One topic based on hard evidence is the relation between reverberation time and speech communication in rooms, as reported by V. O. Knudsen in Paper 3. Knudsen began his career in acoustics under the guidance of Harvey Fletcher, first at Brigham Young University and later at Bell Telephone Laboratories where he was associated with Fletcher's group working on speech research. In this paper, prepared after his move to University of California at Los Angeles, he applies this early experience to architectural acoustics.

In addition to reverberation, which is a convenient way of considering long-term growth and decay of sounds in a room, the short-term sequence of reflected sounds in a hall is known to have a strong influence on a listener's acoustical assessment of a hall. This aspect of hall design has long been recognized, but reliable subjective information has been limited. One of the earliest systematic attacks on the problem was that of C. A. Mason and J. Moir (Paper 4), who were concerned with the quality of sound in cinemas.

THE ROYAL FESTIVAL HALL

The acoustical design of the Royal Festival Hall, London, marked an important landmark in the acoustics of concert halls. Completed in 1951, this was the first important hall to have been built in nearly half a century. Reflecting a new era, with new tastes among listeners and performers and new architectural and economic constraints, the concert hall had a pioneering role to play. The acoustical design, under Hope Bagenal in collaboration with the British Building Research Station, began with a detailed acoustical study of existing halls by P. H. Parkin, W. E. Scholes and A. G. Derbyshire[7]; acoustics was a primary concern throughout the design and construction of the hall; and when it was completed, an experimental concert was staged to provide a subjective appraisal of the final result. The whole enterprise was documented by Parkin, Allen, Purkis, and Scholes (Paper 5).

The result did not match expectations in either physical or subjective terms. But since its completion, the RFH has been the most tested of halls; and its substantial virtues as well as its shortcomings have been recorded in detail. The principal shortcoming, a lack of reverberant sound, has since been rectified by an ingenious electro-acoustic system called "assisted resonance," described in a recent paper by Parkin and Morgan.[8]

The short-term transient response of halls became, in the 1950s, a special interest of Erwin Meyer and his associates at Göttingen University in Germany, and they undertook numerous subjective and experimental studies of hall performance. On the subjective, side H. Niese[9] investigated the effects of irregularities in reverberation decay curves, and considered their influence on the perception of speech. An interesting laboratory study of subjective reactions to a single echo, for several kinds of speech and music, was reported in two papers from Australia by R. W. Muncey, A. F. B. Nickson, and P. Dubout, one of which appears here as Paper 6.

A special objective of the early reflection studies was to investigate the relation between the timing of the various signals reaching the listener and the apparent location of the source of sound. The effect, insofar as it relates to electronic reinforcement systems, was known at least as early as 1935, when it was reported by R. D. Fay[10] and W. M. Hall,[11] although all that appeared in print were two brief abstracts. In 1951, a more complete study was made by Meyer's student, Helmut Haas. A brief English summary of Haas' university dissertation is included here with Paper 7, but a full English translation is also available.[12] Haas' paper has been the starting point for several other studies dealing especially with speech transmission in halls. An example is Paper 8 in which R. H. Bolt and P. E. Doak derive a criterion for auditorium design purposes.

PHILHARMONIC HALL, NEW YORK

There is a striking parallel between the history of the Royal Festival Hall, London, and that of Philharmonic Hall, Lincoln Center, New York, which opened in 1962. As in the case of the Royal Festival Hall, the project began with an exhaustive study of existing halls, reported here by L. L. Beranek in Paper 9 and in more detail in *Music, Acoustics and Architecture*.[13] Beranek had available to him a number of new halls built in the postwar reconstruction period, and also the extensive experimental and theoretical studies stimulated by these new halls. The acoustics of Philharmonic Hall were, nevertheless, disappointing, and prompted many speculations as to the cause; a contemporary report by R. S. Lanier[14] typifies the first reactions. This gave rise to a new wave of studies of both the physical phenomena and the subjective appreciation of halls. Papers by Beranek et al.[15] and by Schroeder et al.[16] are but two of a number relating primarily to Philharmonic Hall.

Renewed attention has since been given to the short-term transient phenomena associated with the first few reflections reaching the listener. One aspect is the *time*-distribution of these early events, which Beranek described in terms of the "initial-time-delay gap." Today, more attention is being given to the *directional* distribution and to the special significance of lateral reflections in providing the optimum acoustical environment for music. This viewpoint is expressed in Paper 10 by A. H. Marshall. A further summation of these ideas is given by M. Barron[17].

Meanwhile the tools and techniques needed to test these concepts, employing the tape recorder, the electronic computer, scale-model simulations of halls, and so forth, have been assembled (see Part II). On the subjective side, perhaps the most important new development is the design of experiments that involve first a separation of the acoustical perception from the extraneous influence of the physical halls. Three approaches of this type are the studies by Nordlund, Kihlman, and Lindblad[18], Yamaguchi[19], and by Schroeder, Gottlob, and Siebrasse[20]. The definitive hall, ideal for all performers, all listeners and all programs, may never be achieved. But at least, capitalizing on these new techniques and criteria, the next major halls should come closer to fulfilling their designers' hopes.

REFERENCES

1. Joseph Henry, "Acoustics Applied to Public Buildings," *Ann. Rep. Smithsonian Inst. Washington, D.C.,* 4:221–237 (1856).
2. T. Roger Smith, *Acoustics of Public Buildings,* in Weale's series of Rudimentary Treatises, 1861.
3. W. C. Sabine, "Reverberation," *The American Architect,* 1900; *Collected Papers on Acoustics,* Dover, New York, 1964, p. 4.
4. F. R. Watson, "Acoustics of Buildings," *J. Franklin Inst.,* July 1924.
5. H. Bagenal and A. Wood, *Planning for Good Acoustics* (Methuen, London, 1931), p. 116.
6. V. O. Knudsen and C. M. Harris, *Acoustical Designing in Architecture* (Wiley, New York, 1931), p. 194.
7. P. H. Parkin, W. E. Scholes, and A. G. Derbyshire, "The Reverberation Times of Ten British Concert Halls," *Acustica* 2(3):97–100 (1952).

8. P. H. Parkin and K. Morgan, "Assisted Resonance in the Royal Festival Hall, London, 1965–1969," *J. Acoust. Soc. Am.* **48**:1025–1035 (1970).

9. H. Niese, "Vorschlag für die Definition und Messung der Deutlichkeit nach subjektiven Grundlagen," *Hochfrequenztechnik und Elektroakustik* **65**:4–15 (July 1956).

10. R. D. Fay, "A Method for Obtaining Natural Directional Effects in a Public Address System," *J. Acoust. Soc. Am.* **7**:239 (1936).

11. W. M. Hall, "A Method for Maintaining in a Public Address System the Illusion That the Sound Comes from the Speaker's Mouth," *J. Acoust. Soc. Am.* **7**:239 (1936).

12. Helmut Haas, "Influence of a Single Echo on the Perceptibility of Speech," (Translation of dissertation from University of Göttingen published by Building Research Station, Great Britain, as Library Communication No. 363, 1949.) Also reprinted in *J. Audio Eng. Soc.* **20** 146–159, March 1972.)

13. L. L. Beranek, *Music, Acoustics and Architecture*, McGraw-Hill, New York, 1962.

14. R. S. Lanier, "What Happened at Philharmonic Hall," *Architectural Forum*, Dec. 1963, p. 119–123.

15. L. L. Beranek, F. R. Johnson, T. J. Schultz, and B. G. Watters, "Acoustics of Philharmonic Hall, New York, During Its First Season," *J. Acoust. Soc. Am.* **36**:1247–1262 (1964).

16. M. R. Schroeder, B. S. Atal, G. M. Sessler, and J. E. West, "Acoustical Measurements at Philharmonic Hall (New York)," *J. Acoust. Soc. Am.* **40**:434–440 (1966).

17. M. Barron, "The Subjective Effects of First Reflections in Concert Halls— The Need for Lateral Reflections," *J. Sound Vib.* **15**:475–494 (1971).

18. B. Nordlund, T. Kihlman, and S. Lindblad, "Use of Articulation Tests in Auditorium Studies," *J. Acoust. Soc. Am.* **44**:148–154 (1968).

19. Kiminori Yamaguchi, "Multivariate Analysis of Subjective Physical Measures of Hall Acoustics," *J. Acoust. Soc. Am.* **52**:1271–1279 (1972).

20. M. R. Schroeder, D. Gottlob, and K. F. Siebrasse, "Comparative Study of European Concert Halls," *J. Acoust. Soc. Am.* **56**:1195–1201, 1974.

Reprinted from pp. 71–77 of *Collected Papers on Acoustics*,
Dover Publications, Inc., New York, 1964, 299 pp.

THE ACCURACY OF MUSICAL TASTE IN REGARD TO ARCHITECTURAL ACOUSTICS

Wallace Clement Sabine

PIANO MUSIC

THE experiments described in this paper were undertaken in order to determine the reverberation best suited to piano music in a music room of moderate size, but were so conducted as to give a measure of the accuracy of cultivated musical taste. The latter point is obviously fundamental to the whole investigation, for unless musical taste is precise, the problem, at least as far as it concerns the design of the auditorium for musical purposes, is indeterminate.

The first observations in regard to the precision of musical taste were obtained during the planning of the Boston Symphony Hall, Messrs. McKim, Mead, and White, Architects. Mr. Higginson, Mr. Gericke, the conductor of the orchestra, and others connected with the Building Committee expressed opinions in regard to a number of auditoriums. These buildings included the old Boston Music Hall, at that time the home of the orchestra, and the places visited by the orchestra in its winter trips, Sanders Theatre in Cambridge, Carnegie Hall in New York, the Academy of Music in Philadelphia, and the Music Hall in Baltimore, and in addition to these the Leipzig Gewandhaus. By invitation of Mr. Higginson, the writer accompanied the orchestra on one of its trips, made measurements of all the halls, and calculated their reverberation. The dimensions and the material of the Gewandhaus had been published, and from these data its reverberation also was calculated. The results of these measurements and calculations showed that the opinions expressed in regard to the several halls were entirely consistent with the physical facts. That is to say, the reverberation in those halls in which it was declared too great was in point of physical measurement greater than in halls in which it was pronounced too small. This consistency gave encouragement in the hope that the physical problem was real, and the end to be attained definite.

Much more elaborate data on the accuracy of musical taste were obtained four years later, 1902, in connection with the new building of the New England Conservatory of Music, Messrs. Wheelwright and Haven, Architects. The new building consists of a large auditorium surrounded on three sides by smaller rooms, which on the second and third floors are used for purposes of instruction. These smaller rooms, when first occupied, and used in an unfurnished or partially furnished condition, were found unsuitable acoustically, and the writer was consulted by Mr. Haven in regard to their final adjustment. In order to learn the acoustical condition which would accurately meet the requirements of those who were to use the rooms, an experiment was undertaken in which a number of rooms, chosen as typical, were varied rapidly in respect to reverberation by means of temporarily introduced absorbing material. Approval or disapproval of the acoustical quality of each room at each stage was expressed by a committee chosen by the Director of the Conservatory. At the close of these tests, the reverberation in the rooms was measured by the writer in an entirely independent manner as described in the paper on Reverberation (1900). The judges were Mr. George W. Chadwick, Director of the Conservatory, and Signor Oresti Bimboni, Mr. William H. Dunham, Mr. George W. Proctor, and Mr. William L. Whitney, of the Faculty. The writer suggested and arranged the experiment and subsequently reduced the results to numerical measure, but expressed no opinion in regard to the quality of the rooms.

The merits of each room in its varied conditions were judged solely by listening to piano music by Mr. Proctor. The character of the musical compositions on which the judgment was based is a matter of interest in this connection, but this fact was not appreciated at the time and no record of the selections was made. It is only possible to say that several short fragments, varied in nature, were tried in each room.

As will be evident from the descriptions given below, the rooms were so differently furnished that no inference as to the reverberation could be drawn from appearances, and it is certain that the opinions were based solely on the quality of the room as heard in the piano music.

The five rooms chosen as typical were on the second floor of the building. The rooms were four meters high. Their volumes varied from 74 to 210 cubic meters. The walls and ceilings were finished in plaster on wire lath, and were neither papered nor painted. There was a piano in each room; in room 5 there were two. The amount of other furniture in the rooms varied greatly:

In room 1 there was a bare floor, and no furniture except the piano and piano stool.

Room 2 had rugs on the floor, chairs, a sofa with pillows, table, music racks, and a lamp.

Room 3 had a carpet, chairs, bookcases, and a large number of books, which, overflowing the bookcases, were stacked along the walls.

Room 4 had no carpet, but there were chairs and a small table.

Room 5 had a carpet, chairs, and shelia curtains.

Thus the rooms varied from an almost unfurnished to a reasonably furnished condition. In all cases the reverberation was too great.

The experiment was begun in room 1. There were, at the time, besides the writer, five gentlemen in the room, the absorbing effect of whose clothing, though small, nevertheless should be taken into account in an accurate calculation of the reverberation. Thirteen cushions from the seats in Sanders Theatre, whose absorbing power for sound had been determined in an earlier investigation, were brought into the room. Under these conditions the unanimous opinion was that the room, as tested by the piano, was lifeless. Two cushions were then removed from the room with a perceptible change for the better in the piano music. Three more cushions were removed, and the effect was much better. Two more were then taken out, leaving six cushions in the room, and the result met unanimous approval. It was suggested that two more be removed. This being done the reverberation was found to be too great. The agreement was then reached that the conditions produced by the presence of six cushions were the most nearly satisfactory.

The experiment was then continued in Mr. Dunham's room, number 2. Six gentlemen were present. Seven cushions were brought into the room. The music showed an insufficient reverberation. Two of the cushions were then taken out. The change was regarded as a distinct improvement, and the room was satisfactory.

In Mr. Whitney's room, number 3, twelve cushions, with which it was thought to overload the room, were found insufficient even with the presence in this case of seven gentlemen. Three more cushions were brought in and the result declared satisfactory.

In the fourth room, five, eight, and ten cushions were tried before the conditions were regarded as satisfactory.

In Mr. Proctor's room, number 5, it was evident that the ten cushions which had been brought into the room had overloaded it. Two were removed, and afterwards three more, leaving only five, before a satisfactory condition was reached.

This completed the direct experiment with the piano.

The bringing into a room of any absorbing material, such as these cushions, affects its acoustical properties in several respects, but principally in respect to its reverberation. The prolongation of sound in a room after the cessation of its source may be regarded either as a case of stored energy which is gradually suffering loss by transmission through and absorption by the walls and contained material, or it may be regarded as a process of rapid reflection from wall to wall with loss at each reflection. In either case it is called reverberation. It is sometimes called, mistakenly as has been explained, resonance. The reverberation may be expressed by the duration of audibility of the residual sound after the cessation of a source so adjusted as to produce an average of sound of some standard intensity over the whole room. The direct determination of this, under the varied conditions of this experiment, was impracticable, but, by measuring the duration of audibility of the residual sound after the cessation of a measured organ pipe in each room without any cushions, and knowing the coefficient of absorption of the cushions, it was possible to calculate accurately the reverberation at each stage in the test. It was impossible to make these measurements immediately after the above experiments, because, although the day was an especially quiet one, the noises from the street and railway traffic were seriously disturbing. Late the following night the conditions were more favorable, and a series of fairly good observations was obtained in each room. The cushions had been removed, so that the measurements were made on the rooms in their original condition, furnished as above described. The apparatus and method employed are described in full in a series of articles in the Engineering Record[1] and American Architect for 1900. The results are given in the accompanying table.

Room Number	Volume	Absorbing Power of Room	Gentlemen Present	Absorbing Power of Clothing	Number of Meters of Cushions	Absorbing Power of Cushions	Total Absorbing Power	Reverberation in Seconds	Remarks
1	74	5.0	0	0	0	0	5.0	2.43	Reverberation too great.
		"	5	2.4	0	0	7.4	1.64	Reverberation too great.
		"	"	"	13	12.8	20.2	.60	Reverberation too little.
		"	"	"	11	10.1	17.5	.70	Better.
		"	"	"	8	7.3	14.7	.83	Better.
		"	"	"	6	5.5	12.9	.95	Condition approved.
		"	"	"	4	3.6	11.0	1.22	Reverberation too great.
2	91	6.3	0	0	0	0	6.3	2.39	Reverberation too great.
		"	6	2.9	0	0	9.2	1.95	Reverberation too great.
		"	"	"	7	6.4	15.6	.95	Reverberation too little.
		"	"	"	5	4.6	13.8	1.10	Condition approved.
3	210	14.0	0	0	0	0	14.0	2.46	Reverberation too great.
		"	7	3.4	0	0	17.4	2.00	Reverberation too great.
		"	"	"	12	11.0	28.4	1.21	Better.
		"	"	"	15	13.7	31.1	1.10	Condition approved.
4	133	8.3	0	0	0	0	8.3	2.65	Reverberation too great.
		"	7	3.4	0	0	11.7	1.87	Reverberation too great.
		"	"	"	6	5.5	17.2	1.26	Better.
		"	"	"	10	9.1	20.8	1.09	Condition approved.
5	96	7.0	0	0	0	0	7.0	2.24	Reverberation too great.
		"	4	1.9	0	0	8.9	1.76	Reverberation too great.
		"	"	"	10	9.1	18.0	.87	Reverberation too little.
		"	"	"	8	7.3	16.2	.98	Better.
		"	"	"	5	4.6	13.5	1.16	Condition approved.

[1] The article in the Engineering Record is identical with the paper in the American Architect for 1900, reprinted in this volume as Part I.

The table is a record of the first of what, it is hoped, will be a series of such experiments extending to rooms of much larger dimensions and to other kinds of music. It may well be, in fact it is highly probable, that very much larger rooms would necessitate a different amount of reverberation, as also may other types of musical instruments or the voice. As an example of such investigations, as well as evidence of their need, it is here given in full. The following additional explanations may be made. The variation in volume of the rooms is only threefold, corresponding only to such music rooms as may be found in private houses. Over this range a perceptible variation in the required reverberation should not be expected. The third column in the table includes in the absorbing power of the room (ceiling, walls, furniture, etc.) the absorbing powers of the clothes of the writer. who was present not merely at all tests, but in the measurement of the reverberation the following night. From the next two columns, therefore, the writer and the effects of his clothing are omitted. The remarks in the last column are reduced to the form "reverberation too great," "too little," or "approved." The remarks at the time were not in this form, however. The room was pronounced "too resonant," "too much echo," "harsh," or "dull," "lifeless," "overloaded," expressions to which the forms adopted are equivalent.

If from the larger table the reverberation in each room, in its most approved condition, is separately tabulated, the following is obtained:

Rooms	Reverberation
1	.95
2	1.10
3	1.10
4	1.09
5	1.16
	1.08 mean

The final result obtained, that the reverberation in a music room in order to secure the best effect with a piano should be 1.08, or in round numbers 1.1, is in itself of considerable practical value; but the five determinations, by their mutual agreement, give a numerical measure to the accuracy of musical taste which is of great interest. Thus the maximum departure from the mean is .13 seconds,

17

and the average departure is .05 seconds. Five is rather a small number of observations on which to apply the theory of probabilities, but, assuming that it justifies such reasoning, the probable error is .02 seconds, — surprisingly small.

A close inspection of the large table will bring out an interesting fact. The room in which the approved condition differed most from the mean was the first. In this room, and in this room only, was it suggested by the gentlemen present that the experiment should be carried further. This was done by removing two more cushions. The reverberation was then 1.22 seconds, and this was decided to be too much. The point to be observed is that 1.22 is further above the mean, 1.08, than .95 is below. Moreover, if one looks over the list in each room it will be seen that in every case the reverberation corresponding to the chosen condition came nearer to the mean than that of any other condition tried.

It is conceivable that had the rooms been alike in all respects and required the same amount of cushions to accomplish the same results, the experiment in one room might have prejudiced the experiment in the next. But the rooms being different in size and furnished so differently, an impression formed in one room as to the number of cushions necessary could only be misleading if depended on in the next. Thus the several rooms required 6, 5, 15, 10, and 5 cushions. It is further to be observed that in three of the rooms the final condition was reached in working from an overloaded condition, and in the other two rooms from the opposite condition, — in the one case by taking cushions out, and in the other by bringing them in.

Before beginning the experiment no explanation was made of its nature, and no discussion was held as to the advantages and disadvantages of reverberation. The gentlemen present were asked to express their approval or disapproval of the room at each stage of the experiment, and the final decision seemed to be reached with perfectly free unanimity.

This surprising accuracy of musical taste is perhaps the explanation of the rarity with which it is entirely satisfied, particularly when the architectural designs are left to chance in this respect.

2

Reprinted from *J. Acoust. Soc. Am.* 1:242–248 (Jan. 1930)

OPTIMUM REVERBERATION TIME FOR AUDITORIUMS

By Walter A. MacNair
Bell Telephone Laboratories

I. Reverberation Time vs. Frequency

There is very little published data in regard to the change in reverberation time with frequency in auditoriums which are considered near ideal. It is often mentioned by engineers and physicists that to secure the best acoustical results, the reverberation time should be the same

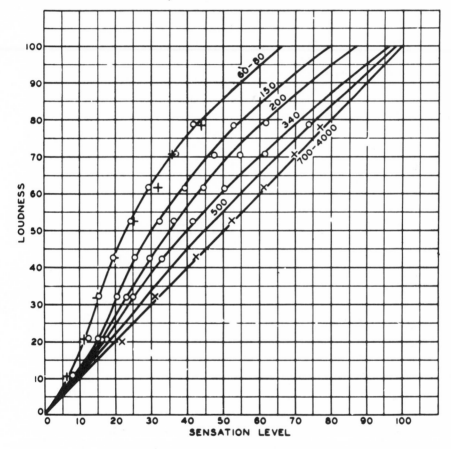

Fig. 1. *Loudness of Pure Tones.*

for all frequencies in any one room. This specifies that the sensation level shall decay at the same rate for all frequencies of interest.

It seems more reasonable, however, to specify that the loudness of all pure tones shall decay at the same rate for all frequencies since it is the loudness of a tone which takes into consideration not only the energy level but also its ultimate effect upon one's brain. In Fig. 1[1] are plotted data which show the relation between the loudness as judged by a considerable number of observers and the sensation level. It will be seen that for frequencies between 700 and 4000 cycles per second these two quantities are equal to each other so that the two points of view mentioned above demand identical conditions throughout this frequency band. Outside of this band, however, any change in the sensation level gives a greater change in the loudness, as may be seen.

The maximum loudness in which we are interested at present is about 73.[2] In the figure the curves may be replaced by straight lines which represent fair approximations to the observed data up to this loudness. This family of straight lines may be represented by the expression

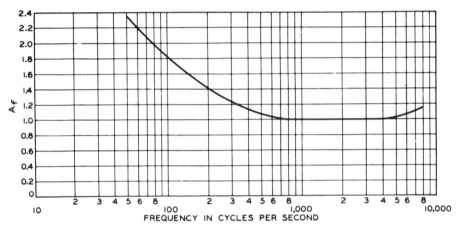

Fig. 2. *Value of A_f vs. Frequency.*

$$L_t = A_f S_t \tag{1}$$

where A_f is the slope of the line adopted to fit the data for the frequency f. The values of A_f chosen from this figure are given by the next, Fig. 2. This approximation simplifies our calculations very much and introduces errors which are not intolerable.

Referring back to Fig. 1, if we wish to adjust the absorption of the room so that the loudness of all pure tones will decay at the same rate, say for the moment 60 units per second, it is seen that the sensation level must drop 60 *db* per second for frequencies between 700 and 4000 cycles and for other frequencies the sensation level must drop $60/A_f$ *db*

[1] This is Fig. 108 from "Speech and Hearing" by H. Fletcher.

[2] This is the loudness that the source chosen in Part II of this paper will produce in a room of 1000 cubic feet having a reverberation time of 0.8 seconds.

per second; or in other words, the reverberation time for frequencies between 700 and 4000 cycles should be one second and outside of this band it should be A_f seconds. Fig. 2, then, which is a plot of A_f vs. frequency now becomes also an illustration of the shape of the reverberation time vs. frequency curve which a room should have in order that the loudness of pure tones of all frequencies shall decay at the same rate.

According to Sabine's well known formula the reverberation time is inversely proportional to the number of absorption units in the room so

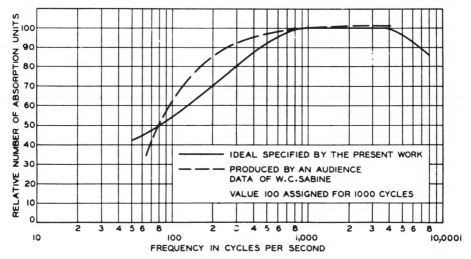

FIG. 3. *Relative Number of Absorption Units vs. Frequency.*

that, if we assume this, we may immediately infer the shape of the curve which represents the number of absorption units necessary at any frequency, referred to the amount required at 1000 cycles, to obtain our required condition. These values are plotted in Fig. 3. If it should happen that the greater part of the sound absorption in a room is caused by one particular kind of surface, then the curve in Fig. 3 is the shape of the absorption curve that this material should have.

A pertinent observation on which every one seems to agree is that if an auditorium has an unusually long reverberation time and consequently is of little use, when empty, it attains excellent acoustic conditions when filled with a large audience. In these cases a very large part of the absorption is caused by the audience. The absorption of an average audience has been measured by W. C. Sabine[3] and his results are also plotted in Fig. 3. The close agreement between this curve and

[3] "Collected Papers on Acoustics" page 86.

the one we have obtained from our hypothesis gives considerable confidence in our general viewpoint.

II. Reverberation Time vs. Volume

It is generally accepted that the best acoustical conditions in a room are obtained when the reverberation time is adjusted to a definite value known as the optimum reverberation time. Observations reported in literature agree that the value of the optimum reverberation time increases with the size of the room in the way shown in Fig. 4 where the curves numbered 1 to 3 are the choices reported by Watson,[4] Lifschitz,[5] and Sabine,[6] respectively. These experimental results have served as the basis of successful adjustment and design of many auditoriums. One naturally seeks the factor which determines a choice of reverberation time of two seconds for a million cubic feet theatre and on the other hand a choice of near one second for a 10,000 cubic foot music room. It is our purpose now to point out the factor which apparently does this.

We will set down a condition which we believe to be this factor and then will show that the requirements demanded by it agree quite closely with the empirical results illustrated in Fig. 4 and mentioned above. The condition is

$$\int_{t_0}^{t_1} L_t dt = -K \tag{2}$$

in which t_0 is the time a sustained source of sound $\bar{E}$ is cut off, t_1 the time the sound becomes inaudible, L_t the loudness of the sound at any instant t, and K a constant. As shown in Fig. 1, the loudness of a one thousand cycle note is equal to the sensation level, that is,

$$L_t = S_t \text{ for } 1000 \text{ cycles.}$$

Since, during the time of decay, S_t decreases uniformly with time, and therefore L_t also, then for a thousand cycle note, evaluating our integral we have

$$L_{t_0} T_1 = 2K \tag{3}$$

or

$$S_{t_0} T_1 = 2K$$

where

$$T_1 = t_1 - t_0.$$

[4] Watson, Architecture, May, 1927.
[5] Lifschitz, Phys. Rev. 27, 618, 1926.
[6] Sabine, Trans. of S.M.P.E. XII 35, 1928.

This last expression is practically in the form in which this condition was first stated by Lifschitz.[5] In (3) there are three unknowns and a fourth is implied, namely, the power of the source, $\overline{E}$.

We now turn our attention to finding the relation between the volume of a room and the reverberation time dictated by the stated condition. Following P. E. Sabine let us take the rate of emission of the source, $\overline{E}$ to be 10^{10} cubic meters (35.3×10^{10} cubic feet) of sound of threshold density per second. Now[7]

$$T_1 = \frac{4V}{ca} \log_e \frac{4 \times 35.3 \times 10^{10}}{c \cdot a}.$$

Where V is the volume of the room in cubic feet.

c is the velocity of sound, 1120 feet per second.

a is the number of absorption units in sq. feet and[8]

$$L_{t_0} = S_{t_0} = 10 \log_{10} \frac{4 \times 35.3 \times 10^{10}}{c \cdot a}.$$

If we should substitute these values in (3) we would obtain a relation between V, a, and K which must be satisfied when condition (2) is satisfied. In other words, (3) specifies the amount of absorption, for a one thousand cycle note, a room should have if it complies with (2). If we assume Sabine's well known formula, namely,

$$T = \frac{.05V}{a}$$

where T is the reverberation tine in seconds we may express this relation in terms of V, T, and K with the result

$$10.40 + \log T_{op} - \log V = \frac{(2K)^{1/2}}{1.283 T_{op}^{1/2}}. \tag{4}$$

Where T_{op} is the value of T imposed by our condition (2) for a thousaud cycle tone.

Referring to Fig. 4 it will be seen that all three observers agree rather closely that the rever'.eration time for an auditorium of 1,000,000 cubic feet should be 2.0 seconds. This value refers to a tone of 512 cycles, the customary frequency used for experimental observation. It has been shown above that the reverberation time for 1000 cycles should be 92.5

[7] See Crandall "Theory of Vibrating Systems and Sounds" page 211.

[5] See Crandall "Theory of Vibrating Systems and Sounds" page 210, and the definition of sensation level.

per cent of the reverberation time for a 512 cycle tone, so that the 2.0 seconds above corresponds to 1.85 seconds for 1000 cycles. We can

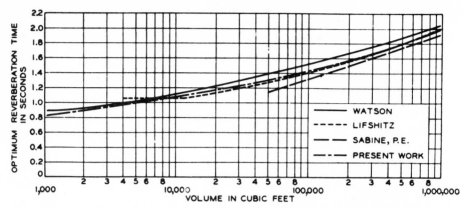

Fig. 4. *Optimum Reverberation Time vs. Volume in Cubic Feet for 512 Cycles.*

evaluate K in (4) by adapting this latter value of T_{op} for a volume of 1,000,000 cubic feet. This gives $K = 32.6$. Substituting this value in (4) we obtain

$$\log V = 10.40 + \log T_{op} = \frac{6.35}{T_{op}^{1/2}}.$$
(5)

From (5) we may obtain T_{op} for 1000 cycles for any volume. See Fig. 5. As mentioned above these values of T_{op} are 92.5% of T_{op} for 512 cycles so that these latter may be easily deduced for comparative purposes. These values are plotted to give curve number 4 in Fig. 4. It is seen that this curve agrees very well with those showing the choices of competent judges.

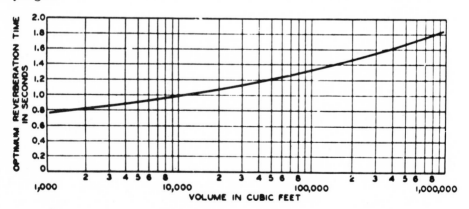

Fig. 5. *Optimum Reverberation Time vs. Volume in Cubic Feet for 1000 Cycles*

III. The More General Hypothesis

Equation (2) may be written as follows, since we have assigned a value to K:

$$\int_{t_0}^{t_1} L_t dt = -32.6 \tag{6}$$

and it will be remembered that we have considered L_{t_0} to be the loudness set up by a certain standard source. Allowing V to vary with f constant (1000 cycles) we have obtained a relation between the optimum reverberation time and volume of rooms for 1000 cycles. We wish to point out now that exactly this same condition (6) with V constant and f variable, will give the same results that we have obtained in Part I of this paper with the only further requirement that for other frequencies than 1000 cycles the strength of the source $\overline{E}$ shall be such that the loudness L_{t_0} set up in the room at the frequency considered shall be exactly the same as the loudness which our standard source would set up at 1000 cycles.

In Part I of this paper our stated condition was that the loudness of all pure tones shall decay at the same rate for all frequencies. Since we have specified that the loudness at the time t_0 shall be the same for all test frequencies and also that the loudness at the time t_1 shall be zero for all frequencies, it is quite evident that the above integral can have the same value at all these frequencies only when the loudness decays at the same rate for all frequencies concerned. In other words, this condition stated as an integral specifies exactly the same requirement on the decay of loudness that we expressed in our statement early in Part I of this paper.

IV. Conclusions

To recapitulate, we have set down an equation, together with a specification of the strength of the virtual source in each case, from which we obtain the value the reverberation time for any frequency tone should have in any sized room according to the condition which apparently controls the choice of observers.

One naturally turns to see what meaning may be attached to this significant expression, namely, the integral of the loudness taken throughout the time of decay to inaudibility. Since this integral has the same values for all auditoriums which are considered ideal, it implies that one's brain is a ballistic instrument which is concerned with not only the maximum value of loudness but also with the effect of loudness integrated throughout a considerable interval of time.

3

Reprinted from pp. 56–59 and 63–75 of *J. Acoust. Soc. Am.* 1:56–78 (Oct. 1929)

THE HEARING OF SPEECH IN AUDITORIUMS

By Vern O. Knudsen
University of California at Los Angeles.

Ever since the monumental work of W. C. Sabine on reverberation there has been a growing tendency, especially in America, to rate the acoustic quality of an auditorium almost solely in terms of its time of reverberation. It is true that reverberation (which determines the rate of growth and decay of sound in a room) has been, and yet is, the most important factor in determining the acoustic properties of a room. However, reverberation is not the only factor affecting the acoustic properties of an enclosure. Thus, the size and shape of the room, and the presence of extraneous noise, all contribute to the resulting acoustic quality. It would seem desirable therefore to evaluate the acoustic merit of a room in terms of all of these factors.

It is not a simple matter to give a quantitative rating to a room which is to be used for music, since so much depends upon the musical taste and dispositon of the listeners. It is, however, a relatively simple matter to give a quantative rating to a room which is to be used for speaking, since our primary concern is how well we hear the spoken words of the speaker. The most feasible scheme for such a rating is probably the one used by telephone engineers for testing speech-transmission over telephone equipment, which goes by the name of "articulation tests." The "percentage articulation' of a telephone circuit sigifies what percentage of typical speech-sounds can be heard correctly when transmitted over the circuit. Thus, if a speaker calls out 1000 meaningless speech-sounds into the transmitting end of the circuit and an observer at the receiving end hears 750 of these sounds correctly the articulation is rated at 75 %. The writer has used this same scheme for investigating the effects of reverberation and noise upon speech-reception in auditoriums.[1] The "percentage articulation" of an auditorium signifies what percentage of typical speech-sounds can be heard correctly by an average listener in the auditorium. A speaker calls out typical monosyllabic speech-sounds, in groups of three, at a rate of three syllables in two seconds. Observers stationed in representative positions in the auditorium write down what they think they hear. If,

---

[1] V. O. Knudsen, Phys. Rev., 26, 287, (1925), and 26, 133–138, (1925). See also The Architect and Engineer, 84, 67–72, (1926) and v. 85, (1927).

on the average, they hear correctly four-fifths of the total number of called speech-sounds, the articulation for this auditorium is rated at 80%. It would seem that such a scheme as this offers a satisfactory means for rating the acoustic quality of an auditorium which is to be used primarily for speaking.

It is obvious that the percentage articulation in an auditorium will depend upon (1) the size of the room, (2) the reverberation character- istics of the room, (3) the amount of disturbing noise in the room, and (4) the shape of the room. It is apparent that, for speaking purposes only, the ideal auditorium is a small room free from all noise, and bounded by perfectly absorbing surfaces. In such a small room the lis- tener will be near the speaker and therefore the speaker's voice will be heard with adequate loudness. Further, there will be no interfering noise, reverberation or delayed reflections. Actual tests conducted in a quiet open space have indicated that with average speakers and lis- teners the articulation in such a room will be about 96%. This figure represents the highest attainable acoustic quality for speech reception in a room. A rating of 100%, that is perfect articulation, can never be attained. A few of the consonantal sounds are sometimes mistaken even under ideal hearing conditions. We are ordinarily unaware of this when we listen to speech because the connotation of the articulated words facilitates the correct interpretations of those words which are not heard distinctly. Even when the speech articulation is as low as 75% the hearing will be regarded as acceptable. An articulation of 96% is, for all practical purposes, about perfect, and therefore there seems to be no necessity for attempting to improve this limited ideal, although it could be done by altering slightly the pronunciation, or even empha- sis, of some of the soft consonantal sounds.

The extension of the size of the room, the use of reflecting materials for the walls and ceiling, and the presence of disturbing noise will all tend to impair the acoustic quality of the room, and thus reduce the articula- tion below 96%. In general, each of the four mentioned factors which affects the acoustics of the room will introduce a distortion or a dis- turbance which can be determined quantitatively. Thus, the resulting percentage articulation in any specified auditorium can be esti.nated by the following equation:

$$\text{Percentage Articulation} = 96 \; k_l \; k_r \; k_n \; k_s, \tag{1}$$

where k_l is the reduction factor owing to the inadequate loudness of the speech, k_r the reduction or distortion factor owing to reverberation,

k_n the reduction factor owing to noise, and k_s the reduction factor owing to the shape of the room. The first three of these reduction factors are fairly well known from existing experimental data. The work of Fletcher[2] has determined the effect of loudness upon speech reception, and the work of the writer[3] has determined the effects of noise and reverberation upon speech reception. The results of these determinations, as they affect the problem of hearing in auditoriums, will now be outlined.

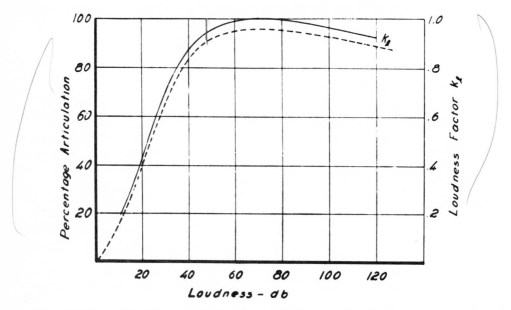

FIG. 1. Curve showing the effect of loudness upon the reception of speech (taken from Fletcher's data). The dotted curve gives the percentage articulation at different loudness levels. The solid-line curve gives the corresponding loudness reduction factor k_1.

The Effect of Loudness upon the Reception of Speech in Auditoriums. Fletcher's data on the effect of loudness upon speech reception are shown by the broken line curve in Fig. 1. The loudness of the speech, expressed in decibels (abbreviated db),[4] was controlled by the gain of a distortionless vacuum tube amplifier. It will be seen that the opti-

[2] Nature of Speech and Its Interpretation, Jour. Frank. Inst., 193, 6, (June, 1922).
[3] Loc. cit.
[4] The loudness of a sound in db is ten times the common logarithm of the ratio of the intensity of the sound to the intensity at the minimal threshold of audibility. This unit is becoming universal as the standard unit for the measurement of loudness. It was recently adopted by the Bell System and the International Advisory Committee on Long Distance Telephony in Europe. In recent years, this same unit has received the designation of transmission unit (TU) or sensation unit (SU).

mal loudness of speech appears to be about 70 db, which corresponds to an intensity which is ten million times the intensity which would be just barely audible. This is somewhat louder than normal conversation which is usually of the order of 50 to 60 db, or less. The data represented by Fig. 1 indicate that if the loudness of undistorted speech be between 50 and 100 db, the articulation is above 90%, which is wholly satisfactory. Below 50 db, the articulation drops off rapidly as the loudness is diminished. Thus, at 30 db the articulation is 66% and 20 db it is only 40%.

The solid line curve shown in Fig. 1 has been derived from the broken line curve. This curve gives the value of the loudness reduction factor, k_l, for different loudness levels from 0 to 120 db. The value of k_l at 70 db, the optimal loudness level, is taken as unity, and the value of k_l at all other loudness levels is the ratio of the percentage articulation at that level to the percentage articulation at 70 db. The solid line curve in Fig. 1 is obviously very useful in connection with equation (1), if the average loudness of a speaker's voice in an auditorium be known.

[*Editor's Note:* Material has been omitted at this point.]

Effect of Reverberation upon the Reception of Speech in Auditoriums

Fig. 2 shows how speech articulation depends upon the time of reverberation[7] in a group of auditoriums having about the same shape

[7] The time of reverberation is here used as defined by W. C. Sabine, that is, the time required for the intensity of a tone of 512 d.v. to decay to one-millionth of its initial intensity.

and volume (200,000 to 300,000 cu.ft.) but different times of reverberation. The small circles in Fig. 2 show the observed values of percentage articulation for the corresponding measured times of reverberation in the auditoriums investigated in this series. The lower curve is drawn to represent the most probable fit with the observed data. It will be noted that, approximately, the articulation decreases 6% for each additional second of reverberation.

The data for the three auditoriums having times of reverberation less than 2.0 seconds, and also the data for six other auditoriums, have been

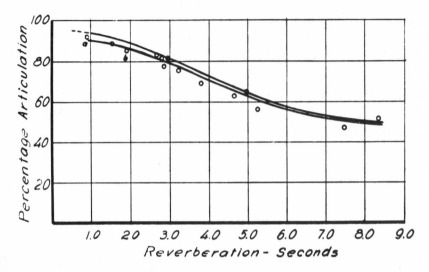

FIG. 2. *Curves showing the interfering effect of reverberation upon the hearing of speech. The lower curve represents the most probable fit with the observed data. The upper curve has been corrected for loudness, and corresponds to a loudness of 70 db.*

obtained during the past three years and therefore were not included in earlier publications. The new data represented in Fig. 2 seem to strengthen the conclusion suggested in the earlier papers, that the optimal time of reverberation for the reception of speech in an auditorium is somewhat shorter than is attained or planned in current practice. The optimal time of reverberation, based upon the combined effects of loudness and reverberation, will be given consideration in a later section in this paper.

The lower curve in Fig. 2. which represents the mean result of the experimental determinations, was not obtained for a constant loudness of speech, because the loudness is dependent upon the amount of absorption in the room. Assuming the power of the speaker's voice to remain constant,[8] the resulting intensity of the speech would be almost inversely proportional to the total amount of absorption in the auditoriums, or directly proportional to the time of reverberation. It was found experimentally that the average loudness of the speakers' voices used in these tests, in an auditorium having a volume of 8440 cu. m. (300,000 cu. ft.) and a time of reverberation of 1.50 seconds, was about 48 db. Using this datum, and the loudness-articulation data given in Fig. 1, it is possible to correct the lower curve in Fig. 2 for variation of loudness. The upper curve in Fig. 2 was obtained by applying such a correction so as to give the percentage articulation for a uniform loudness level of 70 db, which is the loudness level for optimal hearing. This curve has been extrapolated to a time of reverberation of .50 second, as indicated by the dotted portion of the curve. Such an extra-polation is warranted by articulation tests the writer has conducted in a small room in which the percentage articulation increased as the time of reverberation was reduced from 1.0 to .60 seconds.

It is now possible, from the upper curve in Fig. 2, to derive k_r, the reduction factor owing to reverberation, for times of reverberation between .5 and 8.0 seconds. The value of k_r is taken as unity for a time of reverberation of .5 second. The value of k_r for any other time of reverberation is the ratio of the articulation at that time to the articulation for a time of .5 second. The curve in Fig. 3 gives the value of k_r, obtained in this manner, for different times of reverberation. It will be noted that k, decreases almost uniformly as the time of reverberation increases from 1.0 to 6.0 seconds. Above 6.0 seconds the rate of decrease of k_r appears to be less rapid.

The departures of the experimentally determined "percentage articulations" from the values indicated by the smooth curve in Fig. 2 are greater than the experimental errors in determining the time of reverberation and the "percentage articulation." These departures are more probably the result of other factors affecting the reception of speech, such as residual noise and the shape of the room. The tests were conducted during the quiet part of the night, but in many instances there were slight disturbances from passing automobiles or trolley cars. Further, although care was exercised in the selection of auditoriums which would be nearly uniform in shape and size, there

---

[8] This assumption seems more plausible than the alternative one that the speaker maintains a constant loudness level. It seems likely, however, that neither assumption is correct. A speaker generally attempts to raise his voice to the loudness level required for satisfactory hearing, but is limited by the physical characteristics of his speech apparatus.

were certain variations in length, width and height, and in architectural detail. Finally, there are other variations, attributable to the reverberation characteristics of the rooms, which are not included in the independent variable (the time of reverberation) used in Fig. 2. Thus, the dependence of reverberation on pitch and the location of the obsorptive materials in the room both affect the hearing properties. The importance of these two aspects of reverberation has not yet been quantitatively determined, but is under investigation at the present time. Most auditoriums, especially when filled with auditors, are very much more reverberant for the low pitched tones than for the high ones. For example, one of the auditoriums studied

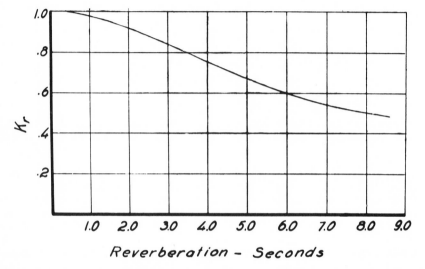

FIG. 3. Curves giving the reverberation reduction factor k_r for different times of reverberation.

in this investigation (the one marked (b) in Fig. 2) had a time of reverberation, empty, of slightly less than two seconds for a tone of 512 d.v., and nearly five seconds for a tone of 128 d.v. The determined "percentage articulation" for this auditorium falls slightly below the smooth curve. On the other hand, another auditorium[*] (the one marked (a) in Fig. 2) had a time of reverberation of .94 second at 512 d.v. and 1.10 seconds at 128 d.v. This auditorium exhibits an un-

---

[*] This auditorium is a sound studio designed by the writer for Metro-Goldwyn-Mayer. The entire ceiling and walls are lined with multiple layers of fibre building board and loose wool separated by air-spaces. This type of absorptive treatment gives an almost uniform time of reverberation for tones of all pitch. The acoustic properties of this room are regarded by the writer and by those who use it as practically perfect, not only for speaking but also for the recording of both speech and music.

usual instance of nearly uniform reverberation for tones of different pitch. The determined articulation for this auditorium lies slightly above the smooth curve of Fig. 1. It is probable that the departures of the points (a) and (b) from the smooth curve in Fig. 2 are actually attributable to the manner in which the reverberation varies with pitch. It is to be noted though, that since these departures are small, it is not likely that the development of absorptive materials that are uniformly absorptive for tones of all pitch will produce a great improvement in the speech-reception properties of rooms. The need for such uniformly absorptive materials appears to be desirable however for naturalness in the quality of speech, and more especially music.

Another characteristic of reverberation in a room is caused by the distribution of the absorptive material used in a room. Watson[10] recently has advocated the use of a highly absorptive space for the listener and fairly reverberant space for the generation of sound. That is, the stage would be left sufficiently reverberant to satisfy the taste of musicians and speakers, and the main part of the auditorium would be treated with very absorptive materials so that there would be little, if any, interfering effect from reverberation. As yet, no quantitative tests have been conducted to determine the effect on speech articulation of such a proposed distribution of absorptive material in an auditorium. The effect, no doubt, is beneficial, especially in large auditoriums where the reflected sound from the stage surroundings ("sounding board") enchances the direct sound wave.

Both of these features just mentioned—the variation of reverberation with pitch and the distribution of absorptive material in the auditorium—will have an effect upon the reduction factor k_r. However, it is not probable that either of these features has a very significant effect upon the "percentage speech articulation" in an auditorium. It is probable therefore that for a first approximation, the value of the time of reverberation for a tone of 512 d.v., as is commonly employed in current practice, can be used for determining, by means of Fig. 2, the appropiate value of k_r for any auditorium. The two features of reverberation under discussion are of unquestioned value in auditorium design, but their most important significance is in relation to the preservation of naturalness of speech and music rather than the improvement of "speech articulation."

[10] F. R. Watson, "Ideal Auditorium Acoustics," Jour. of Am. Inst. of Architects, July, 1928.

Effect of Noise upon the Reception of Speech in Auditoriums

It is a common observation that noise interferes with the hearing of either speech or music; that it produces what is called a "masking effect," the magnitude of which depends upon the loudness of the noise compared with the loudness of the speech or music. The interfering effect of noise upon the reception of speech in auditoriums has been investigated by the writer a number of years ago[11]. The principal

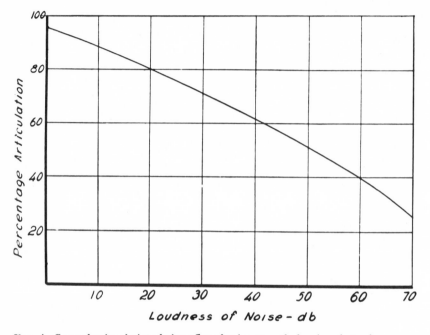

FIG. 4. *Curve showing the interfering effect of noise upon the hearing of speech.*

result of that investigation as applied to the present problem, is represented by the curve in Fig. 4. This curve shows how the articulation is affected by an interfering noise of different loudness levels, for speech which was maintained at a level of 47 db. The tests were conducted in a room which had a volume of 422 cu. m. (15000 cu. ft.) and a time of reverberation of 1.3 seconds. Since the noise tests were conducted in a room having a time of reverberation of 1.3 seconds, there was an interfering effect from reverberation, the magnitude of which can be determined from the curve in Fig. 3. The curve in Fig. 4

[11] V. O. Knudsen, "Interfering Effect of Tones and Noise upon the Reception of Speech," Phys. Rev., 26, 133–138, (1925).

has been corrected for the excess of reverberation above .5 second, and thus the ordinates in this curve are slightly greater than the observed values of the percentage articulation.

The interfering noise was conducted to the observer's ears by means of a pair of telephone receivers which were adjusted on the headband so that each receiver was maintained at a fixed distance of about one inch from the ear. In this manner, the listener observed the speech in the presence of a controllable noise, and further, the speaker was not bothered by the interfering noise. It will be noted that the artic-

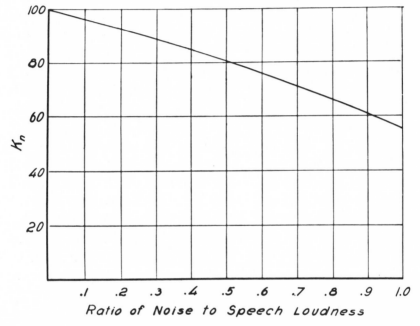

FIG. 5. *Curve showing how the noise reduction factor* k_n *depends upon the loudness of noise. The abscissa gives the ratio of the noise loudness to the speech loudness, where both are expressed in db.*

ulation decreases almost uniformly as the loudness of the noise increases. Further, it will be seen that even a slight noise produces an appreciable impairment. The complete absence of noise is thus seen to be an important factor for ideal hearing.

The harmful effect of actual noises in auditoriums is indicated by some tests which were conducted in two different auditoriums. In a certain high school auditorium, about which the writer was consulted, it was found that one of the chief defects was caused by noise from the ventilating fan and motor. Speech tests conducted in the empty auditorium, with the fan and motor in operation, showed an average speech articulation of 55%. With the fan and motor shut down the average articulation increased to 68%. In another auditorium it was found that the noise from an adjacent corridor was a source of interference. The separation of the auditorium from the corridor by means of a vestibule increased the articulation in the rear portion of the auditorium from 29.7 to 54.5%.

The curve in Fig. 5 is derived from the curve in Fig. 4. It gives the value of k_r for different loudness levels of noise. The loudness of the noise is here represented by the ratio of the noise, in db, to that of the speech, also in db. Thus, when the noise is at the same loudness level as the speech, the abscissa in Fig. 6 is 1.0. The value of k_n for no noise is taken as unity, and all other values of k_n are obtained by taking the ratio of the ordinate (in Fig. 4) for the loudness level in question to the ordinate for zero noise. This method of determining k_n is not strictly rigorous but it gives a close approximation which is sufficiently accurate for practical problems in auditorium acoustics.

The manner of using this curve would be as follows: First determine, by measurement if necessary, the average loudness of the noise in the auditorium. Take the ratio of this loudness level, in db, to the probable loudness level of the speech, in db, and read off from the curve in Fig. 5 the appropriate value of k_n. Thus, if it is found that the average noise level in an auditorium is 10 db and the average speech is 50 db, the value of k_n would be .925. The average noise prevalent in typical auditoriums is rarely lower than 5 db, and may sometimes be as high as 20 to 25 db.

Effect of Shape of Auditorium upon Speech Reception

The auditoriums in which the writer has investigated the effect of reverberation and noise upon the reception of speech were typical school and theatre auditoriums, and were essentially rectangular in shape, having dimensions of about $80' \times 110' \times 32'$ high. No quantitative tests have, as yet, been conducted in auditoriums which differ only in shape. There is undoubtedly some benefit to be gained from the use of suitably designed sounding boards, or from suitably located wall and ceiling surfaces near the speaker, but more data are yet needed to decide the exact benefit derived from such devices. The writer has tested the effect of a large reflecting sheet iron surface placed directly behind the speaker. Speech articulation tests were conducted in an empty auditorium with and without the reflector. The reflector was 10′ high and 12′ wide, and was placed directly behind the speaker, who stood near the middle of the stage floor. With the reflector in position, the average articulation in the auditorium was 62%, and with the reflector removed it was 59%. The observed difference of 3% is not great but it is greater than the experimental error and the reflector therefore is shown to possess some benefit. A similar benefit was observed, with the speaker on the front of the stage, when the

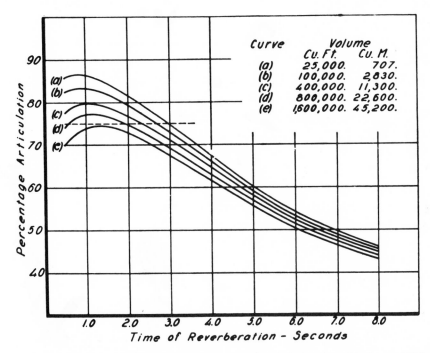

FIG. 6. *Group of curves giving the probable percentage articulation in auditoriums of different sizes and with different times of reverberation. These curves indicate that there is an optimal time of reverberation for the hearing of speech in an auditorium of a certain size.*

asbestos curtain was lowered so as to serve as a reflecting surface directly behind the speaker.

The influence of the shape of an auditorium upon speech reception requires further quantitative investigation. In the auditorium of conventional rectangular shape, it is probable that the k_s (as used in eq. (1)) does not differ appreciably from 1.0. In very large auditoriums, especially with curved surfaces, it is probable that k_s may be reduced to a value as low as .90. It is possible that in small rooms, or in auditoriums designed with properly shaped and located reflecting surfaces,

k_s may reach a value as high as 1.05.   For practical guidance in the design of auditoriums, it probably is advisable to assign a value of 1.0 to k_s, unless the shape of the auditorium is of peculiar design.

Combined Effects of Loudness and Reverberation upon the Reception of Speech in Auditoriums

In the earlier sections of this paper the effects of loudness and reverberation upon speech reception were considered separately.   It is obvious that as the time of reverberation in an auditorium is reduced, the average loudness of speech, assuming a constant power rate for the speaker, will be reduced correspondingly.   This suggests that there may be an optimal time of reverberation for speech in an auditorium. This would occur when a further reduction in the reverberation would

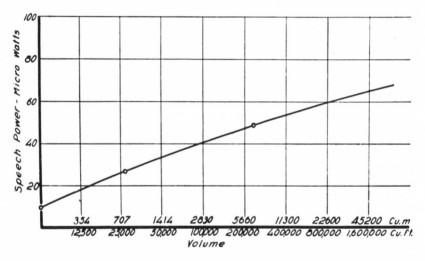

Fig. 7. *Curve showing the probable speech power of the average speaker in auditoriums of different sizes.*

concurrently reduce the loudness to the extent that the impairment occasioned by the diminished loudness would just compensate for the improvement occasioned by the reduction of the reverberation.   The manner in which this occurs is indicated by the series of curves which are plotted in Fig. 6.   These curves give the calculated values of the percentage articulation in auditoriums of different sizes and times of reverberation, for the probable average loudness of speech of the average speaker, based upon the measurements made in this investigation.   The average speech-power of the average speaker in auditoriums of different sizes is obtained from the curve in Fig. 7, which is based upon data obtained in the two auditoriums studied in this investigation and also upon the datum given by Bell Telephone engineers, namely that the average speech-power of the average speaker in a small room is 10 microwatts.[12] Having determined, from Fig. 7, the probable speech-power of the average speaker in a room of a certain size, and assuming that the speaker maintains this power output for different times of reverberation,[13] it is possible to calculate the resulting loudness in an auditorium of any size or time of reverberation. A typical set of calculations for an auditorium is given in table III. The values of k_l and k_r are determined from the curves

TABLE III

Volume of auditorium = 11,330 cu. meters (400,000 cu. ft.).
Average speech-power (from Fig. 7) = 54 microwatts.

Time of Reverberation	Average Speech Intensity*	Average Loudness, db	k_1	k_r	$k_1 k_r$
.50	$.665 \times 10^4$	38.2	.850	1.00	.850
.75	1.00	40.0	.874	.993	.868
1.00	1.33	41.2	.885	.982	.870
1.25	1.66	42.2	.894	.967	.865
1.50	2.00	43.0	.900	.953	.858
2.00	2.66	44.3	.910	.924	.840
3.00	4.0	46.0	.925	.837	.775
4.00	5.3	47.3	.936	.752	.704
6.00	8.0	49.0	.950	.600	.570
8.00	10.6	50.3	.959	.510	.489

* The speech intensity is given in terms of the minimal audible speech intensity.

in Figs. 1 and 3 respectively. It is assumed that the auditoriums are of the typical rectangular shape, so that $k_s = 1.0$. It also is assumed that the rooms are relatively free from disturbing noise, so that the residual noise is only one-tenth as loud as the speech, and therefore k_n will be .96. Eq. (1) then becomes

Percentage Articulation = $.922 \, k_l \, k_r$.

The ordinates of the curves in Fig. 6 were calculated by the use of this equation and a series of tables like Table III.

[12] It is assumed that this speech-power will be maintained by the average speaker in a room having a volume of 175 cu. meters (6200 cu. ft.).
[13] It is possible that a speaker raises the intensity of his voice as more and more absorption is brought into a room. This has not yet been tested, but it is not likely that a large error is is introduced by the assumption of constant power of the speaker's voice in the same room.

It is obvious that, for an auditorium of a certain size, the optimal time of reverberation will be the time for which the product k_lk_r, or the percentage articulation, will be a maximum. Thus the maximal values of the ordinates in the group of curves in Fig. 6 indicate the optimal times of reverberation in auditoriums of different sizes, for speech of about the average loudness that would be commonly used.

The entire series of curves shows very clearly how the reception of speech depends upon the size and time of reverberation of an auditorium. As would be expected, the optimal time of reverberation for speech reception in a small room is as short as .75 seconds. In a very small room the loudness is adequate and therefore speech will be heard more clearly and distinctly the nearer the reverberation approaches zero. In large auditoriums a somewhat longer time is advantageous because it promotes loudness. It will be noted that the peaks of the curves in Fig. 6 are rather broad and flat. This would seem to indicate that there is a considerable allowable variation in the time of reverberation from the optimal time, without appreciable sacrifice in hearing quality. In the design of auditoriums, therefore, it is desirable to so choose the absorptive treatment of the auditorium that the absorption furnished by different sized audiences will make the time of reverberation vary between the limits which determine the approximately flat portion of the curve.

Another factor must also be considered, namely, that the optimal time of reverberation for music is somewhat longer than for speech, and therefore it would be desirable to compromise between the requirements for speech and for music, especially in auditoriums which are to be used both for speech and music.

The optimal time of reverberation for auditoriums has been determined by Watson,[14] Lifshitz,[15] and others. These investigators have arrived at the optimal time primarily by calculating or measuring the time of reverberation in auditoriums which are pronounced good by competent critics. Lifschitz has derived a semi-empirical formula for calculating the optimal times of reverberation[16] for auditoriums of different sizes. This formula yields results which are in fair agreement

[14] F. R. Watson, "Acoustics of Buildings," John Wiley and Sons; Jour. Frank. Inst., July, 1924.

[15] Samuel Lifschitz, Phys. Rev., 27, 618–621, (1926).

[16] This formula is based upon the assumption that the product of the time of reverberation and the loudness, expressed in db, should be constant. Lifshitz calls this product "the energy of musical perception." The assumption seems rather arbitrary but it is fairly consistent with the observed facts.

with Watson's results and with currently accepted optimal times of reverberation. The top curve in Fig. 8 shows the values of the optimal time of reverberation as given by Lifschitz. Although Lifschitz states that his results apply both to speech and music, it is probable that they apply more particularly to music, since the results are based upon the judgments of listeners who regard loudness, resonance, euphony and other qualities as determining factors. The lowest curve in Fig. 8 shows the optimal time of reverberation for speech, based upon the maximal values of the curves in Fig. 6. It would seem that the bottom and top curves in Fig. 8 give the most trustworthy available data for determin-

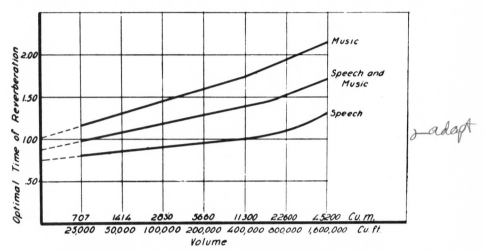

Fig. 8. *Curves showing the optimal time of reverberation for auditoriums of different sizes. The upper curve is taken from the data of Watson and Lifshitz. The lower curve, for speech, is obtained from the maxima in Fig. 6. The middle curve is the arithmetical means of the upper and lower curves, and represents a reasonable choice for both speech and music.*

ing the optimal time of reverberation in auditoriums for either speech or music, where no provision is made for amplifying the power of the voice. If an auditorium is to be used both for speech and music, as is usually the case, it would seem advisable to use the mean value of the two curves. The middle curve is such a mean value curve, and thus gives the optimal time of reverberation for both speech and music.

[*Editor's Note:* Material has been omitted at this point.]

4

Reprinted from *Inst. Electr. Eng. J.* **88**(33), Pt. 3:175–185 (1941)

ACOUSTICS OF CINEMA AUDITORIA*

By C. A. MASON, B.Sc.(Eng.), and J. MOIR, Associate Members.†

(*Paper first received 6th March*, and *in revised form 28th October*, 1940. *The paper was selected for reading and discussion before the* WIRELESS SECTION *on the 4th December*, 1940, *but it was found necessary to cancel the meeting.*)

SUMMARY

The paper records investigations made in an endeavour to discover the reasons for the difference of performance in the reproduction of sound by identical sound equipment in apparently similar cinemas.

Details are given of frequency characteristics and reverberation times, and the requirements of good auditorium design are set out.

The sound levels picked up by a microphone in an auditorium, and recorded on a high-speed level recorder, are shown by means of oscillograms.

INTRODUCTION

Past attempts to improve sound reproduction from films have been directed principally towards improving the frequency response and volume range of the sound equipment and to correcting the reverberation time of the auditoria in order to achieve an agreed optimum rate of decay of the sound. In recent years there has been a growing conviction that other factors may be involved and that these may be of major importance. It has been a point of general experience that identical sound equipment installed in similar auditoria do not produce comparable results. Installation differences have been overcome by making available to the installation engineer testing apparatus capable of checking rapidly the complete electrical and optical characteristics of the equipment from film to loud-speaker, and reverberation time measurements have shown sufficient agreement to remove criticism on this account.

In spite of every care, differences in performance still exist and some cinemas become known for their " good sound," while others apparently identical in every respect are constantly criticized. Faced with this problem the authors commenced the investigation outlined in the following paper. The results emphasize the care necessary in the detail design of the auditorium and in the correct location of the loud-speakers to avoid long-path reflections and the importance of instantaneous sound measurements in obtaining overall performance characteristics of the equipment and hall.

---

* Wireless Section paper.   † British Thomson-Houston Company, Ltd.

The authors feel that no apology is necessary for presenting to an audience of electrical engineers the results of an investigation containing many acoustical problems, because in their opinion further improvement in sound equipment, and particularly in loud-speaker design, is impossible without a thorough appreciation of the effects of the hall in which the apparatus has to be installed.

SUBJECTIVE SURVEY OF CINEMAS

The first step in the investigation consisted of a subjective survey of some 120 cinemas, each having identical sound installations, and the classification of results under the headings of " Average," " Above Average," and " Below Average " sound quality. In order to eliminate as far as possible the personal opinions of the authors a short questionnaire was circulated to sound engineers in charge of the installation and servicing of cinemas throughout the country. A copy of this questionnaire is shown on page 176. The district sound engineers by reason of their experience and contact with the cinemas over many years are well qualified to appraise the sound quality without being led astray by fortuitous changes due to the sound recording on particular films.

It will be seen that the reply to the questionnaire mentions the word " intimacy," and its use in this connection requires explanation. Briefly, intimacy is the close association of sound and picture focus giving the audience the impression that the sound proceeds directly from the screen. The effect is also known as good " presence." Separating the loud-speaker and screen makes the absence of intimacy very noticeable, and if the eyes are opened and closed the separation of sound source and picture can be detected more readily. Even with the loud-speakers mounted behind the screen the sound focus may wander far from the screen and the sense of sound focus may be lost altogether in a cinema where intimacy is poor. It is an effect often confronting the sound engineer in the cinema, but so far as the authors are aware it has received only scant reference in the literature of the art. However, intimacy is extremely important, as theatres are often adversely

Date___*October 8th, 1938.*___

**Questionnaire Concerning Acoustic Properties
of Cinemas**

Name and Address of Cinema_____

Type of Equipment___*35 Watt RPR.*___ Form *RK*

Description of Speakers___*9-Cell H.F.; Standard L.F.*___

General results: ~~AVERAGE~~

~~BELOW AVERAGE~~
ABOVE AVERAGE

(Strike out descriptions which do not apply).

(1) Describe any factors which you think may be responsible.

*It would appear that acoustic conditions in this
hall are excellent, although untreated, intimacy is very good
and ratio of absorption to cubic capacity appears right.*

(2) Are there any acoustic peculiarities in the auditorium?

None.

(3) Is there any indication that the number of audience affects reproduction?

Very slight.

criticized for poor intimacy alone, despite the excellence of the other factors which contribute to good sound. In general, where intimacy is poor the critic is uncertain of the reason for the poor sound, while being perfectly certain that it is unsatisfactory.

Intimacy, or accurate sound focus, must be fully understood and achieved if stereophonic sound is to have practical application in the cinema, because the fundamental requirement of stereophonic sound is that movements of the sound focus across the screen may be appreciated and followed by the audience.

A careful analysis was made of the questionnaire, and extremes in each district were chosen for further tests to ensure that known factors were not primarily responsible for the observed differences. Frequency characteristic and reverberation time were measured at many points in each theatre and the results compared. The methods of measurement involve no new technique and are therefore the subject of a brief Appendix. Results obtained in three typical theatres will serve to indicate the nature of the problem.

In theatre No. 1 sound was excellent throughout and had received very favourable comments from many quarters. Theatre No. 2 was generally below average, while in theatre No. 3 sound was generally very good, but at certain points was poor.

FREQUENCY CHARACTERISTICS

Typical frequency characteristics taken in the three theatres are shown in Fig. 1, while Fig. 2 contains more detailed data on the frequency characteristic at four typical points in theatre No. 3. Curves " a " and " b " in Fig. 2 were recorded at points where the sound quality was considered to be above average, and curves " c " and "d " at points where the sound quality was below average.

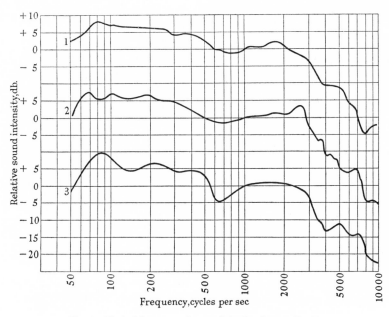

Fig. 1.—Typical frequency characteristics for three auditoria.

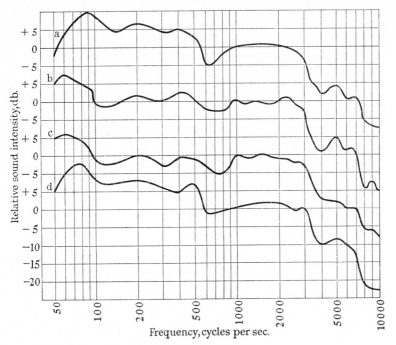

Fig. 2.—Overall frequency characteristics in different parts of auditoria.
Curves a and b.—Sound quality good.
Curves c and d.—Sound quality inferior.

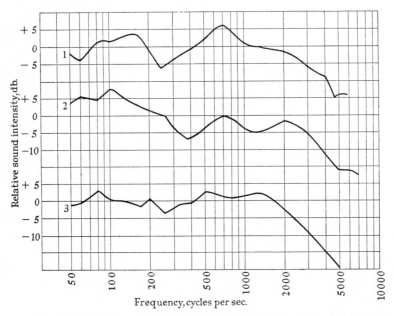

Fig. 3.—Typical frequency characteristics for three leading Hollywood theatres.

Consideration of these results made it evident that there is no significant difference in the frequency response between the good and bad halls of theatres Nos. 1 and 2 (Fig. 1) or between the good and bad position in theatre No. 3 (Fig. 2). Indeed it is worthy of comment that, in the first five cinemas checked, the sound quality was almost inversely proportional to what might have been expected from the critical study of the frequency characteristics alone. This result led to prolonged checking and re-checking of the measuring apparatus without revealing any fault. The accuracy of the technique was further confirmed by the results obtained in three leading Hollywood theatres by American investigators.[1] These are shown in Fig. 3 and it is apparent that the frequency characteristics are comparable with those in Figs. 1 and 2.

It is of interest to observe that none of the characteristics obtained in the cinemas compares even remotely with the normal axial frequency/response curves of the high-frequency loud-speakers obtained in the usual free-space testing condition. In Figs. 1 and 2, for example, the characteristic falls away above 2 000 c./s., while Fig. 4 is

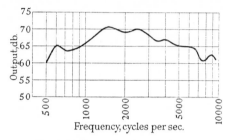

Fig. 4.—Axial characteristic of h.f. speaker.

the frequency characteristic of an identical loud-speaker obtained under test conditions. Early in the investigation the frequency/response characteristics in the cinema were modified until a substantially level acoustic output was obtained to between 5 000 and 5 500 c./s. The aural results were considered unacceptable by the whole of the listening group, and after repeated tests the response characteristics represented in Figs. 1 and 2 were found to give the best aural results. Similar listening tests have been carried out in America, with the result shown in Fig. 5. The method employed in getting this characteristic

was generally the same as that outlined above, modification being made to the equipment until the aural results met the approval of a listening group. In comparing Fig. 5 with Figs. 1 and 2 it should be remembered that the optimum frequency characteristic will depend to some extent on the reverberation-time/frequency curve. This information is not given for the American cinemas and may explain the slight difference in the shape of the curves.

The results in the three cinemas are representative of the results in many cinemas investigated and have led to the conclusion that either frequency response is not the major factor or that the measured frequency characteristic does not represent the characteristic to which the ear responds. Attention was therefore directed towards the acoustic properties of the halls and later to a study of methods of instantaneous sound measurements.

REVERBERATION TIME

Reverberation time is defined as the time taken for a diffuse sound field to be attenuated by 60 db., and is obviously a function of the sound-absorbing property of the interior surfaces of the auditorium. In a uniform rectangular enclosure there is little doubt that a single value represents the reverberation time for the enclosure, but in a hall with a balcony (see, for example, Fig. 6) there is

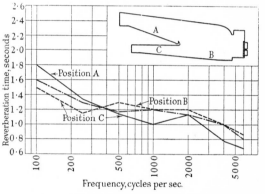

Fig. 6.—Reverberation time at various positions.

reason to doubt whether a single rate of decay exists for the complete enclosure, or whether it might not be more

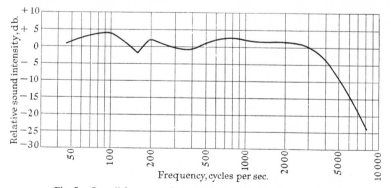

Fig. 5.—Overall frequency characteristic (adjusted for best aural results).
Curve published by American investigators [see Bibliography, (1)].

accurate to consider the enclosure as three separate spaces coupled by the balcony openings. Check tests were made in a typical theatre and the results shown in Fig. 6 suggested that the variations are not sufficient to justify consideration as three separate coupled spaces.

Table 1 gives the reverberation times at 500 c./s. for the

Table 1

500-CYCLE REVERBERATION-TIME DATA ON THEATRES

Theatre	Measured	Optimum	Sound quality
No. 1	1·29	1·38	Above average
No. 2	1·2	1·4	Below average
No. 3	1·13	1·3	Average

Location	500-cycle reverberation time	Sound quality
a	1·11	Above average
b	1·21	Above average
c	1·14	Below average
d	1·11	Below average

three theatres previously discussed, and the reverberation-time measurements in theatre No. 3 at the four positions

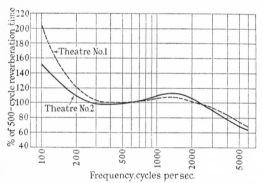

Fig. 7.—Reverberation time as percentage of 500 c./s. value for two theatres.

a, b, c and d, at which the frequency characteristics were measured.

There was still the possibility that the reverberation time varied widely with frequency, although this was unlikely because the furnishing and constructional arrangements of the theatres were generally similar. The possibility was investigated, with the results shown in Fig. 7. This is a record of the reverberation-time/frequency curves for theatres Nos. 1 and 2, and Fig. 8 shows the similar characteristics for the four positions in theatre No. 3. Here again the differences are not sufficient to account for the differences in the aural result.

Considerable data have already been published on the optimum reverberation time as a function of theatre volume and frequency. This information is the result of several theoretical studies and field experience. Fig. 9 represents the authors' optimum time/volume relation together with the recommendation of another large organization,[2] curve B being the 1931 figures and curve C published in 1936, presumably the result of further field experience. An average curve is repeated in Fig. 10 and added to this are the 500-cycle reverberation times of several good and bad theatres. No consistent differences are apparent.

In passing it is of interest to note that on certain assumptions[3] it is possible to derive the optimum shape of the reverberation-time/frequency curve, and while the calculated curve is in general agreement with the measured results in theatre No. 2 below 500 c./s., it departs widely at higher frequencies where, according to the calculated curve, the reverberation time should increase to approximately 150% at 4 000 c./s. The suggested rise in the theoretical curve has little support in the authors' or other investigators' practical experience, which agree in requiring that the reverberation time should decrease steadily above 3 000 c./s.

For obvious reasons the measured times apply to the theatre's empty condition, although the suggested optimum figures are intended to hold with approximately two-thirds of the audience present. It is known that the presence of an audience modifies the acoustic condition, generally leading to improved sound quality, if any effect is at all noticeable. Knowing the reverberation time of the empty theatre it becomes an easy matter to recalculate the probable

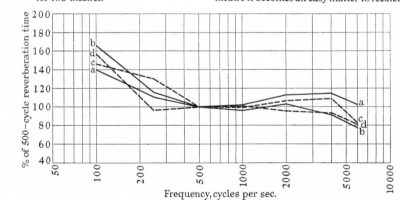

Fig. 8.—Reverberation times at different positions in auditorium.

Curves a and b.—Sound quality good.
Curves c and d.—Sound quality bad.

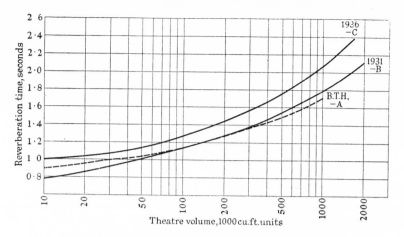

Fig. 9.—(Optimum reverberation time)/(theatre volume) at 500 c./s.

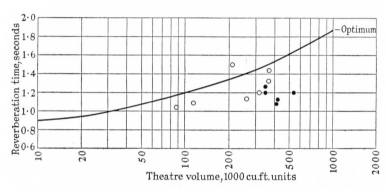

Fig. 10.—(Theatre volume)/(reverberation time)/(sound quality) results.

● ● Sound quality above average.
○ ○ Sound quality below average.

time with the audience present, but, as a check, the effect of an audience in a small theatre was measured, the data being presented in Fig. 11. From these data the average

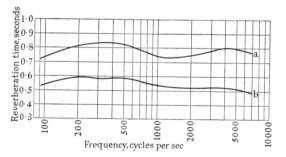

Fig. 11.—Reverberation time in small auditorium.

Curve a.—Empty.
Curve b.—Full.

absorption coefficient per person has been calculated and the data are presented in Table 2, together with Sabine's original figures.

Table 2

AVERAGE ABSORPTION COEFFICIENT OF AN AUDIENCE

Frequency c./s.	Authors' figures	Sabine
100	4·98	3·6
250	5·58	4·3
500	5·37	4·7
1 000	5·86	4·7
2 000	6·25	5·0
4 000	9·7	5·0

It becomes obvious that modern theatres with about two-thirds of the audience present have reverberation times far below the generally assumed figures. The effect, however, is common to all theatres, and cannot contribute to an explanation of observed differences between similar cinemas.

The results of the foregoing measurements, which are typical of many other examples, have convinced the authors that reverberation times and frequency characteristics are not the main criteria, judged by the subjective assessment of sound quality in large auditoria, unless they depart very widely from optimum conditions.

EXTENSION OF INVESTIGATION

The frequency characteristic as measured with normal equipment may not represent, even to an approximate degree, the frequency characteristic which is heard. Measurements are usually made under steady conditions, whereas in practice the ear picks up the sound, which is rapidly changing and is more often composed of transients.

It appears probable that the ear estimates the direction of approach and frequency response by the characteristic of the sound arriving during the first few milliseconds. Existing measuring equipment will average over a much longer period, including in the sum the direct sound and the reflected sound reaching the microphone over a period corresponding to the operating time-constants of the instrument. The frequency characteristic of the reflected sound is entirely different from that of the direct sound, the proportion of the high frequency decreasing at every reflection. On this hypothesis measuring equipment having directional characteristics and time-constants of the ear is required to obtain a true record of the effective response, and it is to be expected that such measuring equipment would show a better response in the high-frequency end than is obtained with the normal recording equipment.

Similarly normal reverberation-time measurements are made under conditions of uniform diffuse sound energy distribution in the auditorium. In practice, sound is propagated from a small area at one end of the auditorium, and in addition is continuously changing in character, so that even approximate uniformity can never exist. These arguments led to further lines of investigation, which are now described.

HIGH-SPEED FREQUENCY CHARACTERISTIC

It was anticipated that if a characteristic could be taken in a time interval which was so short that a signal direct from the loud-speaker covering the complete frequency range could be picked up by the microphone before any reflected sound was picked up, a characteristic would be obtained which approximated more closely to that picked up by the ear. Apparatus was designed to enable a variable-frequency signal to be injected into the theatre equipment covering a range from 40 to 10 000 c./s. in about 0·05 sec. This signal was applied at intervals of about 2 sec., so that all reflected sound would have reached negligible proportions before the application of the next signal.

The characteristic was much too rapid to be recorded by the high-speed level recorder (see Appendix), and a cathode-ray tube was used. A tube using a screen with a long time delay was chosen, so that each image persisted for approximately 2 sec., giving a continuous image for signals at 2-sec. intervals. Whilst quite satisfactory images were obtained, there was considerable evidence to show that even with a signal of 0·05 sec. duration reflected sound was causing serious trouble before the end of the signal. When working from low frequency to high, the recorded level of the high-frequency response was modified by reflected low-frequency sound. The effect was confirmed by reversing the oscillator drive so that the signal

ran from high frequency to low, when similar effects were noted in the reverse order. Attempts were made to reduce the effect by reducing the duration of the signal to 0·02 sec., but even then serious errors were still noticed.

Moreover, in such a short time interval it is doubtful whether the signal applied to the loud-speaker covers adequately the audible range of frequencies. The frequency scale of the oscillator was logarithmically graded, covering about 8 octaves. For the complete sweep of 0·02 sec. each octave occupied 0·0025 sec. This means that the note was changing, for example, from 1 000 to 2 000 c./s. in 1/400th sec., and no signal will exist long enough to be effective as a single frequency. At low frequencies conditions were much worse.

The results of further investigation suggested that the direct and reflected paths may differ by only a few feet, and the problem of sweeping through the frequency characteristic in a time short enough to avoid reflections, but long enough for the effects of the whole band of frequencies to be recorded, became insuperable. The method was therefore abandoned.

INVESTIGATION OF SOUND MEASUREMENT BY IMPULSES

Following the failure of the high-speed trace of the frequency characteristic, a new line of investigation was undertaken in which the paths taken by sounds of short duration in the auditorium are traced out by means of impulses applied to the loud-speakers. The technique is as follows.

The output from a best-frequency oscillator is applied to the theatre amplifiers through a continuously working cam-operated switch. Adjustments are provided enabling the duration of contact to be controlled from 0 to 0·05 sec. The impulse therefore consists of a train of waves the frequency of which is determined by the oscillator setting, and the duration of which is controlled by the adjustment of the cam switch. The impulse is transmitted through the theatre amplifier to the loud-speakers in the auditorium, and a microphone is located in the auditorium to pick up sound. The output from the microphone is amplified and applied direct to the vertical deflector plates of a cathode-ray tube. The horizontal deflector plates are connected to a time base which is synchronized with the rotation of the cam that operates the impulse switch. A diagram of the complete set-up is shown in Fig. 12. Impulses are applied at intervals of about 1 sec., so that reflected sound decays to a negligible value before the next impulse. The horizontal time scale is arranged for a sweep time of 0·5 sec.

Except at very low frequencies, individual waves in the wave train comprising the impulse cannot be seen individually, but the complete impulse appears as a luminous vertical band on the screen of the cathode-ray tube.

A typical impulse injected into the amplifier is shown in Fig. 13(a) (see Plate 1, facing page 182), while Fig. 13(b) shows the same impulse at the output stage of the amplifier. It will be noticed that the impulse is distorted a negligible amount in the amplifier. Fig. 14 (Plate 1) shows the same impulse as picked up by a microphone in front of a high-frequency loud-speaker fitted with a multi-channel horn, in an acoustically dead room. Fig. 14(a) is taken with the microphone located on the horn axis, and Fig. 14(b) with

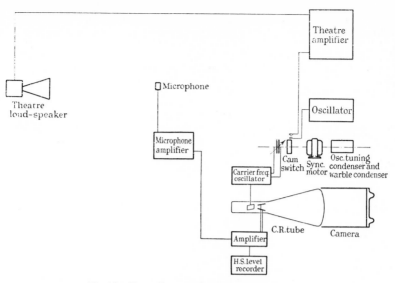

Fig. 12.—Set-up for acoustic survey in auditorium.

the microphone 30° off the horn axis. It will be noticed that the impulse is still not unduly distorted and retains its square-topped form.

The impulses shown in Fig. 14 were taken with the microphone located 6 ft. from the loud-speaker in an acoustically dead room. The length of impulse used was about twice that which has been used for investigations in auditoria. Images picked up by the microphone for typical conditions in a few types of auditoria will now be considered.

IMPULSE MEASUREMENTS MADE IN TYPICAL AUDITORIA

The oscillographic records for a reverberant room show the initial impulse received direct from the loud-speaker, followed by the reflected sound impulses from the walls and ceiling. A study of the time interval between these impulses enables the reflecting surfaces to be located and the sound paths to be determined.

Figs. 15(a) and 15(b) (Plate 1) show two cathode-ray tube images obtained in a large theatre where sound quality was uniformly very good. Diagrams of this auditorium are given in Fig. 16. Sound-absorbent treatment is confined to the rear walls, the floors being heavily carpeted and the seats being of a luxurious type with high sound absorption. It can be seen from Fig. 16 that the shape of the auditorium is such that for a position A there is little chance of anything but directly incident sound being picked up. Direct reflections from side walls or ceiling are not possible. This is illustrated in Fig. 15(a), where the initial impulse is seen to have reflections of a negligible order. Position B (Fig. 16) at the front of the balcony indicates a possibility of some reflection from the ceiling, and this is shown in Fig. 15(b). The reflection, however, is of a small amplitude and is separated from the fundamental by a very small time interval and does not cause any deterioration in sound quality. At the side of the theatre in position C the possibility of multiple re-

flections is high, but the differences in path length are low and the maximum difference between the subtended angles of incidence is small. For this reason the reflected sound has negligible detrimental effects on sound quality. This point is discussed in more detail later.

Fig. 17 shows an auditorium where sound quality was very good, except at a few isolated locations, one of which will be considered. The shape and acoustic treatment

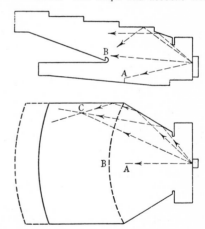

Fig. 16.—Cinema No. 4. Example of theatre of good design.

of the auditorium is such that the same arguments relating to absence of reflections apply as to Fig. 16. Position A (Fig. 17) is typical of this condition, and the corresponding cathode-ray tube image is shown in Fig. 15(c). One isolated bad spot in the theatre occurred under the front of the balcony, and the cathode-ray tube image for this position is shown in Fig 15(d). It shows reflections from the back wall at 80 millisec. delay. A further reflection at 220 millisec. delay, which appears

Plate 1

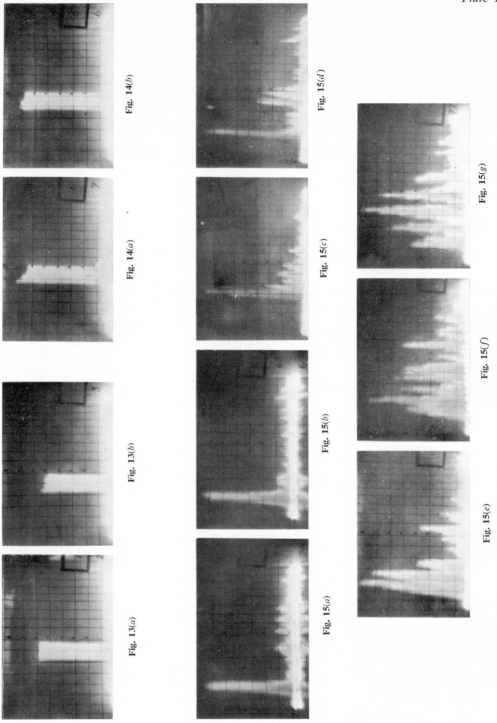

Fig. 14(b)

Fig. 14(a)

Fig. 13(b)

Fig. 13(a)

Fig. 15(d)

Fig. 15(c)

Fig. 15(b)

Fig. 15(a)

Fig. 15(g)

Fig. 15(f)

Fig. 15(e)

Plate 2

Fig. 19

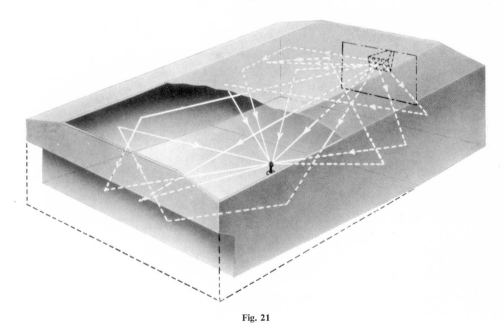

Fig. 21

Plate 3

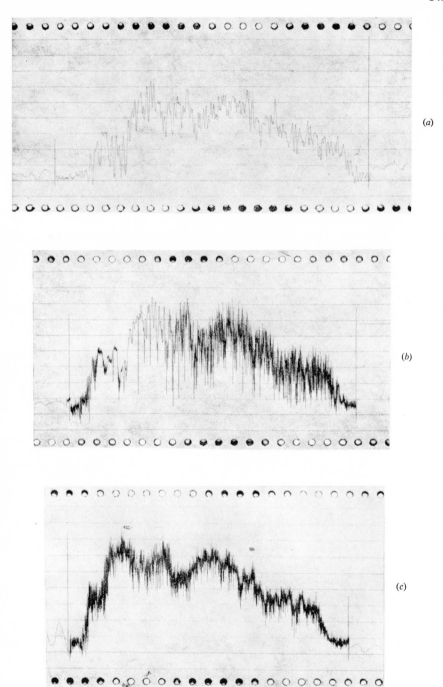

(a)

(b)

(c)

Fig. 23.—Frequency characteristic records.

Plate 4

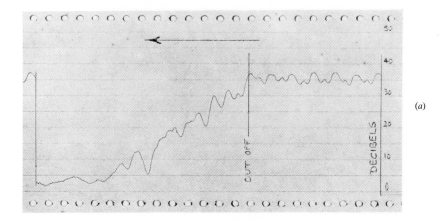

(*a*)

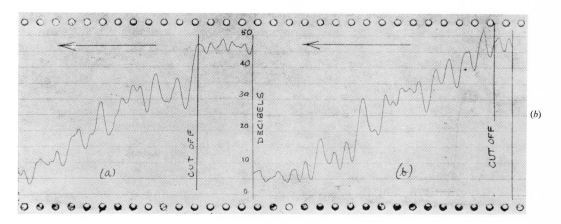

(*b*)

Fig. 24
(*a*) Typical reverberation decay curve.
(*b*) Reverberation decay curves showing differences in initial rate of decay.

to be due to reflection from the rear wall of the balcony on to the ceiling, and then from the curved proscenium arch and sides as indicated in Fig. 17 (position B), is also shown. Experience has indicated that the echo with less than 45 millisec. delay from the back wall can be tolerated, but that the echo with the time delay greater

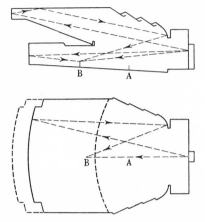

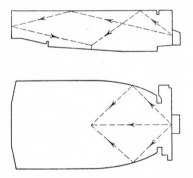

Fig. 17.—Cinema No. 3. Example of theatre showing local sound reflections.

than 50 millisec. leads to a deterioration in sound quality due to lack of intelligibility. It is interesting to note that the frequency characteristics of the equipment taken at positions A and B are both similar and satisfactory, and do not in any way account for the difference in sound quality. These characteristics have been shown in Figs. 2(*a*) and 2(*c*).

Fig. 18.—Cinema No. 5. Example of theatre where intimacy is poor.

Fig. 18 and Fig. 19 (Plate 2) show an auditorium where sound quality was of a medium order, the main complaint being lack of intimacy. The shape of the auditorium, as indicated in Figs. 18 and 19, shows a large expanse of flat roof and walls which are parallel for a considerable distance. Sound-absorbent treatment has been applied to the back wall only. For the microphone position shown in Figs. 18 and 19, the cathode-ray tube photograph shown in Fig. 15(*e*) was obtained. The timing of the reflected im-

pulses indicates that sound reaches the centre of the auditorium by reflection from the side wall, reflection from the ceiling, reflection from the angular space between side wall and ceiling, and a small amount of reflection from the back wall. Since the time delay of all reflections is relatively short, there is little interference with intelligibility, but the directions from which sound emanates is not well marked, and intimacy suffers accordingly. It will be noticed from the reflected paths shown in Figs. 18 and 19 that the solid angle subtended by the incident sound at the centre of the auditorium is far greater than the solid angle subtended in Fig. 16 (position C), which accounts for the much inferior intimacy in this theatre.

The auditorium shown in Fig. 20 and Fig. 21 (Plate 2) is

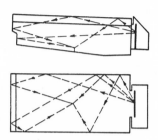

Fig. 20.—Cinema No. 6. Example of theatre with bad acoustic properties.

one where the quality of sound is very inferior, intelligibility and intimacy suffering considerably. At two typical positions in the auditorium, the impulse photographs shown in Figs. 15(*f*) and 15(*g*) were obtained. These indicate large reflections at both short and long time intervals, and the reason for the loss of intelligibility and intimacy is obvious. Both measured frequency characteristic and reverberation time for this auditorium were quite satisfactory.

Impulse measurements have been made in many auditoria in addition to the above and the same general agreement was found to exist between the aural judgments and the results predicted from the impulse photographs.

GENERAL REMARKS ON IMPULSE TESTS

Much still remains to be done in perfecting the technique, but it does appear that quantitative measurements are possible of factors that until now have only been treated qualitatively.

In all the theatres, observations have been made at three positions close together, and at five frequencies—250, 500, 1 000, 2 000 and 3 000 c./s. The measured reflected impulses are then averaged before any conclusions are drawn. The impulse photographs in Fig. 15 are those which approximate most nearly to the average suggested from 15 or 20 observations at each location. There is some change in the type of reflection photograph with changes of frequency and it is thought that this change with frequency is connected with the peculiar discriminating properties in the sound radiated from the loud-speaker in directions at large angles to the normal axis of the loud-speaker.

It is also noticed that the shape of the initial impulse from the loud-speaker as picked up in an auditorium is inferior to that as picked up under free-space conditions or in the acoustically dead room in the laboratory. This point can be seen by comparing Figs. 14 and 15 and remains to be explained.

Emphasis is placed on the value of the reflection photographs in calculating sound paths in an auditorium. In the majority of theatres it has been impossible to estimate the paths from two-dimensional drawings, and the problem has had to be considered largely as one of three dimensions.

Before the investigation was commenced, the significant features in the shape of an auditorium were not fully understood. From architects' plans, or even an examination of the theatre itself, it had been found impossible to deduce the paths taken by reflected sound. The reflection photographs have proved of great assistance in indicating the actual paths taken by reflected sound in typical auditoria. Accordingly, it is now possible to anticipate with fair accuracy the probable sound paths from the plans of a new auditorium and so to forecast the effects of the hall design on intimacy or intelligibility.

REQUIREMENTS OF GOOD AUDITORIUM DESIGN

From the data already collected it appears that further major improvements in sound-film reproduction will require closer co-operation between the equipment designer, the acoustical engineer and the architect in order that the maximum performance be secured from the equipment and the theatre considered as a unit. The authors feel that equipment development is now at such a stage that the predominant factor in obtaining " good sound " is the control of sound reflections in the auditorium, although, of course, this does not imply disregard of the other factors.

It is realized that in many instances the shape and dimensions of a theatre are fixed by the available site, local regulations, the economic viewpoint, etc., but a short discussion of the ideal requirements may be of interest. There are four acoustic factors to which attention must be paid:—

(1) The reverberation time (theatre empty) should approximate to the optimum curve of Fig. 10.

(2) The shape of the reverberation-time/frequency curve should approximate to that of Fig. 7.

(3) The avoidance of reflection paths which exceed the direct path by more than 45 ft. The authors' tests have indicated that reflection paths exceeding this length lead to reduced intelligibility.

(4) The unavoidable reflected sound paths should subtend a small angle at any point in the audience. The test results have indicated that large subtended angles lead to poor intimacy.

With present methods of construction and furnishing, (2) is probably satisfied automatically when requirement (1) is taken care of.

With the present standard of construction and furnishing it appears to be impossible to meet requirement (1) with the audience present, but the authors are inclined to think that reproduction would be improved if this condition could be satisfied. The same remarks apply to requirement (2).

Condition (3) calls for the avoidance of reflection paths which exceed the direct sound path by more than 45 ft., corresponding to a time difference of 40 milliseconds (maximum). This condition must be met in any theatre irrespective of size, and accordingly a standard scale design

with a table of multiplying factors cannot be produced. The majority of theatres built during the last few years fall into the 1 000–1 500 seating capacity class, and to meet this range the design of Fig. 22 is suggested.

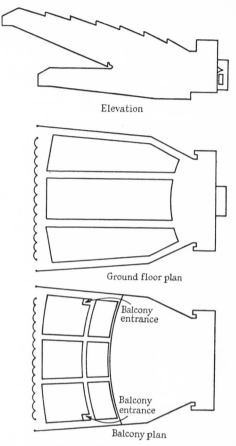

Elevation

Ground floor plan

Balcony entrance

Balcony entrance

Balcony plan

Fig. 22.—Details of proposed theatre.

A balcony type of structure is adopted having the main ceiling stepped down towards the proscenium arch in order to minimize the hall volume and building costs. The reduction of ceiling height towards the proscenium opening has many advantages over the more normal type with the ceiling parallel to the floor. Reflections from the ceiling, back into the audience, follow a path which does not depart seriously from the path of the direct beam of sound from the speaker. In addition, sound incident upon the ceiling is to a large extent scattered by the reflection and diffraction by the ceiling steps which face the stage speaker equipment.

The reverberation times of the three main spaces forming the theatre tend to be equalized by this construction, preventing sound-energy flow from a space having a high reverberation time into a space having a lower time.

The proscenium opening is splayed into the side walls with an included angle of approximately 80° to maintain a sound-energy flow roughly parallel to the splays. As the hall is intended solely for use with a sound reproducer, the side splays are not intended to reinforce the sound source and indeed must not be used for such a purpose. Consequently these splays should be broken up by pillars, panels or other reliefs. Non-parallel side walls are maintained up to the rear wall in order to minimize the time required to produce completely diffuse sound within the enclosure.

The curved rear wall is eliminated and the rear walls are tilted forward in order to reflect incident sound into the rear seating, compensating for the normally lower-intensity level in the rear seats produced by attenuation from the front to the rear of the hall, and preventing the return of echoes of long delay time to the front of the theatre.

Exits and gangways are placed against the side walls in both balcony and ground floor to occupy space difficult to feed with direct sound. The more usually adopted central position for the balcony entrance occupies the equivalent of some 30–40 seats in a position particularly favourably placed for good sound. Curved surfaces should be avoided and where they are absolutely necessary should be kept to a radius not exceeding a few feet.

The question of acoustic treatment is a point of considerable importance. It is the authors' opinion that the theatre design should be such that acoustic treatment is not required, reflection of sound being taken care of by breaking up offending surfaces as part of the decorative scheme. From the data presented in the paper it follows that acoustic treatment is rarely required for the correction of reverberation time owing to the luxurious standard of seating and carpeting in the present-day theatre. Placement of a large proportion of absorbent in the seats is advantageous in minimizing the change in acoustic condition produced by the audience.

When acoustic treatment is required it should be provided, in the form of strips, small panels, etc., of material having an absorption coefficient in the neighbourhood of $0\cdot25$–$0\cdot4$. These strips should be spaced about the side walls. Concentrated areas of material having high absorption coefficient is wrong. Although leading to the same final reverberation time, the initial decay processes are dissimilar and it will be appreciated that the initial stages occupying the first 20–30 db. of the decay period are all-important. From many points of view it would be advantageous to manufacture floor covering with a lower absorption coefficient, the balance being supplied by side-wall treatment of material having a relatively low absorption coefficient ($0\cdot2$–$0\cdot4$). Ceiling treatment is very rarely required and is generally harmful. Sound energy flowing between floor and ceiling is heavily attenuated by the concentration of absorption in the form of seats and carpets, and consequently any further treatment should be confined to the walls.

Back-stage volume should be held to a minimum and the speaker placed as far forward and as heavily draped as possible in order to reduce the sound energy present in the back-stage space.

This combination of requirements has not been found in any one theatre, but the more closely they have been approached the more favourably has the sound been received by all concerned.

Acknowledgments

In conclusion the authors wish to thank Mr. H. Warren, Director of Research of the British Thomson-Houston Co., for permission to publish the results of this investigation, to Mr. Oscar Deutsch and Odeon Theatres for permission to make investigations in theatres of the Odeon circuit, and to Sound Equipment, Ltd., for their active co-operation. Their thanks are also due to Mr. H. L. Webb for assistance in the construction and operation of the test equipment.

REFERENCES

(1) F. Durst and E. J. Shortt: *Journal of the Society of Motion Picture Engineers*, 1939, **32**, p. 169.

(2) W. A. MacNair: *Journal of the Acoustical Society of America*, 1930, **1**, p. 242.

    S. K. Wolf and C. C. Potwin: *Journal of the Society of Motion Picture Engineers*, 1936, **27**, p. 386.

(3) V. O. Knudsen: "Architectural Acoustics" (John Wiley & Sons, New York).

[*Editor's Note:* The appendix and written discussions appearing on pages 185–190 have been omitted due to limitations of space.]

5

Reprinted with the permission of the Controller of Her Britannic Majesty's
Stationery Office from *J. Acoust. Soc. Am.* 25(2):246-259 (1953)

The Acoustics of the Royal Festival Hall, London*

P. H. PARKIN, W. A. ALLEN, H. J. PURKIS, AND W. E. SCHOLES

Department of Scientific and Industrial Research, Building Research Station, Garston, Watford, Herts, England

(Received December 12, 1952)

The Royal Festival Hall was opened in London in May, 1951. This paper describes in some detail its acoustical design, the test concerts held in it before the opening, the objective measurements made in it, and the comments about its acoustics that have been made during the first eighteen months since its opening. These comments show that the "definition" is excellent, but that for some types of music more "fullness of tone" would be desirable. It is concluded that the reverberation time is the only objective measurement which, at the present stage of development, is of practical use. Its value in the Royal Festival Hall when full is 1.5 seconds (at 500 cps), which is 0.2 second shorter than the optimum value given by Knudsen and Harris for a hall of this size. It seems probable that the "fullness" would be adequate if the reverberation time could be lengthened to 1.7 seconds or somewhat longer.

. . . . .every soun,
Nis but of aier reverberacioun,
And ever it wastith lyte and lyt away;
Ther nys no man can deme, by my fay,
If that it were departed equally.

(The Canterbury Tales,
Geoffrey Chaucer,
1340–1400 A.D.)

INTRODUCTION

THE Royal Festival Hall has been built on the south bank of the Thames under the direction of the Architect to the London County Council, Mr. Robert H. Matthew and the Deputy Architect, Dr. J. L. Martin. Mr. Edwin Williams was Senior Architect in charge and Mr. Peter Moro associated architect. The acoustic consultant was Mr. Hope Bagenal, in collaboration with the Building Research Station (Department of Scientific and Industrial Research). The design work started in August, 1948, and the Hall was opened in the presence of Their Majesties King George VI and Queen Elizabeth on May 3, 1951.

Concert hall acoustics is a broad and amorphous subject; it requires a book rather than a paper to deal with it adequately, and this paper is concerned mainly with the practical aspects. The design of the Royal Festival Hall has been described elsewhere[1–4] in rather general terms; this paper puts on record in more detail the acoustical design of the Hall, the objective measurements made in it, and the opinions that have been expressed about its acoustics. The hope is that this information will be of help to other designers; certainly the present writers would have been glad of such detailed information on other recent concert halls.

* *Editor's Note.*—Because the Royal Festival Hall represents the most significant advance in concert hall design in recent years, the editors have invited the authors to publish a shortened account of their work in *The Journal of the Acoustical Society of America*. The complete paper will appear in *Acustica*, vol. 3, No. 1.
[1] Architectural Rev. (Special Issue: Royal Festival Hall) 109, 336 (1951).
[2] P. H. Parkin, Nature 168, 264 (1951).
[3] P. H. Parkin and W. A. Allen, "The acoustic design of concert halls," Building Research Congress Division 3, 15 (1951).
[4] J. L. Martin, J. Roy. Inst. Brit. Architects 59, 196 (1952).

I. DESIGN

A. Objectives

The Royal Festival Hall (abbreviated R.F.H.) was designed to accommodate an audience of about 3000, a symphony orchestra of 120 players, a choir of 250, and an organ (Fig. 1). Its prime purpose was for symphony concerts; that is to say all other considerations, such as its use for speech, were to be subordinate to the acoustical requirements for music.

Our knowledge of the acoustics of large halls has reached the stage where major faults such as echoes can be avoided in the design, or, failing this, can be eliminated by suitable measures in the completed hall. There remain the very considerable problems of assessing and obtaining exactly the musicians' requirements. In the case of the R.F.H. a determined effort was made to consult the professional musicians. This assessment was made by means of a questionnaire sent to musicians,[5] by systematic listening tests in selected concert halls and by discussions of the problems at public[6] and private meetings with musicians and with other workers in the subject. In addition, and most important, there was Bagenal's long and varied experience. It was decided that the most important musical requirements were (a) definition, (b) fullness of tone, (c) balance, (d) blend, (e) no echoes, and (f) a low level of intruding noise. In addition, it was thought important to obtain reasonably uniform acoustics over the whole audience area.

Some description of these musical terms is necessary. "Fullness of tone" is the most difficult to define, although it is easily recognized. Perhaps all we can usefully say about it is that it is the satisfying quality added to the sounds produced by musical instruments (or voices) when in a concert hall as compared with in the open air. Although there may be subtle differences, we must assume for design purposes that musicians mean nearly the same quality when they use such terms as warmth, richness, body, singing tone, sonority or resonance. (It should be remembered that "reson-

[5] Parkin, Scholes, and Derbyshire, Acustica 2, 97 (1952).
[6] J. Roy. Inst. Brit. Architects 56, 70 (1948); 56, 126 (1949).

ance" has different meanings for the musician and the physicist; the Oxford English Dictionary defines it as the reinforcement or prolongation of sound by reflection.) "Definition" has two main characteristics; the first is concerned with hearing clearly the full timbre of each type of instrument so that they are readily distinguished one from another; the second is concerned with hearing every note distinctly so that, for example, it is possible to hear all the separate notes in a very rapid passage. (Speeds of playing of 15 notes per second are not uncommon.) This implies that the sounds from the whole orchestra should be heard well synchronized. "Clarity" is a term commonly used as an alternative to "definition." "Balance" we would define as the correct loudness ratios between the various sections of an orchestra as heard by the audience. "Blend" is another quality difficult to define, but in general terms it is the possibility of hearing a body of players as a homogeneous source rather than as a collection of individual sources.

The basic, obvious, assumptions are that there will be some reverberation and that the audience will receive some of the direct sound from the orchestra. Considering only one aspect of "definition"—speed of playing—it is again obvious and a matter of common experience that a very short reverberation time (abbreviated R.T.) will not prevent a listener hearing all the rapid notes distinctly, while a very long R.T. will. Further, even in the presence of a very long R.T., definition is still maintained when the listener is very close to the source. Assuming for the moment that a very short R.T. is not desirable in a concert hall, it is a simple conclusion that the intensity of the direct sound must be as great as possible. Now the intensity of the direct sound reaching the listener will fall off as the square of the distance, at least for individual instruments if not for the whole orchestra, and in a hall the size of the R.F.H. the difference in intensity between the front and back rows would be about 20 db. (In the Chaucer quotation at the head of this paper, the word "departed" means, in modern English, "divided".) Haas[7] has recently put on a quantitative basis what has long been assumed, namely, that for speech at least, reflections following shortly after the direct sound do not detract from the definition. For example, reflections of the same intensity may be delayed up to about 30 m sec behind the direct sound without any loss of definition. The conclusion is that to maintain definition over the whole audience area, short-path reflections directed towards the rear of the hall should be used to help the intensity of the direct sound.

It is true that the foregoing simple argument is based on geometric acoustics and also takes no account of the many remarkable properties of the ear such as its ability to discriminate against unwanted sounds. Nevertheless in the absence of positive evidence to the con-

[7] H. Haas, Acustica 1, 49 (1951).

Fig. 1. Auditorium of Royal Festival Hall.

trary, these suppositions form a reasonable basis for the design of a hall.

There remain the problems of obtaining "fullness of tone," "balance," and "blend." "Balance" is partly a matter for the conductor, and is discussed below in connection with platform design. As for "fullness of tone," it is very difficult to decide what acoustical factors control this quality. Certainly the only factor we have under any control is the reverberation, and the most reasonable supposition is that the longer the R.T. the more chance there is of obtaining fullness. It may be that, for a given R.T., the lower the intensity of the direct sound the more fullness there will be, but without reflecting surfaces it is certain that we shall have less uniformity over the audience area. Under present conditions it is better to aim for uniformity and therefore to use surfaces close to the orchestra to reflect sound towards the back of the hall and, for fullness, to design for a long R.T.

If the above argument is correct, it follows that there is a conflict between definition and fullness. Obviously the definition will suffer in the presence of a very long R.T. and vice versa, although there may be a range over which fullness can be increased without any noticeable loss of definition. The opinion survey[5] referred to above showed that Liverpool New Philharmonic Hall had by far the best reputation for good acoustics of any concert hall in Britain, and listening tests made in this hall showed that its definition was good. On the other hand, at a meeting of the Acoustics Group[6] the consensus of opinion among musicians was that definition should, if necessary, be secondary to fullness. As it is comparatively simple to shorten the R.T. of a hall, it was decided to design the R.F.H. for as long an R.T. as possible.

"Blend," like "balance," is partly a matter for the conductor, but is probably best helped by surfaces close to the orchestra which reflect some of the sound directly back to the orchestra. These surfaces should

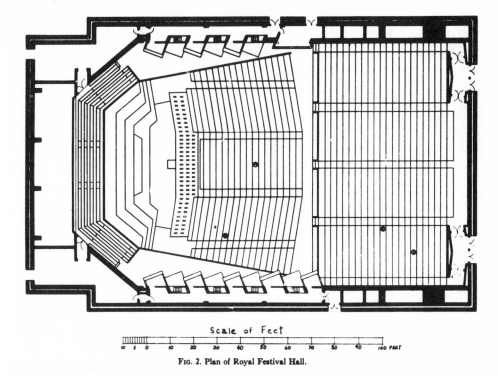

Scale of Feet

FIG. 2. Plan of Royal Festival Hall.

help in two ways: (a) by "mixing" some of the orchestral sound before it reaches the audience and (b) by helping the orchestra to hear itself and thus enabling it to play together.

B. Shape on Plan

The three possible basic shapes of a hall are horseshoe, fan, or rectangular. The horseshoe-shaped hall is theoretically dangerous acoustically because of the concave surfaces involved, although in the past some successful halls have been built in this way, e.g., Usher Hall, Edinburgh. Nevertheless, the dangers involved are too considerable to be ignored, and the choice then is between a fan-shaped or rectangular hall. The main advantage of the fan shape is that the length of the hall is less than that of a rectangular hall seating the same number and of the same width at the orchestra end. The main disadvantage is that the rear wall, balcony front, and seat risers are all curved, causing a serious risk of echoes. The rectangular hall is almost free from this risk, and in addition has a possible advantage that there is more cross reflection between the parallel walls which may give added "fullness." These two considerations, plus the weight of tradition, led to the adoption of a rectangular shape for the R.F.H., although of course the arguments are not conclusive. To overcome the main disadvantage of a rectangular hall—its large

width at the orchestra end—the seating at the front part of the R.F.H. was made fan-shaped at the orchestra level (Fig. 2).

C. Orchestra and Choir Platform

The area of the orchestra platform should be as small as is consistent with adequate room for the players, since the time delays between the various sections of the orchestra are kept short thus aiding definition and homogeneous playing. Further, the design of the reflecting surfaces close to the orchestra is not quite so difficult if the area of the sound source is small. It is more important to keep the depth of the platform to a mininum, rather than the width. This is because the time delays from front to back of the platform are heard by the whole audience: the time delays arising from the width affect a relatively small proportion of the audience, although they do affect instruments widely separated from each other and consequently not easily heard by each other.

In longitudinal section, an orchestra platform can be (a) flat, (b) flat for the front part and raked for the back part, or (c) completely raked from front to back. With designs (a) and (b) the weakest instruments—the woodwind—are screened by the players in the front of the orchestra, and, in design (b), the most powerful instruments—the brass and percussion—have an advan-

59

FIG. 3. Orchestra platform of Royal Festival Hall. (Note: The platform is arranged for a violin recital, with seats for audience. The temporary organ screen referred to in the paper has been replaced by the permanent, openable, screen.)

The raking of the floor area had been calculated[8] on the basis of a free height of 3 in. between successive rows. However, this rake differed so little from a straight line that the considerable expense and inconvenience (e.g., varying step height) were not considered worth while and a straight line was adopted.

The ceiling follows a general line designed to reflect sound towards the rear of the Hall. The small undulations are for lighting, not acoustical reasons. The ceiling was intended to be of solid fibrous plaster, 2 in. thick, but during its construction (it was prefabricated in sections and then brought to the Hall for erection) its thickness was, by mistake, reduced to $\frac{1}{4}$ to $\frac{1}{2}$ in. After erection in the Hall its thickness was made up to the specified 2 in. but using vermiculite plaster instead of the specified fibrous plaster. This slip in the close liaison maintained between the consultants and the architects was a little unfortunate in that, as described later, it probably helped to make the ceiling more absorbent than was intended.

tage which they do not need. Design (c) was adopted for the R.F.H., with all instruments exposed more or less equally (Fig. 3). It is true that the powerful instruments are still unnecessarily exposed, but with all instruments given an equal chance it can then be left to the conductor to achieve the correct balance.

D. Longitudinal Section

In most concert halls the front of the orchestra platform is 3 to 4 ft higher than the front row of audience seats. This arrangement helps to provide good paths for the direct sound to most of the audience, but it had been noticed at listening tests that the screening of the center section of the orchestra by the front rows of the orchestra was noticeable in the front rows of the audience, or over large areas of the audience when the main floor was flat. To overcome this disadvantage, the front of the orchestra platform in the R.F.H. was made only 9 in. higher than the front row of audience seats (Fig. 4).

E. Reflecting Surfaces

The reflector over the orchestra (referred to as the canopy) is the most important reflecting surface. The requirements were to reflect sound towards the rear of the Hall (these reflections to follow the direct sound by as short a time interval as possible) and to reflect some of the sound back to the orchestra to help the players hear themselves and each other. The second requirement can be met by making parts of the canopy horizontal, but the main requirement is complicated by the large area covered by the sound source. It is clear that an angle for the canopy suitable to reflect instruments at the front of the orchestra towards the rear of the hall is not suitable for instruments at the back, and vice versa. The canopy in the R.F.H. is a compromise between the various conflicting requirements. The front leaf is at a slightly greater inclination to the horizontal than the middle leaf, which in turn is more inclined than the back leaf: the general contour is

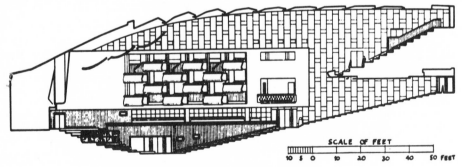

SCALE OF FEET

FIG. 4. Longitudinal section of Royal Festival Hall.

[8] E. Petzold, Deut. Bauhütte 32, 30 (1928).

therefore slightly convex. It is made of wood 2 in. thick: the leaves are fixed by resilient mountings to timber beams which hang by tie rods from the roof trusses. Its weight is 12 tons plus 3 tons of lighting fittings.

The second reflecting surface is the flat floor area between the front of the orchestra platform and the audience seats. This reflector is of slate bedded solid onto concrete, and is intended to reflect sound from the front instruments, i.e., the violins, towards the rear of the Hall.

The third reflecting surface is a wooden screen about 3 ft high which separates the orchestra from the choir seats. The walls splayed on the plan are a fourth group of surfaces; they are made of wood ⅜ in. thick on 4-in. battens with absorbent in the air space. Some diffusion was introduced on these splay walls in the form of protruding triangles.

The organ was not to be installed for some time after the opening of the R.F.H. and in the meantime the opening for the organ was covered by a wooden reflector set at an angle of 15° to the vertical (Fig. 4).

F. Echoes

The three areas considered most likely to cause echoes were the rear wall, the margins where the side walls meet the ceiling, and the side walls near to the orchestra. The risk from the rear wall was slight: the wall was not curved, and its area was comparatively small. The treatment applied to it was therefore only slightly absorbent at mid- and high frequencies, although these areas were used to provide additional low frequency absorbent. The margins were made of unplastered wood-wool slabs on battens, the vertical and horizontal parts each being about 4 ft wide and ran along the whole length of the Hall. The splayed side walls next to the floor level for the first 30 ft away from the orchestra were treated with slit absorbers which had rock wool behind wood facing strips (the so-called "Copenhagen" absorbent from its use in the foyer of the Danish Broadcasting House); for the rest of these side walls the strips were mounted straight on to wood panels.

G. Prevention of Resonances

By resonance is meant the excessive accentuation of one or more discrete frequencies and their associated decay at a much slower rate than the general reverberant sound. These resonances may occur in two ways: by room eigentones between, say, two parallel surfaces with small absorption, or by mechanical resonances of a surface, e.g., a panel. The first type is described by Jordan[9] and was cured in that case by using Helmholtz resonators. The ceiling of the R.F.H. was made with about 1200 2-in. diameter holes for use with resonators

if needed. The surfaces in the R.F.H. which might give rise to the second type of resonance were the ceiling, the wood panels on the side walls, and the orchestra canopy. The ceiling and the canopy were constructed in different sized sections which were then butted together with a strip of soft fiberboard between. A wood panel of the type to be used on the side walls was tested for its decay rate in the laboratory; at its resonant frequency the decay was found to be about 250 db/sec with absorbent in the air-space and about 100 db/sec without absorbent. Both these decay rates were much faster than the expected reverberant sound decay.

H. Reverberation Time

The aim was to achieve as long a reverberation time as possible (at mid-frequencies). At an early stage in the design an R.T. of 2.2 sec was considered, being the value recommended by Bagenal and Wood[10] for a hall of the volume of the R.F.H. As the design developed, and as results of measurements in other British concert halls became available,[5] it was realized that this value was unlikely to be reached; a calculation of the R.T. as the design progressed gave a value of 1.7 sec at 500 cps. This value is that recommended by Knudsen and Harris[11] for a hall of this volume, and as several good British concert halls had been found to lie close to or below the Knudsen and Harris optimum, it was thought to be acceptable.

At high frequencies the R.T. is largely controlled by air and audience absorption, and the designer can do nothing about it. At low frequencies, however, there is a danger that the R.T. may be too long. In traditional constructions there are large quantities of fortuitous low frequency absorption in the form of plaster panels etc.; in modern constructions of reinforced concrete and with little ornamentation this absorption may not be present. The calculation of R.T. at mid-frequencies is uncertain; at low frequencies the uncertainties are worse because of the lack of information on the absorption of surfaces at these frequencies. The major risk was that the R.T. at low frequencies would be too long, the designers would have been satisfied at the design stage, if the R.T. at 125 cps would come in the range 1.7 to 2.5 sec, i.e., between 100 and 150 percent of the value at 500 cps. A conservative estimate of the absorption to be expected from the various surfaces and audience was made for 125 cps; a sufficient area of elm panels spaced from the wall was included in the design to ensure that the R.T. was not too long. As the majority of absorption might thus have been expected to be resulting from these panels, their construction was varied to ensure that absorption was not concentrated over a too narrow band of frequencies. The construction

---

[9] V. L. Jordan, J. Acoust. Soc. Am. 19, 972 (1947).

[10] H. Bagenal and A. Wood, *Planning for Good Acoustics* (Methuen, London, 1931), p. 116.
[11] V. O. Knudsen and C. M. Harris, *Acoustical Designing in Architecture* (John Wiley and Sons, Inc., New York, 1950), p. 194.

and absorption coefficients of these panels have already been described.[12]

It is desirable for the orchestra that the acoustics in a concert hall should not be too different between performance and rehearsal conditions, so the tip-up upholstered seats for the R.F.H. were designed with the underside of the seat perforated, with rock-wool behind the perforations.

I. Sound Insulation

It is intended to describe the sound insulation of the R.F.H. in more detail elsewhere. Briefly, the two major sources of noise at the site were overground electric and steam trains running on the nearby Hungerford Bridge, and the underground trains running directly under the site. The noise from the overground trains was measured at various positions on the site, and a wall and roof construction for the auditorium was devised sufficient to reduce these noise levels to the background noise levels measured in two other concert halls during pauses in quiet passages of music. The wall construction consisted of two leaves of reinforced concrete each 10 in. thick separated by an air-space 12 in. wide, with some absorbent in the air-space. The roof construction consisted of an inner leaf of reinforced concrete 6 in. thick carrying sleeper walls which varied from 2 ft to 4 ft high (depending on the camber of the roof). Over the top of the sleeper walls was draped a 2-in. layer of glasswool and the outer leaf of 4 in. of reinforced concrete rested on this. All doors except two into the auditorium were built with sound locks, and the two exceptions (in the final design) had extra precautions taken in the foyers leading to them.

At this stage of the design it was necessary to give the architects as much freedom as possible in the positioning of the auditorium, and the constructions specified were intended to give adequate insulation on their own. In the event, most of the auditorium was enclosed by foyers, corridors, etc., all of which were designed to give extra insulation, both as an added safeguard and as an aid to reducing the noise level below that measured in the two other halls.

The ground vibrations arising from the underground trains were measured and were thought to be sufficiently small not to produce troublesome noise in the auditorium. As a safeguard the auditorium was placed as high above ground level as possible.

The inlet and outlets ducts for the ventilation plant were treated to reduce the external noise by the same amount as the walls and roof, and the plant itself was designed for minimum noise.

J. Capacity

The final figures for the capacity of the R.F.H. are as follows:

---

[12] P. H. Parkin and H. J. Purkis, Acustica 1, 81 (1951).

Seating		
Orchestra	about	125
Choir		248
Stalls		663
Terrace Stalls		1120
Grand Tier		616
Side Balconies		152
Boxes		200
Standing		
Side Balconies		120
Sides of Terrace Stalls		80
Behind Terrace Stalls		24
Behind Grand Tier		56
		———
Total		3404

The volume is 777 000 cubic feet so that the volume per person when the Hall is completely full is 225 cubic feet.

II. TEST CONCERTS

The first purpose of the test concerts which were held in the R.F.H. prior to its opening was to test for major faults such as echoes. Reliance was placed mainly on subjective listening tests; only limited objective measurements were possible with an audience present, and in any case even obvious faults such as echoes can not be assessed with much reliability from objective measurements. At the first test concert there were two classes of listeners in addition to the normal audience. The first class consisted of 13 groups of 20 listeners each distributed over the whole seating area. These were classified as ordinary concertgoers, and had all been to at least six concerts in the previous year. They answered questionnaires which dealt with such subjects as noise, echoes and loudness. The second class consisted of six groups of two or three people each who had some knowledge of acoustics and most of whom had already taken part in similar tests in other concert halls. Each group of "specialists" sat in different positions in the Hall for each of the three parts into which each test concert was divided. These positions were B, C, and D, as shown in Fig. 2, and symmetrical positions on the other side of the center-line of the Hall. Their function was to give more detailed reports on the acoustics and on the variations between positions; for example if echoes had been heard they would have been expected to estimate the probable source.

It was clear at the first test concert that there was only one major fault. The group of 20 at the extreme back of the Grand Tier had behind them an absorbent gangway which had a wood-wool ceiling and the leather-paneled rear wall; this group gave less favorable answers, on the whole, than the other groups. As examples, in reply to the question "Was it easy to listen attentively?" 95 percent of all the other groups answered "Yes" compared with 70 percent for this group; in reply to the question "Were crescendos effective?" 80 percent of all the other groups answered "Yes" compared with 35 percent for this group. Before the next test concert the wood-wool ceiling of this gangway was plastered over; this group then gave answers similar on the average to the other groups.

Attention was subsequently concentrated on the proper balance between fullness of tone and definition. The majority of both classes of listeners at the first test had said that more fullness was required; accordingly some absorbent was removed before the second test concert. At this second test and at the final test a third class of listeners was used, consisting of professional music critics who, like the second class, were divided into three groups sitting in different positions (*B*, *C*, and *D*) for each part of the tests.

It will be simplest if the main result of these subjective tests is stated first. This was that the absolute opinions on the acoustics were not consistent. For some reason, possibly because of the effect of listening mainly in the Royal Albert Hall for ten years, most of the listeners found the acoustics of the Royal Festival Hall rather surprising at a first hearing. This is best illustrated by considering the replies of the professional critics. At the first concert they attended (actually the second test concert) 10 out of 14 critics wanted more fullness of tone, three wanted more definition and one wanted the Hall left as it was. At the next concert nine of these 14 wanted the Hall left as it was, three still wanted more fullness and two wanted more definition. The measured R.T. was practically the same at these two concerts and although the R.T. is, of course, not the only factor in assessing the acoustics (e.g. the amount of diffusion might have changed between the two concerts while leaving the R.T. unchanged), it is reasonably certain that it was the effect of a second hearing on the critics that made them change their minds. This is confirmed by the fact that of another eight critics who were in the Hall for the first time at the final test concert seven wanted more fullness of tone. This result was not unexpected; Bagenal had forecast at an early stage that it would take several months for musicians used to more traditional halls to get used to the acoustics of the R.F.H.

In view of this unreliability there is little point in giving any further results of these subjective assessments, except to state that, at the final test, of the second class of listeners (the "specialists") who had been to all the test concerts, one gave *B* as the best position, two gave *C*, three gave *D*, and four expressed no preference; of the professional music critics who had then been twice to the Hall, three preferred position *B*, two position *C*, nine position *D* and two expressed no preference.

Although these test concerts were not useful for absolute judgments, it should be emphasized that they were invaluable for three main reasons. The first was that the removal of the absorbent areas undoubtedly caused an improvement in the acoustics. It would not have been practicable to remove them when the Hall was in use, and if they had not been removed the R.T. would have been shorter than it is now: it is shown below that

the R.T. should not be any shorter. At the same time it was possible to check (by using the groups of listeners) that this removal of absorbent did not give rise to any serious echoes. The second reason was that the only major fault noticed—the bad acoustics at the back of the Grand Tier—was corrected before the opening. If other and more serious faults had occurred, e.g., the sort of resonances found by Jordan,[9] these too could have been eliminated. Any such faults not corrected before the opening would have been very difficult to deal with when the Hall was in use and might have affected its reputation seriously. The third reason was that under the conditions of the test concerts, it was possible to make measurements, both objective and subjective, which would have been impracticable at normal concerts.

III. OBJECTIVE MEASUREMENTS

A. Technique

Western Electric type 640AA condenser microphones were used for most of the measurements. The R.T.'s were measured in octave bands using a 0.45-in. Colt pistol as the sound source, fired from the front center of the orchestra platform. Unless otherwise stated, the average value of the decay between −5 and −35 db following the maximum deflection on the logarithmic recorder was taken to represent the R.T. The four microphone positions used are shown in Fig. 2. Position *A* was half-way between the floor and the ceiling; positions *B*, *C*, and *D* (in the Stalls, Terrace Stalls, and Grand Tier, respectively) were all at ear height.

If the conditions are carefully controlled it is possible to measure R.T.'s with great relative accuracy. For example, the coefficient of variation (the standard deviation divided by the arithmetic mean) obtained from firing 101 pistol shots with the microphone in one position was found to be between 1 and 3 percent for all octaves. However, under most conditions of measurements this accuracy can not easily be maintained (largely because of difficulties of "interpretation" of the traces) and the coefficient of variation is probably about 5 percent for the R.T. results presented here. All values given for the R.T.'s are the mean of at least six readings, so that we can expect differences of 0.1 sec

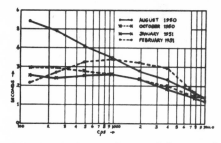

Fig. 5. Reverberation times at various stages during construction.

between mean values to be significant (in the statistical sense if not subjectively).

To check the effect of the source of sound on the measurement of R.T., the results using (a) pistol, (b) warble-tones (±10 percent about the mean frequency up to 2000 cps and ±200 cps at higher frequencies) from loudspeakers, and (c) an orchestra playing staccato chords were compared, and it was found that the pistol and the orchestral chords gave the same results within the limits of experimental error. The warble-tones also gave the same results except at the highest frequencies where they were rather shorter. This was due to the fact that octave bands are too wide for ranges where the R.T. is changing rapidly with respect to frequency. Nevertheless, the values obtained from octave analysis will be used since the pistol and orchestra had to be used as sources for most of the measurements.

B. Reverberation Time During Construction and Test Concerts

The R.T. of the R.F.H. was measured at various stages during the construction (Fig. 5). For all these measurements the microphone was at position A. The first measurement was made in August, 1950; the roof and structural walls were complete but the false floor which carried the seats was not installed, nor were any of the finishings. At just below ceiling height was a layer of planks laid loose on scaffolding covering the whole ceiling area. At the time of the next measurement (in October, 1950) the ceiling, the wood-wool margins, and the false floor had been installed; the layer of scaffolding planks just below the ceiling was still in position. Otherwise the conditions in the hall were very similar to the previous measurement. It is seen that the R.T. at 125 cps had dropped from 5.4 to 2.9 sec and at 500 cps from 4.1 to 2.8 sec. If we allow for the absorption of the wood-wool and make a rough guess that of the additional absorbent 80 percent was due to the ceiling and 20 percent was due to the false floor, we get absorption coefficients of about 0.3 for the ceiling and 0.1 for the floor at 125 cps, and 0.1 for the ceiling and 0.05 for the floor at 500 cps. These very approximate calculations show that the ceiling was acting as quite an efficient panel absorbent, presumably partly owing to the use of vermiculite plaster described above. At the time of the next measurement, in January, 1951, all the elm panels on the side walls had been installed. Some of the other finishes, e.g. the "Copenhagen" absorbent, had also been installed, but the layer of scaffolding planks below the ceiling had been removed. Most of the rock-wool was in position on the rear and side walls ready to be covered by leather-faced panels. The net result was that at low frequencies the R.T. was shorter, but at higher frequencies (1000 cps and upwards) the R.T. was practically unchanged. If we make another rough calculation for 125 cps and guess that 80 percent of the extra absorption was due to the elm panels,

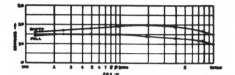

Fig. 6. Reverberation time of completed Hall (microphone at position A).

we get a coefficient for them of about 0.35; this value is close to the laboratory figure of about 0.4.[12] The fourth curve of Fig. 5 shows the R.T. of the Hall completed except for the seats. The big increase in R.T. at mid-frequencies was presumably the result of the removal of all the builder's material that had previously been lying about in the Hall and to the covering over of the rock-wool on the rear and side walls.

As mentioned in Part II, some absorbent areas were removed between the first and second test concerts and some further areas were removed between the second and third tests. However, the permanent upholstered seats were being installed throughout the test period (at the first test the audience sat on the floor) and the net effect was that the R.T. (in the full Hall) throughout the test period was very little changed at mid- and high frequencies and was slightly shortened at the low frequencies.

C. Reverberation Times in Completed Hall

During the test concerts none of the audience was standing, and the choir seats were unoccupied. For normal concerts and when at least 80 percent of the seats are sold, the conditions are different. The choir seats are always occupied, either by a choir or by audience, and there are about 120 standing audience in the side-balconies, while some of the seats may be empty. This is because of the relative prices of the seats. When all the seats are sold there are usually another 150 standing at the sides and back of the Hall. Thus the R.T. of the full, completed Hall is shorter than that measured at the last of the test concerts and on which the previously published[2] figures of the R.T. were based.

The R.T. has been measured at several concerts when the seated audience has varied between 80 and 100 percent capacity and the choir and standing audience between 250 and 500. No significant variation (i.e., no difference as great as 0.1 sec) was found in the measured R.T. over this range of audience size, and only slight differences were found between the microphone positions A, B, C, and D. There were not sufficient measurements to be able to decide on the exact differences between positions; the average value for position A (corresponding to the microphone position used in other halls) has therefore been selected to give the R.T. characteristic for the full Hall (Fig. 6).

In the empty Hall, the R.T. at position B was slightly longer at mid- and low frequencies compared

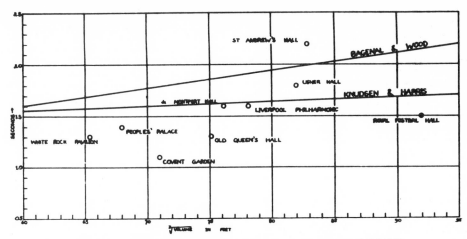

FIG. 7. Reverberation times (500 cps) of eight good Halls (full) and of Royal Festival Hall (full).

with position A; at positions C and D it was slightly longer at mid-frequencies only. At high frequencies the R.T. at B, C, and D was slightly shorter than at A.

The R.T. in the empty Hall at frequencies between 50 and 100 cps was investigated in some detail both with the pistol plus one-third octave filters and with warble-tones. The decays in this range were very irregular and often inconsistent, and it was not possible to make any sense out of the results. This erratic behavior was probably the result of the space between the false floor which carries the seats and the structural floor. This space is about 3 ft high, and is connected to the auditorium by a large number of ventilation openings; although the floor of the space is covered with 2 in. of rock-wool, the R.T. measured inside it was 6 sec at 50 cps falling to 4 sec at 125 cps.

D. Noise Levels

As mentioned above, a detailed description of the sound insulation and noise levels is to be given elsewhere, but a brief statement of the noise levels in the finished auditorium is of interest here. The overground trains were found to be completely inaudible in the auditorium, but the underground trains produced a low frequency noise which was audible for about 5 seconds for each train. At the time of writing, the sound pressure levels arising from the underground trains have not been measured, but subjective measurements using a Barkhausen meter, while very approximate, showed that the noise level reached a maximum in the empty Hall of about 30 phons. At one of the test concerts, the test groups were asked to listen for this noise; out of the total of about 240 listeners, 18 were able to detect them but only, of course, when the trains coincided with a pause or a very quiet passage in the music.

The ventilation plant was audible at a very few places

(e.g., behind the boxes), where the inlet or outlet grilles were close to the listeners. The noise level at these positions appeared to be about the same as the underground trains, i.e., 30 phons. Over the major part of the seating area the plant noise was not distinguishable, and the total background noise was probably of the order of 20 phons.

E. Discussion of Objective Measurements

Certainly the most useful measurement from the practical point of view is the R.T. The measurements made in the R.F.H. during construction and during the test concerts provided a firm basis for the discussions on what alterations to make. Further, in the completed Hall it is the only well-established criterion for comparison with other halls (Fig. 7). For R.T. measurements in a full hall the orchestra is normally used as a source, and analysis into octave bands is sufficient over most of the frequency range. The use of narrower bands reduces the accuracy when, as happens in practice, only a limited number of measurements are possible, by increasing the spread of results in a given band resulting from the variations in the spectrum of the orchestral sound. At low and high frequencies, however, where the R.T. may be changing comparatively quickly with respect to frequency, octave-band analysis can usefully be supplemented by one-third octave analysis. In empty halls it seems logical to use a similar method of measurement, e.g., a pistol and octave analysis in preference to, say, warble-tones. In the R.F.H. no difference was found between pistol shots and warble-tones but this is not always the case. As to the desirable standard of accuracy, there seems to be little point usually in presenting results with any greater accuracy than to the nearest 0.1 sec; changes in the conditions, such as the number of audience present, would nullify any greater accuracy,

65

and it is unlikely that subjective impressions would be affected by changes smaller than 0.1 sec.

All objective measurements other than the R.T. are still tentative, and the authors can do no more at the present stage than indicate what, in their view, are the most promising lines of development. To be of practical use, objective measurements should, in the long-term, be capable of fulfilling three conditions. First, it should be possible to state the results in a quantitative metrical form. Secondly, it should be possible to correlate the objective measurements with subjective impressions either as being typical of a hall as a whole or as relating to particular positions in a given hall. This does not necessarily mean that one particular type of objective measurement need be directly related to one particular subjective quality. If a measurement is to refer to a particular position, its value should not change too quickly with position. In other words, a measurement which gives one value at a given seat is of little use if its value is quite different only two or three seats away, because obviously a listener would hear no difference between the two positions. Thirdly, it should be possible, ideally, to make the measurements in full halls, but as this is unlikely, the type of measurement should be such that the results are not too dependent on the full or empty state of the hall.

As we are concerned here with practical problems, it appears that most progress can be expected by working in terms of geometric acoustics. That is to say, we should examine at a given point in a hall the variation with time, direction and frequency of the pressure received when an impulsive or a continuous sound is emitted from the source. There would be a hope of correlating such measurements with subjective impressions, perhaps not directly but through intermediate laboratory experiments.

The measurements needed in a concert hall would require first a nondirectional source for impulsive and steady sounds. For the impulsive sounds either pulses of tone from a loud-speaker or explosive sources such as a pistol or an electric spark could be used. If a loud-speaker source is used it must be made nondirectional, e.g., by the use of several loud-speakers mounted on a spherical or hemispherical baffle, and the pulses must be very short. Certainly a pulse length of 25 m sec is too long; measurements made in the R.F.H. using this pulse length showed very large variations in the received signal when the microphone was moved three seats and thus did not satisfy the second criterion for objective measurements. Pulses of only 2 to 15 m sec have been shown[13,14] to give better results, but with such short pulses the transient response of the loud-speaker

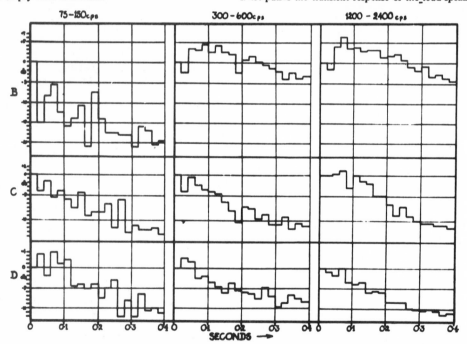

Fig. 8. Transient responses at positions, B, C, and D, averaged over 20 m sec intervals.

[13] R. H. Bolt and P. E. Doak, J. Acoust. Soc. Am. 22, 507 (1950).
[14] Bolt, Doak, and Westervelt, J. Acoust. Soc. Am. 22, 328 (1950).

must be allowed for, and it is difficult to put sufficient low frequency energy into the room. An explosive source is better in these respects; using the 0.45 pistol in the R.F.H. measurements were possible over most of the frequency range and the received signal varied only slowly with position. The transient responses recorded on an omnidirectional microphone at the three positions B, C, and D when the 0.45 pistol was fired from the front center of the platform at the last test concert are shown in Fig. 8, where the C.R.O. photographs have been "smoothed" by taking the mean rectified pressure for successive 20 m sec bands and plotting the results on a logarithmic amplitude scale referred to the mean pressure in the first 20 m sec band. The outstanding feature of these results is that at position B (in the Stalls) it is some 200 m sec before the pressure level starts to decay. This is a reasonable result because of the reflections arriving from the canopy and the organ screen. However, this technique requires further development; rather different results are obtained from a less intense source, and it is possible that there may be some nonlinearity in the behavior of the various surfaces. Further, the pressure per octave-band from a pistol shot increases by about 10 db per octave and it might have been better to insert an equalizing network before the octave analysis.

If these transient measurements with an omnidirectional microphone were to be extended using a directional microphone and supplemented by steady-state measurements, we could expect to obtain results which would define completely the behavior at given points in the room. This discussion has only been concerned with tentative methods, but it is hoped that enough has been said to indicate the most promising lines of development.

IV. SUBJECTIVE ASSESSMENTS

A. Press Comments

A fairly comprehensive review of the comments in the national press (not the musical press) on the acoustics of the R.F.H. has been made throughout the first eighteen months since the Hall's opening (omitting comments on the test concerts). These press comments may be interpreted in terms of the requirements set out in Part I. There were about 100 comments, and the numbers of comments on the particular acoustical qualities are shown in Table I.

In addition, there were 20 favorable and 11 unfavorable comments on the sense of climax for loud passages, 8 favorable and 4 unfavorable comments on the string tone, 4 favorable and 1 unfavorable comments on the brilliance, and 7 comments that the acoustics were particularly good for small bodies of players.

B. Stokowski Concert

Dr. Leopold Stokowski conducted a concert in the R.F.H. on June 9, 1951, with all the strings arranged on one side so that they were all more or less facing the

TABLE I. Press comments in first eighteen months.

Quality	Number of Comments Favorable	Unfavorable
Definition or Clarity	24	3
Fullness or Resonance or Singing Tone	15	10
Blend	8	3
Balance	15	7
Echoes	2 comments on absence of echoes	

audience. The purpose of this arrangement was to see if it had any effect on the string tone. Of six music critics who were all at about the middle of the Hall, five commented that the string tone was better than they had ever heard in this Hall, although three out of the five qualified this remark by stating that it was probably as much the result of the excellence of the conducting as of the orchestral layout. Three said that the balance was excellent, one said that the balance was fair, and one said the balance was poor in the Beethoven and Bach works in the program but good in the Stravinsky work. In general, the comments were very favorable.

C. Foreign Scientists

A concert in the R.F.H. on September 16, 1951, was attended by 18 foreign scientists all of whom were engaged in acoustical research. They were divided into three groups and sat successively in different positions in the Hall. These positions were: Front Stalls (seventh row, some in the center and some towards the left), Terrace Stalls (seventh row, in the center) and Grand Tier (seventh row, in the center). They were given a questionnaire and their answers were as follows:

1. Four preferred the Front Stalls, nine the Terrace Stalls, three the Grand Tier, and two expressed no preference.
2. One wanted more definition, six more fullness, and eleven no change.
3. Three knew better halls and two others knew concert halls which were better in some respects but not on the whole. Nine stated definitely that they did not know any better hall, and four did not commit themselves.

The fourth question was: "Apart from your own preference, if you were asked by the musicians to provide (a) more definition or (b) more fullness of tone, how would you do it?"

(This question was badly worded, as "musician" was not defined and was queried in a few replies as referring to the players or the listeners. However, most replies seemed to understand the question—as was intended —as referring to musician listeners.) To increase the definition, six suggested more reflecting surfaces closer to the orchestra, and three said lower the canopy. One said raise the height of the platform and two said shorten the R.T. To provide more fullness, seven said increase the R.T. (of which seven, five particularly mentioned the low frequency R.T.); one said more diffusion; one said reduce and one said increase the reflecting surfaces close to the orchestra.

From the above answers and from the general comments made by these listeners, it is possible to divide their opinions into two broad classes. Thus thirteen of the listeners would put this Hall in the "excellent to very good" class and the other five in the "good" class.

D. Acoustics Group Meeting

The Acoustics Group of the Physical Society held a meeting[15] on November 23, 1951, to discuss the acoustics of the R.F.H. Most of the speakers had been previously invited by the Acoustics Group to contribute, largely on the grounds of their known interest in the subject, and the contributors can not be taken as a scientifically selected cross section of the musical world. Nevertheless, they were probably more fully representative of musical opinion than the comments in the general press, and this meeting is certainly the most authoritative musical review of the acoustics that has as yet taken place.

It is convenient to divide the contributors into two classes, performers and listeners. Of the performers, the conductors (Beecham, Blech, Boult, Krips, Sargent, Stokowski) were on the whole very pleased with the acoustics and three of them (Krips, Sargent, Stokowski) said that it was the best hall they knew. However, all except Krips qualified their praise by a request for a little more "resonance." Two other performers (Miss Joan Hammond and Mr. Denis Matthews) also were very pleased with the acoustics, but again one (Matthews) wanted a little more resonance. Another performer (Heifetz) said, by proxy, that it was the finest hall in the world.

The three chairmen or secretaries of the main London orchestras (Royal Philharmonic, London Philharmonic and London Symphony) can, from their remarks, be classified as listeners. One made three points: that the smaller the number of performers the better it sounded; that there was a weakness under the Grand Tier but that otherwise it was good; and that there was a cross echo at the front from trumpets. The other two said that their orchestras were coming more and more to like playing in the R.F.H., but both stressed the need for more resonance for the listeners and one said that the players needed a little more help in hearing each other. Of the nine other listener contributors (three of them professional music critics) five wanted more fullness, and two did not; two complained that the middle sections of the orchestra were sometimes lost, but one did not agree; two suggested that more orchestral layouts should be tried to overcome some of the difficulties.

Throughout the whole discussion there was general agreement that the definition and clarity in the R.F.H. were outstandingly good, and there were several comments that this has led to an improvement in orchestral performance.

[15] J. Roy. Inst. Brit. Architects 59, 47 (1951).

E. Casual Observations

From casual observations made by the authors, three points are worth mentioning. The first is that under normal concert conditions it is possible to hear a change in the quality of the orchestral sound as one moves to the extreme back of the Terrace Stalls, i.e., under the Grand Tier. This "under-balcony" effect is, of course, quite common and is not in this case very great. However, on one occasion an orchestra was playing on a flat platform specially erected for ballet performances, and which was 5 ft high compared with the 9-in. height of the front of the usual platform. Under this condition it was not possible to hear any change in the sound between the back and the front of the Terrace Stalls, although it should be mentioned that all these seats were empty.

The second point is that in the boxes and the side stalls, it is possible to hear slight high frequency echoes, particularly from the brass. These presumably arise from the removal of the absorbents from behind the "Copenhagen" strips and to the plastering-over of most of the wood-wool margins.

The third point is the unusual loudness of the sound at the back of the Grand Tier. This is particularly noticeable with solo instruments or voices, so much so that there is a conflict between sight and sound. The performer from this distance (135 ft) looks, of course, quite small yet the sound is "unnaturally" loud.

F. Discussion of Subjective Assessments

The opinions expressed about the acoustics of the R.F.H. have been given in some detail because the science has now reached the stage where we can attempt to obtain in the design—if not to measure—definite musical qualities, as distinct from the broad classifications of good or bad acoustics. We can not hope to get what might be called absolute opinions; in other words we can not get opinions under controlled conditions, such as comparing directly slightly different acoustical conditions. This, however, is not a serious criticism of the opinions presented here. The fact is that these opinions are what people have said and what they believe; this definition of the validity of the opinions is sufficient for the present writers.

One difficulty with subjective assessments is the use by musicians of a large number of terms, and to bring order into the problem it has been necessary to translate some of their opinions into our own terms. It does appear, however, that there is not much difference between many of these terms, certainly not enough to be of any consequence when dealing with design problems.

The general opinion about the R.F.H. is that its acoustics are very good, several distinguished individuals going so far as to say that it is the best of its size in the world. Dealing with the acoustic qualities in detail, it is obvious that the definition in this Hall is exceptionally good. It is also established that there are no echoes nor intruding noise sufficient to call for comment. There is not such universal agreement on the other qualities, although there have been more favorable than unfavorable comments about all the qualities. The blend has been criticized on some occasions but this is not entirely an acoustical phenomenom; on occasions one orchestra has been criticized for lack of blend and, a few days later and by the same critic, another orchestra has been praised for the excellent blending. This remark also applies to a certain extent to balance; the platform has been designed to give all instruments as equal a chance as possible and, in the authors' opinion, it is now the responsibility of the conductors to achieve good balance. It would be interesting to see some more experiments made similar to the Stokowski concert.

The most important criticisms have been concerned with fullness of tone and, what is presumably connected with it, the sense of climax at loud passages, particularly with the extreme romantic composers such as Wagner. Most, not all, of the musicians want more fullness but only a little more. On the other hand, of the 18 scientists six wanted more fullness and one more definition, but 11 wanted no change. It should be noted that there appears to be no lack of fullness for the quieter and more classical pieces of music; in fact there have been several comments that the Hall is very satisfactory for small bodies of players and several chamber music concerts have been given in it with complete success.

V. CONCLUSIONS

The main conclusions that the authors would draw from the above objective and subjective results are most usefully discussed in terms of what they would alter if they had again to advise on the acoustical design of the Hall. On all major points they would not change the design. The R.F.H. is a good hall acoustically with no serious faults; the rectangular plan is still favored because of the smaller risk of echoes; the good sound insulation may be helping the definition, has made possible the use of the Hall for chamber music and, by enabling orchestras to play very quietly, has increased the dynamic range possible; the canopy and the raking of the seats have produced excellent definition and uniform conditions.

The only criticisms which have been at all serious are those concerning lack of fullness. The authors hold to their original view that this lack would be overcome by a longer R.T.; the value of 1.5 sec at 500 cps is 0.2 sec below the Knudsen and Harris optimum (Fig. 7). It may be that an increase of 0.2 sec would be sufficient, but a greater increase might be better, although the upper limit is indicated by the fact that when listening to rehearsals during the test period (the R.T. was about 2.3 sec) the definition was appreciably worse. In any new design, it would be desirable to make every effort to get the R.T. at mid-frequencies as long as

possible; if it were found that it was too long it would be a simple matter to shorten it, e.g., by introducing carpets or by perforating wood panels. Over half the total absorption at midfrequencies in the full Hall is due to the seats and the audience, which indicates that an increase in the volume per seat would be the most important method for lengthening the R.T. It is still thought essential to ensure that the R.T. at low frequencies is not too long; this would be a major fault as compared with the minor defect of a too short R.T. However, in the R.F.H. as it is at present any increase in the R.T. at mid-frequencies should be supplemented by a corresponding increase in low frequencies. It might be that a greater increase at low frequencies, or even an increase only at low frequencies, would be the most effective way of increasing the fullness. The criticisms of lack of fullness have applied only to the louder passages of romantic and choral music, although there have been comparatively few concerts with large choirs. This lack does not always occur; it may be simply that, for this class of music, most orchestras are a little too small for a hall of this size, or it may be that the absorption of some of the surfaces varies with intensity.

On minor points, the authors would not change the design of the canopy except that it might be better to obtain more blending by a general "closing-up" of all the surfaces round the orchestra; this could not be done in the R.F.H. because of the very large organ opening (60 ft wide by 30 ft high) required by the organ consultant. The organ is always a difficulty (acoustically) in a concert hall, and it is debateable how far the orchestral and choral conditions should be sacrificed in order to help the organ. The canopy in the R.F.H. has, for the sake of the organ, been made higher than the authors would have wished, and although this does not appear to have hindered the definition, it has probably detracted from the blend.

The danger of the screening of the middle instruments by the front instruments was probably exaggerated; it occurs mainly in halls with flat floors and it might have been better in the R.F.H. to have had a higher platform front while still keeping the platform rake. Any screening would only have affected the front two or three rows, and several rows at the back of the Terrace Stalls would have been helped. Also, it would have been better to keep to the calculated rake; at the back of the Terrace Stalls the total difference between the calculated rake and the straight line adopted was about 2 ft 6 in., but the greater height, small as it is, might have helped considerably. Apart from this slight weakness at the back of the Terrace Stalls, the uniformity of the acoustics is good. This is particularly so at the back of the Grand Tier and it would have been possible to go further back here with little ill effect acoustically, although, in a sense, the limit on visual grounds has just about been reached.

It should be emphasised again that this paper has

been concerned with the practical aspects of concert hall design. A large number of factors enter into a full discussion of concert hall acoustics, factors such as new halls affecting opinions, the change of music with time, the reactions of the ordinary public, and economic considerations. This paper represents the views of the authors, based largely on their experience with the R.F.H. but also influenced by two other new concert halls (Colston Hall, Bristol, and Free Trade Hall, Manchester) with which they have been connected. For those who disagree with these views they would end with another quotation from Chaucer:

"And who so sayth of trouthe I varye, Bid hym proven the contrarye."

The authors wish to record the pleasure it has given them to work with Mr. Bagenal and with the L.C.C. architects and staff. It will be obvious from this paper that the acoustic consultants enjoyed an extremely close collaboration with the architects. This collaboration began at the earliest stages and continued in detail throughout the period of construction and testing up to the present time, when modifications are still under consideration. The willingness of the architects to meet the numerous acoustical requirements has made the work of the consultants a pleasure and, of course, has contributed greatly to the success of the Hall.

The authors wish to thank Professor R. H. Bolt, Mr. F. Ingerslev, and Professor Erwin Meyer for helpful discussions during the design of the R.F.H., and the last named for some of the ideas discussed in part III E. The following also helped in the design and testing: Miss E. S. J. Stedeford and Mr. H. Creighton, Mr. R. O. B. Hinsch (of the Argentine), Mr. H. R. Humphreys, Mr. C. A. G. Pursell, Mr. T. G. C. Spiers, and Mr. E. F. Stacy.

The work described here is part of the research program of the Building Research Board of the Department of Scientific and Industrial Research and is published by permission of the Director of Building Research.

6

Reprinted from *Acustica* 4(4):447–450 (1954)

THE ACCEPTABILITY OF ARTIFICIAL ECHOES
WITH REVERBERANT SPEECH AND MUSIC

by A. F. B. Nickson, R. W. Muncey and P. Dubout

Division of Building Research, Commonwealth Scientific and Industrial Research Organization, Australia.

Summary

Groups of subjects seated in a room of short reverberation time (0.15 s) assessed whether or not a single artificial echo added to speech or music was disturbing. The artificial echo varied 50 dB in intensity level and up to 600 ms in delay. The acceptable-echo-level/delay relation is shown to consist of three parts; the initial period when the Haas effect is prominent, an interim period determined by the decay time of the sound type, and a final period when the level of acceptable echo scarcely changes with the delay, being determined by the dynamic range of the programme material.

Sommaire

Des groupes de sujets assis dans une salle ayant une faible durée de réverbération (0,15 s), ont été invités à dire si un écho artificiel simple ajouté à de la parole ou de la musique était ou non gênant. Cet écho artificiel avait un niveau d'intensité variant jusqu'à 50 dB, et son retard atteignait 600 ms. On montre que la relation entre l'intensité d'écho admissible et le retard, se compose de trois portions: une portion initiale, où l'effet Haas est prédominant, une portion inter-médiaire déterminée par la durée d'affaiblissement du son employé, et une portion finale où le niveau de l'écho admissible varie à peine avec le retard, étant déterminé par le champ du dispositif de reproduction.

Zusammenfassung

Versuchspersonen, die in einem Raum mit kurzer Nachhallzeit (0,15 s) saßen, stellten fest, ob ein einzelnes künstliches Echo, das zu Sprache oder Musik hinzugefügt wurde, störend wirkte oder nicht. Das künstliche Echo wurde 50 dB in der Schallintensität und bis zu 600 ms in der Verzögerungszeit geändert. Es wird gezeigt, daß die Beziehung zwischen der zulässigen Echo-intensität und der Verzögerungszeit aus drei Bereichen besteht: dem Anfangsbereich, in welchem der Haas-Effekt vorherrscht, einem Zwischenbereich, der durch die Nachhallzeit der Schallart bestimmt wird, und einem Endbereich, in welchem die zulässige Echointensität sich kaum mit der Verzögerungszeit ändert, und der durch den Dynamikbereich der Wiedergabe bestimmt wird.

1. Introduction

All sounds as commonly heard consist of an initial direct sound followed by echoes. Workers [1], [2] have studied the acceptability of artifi-cially added echoes; the results can be expressed graphically as the echo level intensity at various delay intervals that are acceptable to a given percentage of the audience at the test. Presented in this form three parts are apparent: the first occupies the initial period to about 50 ms, within which period an echo from as loud as the original sound to as much as 10 dB greater is acceptable, the second occupies the next few hundred ms when the tolerability of the echo decreases almost linearly with increasing delay, and the third when the level of acceptable echo scarcely changes with the delay period.

This report is concerned with listening experi-ments using the previous technique of listeners seated in a comparatively dead room [2] but the sounds used in this study have been recorded in reverberant spaces so that in addition to the artificial echoes there are natural echoes in the sound. The results lead to a hypothesis which explains them in terms of the characteristics of the original sound.

2. Experimental technique

The equipment used in this work is essentially that used previously [2]. The endless tape system is unchanged but the amplifiers and the control system have been reconstructed to provide in-creased operating facility and somewhat improv-ed sound crispness and quality. A new room for listening has been constructed; it measures about 6 m × 5 m × 3 m and has a reverberation time/frequency characteristic as shown in Fig. 1, room A. Two loudspeakers placed side by side were used for the initial and echo sounds and the number of listeners who answered the question "are you disturbed by the echo?" in the affirma-tive was recorded electrically as before.

Several sound types were used; three were one excerpt of 20 s length read at 5 syllables per

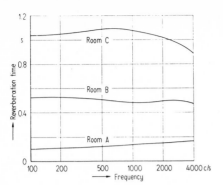

Fig. 1. Room reverberation times.

second in the three rooms numbered A, B and C in Fig. 1, the fourth was string music recorded from a quartette playing outdoors, the fifth this string music played in room C and recorded from a microphone, and a sixth was the speech recorded in room A re-recorded concurrently with random noise onto the endless tape. As results show, it is unlikely that the actual words and music are significant; they are therefore not reported.

The audience at each listening test was normally ten; occasionally the whole ten were not available and the corresponding adjustment was made in the evaluation of the results. The eight persons of the acoustics group of this Division were included in the ten and, from the previous paper, it can be expected that their results will be representative generally.

The sound level used was the same as in the earlier test, being about 80 dB (peak) above 0.0002 dynes/cm², as desired by the subjects for comfortable listening. In addition, for the speech from room A a further test was conducted at a level 20 dB less to find the effect of intensity on the results.

Finally, the overall level of the non-echoed sound of each type was recorded by a Brüel and Kjær type 2301 high-speed level recorder adjusted to have an instrument decay equivalent to a reverberation time of 0.14 s, being close to that of the ear [3]. The rate of sound decay after each syllable of speech or music has been termed the programme reverberation time.

3. Results

The results obtained from the various sound types are shown in Figs. 2 and 3. Initially a plot was made of the percentage disturbed by the echo against the amount of the delay and from these plots the echo level/delay relation acceptable to 20 and 50 percent of listeners was determined. The precision of the results decreased

with increasing reverberation time on the original sound and was greater for speech than for music. The standard error for a single observation for speech recorded in room B was about 12 per cent for any particular level/delay combination. Measurements were repeated to reduce the error of the mean to about 5 per cent for speech and 10 per cent for music.

The records of the sound level were analysed to measure the level difference between the peaks and the troughs (roughly within one reverberation time) and the minimum noise level and also the programme reverberation time measured as each syllable or sentence of speech or music died away. These are recorded in Table I together with details of the equivalent reverberation time (obtained as a tangent to the echo level/delay time curve), and the level below which echo acceptability is independent of delay taken from Figs. 2 and 3.

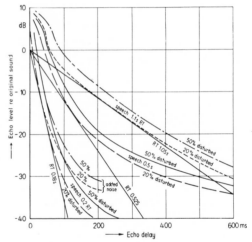

Fig. 2. Acceptable echo levels for speech.

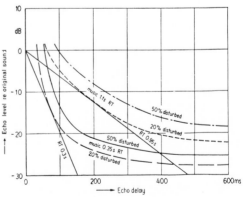

Fig. 3. Acceptable echo levels for string music.

As might be excepted, the result of the test at a level 20 dB below that normally used gave results which did not differ essentially from those at the usual level.

Table I

Sound type	Level records			Listening test results (20% disturbed)	
	Programme reverberation time	Ratio between peaks and troughs	Ratio between peaks and tape noise	Equivalent reverberation time	Level for echo independent of delay
	s	dB	dB	s	dB
Speech room A	0.20	42	42	0.18	—45
Speech room A with added noise	0.20	28	28	0.18	—35
Speech room B	0.5	35	35	0.52	—30
Speech room C	1.1	32	32	1.05	—30
String music outdoors	0.35	23	38	0.3	—25
String music room C	1.1	20	32	0.95	—20

4. Discussion

It is quite apparent that the nature of the original sound has a profound effect on the results obtained in listening tests such as described here. The agreement of the programme reverberation times and maximum intensity changes, measured from level records of the sound, with the equivalent reverberation times and minimum echoes corresponding to the more critical section of listeners (20 percent disturbed) is quite striking. In fact it appears almost certain that it is the former two parameters that determine the shape of the curve after the initial period when the "HAAS" effect operates i. e. when the ear is capable of accepting echoes greater than those given by a linear reverberation decay. The tentative criterion for auditoria proposed by BOLT and DOAK [4] follows this idea.

As a further check on this hypothesis the recorded sounds used in the previous study [2] have been examined and the results are compared in Table II. Again the agreement is very good with the exception of the slow music. For all cases the ratio between peaks and troughs of the sound was measured also after passing through a tuned circuit: for the slow string music the ratio was then about 28 dB and for the slow organ music 20 dB, whereas it was practically

unchanged for the other sounds. This suggests that the frequency sensitivity of the ear is being used to detect low level echoes in slow music. Further, many of the subjects complained of the difficulty of detecting echoes with slow music and about 30 percent reported being disturbed by an echo even when the sound was presented without echo. These results would appear to be less reliable than the others.

An independent confirmation is supplied by MEYER and SCHODDER [5]. Study of their Fig. 8 shows that for echoes to pass unnoticed by the more critical section of an audience they must fall close to the level expected from a linear decay at a rate corresponding to the reverberation time of the listening room.

Table II

Sound type	Level records		Listening test	
	Programme reverberation time	Ratio between peaks and troughs	Equivalent reverberation time	Level for echo independent of delay
	s	dB	s	dB
Speech fast	0.20	40	0.18	—35
slow	0.28	27	0.28	—34
String music fast	0.35	30	0.41	—32
slow	0.4	18	0.75	—30
Organ music fast	0.6	33	0.76	—31
slow	0.8	8	0.65	—30

The results from listening in a dead space to a sound recorded in a live space are qualitatively comparable with those previously reported for listening in a live space to material recorded in a dead space. For example, replotting HAAS' Fig. 4 in the form used here suggests (from the slope of the tangent) an equivalent reverberation time of about 0.8 s. In both instances then, the tolerance for echoes is largely defined by the masking echoes introduced by the room. This leads to the conclusion that, apart from short delay echoes, the listening tests are only a complex method of measuring phenomena connected with reverberation.

5. Conclusions

The relation between acceptable echo level and echo delay in artificial echo listening tests depends in the middle period on the natural echoes present in the recording room, and in the long delay period on the difference in sound level between the loud and quiet portions of the programme material. Any attempt to translate results of listening tests into criteria for examination of pulse measurements from rooms would therefore produce merely a reflection of the test conditions and not be universally accurate.

(Received 15th December, 1953.)

References

[1] HAAS, H., Über den Einfluß eines Einfachechos auf die Hörsamkeit von Sprache. Acustica **1** [1951], 49···58.

[2] MUNCEY, R. W., NICKSON, A. F. B. and DUBOUT, P., The acceptability of speech and music with a single artificial echo. Acustica **3** [1953], 168···173.

[3] STEVENS, S. S. and DAVIS, H., Hearing; its psychology and physiology. Wiley & Sons, New York 1948; 3rd imp., p. 222.

[4] BOLT, R. H. and DOAK, P. E., A tentative criterion for the short-term transient response of auditoriums. J. acoust. Soc. Amer. **22** [1950], 507···509.

[5] MEYER, E. and SCHODDER, G. R., Über den Einfluß von Schallrückwürfen auf Richtungslokalisation und Lautstärke bei Sprache. Nachr. Wiss. Ges. Göttingen **6** [1952], 31···42.

Reprinted from *Acustica* 1(2):49–58 (1951)

ÜBER DEN EINFLUSS EINES EINFACHECHOS AUF DIE HÖRSAMKEIT VON SPRACHE

Von HELMUT HAAS

(III. Physikalisches Institut der Universität Göttingen)

[*Editor's Note:* In the original, material precedes this excerpt.]

1. Einleitung

Bei Schalldarbietungen in geschlossenen Räumen treten an Begrenzungsflächen und Einrichtungsgegenständen zahlreiche Schallreflexionen auf. Der reflektierte Schall kommt am Beobachtungsort später an als der direkte und kann bei größeren Wegunterschieden bzw. Laufzeitdifferenzen zu einer Störung des Klangeindruckes führen.

Ist der Wegunterschied genügend groß, so können der direkte und der reflektierte Schall gegebenenfalls getrennt wahrgenommen werden, und man spricht dann von einem Echo. Wir wollen in dieser Arbeit jedoch jede Schallreflexion unabhängig von der Größe ihrer Laufzeit als Echo bezeichnen.

Der Schwellenwert der Laufzeit, oberhalb dessen sich eine merkbare Verschlechterung des Klangeindruckes ergibt, wurde von PETZOLD [1] als Verwischungsschwelle bezeichnet und für Sprache mit

$$t = 0{,}05 \pm 0{,}01 \text{ s}$$

angegeben, was einer Wegdifferenz in Luft von $s = 17 \pm 3$ m entspricht. Für kleinere Räume, bei denen zur Erreichung dieser Laufzeitdifferenz viele Reflexionen notwendig sind, ist mit einer Echostörung nicht zu rechnen, da einerseits infolge der Schallschluckung der reflektierenden Flächen die Intensität des mehrfach reflektierten Schalles nach 50 ms sehr stark abgesunken sein wird, und andererseits das Ohr eine Vielzahl von kurz aufeinanderfolgenden kleiner werdenden Reflexionen erhält, die es gar nicht voneinander getrennt wahrnimmt. Das allmähliche Abklingen des Schalles in Räumen wird als Nachhall bezeichnet. Eine gewisse Größe der Nachhallzeit ist für die Hörsamkeit sogar vorteilhaft, da sie neben einer Schallverstärkung eine Beeinflussung des Klangbildes hervorruft, die wir als angenehm empfinden.

In großen abgeschlossenen Räumen kann es jedoch sehr leicht vorkommen, daß nach einer oder nach wenigen Schallreflexionen bereits die kritische Laufzeitdifferenz überschritten wird. Ist die Energie des reflektierten Schalles dann noch so groß, daß einzelne Schallrückwürfe sich sehr stark aus der das allmähliche Abklingen des Schalles charakterisierenden Nachhallkurve herausheben, so werden wir u. U. eine Beeinträchtigung der Darbietung in ungünstigem Sinn erhalten, im Extremfall ein ausgesprochenes Echo hören.

Für den Entwurf und beim Bau großer Räume ist es von Bedeutung, die Größe der durch Echoerscheinungen möglicherweise hervorgerufenen Störungen im voraus abschätzen zu können. Auch bei künstlicher Schallverstärkung in Räumen oder bei Schallübertragung durch mehrere räumlich voneinander entfernte Lautsprecher können so große Laufzeitunterschiede entstehen, daß störende Echoerscheinungen auftreten. Schließlich sei noch darauf hingewiesen, daß auch in der Nachrichtentechnik u. U. Echoerscheinungen auftreten können, die bei Sprechverbindungen Anlaß zu Störungen geben. Es soll im folgenden untersucht werden, welchen Einfluß Echos in Abhängigkeit von verschiedenen Parametern wie Laufzeit, Intensität, Klangfarbe, Richtung usw., bei binauralem Hören auf die Hörsamkeit von Sprache ausüben. Um übersichtliche Verhältnisse zu erhalten und den technischen Aufwand zu begrenzen, werden

[*Editor's Note:* An English summary of this article prepared by T. D. Northwood follows.]

die Untersuchungen mit einem Einfachecho durchgeführt, d. h., dem direkten Schall folgt eine einmalige Wiederholung. Dabei werden vor allem Messungen vorgenommen, deren Ergebnisse für die Praxis von Bedeutung sind. Untersuchungen auf diesem Gebiet unter Hinzuziehung einer großen Anzahl von Beobachtern sind bisher noch nicht angestellt worden. H. DECKER [2] hat die bei Fernsprechverbindungen über lange Kabelleitungen auftretenden Echoerscheinungen untersucht und gefunden, daß sich bei Ferngesprächen für den Fall, daß die Echointensität gleich der Primärschallintensität ist, $50^0/_0$ seiner 15 Beobachterpaare bei einer Echolaufzeit von 100 ms im Flusse ihrer Unterhaltung gestört fühlen, während für den Fall, daß die Echoamplitude nur $60^0/_0$ der Primärschallamplitude beträgt, die kritische Laufzeitdifferenz auf 150 ms ansteigt.

H. STUMPP [3] untersuchte den Einfluß verschiedener Einfallsrichtungen eines Einfachechos. Er fand für den Fall, daß Primärschall und Echo gleiche Intensität haben und den Beobachter seitlich aus der gleichen Richtung treffen, für Sprache im Freien eine kritische Laufzeitdifferenz von 80 ms; kamen Primärschall und Echo seitlich aus entgegengesetzten Richtungen, dann betrug die kritische Laufzeitdifferenz 50 ms.

2. Beschreibung einer Apparatur zur künstlichen Erzeugung von Echos

Um die Eigenschaften des reflektierten Schalles, also des Echos, genau definieren zu können, war es erforderlich, eine Apparatur zur künstlichen Erzeugung von Echos zu erstellen. Sie hat die Aufgabe, einen Schallvorgang nach Ablauf einer bestimmten Zeit zu wiederholen.

Für die Erreichung der Zeitverzögerung wurde im Hinblick auf hohe Übertragungsqualität das Magnettonverfahren angewendet. Es besteht im wesentlichen darin, daß man den zu verzögernden Schallvorgang auf einen Träger magnetisch aufzeichnet, von dem man anschließend direkten Schall und Echo mit praktisch jeder gewünschten Verzögerungszeit wieder abnehmen kann.

Abb. 1 stellt die ausgeführte Anlage im Blockschaltbild dar. Mittels eines Mikrophons M wird der Schallvorgang, wir wollen uns im folgenden auf Sprache beschränken, in elektrische Spannungsschwankungen umgeformt; diese werden in einem nachfolgenden Verstärker verstärkt und den Aufsprechentzerrern der Magnetophonapparatur zugeleitet. Will man für besondere Messungen immer wieder einen bestimmten, gleichbleibenden Text verwenden, so kann man diesen mit einem handelsüblichen Magnetbandgerät aufnehmen und wiedergeben. Der nach dem Mikrophonverstärker eingezeichnete Umschalter dient der Wahl zwischen direkter und indirekter Wiedergabe. Mittels eines Aussteuerungsmessers wird der Pegel überwacht und konstant gehalten. Zur eigentlichen Schallverzögerung dient eine Endlos-Band-Magnetophonanlage, die in Abb. 1 oben, skizziert ist. Die Aufzeichnung erfolgt nach dem Hochfrequenzverfahren. Eine Schiene aus kräftigem U-Profileisen trägt im Abstand von ca. einem Meter zwei drehbar gelagerte Leitrollen R mit einem Durchmesser von 150 mm, um die eine Schleife aus Magnetophonband gelegt ist. Der Bandantrieb erfolgt in Pfeilrichtung durch die rundgeschliffene Welle eines Synchronmotors ($n = 1500$ Umdr./min), an die das Magnetophonband von einer Gummiandruckrolle gepreßt wird, mit einer Geschwindigkeit von 77 cm/s. Auf der Schiene einer optischen Bank sind die durch Kreise angedeuteten Löschköpfe L_I, L_{II}, die Sprechköpfe S_I, S_{II}, und die Hörköpfe H_I und H_{II} befestigt, von denen H_{II} in Bandrichtung verschiebbar ist. Die Wirkungsweise der Apparatur ist folgende:

L_I und L_{II}, die mit Hochfrequenz (ca. 80 kHz) aus den Aufsprechentzerrern gespeist werden, löschen die auf dem Band befindlichen magnetischen Aufzeichnungen. S_I und S_{II}, die aus den eingangsseitig parallel geschalteten Aufsprechentzerrern I und II erregt

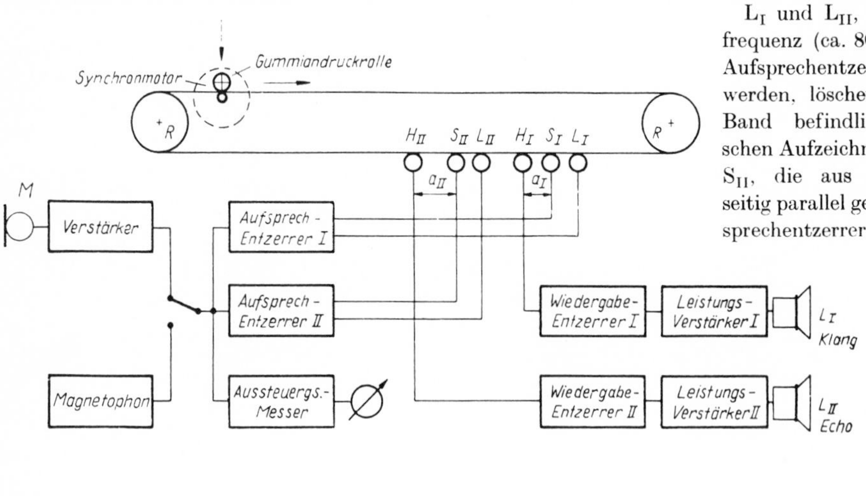

Abb. 1.
Echogerät,
Blockschaltbild.

werden, zeichnen gleichzeitig den gleichen Schallvorgang auf das Band auf. Mittels der Hörköpfe H_I und H_{II} wird dieser Vorgang dann wieder abgetastet. Die Laufzeitverzögerung Δt ist gegeben durch

$$\Delta t = (a_{II} - a_I)\, v \,.$$

wenn a_I den Abstand der Spaltmitten von $S_I — H_I$, a_{II} den entsprechenden Abstand $S_{II} — H_{II}$ und v die Bandgeschwindigkeit bedeuten. Ist $a_I = a_{II}$, dann ist die Laufzeitdifferenz $\Delta t = 0$, d. h., die Schallvorgänge werden genau so gleichzeitig abgetastet, wie sie aufgenommen wurden. Die Einstellung der verschieden großen Echolaufzeiten erfolgt durch Verschieben von H_{II} längs einer in ms geeichten Skala.

Die in H_I und H_{II} vom bewegten Magnetophonband induzierten Spannungen werden zwei getrennten Wiedergabeentzerrern I und II zugeführt, anschließend verstärkt, und man erhält aus dem Lautsprecher L_I den eigentlichen Schallvorgang, während L_{II} denselben Schallvorgang zeitlich verzögert, also das Echo wiedergibt.

Es wird somit auch der erste Schallvorgang, im folgenden „direkter Schall" oder „Primärschall" genannt, über einen Lautsprecher wiedergegeben, was zur Vermeidung akustischer Rückkopplung und zur Ausschaltung einer Beeinträchtigung des Sprechers durch sein Echo zweckmäßig erschien.

Die objektive Messung der Güte der Übertragungsapparatur ergab, daß bei konstant gehaltenem Schalldruck am Aufnahmemikrophon die Schalldruckschwankungen, im Abstand von 2 m in Achsrichtung vom Wiedergabelautsprecher gemessen, im Frequenzgebiet zwischen 100 Hz ... 10 kHz kleiner als ± 7 db waren. Dabei war der Lautsprecher in einer quadratischen Schallwand von 40×40 cm^2 montiert. Die geringen linearen Verzerrungen der Anlage können in unwesentlichem Ausmaß die Klangfarbe des Sprechers ändern, doch ist dies für die Durchführung der folgenden Messungen belanglos, da die Verständlichkeit dadurch nicht herabgesetzt wird.

Da Verstärker und Lautsprecher für Sprachwiedergabe in natürlicher Lautstärke genügend überdimensioniert waren, wird die Größe der nichtlinearen Verzerrungen nur von der Magnetophonapparatur bestimmt. Der Klirrfaktor des Magnettonverfahrens wird in der Literatur bei 800 Hz mit $<3^0/_0$ angegeben.

Die angestellten subjektiven Untersuchungen zeigten, daß die Silbenverständlichkeit über die gesamte Übertragungsanlage mit einem geübten Hörtrupp $96^0/_0$ betrug, also bereits das erreichbare Maximum darstellt, welches auch bei direkter Sprachübertragung von Mund zu Ohr nur erzielbar ist.

3. Untersuchungen bei kleinen Laufzeitdifferenzen

Bei Schalldarbietungen in kleineren Räumen sind stets Echos mit kleiner Laufzeitdifferenz gegenüber dem direkten Schall vorhanden, die uns jedoch erfahrungsgemäß nicht stören; wir empfinden vielmehr eine Unterhaltung als anstrengender und unnatürlicher, wenn diese kurzzeitigen Reflexionen fehlen, wie beispielsweise im Freien bei Vorhandensein einer starken Neuschneedecke oder in schallgedämpften Räumen.

Diese Beobachtung zeigt, daß unser Gehörorgan, das fast immer neben direkten Schalleindrücken auch deren kurzzeitig folgende Reflexionen verarbeitet, so eingerichtet ist, daß es diesen Zustand als natürlich empfindet. Welchen Unterschied können wir nun überhaupt zwischen einem reflexionsfreien Schallvorgang und einem solchen, der mit einem kurzzeitigen Echo behaftet ist, gehörmäßig feststellen? Zur Klärung dieser Frage wurde folgende Versuchsanordnung aufgebaut:

Da kein schallgedämpfter Raum zur Verfügung stand, mußte die Untersuchung im Freien ausgeführt werden. Als hierfür geeignet erwies sich das flache Dach eines freistehenden Gebäudes. Zur Messung wurden zwei gleichartige Lautsprecher in einer Entfernung von 3 Meter unter einem Winkel von 45^0 halblinks und halbrechts vor dem Beobachter aufgestellt. Diese Aufstellung wurde lediglich deshalb gewählt, weil sie sich für die Durchführung der Messung insofern als günstig erwies, als der Abgleich so am besten erfolgen konnte. Die Lautsprecherachsen waren auf den Kopf des Beobachters gerichtet, um keinen Abfall der hohen Frequenzen infolge der Richtcharakteristik der Lautsprecher zu bekommen. Primär- und Echolautsprecher strahlten mit der gleichen Intensität, die Lautstärke am Ort des Beobachters betrug ca. 50 Phon, was durch Lautstärkevergleich mit einem 1000 Hz-Ton und einem geeichten Mikrophon nachgeprüft wurde.

Vor Beginn der Messung wird an der Apparatur die Verzögerung Null eingestellt und der Kopf des Beobachters in eine solche Lage gebracht, daß Mitteneindruck entsteht, d. h., der Schall direkt von vorn zu kommen scheint. In dieser Stellung verbleibt die Versuchsperson während der folgenden Einstellungen. Bei sehr kleinen Laufzeitdifferenzen zwischen 0 und 1 ms beobachtet man ein allmähliches Wandern der fiktiven Schallquelle, in Richtung auf den Primärlautsprecher hin. Wir empfinden also bei sehr kleinen Laufzeitdifferenzen eine Änderung des Richtungseindruckes. Nun wird beispielsweise die Sprache aus dem Echolautsprecher gegenüber dem Primärlautsprecher um 10 ms verzögert, was einem Laufwegunterschied in Luft von etwa 3,4 m entspricht, und der auffallendste Effekt ist nun der, daß man den Echolautsprecher überhaupt nicht mehr hört, obwohl er mit derselben Energie strahlt wie der Primärlautsprecher, und man den Eindruck hat, die Sprache käme allein aus diesem. Eine Erklärung für diesen Effekt ist noch nicht bekannt. Er beruht wahrscheinlich auf einer Funktion unseres Zentralnervensystems und wäre vielleicht so zu erklären, daß bei Eintreffen eines Schallvorgangs ein Sperrvorgang ausgelöst wird, der verhindert, daß wir kurzzeitige Wiederholungen desselben Schalleindruckes getrennt wahrnehmen können. Es wäre denkbar, daß die Anstiegsflanke dieses Sperrvorganges zum Richtungshören beiträgt.

Die Vorgänge, die es uns rein akustisch ermöglichen, die Richtung, aus der wir einen Schalleindruck empfangen, festzustellen, wurden bereits eingehend von G. v. BÉKÉSY [4], M. REICH und H. BEHRENS [5], K. DE BOER [6] und H. WARNCKE [7] untersucht. Diese Autoren haben gefunden, daß der Richtungseindruck durch Laufzeit- und Intensitätsunterschiede, mit denen ein Schalleindruck unsere beiden Ohren trifft, hervorgerufen wird, doch wurden bei diesen Untersuchungen hauptsächlich Laufzeitunterschiede zwischen $0 \ldots 0,62$ ms betrachtet, die einer Wegdifferenz in Luft bis zu 21 cm entsprechen. Das ist etwa der größte Wegunterschied, mit dem ein von einer einzigen Schallquelle herrührender Schalleindruck unsere Ohren treffen kann. Wird die Laufzeitdifferenz über dieses Maß hinaus vergrößert, so tritt keine Änderung der Richtungsempfindung mehr auf. Welche Beobachtungen kann man nun machen, wenn die Laufzeitdifferenz weiter vergrößert wird?

Bei kleinen Laufzeiterhöhungen zwischen $1 \ldots 30$ ms ist gegenüber dem Fall, daß der Echolautsprecher abgeschaltet wird, eine Änderung des Klangeindruckes und eine Erhöhung der Lautstärke festzustellen, auf die später noch ausführlicher eingegangen werden soll. Man hat den Eindruck, daß der Klang voller wird und die Schallquelle voluminöser zu werden scheint.

Dieser „pseudo-stereophonische" Effekt ist schon seit längerer Zeit bekannt und fand seine praktische Anwendung bereits 1926 bei der Konstruktion des Ultraphons [8].

Derselbe Effekt läßt sich auch bei gleichzeitigem Betrieb zweier Lautsprecher erzielen, wobei die erforderliche Laufzeitdifferenz durch verschieden große Abstände des Beobachters von den beiden Lautsprechern erreicht wird.

Man lokalisiert zwar eindeutig auf den zuerst ertönenden Lautsprecher, hat aber bei Vorhandensein eines kurzzeitigen Echos nicht mehr den Eindruck eines typischen Lautsprecherklanges, der auch bei guter Wiedergabe vorhanden ist, d. h., man hat nicht mehr die Empfindung, daß die Ausdehnung der Schallquelle nur auf die Abmessungen des Lautsprechers beschränkt bleibt.

Dieser Zustand bleibt bei Laufzeitunterschieden zwischen $1 \ldots 30$ ms ziemlich unverändert erhalten. Erst bei Laufzeiten in der Größenordnung um 40 ms bemerkt man, daß auch der Echolautsprecher Schall abstrahlt, doch lokalisiert man die Schallquelle nach wie vor auf den zuerst ertönenden Primärlautsprecher. Wird die Laufzeitdifferenz weiter über 50 ms hinaus vergrößert, so beginnt man ein getrenntes Echo zu hören, doch bleibt der Schwerpunkt der Schallabstrahlung nach wie vor auf dem Primärlautsprecher liegen. Um eine quantitative Erfassung des bei kleinen Laufzeitdifferenzen auftretenden subjektiven „Unterdrückungseffektes" von Echos in Abhängigkeit von der Laufzeit zu ermöglichen, wurde folgendermaßen vorgegangen:

Primär- und Echolautsprecher strahlten in der beschriebenen Anordnung mit gleicher Intensität fortlaufenden Sprachtext. Die Laufzeitdifferenz zwischen den beiden Lautsprechern wurde statistisch im Bereich von $1 \ldots 50$ ms sprunghaft geändert. Der Beobachter hatte die Möglichkeit, mittels eines in db geeichten Dämpfungsgliedes die Intensität des Primärlautsprechers so weit zu vermindern, bis er den Eindruck hatte, Primär- und Echolautsprecher gleich laut zu hören. Man hat dabei keineswegs einen Mitteneindruck, wie man ihn bei einer gegen Null gehenden Laufzeitdifferenz feststellen kann, sondern das Empfinden, gleichzeitig zwei Schallquellen aus verschiedenen, der Aufstellung der Lautsprecher entsprechenden Richtungen zu hören. Die Einstellung auf subjektiv gleich empfundene Lautstärke machte den Beobachtern keine Schwierigkeiten und ergab verhältnismäßig kleine Streuungen. Das Ergebnis dieser Messung, die mit 15 verschiedenen Beobachtern durchgeführt wurde, zeigt Abb. 2.

Es ist die Intensitätsdifferenz in db, um die die Intensität des Primärlautsprechers geschwächt werden mußte, damit beide Lautsprecher gleich laut empfunden wurden, in Abhängigkeit von der Laufzeit aufgetragen. Die stark ausgezogene Kurve stellt den Mittelwert aus allen Meßpunkten dar, die beiden ähnlich verlaufenden gestrichelt gezeichneten Linien deuten die Streugrenzen an. Um einen Überblick über die bei den einzelnen Beobachtern auftretenden Streuungen zu geben, sind die Meßpunkte, die bei zwei verschiedenen Beobachtern erhalten wurden, besonders eingetragen. Wie man aus Abb. 2 erkennt, muß die Intensität des Echolautsprechers die des Primärlautsprechers bei Laufzeitdifferenzen zwischen $5 \ldots 30$ ms etwa zehnmal überwiegen (10 db), um den

Eindruck gleicher Lautstärke hervorzurufen. Bei Echolaufzeiten über 30 ms tritt ein leichtes Absinken der Intensitätsdifferenz auf und die Streuung der Meßpunkte wird größer.

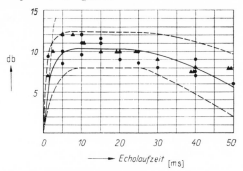

Abb. 2. Echounterdrückungseffekt in Abhängigkeit von der Laufzeit bei Sprache.
(— — — — Meßergebnisse von DE BOER [6] bei 0...3 ms)

Es kann angenommen werden, daß dieser im vorstehenden beschriebene Effekt der gehörmäßigen Unterdrückung von Echos mit kleiner Laufzeitdifferenz uns das Richtungshören in geschlossenen Räumen ermöglicht, wobei zur Richtungsbestimmung von unserem Gehörorgan immer nur der erste Klangeindruck des direkten Schalles ausgewertet wird. Es wurde weiter mit Hilfe von Sprache untersucht, ob die Richtung, aus der das Echo kommt, einen Einfluß auf den Unterdrückungseffekt hat. Es stellte sich heraus, daß kein über die Streugrenzen in Abb. 2 hinausgehender Einfluß festzustellen war. Das stimmt mit der praktischen Erfahrung überein. In Räumen können kurzzeitige Echos aus allen Richtungen kommen, ohne als störend empfunden zu werden.

Es soll in diesem Zusammenhang nicht versäumt werden, auf eine technische Anwendung des Unterdrückungseffektes für kurzzeitige Echos hinzuweisen, auf die L. CREMER [9] bereits aufmerksam gemacht hat. Es ist nämlich durchaus möglich, eine künstliche Schallverstärkung um 10 db oder vielleicht noch mehr vorzunehmen, ohne daß man das Vorhandensein der verstärkenden Schallquelle merkt, wenn man zusätzlich noch durch einen weiteren, beispielsweise optischen Eindruck abgelenkt wird.

Wir wollen uns nun der Betrachtung der durch ein kurzzeitiges Echo hervorgerufenen Lautstärkeerhöhung zuwenden. In unserem Fall, bei gleicher Intensität des Echos, wird dem Gehörorgan, um die Laufzeitdifferenz verzögert, zweimal eine bestimmte, gleich große Energie zugeführt. Unser Schallempfindungsvermögen ist jedoch mit einer gewissen Trägheit behaftet. Die empfundene Lautstärke er-

reicht beispielsweise nach Einschalten eines Tones ihren Endwert erst, wie v. BÉKÉSY [10] gefunden hat, nach einer Zeit von 200 ms. Umgekehrt fällt nach Untersuchung von U. STEUDEL [11] die subjektiv empfundene Lautstärke eines Tones nach dessen plötzlichem Abschalten nur allmählich ab. BÜRCK, KOTOWSKI und LICHTE [12] haben im Zusammenhang mit der Entwicklung eines objektiv anzeigenden Lautstärkemessers aus den Messungen von v. BÉKÉSY [10] und STEUDEL [11] die Zeitkonstanten des subjektiven An- und Ausklingvorganges bestimmt. Es ergab sich bei v. BÉKÉSY im Mittel ein Wert von 130 ms, bei STEUDEL ein solcher von 50 ms.

Daraus kann man schließen, daß unser Gehörapparat die Schallintensitäten über kurze Zeiträume hinweg integriert, ähnlich etwa wie ein ballistisches Meßinstrument.

Wenn man nun für unseren Fall eines Echos kurzer Laufzeitdifferenz das Energieadditionsgesetz anwendet, so würde sich die Lautstärke unter Berücksichtigung der angenähert logarithmischen Ohrempfindlichkeit bei doppelter Schallintensität um 3 Phon erhöhen. Dies wurde auch durch von AIGNER und STRUTT [13] vorgenommene Messungen mit Sprache und Musik für den Fall bestätigt, daß die Klangfarben von Primärschall und Echo gleich waren. Unterschieden sich jedoch die Klangfarben wesentlich, so erhielten diese beiden Autoren eine um 6...9 Phon über das Energieadditionsgesetz hinausgehende subjektiv empfundene Lautstärkeerhöhung. E. LÜBCKE [14], der die von AIGNER und STRUTT in diesem Zusammenhang angestellten Versuche mit Geräuschen wiederholte, fand auch bei gleicher Klangfarbe und kleinen Schallwegunterschieden für den Fall, daß Primärschall und Echo gleiche Intensität hatten, eine subjektive Lautstärkeerhöhung um 5...6 Phon, also um 2...3 Phon mehr, als nach dem Energieadditionsgesetz zu erwarten wäre.

Die beiden erstgenannten Autoren verwendeten als Schallquellen Rundfunkempfänger. Die Laufzeitdifferenz wurde entweder durch Aufstellung der Empfänger in verschiedenem Abstand vom Beobachter erzielt oder bei gleichem Abstand durch eine doppelte verzögerte Schallplattenabtastung erreicht. Die subjektive Messung der Lautstärke erfolgte in beiden Arbeiten mittels eines Geräuschmessers nach BARKHAUSEN [15], d. h. durch Lautstärkevergleich mit einem 800 Hz-Klang.

Obwohl bereits AIGNER und LÜBCKE angegeben hatten, daß die Größe des von ihnen festgestellten Effektes nur unwesentlich von der Laufzeitdifferenz abhängt, solange diese unter der Verwischungsschwelle bleibt, sollte im Zusammenhang mit dieser Arbeit der Einfluß verschieden großer Laufzeiten untersucht werden. Die Lautstärkeerhöhung wurde im Gegensatz zu anderen Autoren nicht mit einem Lautstärkemesser, sondern durch direkten Vergleich eines Schalleindruckes ohne Echo mit demselben Schalleindruck bei Vorhandensein eines Echos gemessen. Überraschenderweise konnte mit dieser Versuchsanordnung keine größere Lautstärkeerhöhung als die bei Anwendung des Energieadditionsgesetzes zu erwartende, nämlich 3 Phon, festgestellt werden. Es wurde die Laufzeit innerhalb der Verwischungsschwelle geändert und die Einfallsrichtungen von direktem Schall und Echo variiert. Es wurde die Klangfarbe des Echos geändert, wobei darauf geachtet wurde, daß Primär- und Echolautsprecher für sich allein denselben Lautstärkeeindruck hervorriefen. Das Experiment wurde sowohl im Freien als auch in einem geschlossenen Raum mit verschiedenen Beobachtern durchgeführt. Immer aber blieb die durch ein Einfachecho gleicher Intensität hervorgerufene Lautstärkeerhöhung 3 Phon.

Eine Erklärung für den Unterschied dieses Ergebnisses gegenüber den Resultaten früher angestellter Untersuchungen ist vielleicht in der Art der Feststellung des Lautstärkezuwachses zu suchen. Während hier die Klangeindrücke direkt miteinander verglichen werden konnten, wurde dort ihre Lautstärke einzeln durch Vergleich mit einem Geräusch, dessen Intensität sich nur in Stufen von 5 zu 5 Phon variieren ließ, festgestellt. Weiter fiel bei der Lektüre früherer Arbeiten auf, daß keiner der Autoren erwähnte, daß der Echolautsprecher akustisch nicht lokalisierbar war. Möglicherweise ist auch der Grund der früheren Ergebnisse z. T. in der Unvollkommenheit der damaligen elektroakustischen Übertragungsanlagen zu suchen.

Zusammenfassend kann auf Grund der vorstehenden Untersuchungen gesagt werden, daß ein Einfachecho mit kleiner Laufzeitdifferenz infolge der Trägheit unseres Gehörapparates nicht als Echo wahrgenommen werden kann. Wir empfinden lediglich eine Zunahme der Lautstärke, die dem Energieadditionsgesetz entspricht, und eine angenehme Änderung des Klangeindruckes, d. h. eine scheinbare Verbreiterung der Schallquelle.

4. Untersuchungen bei großen Laufzeitdifferenzen

Erreicht die Laufzeitdifferenz des Echos einen bestimmten Wert, so beginnen wir Primärschall und Echo getrennt wahrzunehmen. Bei weiterer Vergrößerung des Laufzeitunterschiedes leidet die Deutlichkeit des Klangbildes, d. h. bei Sprache die Verständlichkeit. Wir können zwar die Störung infolge unserer Konzentrationsfähigkeit bis zu einem gewissen Grad unterdrücken, doch wirkt dann längeres Zuhören ermüdend. Wird die Echolaufzeit weiter vergrößert, sinkt die Verständlichkeit stark ab.

Es war naheliegend, den Einfluß eines Echos auf die Verständlichkeit von Sprache nach dem Silbenverständlichkeits-Meßverfahren zu untersuchen [16]. Es stellte sich jedoch heraus, daß dieses Verfahren ungeeignet war. Auch mehrsilbige Logatome oder die Überlagerung eines Störgeräusches brachten nicht den gewünschten Erfolg.

Es wurde daher eine neue Methode zur Beurteilung der durch Echoeinflüsse hervorgerufenen Störung entwickelt, die den Verhältnissen der Praxis weitgehend entspricht und die selbst mit einer geringen Anzahl von Versuchspersonen noch brauchbare Ergebnisse liefert. Den Beobachtern wurde fortlaufender Sprachtext zu Gehör gebracht, der durch Echoeinwirkung verschieden stark gestört war, wobei das Ausmaß der Störung sprunghaft geändert wurde. Die Zuhörer hatten durch Ja-Nein-Entscheidung zu vermerken, ob sie sich durch das Echo gestört fühlten oder nicht. Als „störend" wurde ein Echo dann bezeichnet, wenn es das Zuhören unangenehm anstrengend machte, wobei der Text u. U. noch voll verständlich sein konnte.

Um statistisch zuverlässige Ergebnisse zu erhalten, wurde die Anzahl der Beobachter groß gewählt. Sie schwankte zwischen 50 und 100 Personen. Aus technischen Gründen mußte auf eine Durchführung im Freien verzichtet und die Untersuchungen in einem geschlossenen Raum vorgenommen werden. Seine mittlere Nachhallzeit in besetztem Zustand betrug 0,8 s. Um das Energieverhältnis und die Laufzeitdifferenz von Primär- und Echolautsprecher an allen Punkten des Raumes möglichst gleich groß zu erhalten, wurden die Lautsprecher in geringem Abstand voneinander an der Stirnseite aufgestellt und waren auf die Zuhörer gerichtet. Der Lautsprecherabstand bis zur ersten Beobachterreihe betrug 4 m, so daß die Lautstärke für alle Versuchspersonen praktisch konstant war. Sie betrug ca. 55 Phon. Als Meßtext kam fortlaufende Sprache mit einer Sprechgeschwindigkeit von 5,3 Silben/s zur Verwendung.

a) Einfluß der Sprechgeschwindigkeit

Intensität und Klangfarbe von Primärschall und Echo waren einander gleich. Es wurden drei Meßreihen mit verschiedenen Sprechgeschwindigkeiten durchgeführt, und zwar mit 3,5 Silben/s, 5,3 Silben/s und 7,4 Silben/s. Innerhalb dieses Bereiches liegen normalerweise die Sprechgeschwindigkeiten.

Abb. 3 stellt das Ergebnis dieser Messungen dar. Es ist die prozentuale Anzahl der sich gestört fühlenden Beobachter in Abhängigkeit von der

Echolaufzeitdifferenz aufgetragen. Bei Kurve II, wie sie für eine Sprechgeschwindigkeit von 5,3 Silben/s gefunden wurde, sind zur Veranschaulichung der Größe der Streuungen die einzelnen Meßpunkte, die zwei unter gleichen akustischen Bedingungen durchgeführte Messungen mit verschiedenen Beobachtern lieferten, dargestellt durch kreis- und dreieckförmige

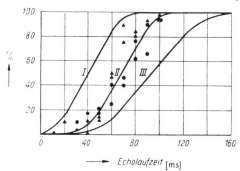

Abb. 3. Echostörung in Abhängigkeit von der Sprechgeschwindigkeit;
   I. 7,4 Silben/s,
   II. 5,3 Silben/s,
   III. 3,5 Silben/s.

Punkte, eingetragen. Man erkennt, daß bei Laufzeitunterschieden über 40 ms der Anteil der sich durch das Echo gestört fühlenden Beobachter schnell zunimmt und bei 100 ms Laufzeitdifferenz 100 % erreicht.

Die Auswertung der Meßpunkte erfolgte so, daß durch sie in möglichst guter Annäherung eine in ihrem mittleren Teil geradlinig verlaufende Kurve gelegt und als kritischer Wert der Laufzeitdifferenz derjenige angesehen wird, bei dem 50 % der Beobachter eine Störung empfinden. Er beträgt somit bei Kurve II 68 ms, und es kann angenommen werden, daß dieser Wert die Verhältnisse recht gut kennzeichnet. Versucht man, durch die kreisförmigen und dreieckförmigen Meßpunkte in Abb. 3 zwei verschiedene Kurven zu ziehen, so erkennt man, daß dadurch der Wert der kritischen Laufzeitdifferenz nur unwesentlich beeinflußt wird. Wenn man bedenkt, daß bei durchschnittlich 80 Beobachtern und 25 Meßpunkten in einer Kurve 2000 Urteile enthalten sind, so kann man annehmen, daß den Erfordernissen der Statistik weitgehend Rechnung getragen wurde.

Die kritische Laufzeitdifferenz ist von der Sprechgeschwindigkeit abhängig und beträgt für schnell gesprochenen Text (7,4 Silben/s) 40 ms, für normale Sprechgeschwindigkeit (5,3 Silben/s) 68 ms und für getragene Sprache (3,5 Silben/s) 92 ms. Sie ist etwa der Sprechgeschwindigkeit umgekehrt proportional. Aus dem Ergebnis dieser Messung erkennt man, daß

in einem Raum, in dem der Zuhörer langsam vorgetragener Sprache noch ungestört folgen kann, bei steigender Sprechgeschwindigkeit bereits Echostörungen auftreten können.

b) Einfluß der Echointensität

Die Sprechgeschwindigkeit betrug 5,3 Silben/s, die Klangfarben von Primärschall und Echo waren gleich. Die Intensität des Echos wurde stufenweise variiert, und zwar betrug der Unterschied zur Intensität des direkten Schalles + 10 db, 0 db, — 3 db, — 6 db und — 10 db. Das Ergebnis dieser Messungen ist in Abb. 4 dargestellt. Daraus erkennt man quantitativ, wie sich die Echostörung durch eine Schwächung der Echointensität vermindern läßt, was man bekanntlich in Räumen durch Anbringen von Schallschluckstoffen auf reflektierende Flächen erzielen kann.

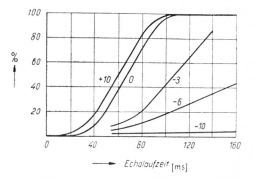

Abb. 4. Echostörung in Abhängigkeit von der Intensität des reflektierten Schalles (die eingetragenen Zahlen geben die Intensitätsunterschiede des Echos gegenüber dem direkten Schall in db an).

Während für den Fall, daß die Echointensität gleich der Primärschallintensität ist, sich wieder eine kritische Laufzeitdifferenz von 68 ms ergibt, steigt diese schon bei einer Schwächung des Echos um 3 db auf 108 ms an, bei einer Schwächung um 6 db auf ca. 175 ms und bei einer Schwächung um 10 db tritt praktisch überhaupt keine Störung mehr auf.

Eine Steigerung der Echointensität um 10 db, wie sie beispielsweise durch Schallkonzentrierung infolge gekrümmter Raumbegrenzungsflächen auftreten kann, z. B. bei Kuppeln oder Planetarien älterer Bauart, verkleinert die kritische Laufzeitdifferenz verhältnismäßig wenig unter den Wert bei gleicher Echointensität. Sie verringert sie nur von 68 ms auf 60 ms.

c) Einfluß der Echoklangfarbe

Die Sprechgeschwindigkeit des Textes war 5,3 Silben/s, Primär- und Echolautsprecher hatten gleiche Intensität. Die Klangfarbenänderungen des Echos wurden auf elektrischem Wege dadurch erzeugt, daß die lineare Frequenzcharakteristik eines im Echokanal liegenden Verstärkers (Abb. 1) verzerrt wurde.

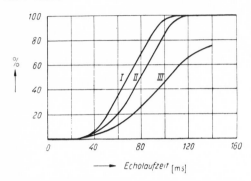

← Echolaufzeit [ms]

Abb. 5. Echostörung in Abhängigkeit von der Klangfarbe des Echos;
 I. Primärschall und Echo haben gleiche Klangfarbe,
 II. Tiefe Frequenzen des Echos sind gedämpft,
 III. Hohe Frequenzen des Echos sind gedämpft.

Das Meßergebnis zeigt Abb. 5. Kurve I gilt für den Fall gleicher Klangfarbe von Primärschall und Echo. Die kritische Laufzeitdifferenz ist hier wie früher auch 68 ms. Kurve II mit einer kritischen Laufzeitdifferenz von 80 ms wurde bei einer Schwächung der tiefen Frequenzen des Echos erhalten (allmählicher Dämpfungsanstieg der Frequenzen unter 300 Hz, Dämpfung bei 100 Hz: —5 db). Das Ansteigen der kritischen Laufzeitdifferenz ist hier zum Großteil auf die mit der Dämpfung der tiefen Frequenzen verbundene Energieabnahme des Echos zurückzuführen. Bei Dämpfung der hohen Echofrequenzen (allmählicher Dämpfungsanstieg der Frequenzen über 1000 Hz, Dämpfung bei 10 kHz: —15 db) ergab sich Kurve III mit der kritischen Laufzeitdifferenz von 105 ms. Die Erhöhung der kritischen Laufzeitdifferenz ist hier allein auf das Fehlen der hohen Echofrequenzen zurückzuführen. Eine Verringerung der Echolautstärke war nicht feststellbar, da die Frequenzen über 1000 Hz nur noch sehr wenig zur Gesamtenergie der Sprache beitragen. Abb. 5 liefert somit die quantitative Grundlage für die bekannte Erfahrung, daß man zur Verringerung von Echostörungen die störenden Reflexionsflächen mit porösen Schluckstoffen belegt, um dadurch vor allem eine große Dämpfung der hohen Frequenzen zu erzielen.

Der Grund für die frequenzabhängige Störempfindlichkeit des menschlichen Gehörorgans durch Echos ist noch nicht bekannt. Vielleicht liegt er in einer frequenzabhängigen Größe der Zeitkonstanten für An- und Abklingvorgänge.

d) Einfluß der Lautstärke

Die Sprechgeschwindigkeit bei dieser Messung war 5,3 Silben/s, Primär- und Echolautsprecher hatten gleiche Intensität und Klangfarbe. Die Intensität beider Lautsprecher wurde gleichzeitig stufenweise variiert, und zwar, um den Verhältnissen der Praxis nahezukommen, nach kleineren Lautstärken hin, zwischen 55, 45 und 35 Phon. Es ergab sich bei diesem Versuch, daß das Verändern der allgemeinen Lautstärke im Bereich zwischen 55 ... 35 Phon keinen Einfluß auf das Ausmaß der subjektiv empfundenen Echostörung hatte.

e) Einfluß der Herkunftsrichtung des Echos

Primär- und Echolautsprecher hatten gleiche Intensität, die Sprechgeschwindigkeit betrug 5,3 Silben/s. Diese Untersuchung wurde mit mehreren kleinen Gruppen von sechs Personen angestellt, um zu garantieren, daß die Herkunftsrichtung des Echos für alle Beobachter gleich war, da der Lautsprecherabstand nicht beliebig groß gemacht werden konnte. Die Messungen wurden, um saubere Versuchsbedingungen zu erhalten, im Freien durchgeführt. Dabei erhielten die Beobachter den direkten Schalleindruck immer von vorn, während die Echos, deren Laufzeit sprunghaft geändert wurde, aus elf verschiedenen Richtungen kamen. Es ergab sich, daß die Störung nur in verhältnismäßig geringem Ausmaß von der Echorichtung abhängig ist. Das geringe Ansteigen der kritischen Laufzeitdifferenzen bei seitlichen Echoeinfallsrichtungen dürfte auf die dabei auftretende Intensitätsabnahme am abgewandten Ohr gegenüber dem direkten Schalleindruck zurückzuführen sein.

f) Einfluß des Raumnachhalls

Bei den vorhergehenden Messungen im Freien ergab sich im Mittel eine kritische Laufzeitdifferenz von 44 ms im Gegensatz zu 68 ms bei früheren Untersuchungen im geschlossenen Raum. Dies läßt darauf schließen, daß die Größe der Raumnachhallzeit ebenfalls von Einfluß auf die kritische Echolaufzeitdifferenz ist. Um festzustellen, ob eine weitere Vergrößerung der Raumnachhallzeit auch eine Zunahme der kritischen Laufzeitdifferenz bewirkt, wurde eine Messung in einem Raum mit einer mittleren Nachhallzeit von 1,6 s vorgenommen.

Primär- und Echolautsprecher strahlten mit gleicher Intensität und Klangfarbe so, daß die Lautstärke am Beobachtungsort ca. 55 Phon betrug. Die Sprechgeschwindigkeit war 5,3 Silben/s. Bei der so durchgeführten Messung ergab sich eine kritische Laufzeitdifferenz von 78 ms. Abb. 6 stellt den Einfluß der Nachhallzeit auf das Ausmaß der subjektiv empfundenen Echostörung dar und ist aus den Ergebnissen dieser und früherer Messungen zusammengestellt. Man erkennt, daß bei zunehmender Nachhallzeit die kritische Laufzeitdifferenz vergrößert wird.

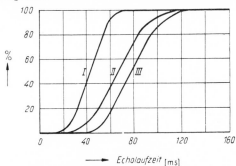

Abb. 6. Echostörung in Abhängigkeit von der Raumnachhallzeit;
   I. Nachhallzeit 0 s,
  II. Nachhallzeit ca. 0,8 s,
 III. Nachhallzeit ca. 1,6 s.

Offenbar verdeckt der Nachhall etwas die Erkennbarkeit der Echos bei kleinen Laufzeitdifferenzen. Es ist allerdings zu berücksichtigen, daß größere Nachhallzeiten allein bereits einen ungünstigen Einfluß auf die Verständlichkeit von Sprache haben.

Zum Schluß dankt der Verfasser Herrn Prof. Dr. E. MEYER für die Anregung zur vorliegenden Arbeit und das fördernde Interesse, das er ihr entgegenbrachte.

Dank gebührt auch seinen Mitarbeitern für manch wertvollen Hinweis und den Zeitaufwand bei der Teilnahme an den Echountersuchungen. Schließ-

lich sei noch ein Wort des Dankes an die zahlreichen Beobachter für die Mitwirkung bei den Messungen gerichtet.

(Eingegangen am 21. August 1950.)

Schrifttum

[1] PETZOLD, F., Elementare Raumakustik. Bauwelt Verlag, Berlin 1927, S. 8.

[2] DECKER, H., Eine Verzögerungsleitung für Messung und Vorführung von Laufzeitwirkungen in Fernmeldesystemen. Elektr. Nachr. Techn. 8 [1931], 516.

[3] STUMPP, H., Experimentalbeitrag zur Raumakustik. Beihefte zum Gesundheitsing., Reihe II, H. 17 [1936].

[4] V. BÉKÉSY, G., Über das Richtungshören bei einer Zeitdifferenz oder Lautstärkeungleichheit. Phys. Z. 31 [1930], 824, 857.

[5] REICH, M. u. BEHRENS, H., Das Richtungsempfinden bei Tönen und Klängen. Z. techn. Phys. 14 [1933], 6.

[6] DE BOER, K., Plastische Klangwiedergabe. Philips' techn. Rdsch. 5 [1940], 108.

[7] WARNCKE, H., Die Grundlagen der raumbezüglichen stereophonischen Übertragung im Tonfilm. Akust. Z. 6 [1941], 174.

[8] LINON, A., Das Küchenmeister-Intervall und das Ultraphon. Z. VDI 70 [1926], 33.

[9] CREMER, L., Geometrische Raumakustik. S. Hirzel Verlag, Stuttgart 1948, S. 126.

[10] V. BÉKÉSY, G., Zur Theorie des Hörens. Phys. Z. 30 [1929], 118.

[11] STEUDEL, U., Über Empfindung und Messung der Lautstärke. Hochfrequenztechn. u. Elektroak. 41 [1933], 116.

[12] BÜRCK, W., KOTOWSKY, P. u. LICHTE, H., Die Lautstärke von Knacken, Geräuschen und Tönen. Elektr. Nachr. Techn. 12 [1935], 278.

[13] AIGNER, F. u. STRUTT, M. J. O., Über eine physiologische Wirkung mehrerer Schallquellen auf das Ohr und ihre Anwendung auf die Raumakustik. Z. techn. Phys. 15 [1934], 355.

[14] LÜBCKE, E., Über die Zunahme der Lautstärke bei mehreren Schallquellen. Z. techn. Phys. 16 [1935], 77.

[15] BARKHAUSEN, H., Ein neuer Schallmesser für die Praxis. Z. techn. Phys. 7 [1926], 599.

[16] PANZERBIETER, H. u. RECHTEN, A., Subjektive Bestimmung der Güte von Fernsprechverbindungen. Arch. techn. Messen V 3719—3, Dez. 1942.

THE INFLUENCE OF A SINGLE ECHO ON THE PERCEPTIBILITY OF SPEECH

Helmut Haas

This English summary was prepared expressly for this
Benchmark volume by T. D. Northwood, National Research
Council of Canada, from "Uber den einfluss eines einfachechos
auf die horsamkeit von sprache," Acustica, 1:49–58 (1951).

The perception of transient sounds such as speech in rooms is complicated by the time delays between the sound reaching the listener directly and the various reflected sounds. It is of interest to determine the conditions under which these delayed sounds have a beneficial or harmful effect on the perceptibility of speech. Earlier studies have examined particular parts of the subject, including very short or very long delays [1, 2, 3]. The present paper considers delays ranging from zero to 160 milliseconds, corresponding to early reflections in typical rooms, and examines a number of parameters, viz: echo level relative to direct sound; rapidity of speech; frequency spectrum; room reverberation time.

The apparatus (Figure 1) consisted of an endless-loop tape recorder/reproducer that permitted an input signal to be reproduced in two channels separated in time by an adjustable interval. These two channels terminated in two separate loudspeakers, typically mounted on axes 45 degrees to the right and left of the subject. Three test environments were used: outdoors, a fully occupied auditorium, and the same auditorium empty except for a group of six subjects.

Several possible measures of the subjective effect of a delayed echo on the reception of speech were considered, including the scores resulting from various types of intelligibility test. Finally, however, the subjects were simply asked to indicate whether or not the presence of the echo was disturbing in their effort to understand continuous speech. The percent disturbance, used as a parameter throughout the study, is the percentage of observers who reported being disturbed in this sense.

CONCLUSIONS

(1) Tests with small echo-delay differences, between 1 msec and 30 msec, showed an increase in loudness, in agreement with the law of the addition of energies, and a pleasant modification of the sound impression in the sense of a broadening of the primary sound source; within a certain range of relative intensities, the echo source was not perceived acoustically. The magnitude of the auditory "suppression effect" for echoes with 1 msec to 30 msec delays was found to be 10 dB, i.e., the intensity of the echo must exceed that of the primary sound by about 10 dB in order to make the echo separately perceptible in the said range of delay differences (Figure 2). More explicitly, this is the level to which the echo must be raised in order to be perceived as being equal in loudness to the first arrival.

(2) Echoes with greater delay differences above a well-defined threshold value cause the perception of speech to be disturbed, even to complete unintelligibility. The concept of a "critical delay difference," beyond which disturbance to the listener rises rapidly, is introduced as a means of gauging the degree of disturbance, and of comparing measurements made under different circumstances.

- (a) Its value is approximately inversely proportional to the speed of speech in the range examined. (Figure 3, Curve I–7.4, II–5.3, III–3.5 syllables/sec.)
- (b) The intensity of an echo exerts an important influence on the critical delay difference. An attenuation of the echo intensity by only 5 dB doubles the critical difference. Echo intensities more than 10 dB below that of direct sound do not disturb at all the reproduction of continuous speech (Figure 4).
- (c) The high frequencies of the echoes determine the amount of the subjective disturbance. Their attenuation (Figure 5, Curve III) makes possible a considerable raising of the critical difference, with almost no noticeable reduction of the loudness of the echo, whereas attenuation of low frequencies (Curve II) makes little difference.
- (d) The magnitude of echo disturbance does not depend on the absolute loudness over the loudness range of speech.
- (e) The direction of incidence of the echo does not significantly affect the critical delay difference, as long as the direct sound is incident from the front.
- (f) A longer reverberation time in a room produces a greater critical delay difference. (Figure 6: Curve I–R.T. = 0 sec, II = 0.8 sec, III = 1.6 sec.)

Reprinted from *J. Acoust. Soc. Am.* **22**(4):507–509 (1950)

A Tentative Criterion for the Short-Term Transient Response of Auditoriums*

R. H. BOLT AND P. E. DOAK

*Acoustics Laboratory, Massachusetts Institute of Technology,
Cambridge, Massachusetts*

April 5, 1950

Studies of the response of an auditorium to a short tone burst have indicated that the character of the first 20 or 30 db of sound decay, the "short-term" response, is closely related to the subjective "hearing quality" of the room. Recently Haas has investigated the effect of a single echo on the subjective hearing of speech. On the basis of some of Haas' results and other information on the hearing of speech in rooms, a tentative criterion for the short-term response has been formulated as curves of amplitude *vs.* time delay of each reflection relative to the direct sound. The criterion correctly rank-orders subjective quality observations reported by others, and agrees quantitatively with some subjective judgments reported here.

IN recent years many acousticians have become increasingly convinced that the character of the first 20 or 30 db of sound decay in a room, i.e., the "short-term" transient response, is closely related to the subjective "hearing quality" of the room. For example, it is common experience that two rooms with nearly identical long-term reverberation times often are markedly different in subjective hearing quality.

The studies of Mason and Moir,[1] Somerville[2] and others have indicated that certain features of the subjective hearing quality can be correlated with cathode-ray tube oscillographs of the response of the room to a single, short (10 to 50 msec.), tone burst. Mason and Moir found that reflection paths longer than 45 ft. (corresponding to a maximum time delay of 40 msec.) are to be avoided, and their published data indicates that an echo with a time delay of from 70 to 100 msec., for example, should have an intensity level at least 10 db below the direct sound.

Independently of the British pulse studies, Haas[3] has recently investigated the effect of a single echo on the subjective hearing of speech, as a function of echo delay, amplitude, and spectrum. Haas observed that, in the case of continuous speech, an echo of from one to 35 msec. delay is "masked" by the direct sound (i.e., is not heard as a separate distinct echo) as long as the echo intensity level is less than +10 db *re* the intensity level of the direct sound. Further, echoes in this masked region were not detrimental to the sound quality but rather enhanced it. When the direct sound was accompanied by such an echo of loudness equal to that of the direct sound, listeners' judgments showed that there was an increase in subjective loudness of 3 db *re* the loudness in the absence of the echo.

Haas also investigated the effect of an echo of from 40 to 160 msec. delay, which, if it had a suitable amplitude, could be perceived as a distinct echo. He found that comparison of articulation scores with and without the echo gave inconclusive results. Hence he evaluated the effect of the echo in terms of a "disturbance" criterion. Listeners were presented with one minute samples of continuous speech having different echo time delays, amplitudes, and spectra. The listeners were asked to answer "yes" or "no" to the question: "Are you disturbed by the echo?" The percent of the listeners disturbed was taken as a measure of the influence of the echo on the quality of the sound transmitted. The percent disturbance was found to increase with increasing delay time and with increasing echo amplitude. On the other hand, the percent disturbance decreased with decreasing speech rate (syllables/sec.), and with decreasing high frequency content ($f > 1000$ c.p.s.) in the echo spectrum. Figure 1 shows Haas' percent disturbance as a function of time delay for various echo amplitudes, the echo spectrum being the same as that of the direct sound. The curves in Fig. 1 are for the average German speech rate of 5.3 syllables/sec.

Now suppose that, using Fig. 1 and Haas' data for the dependence of percent disturbance on speech rate, we plot contours of constant percent disturbance on a graph whose coordinates are echo time delay and echo intensity level in db *re* direct sound intensity level. Such a plot, somewhat modified and extrapolated, is shown in Fig. 2, for a speech rate of about 4.5 syllables/sec. (about average for English).

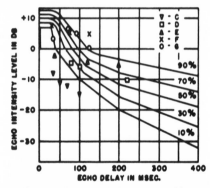

FIG. 2. The proposed tentative criterion, as represented by contours of constant percent disturbance plotted on echo intensity level *vs.* echo delay coordinates. The points indicated correspond to echoes observed by Mason and Moir.

The extrapolations and modifications consist of the following. The spacing of the disturbance contours for echoes of intensity levels greater than the direct sound and time delays less than 125 msec. was adjusted to correspond to Haas' results on echo masking for one to 40 msec. echoes. The spread between the 10 and 90 percent contours in the one to 40 msec. region corresponds to the scatter in the data for this experiment. The 10 percent disturbance contour is extrapolated to −20 db at 200 msec., since the average English syllable length is around 230 msec. and the weaker consonants which usually initiate or terminate syllables have intensities of the order of −20 db *re* the vowel sound intensity of the syllable. From 200 to 400 msec. the extrapolation is based on optimum reverberation time data. A reverberation time of one second is a little below the usual optimum time for speech for auditoriums of volume greater than 100,000 cu. ft., and hence seems to be a reasonable guess for the 10 percent disturbance contour. Similarly a slope corresponding to a reverberation time of 2.5 sec. seems reasonable for the 90 percent contour.

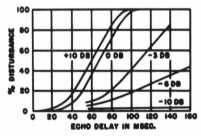

FIG. 1. Percent disturbance caused by a single echo, as measured by Haas for several echo intensity levels *re* the direct sound intensity level. The echo has the same spectrum as the direct sound. Speech rate was 5.3 syllables/sec.

These adjustments and extrapolations of Haas' data are obviously fairly arbitrary and intuitive. Figure 2 is thus not a strict replotting of Haas' data.

We now suggest that a disturbance plot such as Fig. 2 may provide a criterion for some acoustical properties of a room which are closely related to the subjective hearing quality. Specifically, the disturbance plot may be useful in interpreting certain features of pulse response photographs.

The points shown on Fig. 2 represent reflections observed by Mason and Moir. Each point gives the time delay and intensity level of an echo re the direct sound. The points on curve C correspond to a short-time echo structure photographed at a typical location in a motion picture theater where the sound quality is described as "very good."[1] Points D show the echo structure for "one isolated bad spot in the theater . . . under the balcony."[1] Points E represent a typical response in "an auditorium where sound quality was of a medium order, the main complaint being lack of intimacy."[1] Points F and G show typical pulse responses in an auditorium "where the quality of sound is very inferior, intelligibility and intimacy suffering considerably."[1] Mason and Moir point out that for all three auditoriums the measured frequency characteristics and reverberation times were quite satisfactory. It is clear from Fig. 2 that the percent disturbance for each location as read from this plot rank order with the subjective hearing quality judgments reported by Mason and Moir.

Some pulse studies of a motion picture theater have been made at the M.I.T. Acoustics Laboratory as part of a general study of acoustical properties of existing motion picture theaters and auditoriums. At the front of the main floor of this theater a distinct slap echo from the rear wall could be heard. As an observer walks back down the aisle toward the rear wall, the slap echo comes closer in time to the direct pulse, until at a point about 35 ft. from the rear wall the slap echo is no longer heard. Figure 3 shows envelope tracings of pulse response photographs at different positions along the aisle, overlaid by the disturbance plot. The pulse duration was 14 msec. and the carrier frequency 700 c.p.s. The direct pulse is indicated by A and the time of arrival for a

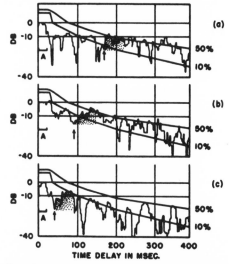

FIG. 3. Envelope tracings of pulse response photographs showing the disappearance of a slap echo. In Fig. 3(a) the microphone was in the right center aisle at the front of the theater; in Fig. 3(b) about halfway back toward the rear wall; and in Fig. 3(c) about 35 ft. from the rear wall. Note that the audible slap echo appears as a hump in the envelope rising above the 10 percent contour.

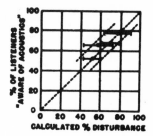

FIG. 4. Graph showing correlation between the percentage of observers "aware of the acoustics of the theater" and the percent disturbance as calculated from pulse photographs using the disturbance contours shown in Fig. 2.

reflection from the rear wall by the arrow. The hard plaster rear wall is broken up by stairways and other deep indentations so that a clean reflection would not be expected. The stippled areas indicate approximately the regions containing those components that have undergone a single reflection from some section of the rear wall. The location of the microphone for Fig. 3(c) is that at which several observers judged that the slap echo could no longer be heard. At this point the echo peaks just pass below the 10 percent disturbance contour out to 180 msec. time delay. It is evident that the subjective "slap echo" is actually not a sharp well-defined peak but an irregular hump, comprised of components from a number of image sources, which rises above the 10 percent disturbance contour. The subjective aural interpretation of the pulse response is thus quite different from the visual impression produced by the photograph. Unless a pulse photograph is analyzed by means of a disturbance plot, such a hump might appear insignificant.

A standard listening test reel was played in this theater for a group of 21 observers as part of the over-all study of the acoustical properties of the theater. The observers were in four groups and listened to the test reel successively at four different locations. The first question on a questionnaire which the observers filled out was "Were you aware of the acoustics of the theater?" Subsequent questioning of the observers indicated that they interpreted "awareness" as awareness in the sense of being disturbed. Hence one could hypothesize that this question was similar to Haas' "disturbance" question. Pulse response photographs for the four different observer locations were analyzed as follows. The reflected sound out to 400 msec. delay was considered as one long single echo. By the use of an overlay corresponding to Fig. 2, different percent disturbances for each echo peak on a photograph could be read. The largest percent disturbance for each photograph was thus determined and recorded. For each location an average percent disturbance was calculated by averaging this maximum percent disturbance for pulse carrier frequencies of 350, 500, and 720 c.p.s. (pulse length 20 msec.), and 1000, 1500, 2000, 2800, and 3500 c.p.s. (pulse length 5 msec.). Figure 4 is a plot of the percent disturbance calculated in this way against the percent of the observers who were "aware of the acoustics of the theater," the standard deviations of the calculated values being indicated. The correspondence is evidently very nearly one to one.

The general qualitative and quantitative agreement between this tentative "percent disturbance" curve and these experimental results seems very encouraging and perhaps is not entirely fortuitous. It therefore appears reasonable to propose this disturbance plot as a measurable tentative criterion for the short-term response. According to this criterion, for example, an echo structure having no peaks above the 10 percent disturbance contour would indicate good hearing quality, and an echo above the 10 percent contour would indicate a certain percent disturbance.

Obviously, to establish the criterion on a more accurate quantitative basis much more experimental work must be done. Among questions to be examined more fully are: the psychological meaning and validity of the "disturbance" question and its relation to hearing quality; the influence on subjective hearing quality of multiple echoes of various time delays, amplitudes and spectra; and the influence of echo directionality. Haas' results indicate a strong dependence of disturbance on high frequency content of the echo spectrum and a weak dependence on directionality.

In regard to evaluation of pulse photographs by a plot such as Fig. 2, it should be emphasised that pulses of different lengths and carrier frequencies in general give quite different pulse pictures because of the different interference conditions. Interference can be kept at a minimum by making pulse lengths as short as possible. However, shortening the pulse decreases the low frequency content for a given carrier frequency, and increases the high frequencies. In any case, the value of the carrier frequency, and practical considerations of loudspeaker behavior place a lower limit on pulse length. Thus, formulating a definition of a "standard pulse" is of primary importance. The definition of such a pulse should most certainly depend on the results of further subjective listening tests.

We wish to thank Professor L. L. Beranek for the use of the results of the listening test, and Mr. J. A. Kessler for his help in making the pulse measurements.

* This work was supported in part by the ONR, Department of Navy, under Contract NObs 25391, Task 7.
¹ C. A. Mason and J. Moir, "Acoustics of cinema auditoria," J. Inst. Elec. Eng. 88, III, No. 3, 175 (1941).
² British Broadcasting Corporation, Engineering Division, Research Department Reports B.027, B.035.
³ H. Haas, "Ueber der einfluss eines einfachechos auf die hoersamkeit von sprache," dissertation, Goettingen (1949). (English translation: Library Communkation No. 363, December, 1949, Department of Scientific and Industrial Research, Building Research Station, Watford, Herts., England.)

9

Reprinted from *Int. Congr. Acoust., 4th, Copenhagen, 1962, Proc.,* pp. 15-29

Rating of Acoustical Quality of Concert Halls and Opera Houses[1]

Leo L. Beranek

Bolt Beranek and Newman Inc., 50, Moulton Street, Cambridge, Massachusetts.

I. Introduction

A detailed study of the acoustics of concert halls and opera houses was launched in 1955 for the purpose of discovering why most halls built in the twentieth century have failed to satisfy musicians, music critics, and audiences. This investigation has ranged over twenty nations and from it detailed architectural and acoustical data have been obtained on 54 halls for music. Twenty three internationally known conductors and musicians were interviewed in depth—some interviews taking up to three hours. [2] Twenty one music critics of the United States, Canada, Great Britain, and Germany were interviewed in the same manner. [3] A spread of ages and backgrounds were represented.

The acoustical measurements that were performed and technical details assembled include (a) reverberation time as a function of frequency with the audience and orchestra present, (b) reverberation time as a function of frequency with the hall empty, (c) pulse measurements (in some halls only) designed to show the decay of sound in detail at several positions

1. The material in this paper is abstracted from a book by the author, *Music, Acoustics, and Architecture,* John Wiley and Sons, Inc., New York and London (1962).
2. The artists include: Sir John Barbirolli, Sir Adrian Boult, Leonard Bernstein, Eleazar de Carvalho, Alexander Gibson, Tauno Hannikainen, Irwin Hoffman, Herbert von Karajan, Erich Leinsdorf, Igor Markevich*, Dimitri Mitropoulos, Pierre Monteux, Charles Munch, Eugene Ormandy, Fritz Reiner, Sir Malcom Sargent, Hermann Scherchen, Stanislaw Skrowaczewski*, Izler Solomon*, William Steinberg*, Isaac Stern, Leopold Stokowski, and Bruno Walter. (*Interviewed by Russell Johnson.)
3. The music critics and their newspapers are: Thomas Archer*, Montreal *Gazette;* Roger Dettmer, Chicago *American;* Cyrus Durgin. Boston *Globe;* Alfred Frankenstein, San Francisco *Chronicle;* Albert Goldberg, Los Angeles *Times;* Jay Harrison, *Music Magazine;* Peter Heyworth, London, *The Observer;* Frank Howes, London, *The Times;* Paul Hume, Washington *Post;* Irving Kolodin, New York *Saturday Review;* Paul Henry Lang, *New York Herald Tribune;* Eric McLean*, *Montreal Star;* Robert C. Marsh*, Chicago-*Sun Times;* Colin Mason, Manchester, *The Guardian;* Antonio Mingotti, Munich, *Abendzeitung;* Harold Rogers, *The Christian Science Monitor;* Max de Schauensee, *Philadelphia Bulletin;* Harold Schonberg, *New York Times;* Desmond Shawe-Taylor, London, *The Sunday Times;* Howard Taubman, *New York Times;* Jules Wolffers, *Boston Herald.*

in the hall, (d) accurate architectural drawings, (e) dimensional measurements of stage, pit, stage enclosure, and the hall itself, generally made from existing architectural plans, with portions checked to insure accuracy, (f) the number of seats on each floor level, number of standees and number in boxes, and (g) details on seats, carpets, wall and seating surfaces, stage and hall floors, draperies, curtains, proscenium opening, etc.

Much information was collected during personal visits as the architectural drawings seldom show up-to-date detail. Also, it was found that many halls have been modified: the seats changed, balconies rebuilt, stairways relocated, stage extended or shortened. As a result, great care was used to check all factors rather than assuming blindly that the hall was in conformance with existing drawings. All technical material was sent to appropriate people for checking for accuracy.

II. Results of the Interviews and Listening Judgments

In order for acoustical science to provide directions to architecture in the interests of music, a scale is needed against which the quality of a hall can be gauged, and the importance of a particular acoustic attribute weighed. The evaluations of 53 of the halls were limited to two types of performances, namely, symphony orchestra concerts, or non-Wagnerian opera, or both. With the help of the interviews with musicians and critics and my own notes made while listening in almost all the halls, I have been able to assign them into five categories of quality: A+, excellent; A, very good to excellent; B+, good to very good; B, fair to good; and C+, fair. The 54th hall, Bayreuth Festspielhaus, is especially suited to Wagnerian opera and must be rated on a different basis. This study has purposely eliminated poor halls as their gross defects are easily identified.

It is probable that any particular hall could be ranked one category higher or one category lower than its present assignment. An error of three categories is almost impossible, and an error of two categories is highly unlikely.

But, the initial assignment of the halls into categories was not depended on alone. After the compilation was completed, it was sent with a request for criticism to most of the men interviewed. A small number of shifts to an adjacent category resulted from their comments, but the majority wrote as did Howard Taubman of the *New York Times*, "On the whole I agree with you about the halls that I know anything about."

A. *Definitions of the Categories*

Category A+: The halls in this category were rated "excellent" for symphony orchestra concerts or non-Wagnerian opera by nearly every musician or critic interviewed. The halls received very few adverse comments. Six concert halls and seven opera houses fell into this category.

Category A: The halls in this group were generally rated "very good" to "excellent." They are all deficient to some extent in one or two of the important acoustical attributes. Each hall is highly treasured by the community it serves and is spoken of favorably by the majority of those musicians who perform there. There are nineteen concert halls and eight opera houses in this category.

Category B+: The halls in this category were rated "good" to "very good." They received more favorable than adverse criticism. These halls are not mediocre; any negative criticism received came in response to a request for an assessment of both the good and bad qualities. Some of the halls are better suited to the music of the Baroque period that to that of either the Classical or Romantic period. Some are very large. All but three were built for all types of use—music, speech, dance, and so forth—calling for compromises that generally degrade the acoustics of the hall for concerts or opera. There are fourteen concert halls and four opera houses in this category.

Category B: The halls in this group were rated "fair" to "good." They received more adverse than favorable criticism in the interviews, but they definitely are not poor or unsatisfactory halls. Of the seven concert halls in this category all but two were built to serve many different purposes.

Category C+: There is only one hall in this category. The chief factor that adversely affects its acoustics are its large lateral dimensions and its huge cubic volume. It is quite satisfactory for organ and choral music of the cathedral type and for ceremonial orchestral and band music.

B. *Cubic Volume and Age*

During the interviews, musicians often commented that: (1) small halls generally sound better than large ones; (2) halls built to serve many purposes are inferior to halls built especially for concert or opera; and, (3) old halls sound better than new ones. Investigation reveals that the median cubic volume of the ten best-liked concert halls is 620,000 cu ft (17,600 cu m) compared to 770,000 cu ft (21,800 cu m) for the 10 least-liked concert halls. This difference of 25% in cubic volume is hardly significant by itself.

In opera houses, size is very important. An artist can sing more easily in a small house than in a large one. Singers who sound one way at the Metropolitan Opera House in New York, sound different in smaller European theaters. In small houses, singers are less inclined to force their voices and so are more relaxed. Obviously, for equal reverberation times, a voice is louder in a small house than a large one.

In regard to age, neither acoustical data nor reviews by music critics indicate that any hall has changed acoustically with time unless architectural changes were made. The eight least-liked 20th Century halls of this study differ from the seven best-liked 19th Century halls in the following ways: The median floor area occupied by seats and orchestra is 65 per cent greater in the newer halls than in the older halls. The median ceiling height is 8 ft lower. The median mid-frequency reverberation time is 0.2 sec lower. The interiors in the new halls, in general, contain thin wood (two contained large areas of sound-absorbing tiles). Most damaging of all to the 20th Century halls, the median width is 83 per cent greater, yielding a greater difference between the times of arrival at listeners' ears of the direct sound and the first reflections. Architectural differences, not age, are clearly responsible for the acoustical differences.

C. *Importance of Reverberation Time*

Let us see if there is a high correlation between the subjective rank orderings of concert halls and their mid-frequency reverberation times (See Table 1). In Category A+ there are six halls with reverberation times between 1.7 and 2.05 seconds—a median of 1.9 seconds. In the three categories of A, B+, and B the reverberation times range between 1.0 and 2.0 seconds, with median values of 1.5 and 1.6 seconds.

Table I

Correlation Between Mid-Frequency Reverberation Times and Rank-Order Categories for 47 Fully-occupied Concert Halls

Rank order category	Mid-Frequency reverberation times—sec.	Median sec.
A+	1.7, 1.8, 1.8, 2.0, 2.05, 2.05	1.9
A	1.2, 1.3, 1.4, 1.4, 1.5, 1.5, 1.5, 1.6, 1.6, 1.6, 1.6, 1.7, 1.7, 1.7, 1.7, 1.7, 1.8, 1.9, 2.0	1.6
B+	1.0, 1.3, 1.4, 1.4, 1.5, 1.5, 1.5, 1.5, 1.5, 1.6, 1.7, 1.7, 1.7, 1.9	1.5
B	1.3, 1.4, 1.5, 1.6, 1.6, 1.6, 1.8	1.6
C+	2.45	2.45

Four tentative conclusions can be drawn from Table 1.

1. In order for a concert hall to fall in the A+ category, its reverberation time must be at least 1.7 sec.

2. A reverberation time as long as 1.7 sec does not ensure that a concert hall will fall in category A+ and, therefore, reverberation time alone does not distinguish an excellent hall from an inferior hall.

3. Let us assume that the only reason the seven halls in Category A whose reverberation times are shorter than 1.6 sec do not fall in A+ is their short reverberation time since, if they were deficient in additional ways, they would fall in a still lower category. Then it follows that the difference between the median of the A+ group (1.9 sec) and the median of the seven halls with short reverberation time (1.4 sec) is one rating category.

4. Thirteen halls in Categories A, B+, and B have reverberation times between 1.7 to 2.0 seconds. Perhaps deviation in only one acoustical attribute is primarily responsible for their falling into a lower category below A+.

III. The Positive Acoustical Factors Affecting Musical Quality

Let us examine Item 4 above to see if there is one acoustical attribute that, when the other attributes are alike, is responsible for the observed differences in the rank orderings.

A. *Initial-Time-Delay Gap*

The acoustical data for the 20 concert halls of Table 1 with reverberation times at least as great as 1.7 seconds reveal that all but three halls have about the same ratios of low frequency and high frequency reverberation times to those at mid-frequencies; there is no observable tonal distortion; there is almost no noise; balance and blend are good; sound diffusion is adequate; and those listeners in the middle of the main floor are about 60 ft from the concert master. The only clearly observable variable among the remaining 17 concert halls is the time difference between the arrival (at the listeners' ears) of the direct sound and the first reflection. Let us call this difference the initial-time-delay gap (See Figs 1 and 2).

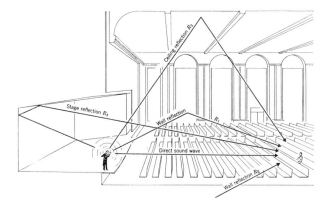

Fig. 1. Sketch showing direct and four reflected sound waves in a concert hall.

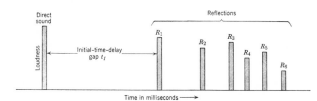

Fig. 2. Time diagram showing that at a listener's ears, the sound that travels directly from the performer arrives first, and after a gap, reflections from the walls, ceiling, stage enclosure and hanging panels arrive in rapid succession. The height of a bar is related to the intensity of the sound. The initial-time-delay gap t_1 is indicated.

Concert Halls: The initial-time-delay gap at the ear of a listener on the main floor of a concert hall is easily measured from architectural drawings. The initial-time-delay gaps for the 47 concert halls of this study are plotted within the blocks with solid boundaries on Fig. 3. The 17 halls that differ primarily in initial-time-delay lie within the blocks with solid boundaries. One other hall with slightly low bass reverberation also falls in this group. Of the 18 halls, the six in Category A+ have initial-time-delay gaps, measured at the center of the main floor, of between 8 and 21 milliseconds; the 8 in A, between 22 and 34 milliseconds; the 2 in B+, between 35 and 46 milliseconds; the one in B, between 47 and 58 milliseconds; and the one in C+ between 59 and 71 milliseconds. Each solid block is about 12 milliseconds

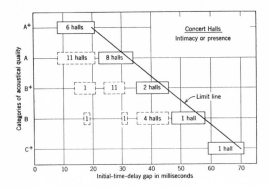

Fig. 3. Chart of rank-order categories for 54 halls as a function of initial-time-delay gap. The heavy diagonal line indicates the highest category that a hall with a given initial-time-delay gap can achieve.

long. The remaining 29 concert halls are plotted in blocks with dashed boundaries. These 29 halls fall one or more categories below what would be expected if the initial time delay gap alone could determine the excellance of a concert hall. Apparently, these 29 halls have deficiencies in some or all of other acoustical attributes.

Of the 26 Concert Halls of Fig. 3 that fall one category below the solid blocks, 9 have short reverberation times at all frequencies, 7 are slightly deficient in a number of acoustical attributes, 5 have somewhat short reverberation times at mid and high frequencies and are somewhat deficient in the ratio of bass reverberation to mid-frequency reverberation, 4 have distortion, combined with fairly short reverberation times at mid and high frequencies; and one is seriously deficient in the ratio of bass reverberation to mid frequency reverberation.

Of the 2 Concert Halls of Fig. 3 that are two categories below the solid blocks, both have short reverberation times at all frequencies, both have low ratios of the reverberation times at low frequencies to those at mid frequencies, and both are wanting in balance and blend. The 1 Concert Hall of Fig. 3 that falls three categories below a solid block has a short reverberation time, a deficiency in bass reverberation, poor balance and blend, some tonal distortion, and some echo.

From this recital of deficiencies one can make certain deductions about concert halls:

1. A relatively short mid-frequency reverberation time, (the average of R.T.'s at 500 and 1000 cps), of about 1.4 sec with full occupancy, causes a drop of about one rating category. This was also suggested by tentative conclusion 3, relating to Table 1.

2. A serious deficiency in ratio of bass to mid-frequencies reverberation produces a drop of one category.

3. Poor balance and blend account for a drop of about one half a category.

4. Distortion to the extent found in these halls causes a drop of about one-half a category.

Obviously, a short initial-time-delay gap is necessary to assure excellance in a concert hall, but it alone is not sufficient to guarantee it.

Opera Houses: A similar graph for opera houses [4] has been drawn. It is found that about 20 per cent longer initial-time-delay gaps than those shown on Fig. 3 appear to be permissible in each rating category.

In Fig. 3, a heavy, diagonal line intersects the right-hand end of each of the five solid blocks. This line delimits the highest category into which a concert hall can fall for any initial-time-delay gap. It appears to have basic psychoacoustic significance.

Haas [5], and Muncey, Nickson and Dubout [6] have investigated the addition of a delayed reflection to original music. The Australian experiments were performed by playing music through loudspeakers to a group of twenty listeners and adding to it the same music delayed in time by up to 1000 milliseconds. The subjects, laboratory personnel, were asked to decide whether they were disturbed by the added reflection. The results showed they were most disturbed by reflections added to fast string music, the kind of music that forms the foundation of present-day symphonic concerts. When the intensity of reflections was not more than 5 decibels below the intensity of the direct sound, fewer than 20 per cent of the subjects were disturbed by delays of 15 to 30 milliseconds. Nearly all the listeners were disturbed by reflections that were delayed by 100 milliseconds.

If we assume that those 10 to 20 per cent of the group who found delays of 15 to 30 milliseconds disturbing were critical listeners, then it should be reasonable to conclude that an initial-time-delay gap in excess of about 20 milliseconds detracts from musical quality. The result of the tests imply also that, the greater the initial-time-delay gap the less pleasant the music becomes. It seems safe to conclude that acoustical intimacy cannot survive an initial-time-delay gap of 70 milliseconds, particularly if the first reflection has an intensity within 10 decibels of the direct sound. Hence, the findings of the Australian group and the results shown in Fig. 3 are in general agreement.

According to Fig. 3, since an initial-time-delay gap of 70 milliseconds or more corresponds to a ratio of C+ or lower, it contributes no points to the rating of a concert hall. How many points should be assigned to a change of one category? It appears from the interviews and the author's judgment that those halls in Category B are subjectively about 70 per cent as good as those in A+. Hence, it seems that the right number is about 10 points per category on a 100 point scale. (See Fig. 4) Hence, if an initial-time-delay gap of 70 milliseconds is zero, then one of 20 milliseconds or less, being four categories higher, must equal 40 points. Rating scales so derived, for the attribute of initial-time-delay gap for concert halls and opera houses, are shown in Fig. 5.

B. *Reverberation Times at Middle and High Frequencies*

The interviews clearly show that for today's typical orchestral repertoires, musicians and most music critics consider that reverberation times at mid-frequencies of under 1.6 seconds are too short. The optimum reverberation times, for full occupancy, at mid-frequencies

4. Twelve of the halls used for deriving rating scales for concerts are also used for operas. The musicians' and conductors' ratings of a hall are generally different for opera than concerts and this fact was taken into account in the ratings.
5. H. Haas, *Acoustica, 1*, pp. 49–58 (1951)
6. R. W. Muncey, A. F. B. Nickson, and P. Dubout, *Acoustica 3* pp. 168–173 (1953)

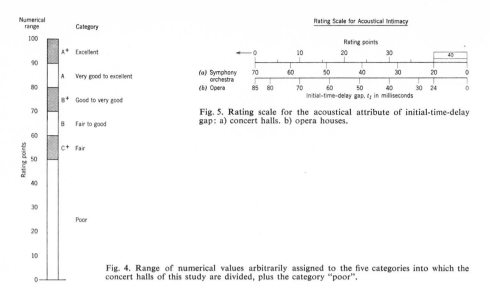

Fig. 5. Rating scale for the acoustical attribute of initial-time-delay gap: a) concert halls. b) opera houses.

Fig. 4. Range of numerical values arbitrarily assigned to the five categories into which the concert halls of this study are divided, plus the category "poor".

(average of R.T.'s at 500 and 1000 cps), are found to be 2.1 to 2.3 sec for music of the Romantic period; 1.6 to 1.8 for music of the Classical period; and 1.4 and 1.6 for music of the Baroque period. The compromise optimum reverberation time for today's symphonic repertoire is found to be 1.8 to 2.0 sec. These times seem independent of cubic volume, for the range of volumes studied, (200,000 to 1,500,000 cu ft.).

Measurements in the 54 halls showed that, if there were no significant amounts of draperies or other sound-absorbing materials on the ceiling or walls, the reverberation times at 2000 cps was about 0.9 that at mid-frequencies; that at 4000 cps was about 0.8 and that at 6000 cps was about 0.7 [7]. Reverberation times apply to full occupancy.

We learned from the previous section that in a concert hall when the reverberation time is low (about 1.4 sec), compared to its optimum of about 1.9 sec, the rating decreases by about one rating category, equivalent to 10 rating points. Also, it is well accepted that the reverberation in a concert hall adds very little to symphonic music if it is shorter than about 1.1 sec. One judges that if a change in the region from 1.9 sec down to 1.4 sec is 10 points, then an additional drop of 0.3 sec to 1.1 sec amounts to about five points. The overall range, therefore, of the scale for mid-frequency reverberation times is 15 points.

From the evidence at hand, it seems that an opera house for Italian music should have about the same reverberation time as a concert hall for Baroque music and one for Wagnerian music should have about the same R.T. as a concert hall for Classical music.

Rating scales for mid-frequency reverberation time for concert halls and opera houses are given in Fig. 6.

7. Leo L. Beranek, *J. Acoust. Soc. Am., 32,* pp. 661–670 (1960)

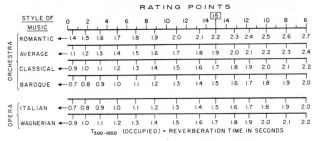

Fig. 6. Rating scale for the acoustical attribute of mid-frequency reverberation time for fully occupied halls.

T_{MID} = AVERAGE OF REVERBERATION TIMES AT 500 AND 1000 CPS FOR FULLY OCCUPIED HALL

C. *Bass to Mid-Frequency Reverberation Time Ratio:*

Many concert halls were criticised by musicians and critics as being deficient in bass. If one divides the concert halls into four groups: (1) excellent bass, (2) good bass, (3) fair bass, and (4) poor bass, as judged by those interviewed, he finds that the *bass ratio*, the average of the reverberation times at 125 and 250 cps divided by the average of the R.T.'s at 500 and 1000 cps, is for (1) between 1.2 and 1.25, for (2) about 1.05; and (3) about 0.96, and for (4) about 0.88. In no hall was there a specific complaint about too much bass even though the bass ratios are as high as 1.4.

In Item 2 of the section on initial-time-delay gap, we said that a serious deficiency in ratio of bass to mid-frequency reverberation corresponds to a drop of about one category. Also, it seems that the bass ratio is about of the same importance as the mid-frequency reverberation time in setting the musical quality of a hall. Hence, let us assign, for concert halls, 15 points to an average bass ratio of between 1.2 and 1.25. For opera houses, the bass ratio need not be as great as in concert halls, because of the relative unimportance of the bass singing part.

Rating scales for concert halls and opera houses, as derived from this study for the attribute of bass ratio are shown in Fig. 7.

D. *Intensity of the Direct Sound*

There is an optimum listening level for music and this applies separately to the direct sound and the reverberant sound. If the direct sound is too weak, it may get lost in the reverbera-

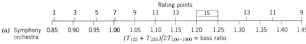

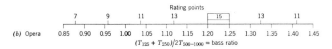

Fig. 7. Rating scale for the acoustical attribute of bass ratio, the average of the reverberation times at 125 and 250 cps to the average of the R.T.'s at 500 and 1000 cps (mid frequencies), for: a) Symphony Orchestra. b) Opera.

tion or in general audience noise. Reflecting surfaces enhance the level of the direct sound by shortening the initial-time-delay gap—but this contribution has already been rated. Furthermore, the design of the stage affects the balance and blend of the music, a contribution that is to be rated later.

Investigation of people's preferences for seats on the main floors of halls for music, indicate that the optimum distance for an auditor from a large symphony orchestra is about 60 feet measured from the concertmaster or, in an opera house, from the average singers' position. In halls with a very large main floor and no balcony overhang (e.g., Tanglewood Music Shed at Lenox, Massachusetts [8]), the direct sound decreases noticeably in intensity at the rear of the hall. It is the author's observation that at about 160 ft on the main floor, the intensity of the direct sound is weak enough to have deteriorated the musical quality by about one category or 10 rating points. In balconies, the situation is different. The sound does not have to travel directly over the heads of as many people to reach an auditor, and, further, additional reflecting surfaces usually come into play. Hence, in balconies, corrections to the main floor ratings are necessary.

The rating scale for intensity of the direct sound on the main floor of a hall for music is shown in Fig. 8. The correction chart for balconies is given in Table 2. It should especially be noted that the rating system cannot be used in any situations where the direct sound does not travel in a customary manner from the stage to the listeners. For example, the method will not rate the quality of the acoustics of seats at the front of a hall and to one side of the proscenium where the listeners are cut off from the direct sound.

Table 2

Corrections to the Rating Scale for Intensity of Direct Sound for Seats in Balconies

Conditions	Incremental Rating Points
Favorable reflections (under 35 msec delay) From a pair of balcony fronts	2
Favorable reflections (under 35 msec delay) From a pair of side walls	4
Highly directive ceiling	4
Maximum possible increment	8

Note: The combined direct sound rating, i.e., the sum of the rating from fig. 8 and the correction above, must not exceed 10.

8. F. R. Johnson, L. L. Beranek, R. B. Newman, R. H. Bolt, and D. L. Klepper, *J. Acoust. Soc. Am., 33* 475–481 (1961)

Fig. 8. Rating scale for intensity of the direct sound. This scale is applicable to seats on the main floor only. Balcony seats must be handled by a combination of the scale above and correction numbers from Table 2. The combination, however, must not exceed a total rating of 10 points.

E. *Loudness of the Reverberant Sound*

The reverberant sound is too loud in some halls and too weak in others. Following L. Cremer [9] the loudness is related to,

$$L = (T/V) \times 10^6$$

where T is the mid-frequency reverberation time with full occupancy and V is the volume in cubic feet.

Using 14 halls with acceptable loudness as examples, it is found that optimum loudness, L, should have a value between 2.5 and 3.5. A rating scale, based on the available data, is given in Fig. 9.

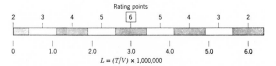

Fig. 9. Rating scale for loudness of the reverberant sound *vs* the quantity $L = (T/V) \times 1{,}000{,}000$, where T is the reverberation time in seconds and V is the volume in cubic feet.

The rating system does not apply to seats that are under balconies or are deep in boxes where the listener is shielded from the reverberant sound in the hall.

F. *Diffusion*

It has been recognized that diffusion is of importance in a concert hall [10]. The listener observes that with adequate diffusion, sound arrives at his ears from all directions. Without diffusion, sound seems to come straight at him from the front of the hall. The present study reveals that except for its strong relation to reverberation time, diffusion *per se* is not a very large factor in the rating of a hall, so that visual observation of the nature of wall and ceiling irregularities is an adequate measure. The rating scale for diffusion in concert halls is given in Fig. 10.

Fig. 10. Rating scale for sound diffusion in concert halls. Since sound diffusion is not important in opera houses, this scale is not used in rating them.

These studies indicate that diffusion is unimportant in opera houses. Hence, it receives no consideration in the rating scheme for them described here.

9. L. Cremer, *Statistische Raumakustik*, S. Hirzel, Stuttgart, p. 231 (1961)
10. E. Meyer, *J. Acoust. Soc. Am. 26*, 630–636, (1954)

G. *Balance and Blend*

By balance in concert halls, we mean the relative loudness of the various sections of the orchestra as heard in the hall. In opera houses, we mean the relative loudness of the singers *vs* the orchestra. By blend we mean the combination of the sound from the sections of the orchestra into a pleasant whole. If the stage or pit for the orchestra is too wide or too deep, time delays destroy the blend for the listeners who sit off the centerline of the hall. No numerical measure of balance and blend has been derived. An experienced observer can judge the quality of the balance and blend from architectural drawings. A musician or other trained listener can hear the quality of the balance and blend immediately.

In concert halls, the attribute of blend and sectional balance is rated a maximum of 6 points. In opera houses, balance and blend between singers and orchestra is so important that this attribute is rated a maximum of 10 points. The rating scales are given in Fig. 11.

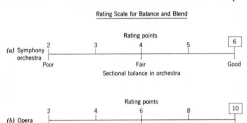

Fig. 11. Rating scales for balance and blend:
a) Between sections of symphony orchestra.
b) Between opera singer and pit orchestra.

H. *Ensemble*

Good ensemble results when the performers play together in perfect unison. Ensemble is partly a matter of the skill of the conductor and the performers and partly a matter of the design of the stage enclosure or the reflecting surfaces to the sides and above the stage. Insofar as the hall, (as distinct from the performers) contributes to ensemble the rating scale is that shown in Fig. 12.

Fig. 12. Rating scale for ensemble, i.e., performers' ability to hear each other.

We still have to take into account acoustical blemishes. These comprise echo, noise, tonal distortion and hall non-uniformity and are handled as deductions. They contribute nothing positive to the quality of the music.

IV. *Acoustical Blemishes*

A. *Echo*

In this paper, an echo is defined as a reflection delayed by more than 70 milliseconds, sufficiently intense to be annoying. Echo arises from a rear wall whose smooth surface is not broken up by balconies or diffusion. It is intensified by inadequate ceiling and side wall diffusion and low reverberation time. Without ceiling and side wall irregularities, the first reflection that returns to the front part of the hall from anywhere is from the rear wall and will stand out in the front part of the hall like a black spot on a white surface. The worst potential source of echo is a long curved wall at the rear of a wide fan-shaped hall. Two procedures for control of echos in such halls are published elsewhere. [8, 11]

B. *Noise*

Noise may arise from ventilation openings, machinery, subways, trains, aircraft or traffic. Vibration from machinery of subways may also be troublesome. Criteria for acceptable noise levels in concert halls and opera houses (measured without audience) are shown in Fig 13. Only if economics is a serious factor, should the levels be permitted to approach the upper curve.

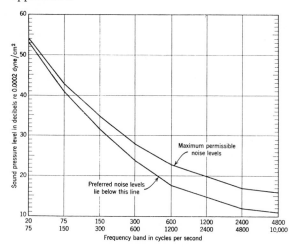

Fig. 13. Criteria for acceptable noise levels in concert halls and opera houses. The lower curve (NC-15) is recommended. Economy considerations should not permit the levels to exceed the NC-20 curve under any circumstances. Measurements are made with an American Standard Sound Level Meter and Octave Band Analyzer. The noise levels are measured without audience.

C. *Tonal Distortion*

Tonal distortion may arise from several sources: (1) selective sound absorption in the hall, e.g., wave coincidence effects in thin, stiff materials, etc.; (2) sympathetic ringing tones, e.g., decorative grilles in front of pipe organs; (3) diffraction grating effects, e.g., from orderly vertical strips on side walls; (4) flutter echo, e.g., as between parallel surfaces; and (5) poor balance among sections of an orchestra due to improper design of stage enclosures.

D. *Hall Nonuniformity*

Nonuniformity of sound in a hall may arise from balcony overhangs that are too deep, boxes with too small openings, better reflections of sound into a balcony than to the main

11. J. B. C. Purcell, NOISE REDUCTION, L. L. Beranek, Editor, McGraw-Hill Book Company, New York, (1960), pp. 408–409

floor, or vice versa, poor sight lines (the interfering heads block off sound), seats near the front of a wide hall and to one side of the stage, and seats too near the stage.

E. *Effect of Blemishes on Rating*

Table 3 gives a suggested means for de-rating a hall if it possesses one or more of the three blemishes shown. Hall nonuniformity is handled by performing the overall calculation for a number of seats and then striking an average. The calculation scheme is not valid for seats under balconies, deep in boxes, or to one side of the proscenium near the front of the hall.

V. Validation of the Rating System

The major part of the job of rating the musical quality of a hall has now been completed. A perfect hall would rate 100 points. A hall that rates under 50 points is unsuited for musical performances.

To validate the rating system, the physical data on each hall were assembled. Interviews had already placed the halls in their respective rating categories, as explained earlier. The calculations of acoustical quality were carried out for two positions in the hall (1) a seat to one side of the center line of the main floor and about half way between the most protruding balcony front (if any) and the stage and (2) a seat in the balcony (if any) and about half way between the front rail and the rear balcony. The ratings for the two seats were averaged to yield the overall rating for the hall.

Table III

Corrections to the Rating Scales for the Negative Attributes, Echo, Noise, and Distortion.

A Separate Correction is Applied for Each Blemish to the Rating.

Amount of Negative Attribute	Correction to Rating Points
None	0
Some	−5
Substantial	−10
Bad	−15 to −50

The best concert hall of the A+ group in this study calculates 96 points on the main floor and 95 points in the balcony. The worst hall in the B group calculates 58 points on the main floor and 71 points in the balcony for an average of 65 points. The one hall in the C+ group is difficult to rate because it is very non-uniform and very large.

Comparison of the subjective ratings with the calculated ratings showed that 50 of the 54 halls were calculated to fall in the correct one of the five subjective categories, A+, A, B+, B, and C+. The calculations for the four missed their categories by only a small amount. This rating system is probably not the ultimate one although it rank orders existing halls well. Experience and laboratory experiments will suggest refinements. The author will be satisfied if these studies lead to better concert halls and opera houses immediately and stimulate further research throughout the acoustical world.

VI. Acknowledgment

The author wishes to acknowledge the help of more than two hundred people—scientists in many countries, musicians, orchestra and opera managers, music critics, architects, hall owners, colleagues at Bolt Beranek and Newman Inc. and others who supplied me with data and judgments and criticised my efforts along the way.

Author's Note Added in Galley Proof:

Recent experience shows that the attribute of acoustical *warmth* is effected by the ratio of the bass energy to mid-frequency energy in the sound reflections that arrive within 50 milliseconds of the direct sound, as well as by the reverberant bass ratio defined in III-C. Thus the rating scale of Fig. 7 must be modified in those cases where the early reflections are deficient in bass energy.

10

Reprinted with the permission of Research Publications Pty. Limited
from *Archit. Sci. Rev. (Australia)* 11:81–87 (Sept. 1968)

Acoustical Determinants
for the Architectural Design
of Concert Halls

BY

A. H. MARSHALL

B.Arch., B.Sc., Ph.D., A.N.Z.I.A.

Senior Lecturer in Architectural Science, The University of Western Australia

Introduction

What formal guides are there for the architect of a concert hall? Is it necessary for him to delegate responsibility for the interior shape of the hall to his consultants? Need he depend solely or mainly upon imitation of the shape of an earlier concert hall with "good" acoustics in order to produce similar desired qualities of sound in a new hall? This paper is intended, while fitting in with established acoustical procedures, to give a clear direction for architectural development in the room form, and to provide a means for asessing the subjective significance of such details as raked floors, diffusing elements and stage reflectors. Such direction has been notably lacking in the past even though sophisticated techniques for predicting acoustical quality in a given hall have been developed, for example, by Prof. Spandöch at Munich. As Lochner and Burger put it . . . "it is clear that architectural design (of concert halls) is, in spite of all efforts that have been made, still largely based on empirical rules".

(ref. 5, p.443).

The paper is based upon the supposed equivalent of the subjective experience of simulated room-acoustical conditions produced in an anechoic room using an electroacoustical installation, and the subjective experience of the acoustical conditicns in a corresponding real room. As far as I know, this equality has never been conclusively established, and the limits within which the spatial distribution of the source and spectral and directivity discrepancies in the corresponding radiation from loudspeakers may be reasonably held to be equivalent, are of obvious significance in the following. A research project which it is hoped to mount shortly at the University of Western Australia is aimed at resolving some of these doubts. Never-theless it does appear that gross effects in the two situations are equivalent, and it is a gross effect that this paper is concerned with.

Definitions

All the available psychophysical data upon the effect of reflections in auditoria are derived from electroacoustical room representations. From experience gained in constructing such representations at Goettingen and elsewhere (refs. 1, 2), one may usefully distinguish between the following components of the sound which arrives at a listener in an auditorium. The characteristic a c o u s t i c a l property of architectural enclosure is, of course, *Reverberation*, but this term is too wide to be useful. Instead, we identify: the direct sound, the reflections, the echoes, and field-reverberation (these being equivalent to the German *Direkt, Rückwürfe, Echos and Nachhall*). Briefly, they are distinguishable as follows: For about 100 MS after the direct sound, sound which arrives as *reflections* is integrated by the hearing process with the impression of sound from the source, contributing to acoustical quality, localisation, intelligibility (for speech), and loudness. The closer to the direct sound the reflection occurs the greater the proportion of its energy' that can be integrated. In adding such reflections to a simulated reflection sequence, it has been found that the direction from which they come relative to the direct sound is of fundamental importance in their effectiveness, as is their precise time location in the sequence.

Field-reverberation, on the other hand, is added without directionality, the time at which it is injected into the sequence is of little significance within relatively wide limits, nor is the shape of its rise, and it contributes only an effect of exponential decay. Without it, however, there is no eflect of enclosure ("Raumeindruck") (ref. 3), and its significance in the perception of the concert hall as a room is thus very considerable.   The transition from reflection to field-reverberation depends upon the room (and so upon the number and direction of reflections), the type of source, and the general level of reverberation, in a complex manner and a rigid distinction between the the two is not implied.

Under certain circumstances a reflection may be perceptible as an acoustical event which interferes with the unified (integrated) perception of a sound in a room.   Such reflections are classed as *echoes*.   The level and delay at which this occurs are simply related and have been widely studied for various sound sources (refs. 4, 5, 6, 7, 8).

The Premium Quality

Nickson & Muncey (ref. 9) in 1960 summarised auditorium conditions which they considered to be both necessary *and sufficient* for concert halls to be rated by the public and performers as excellent. The five conditions they listed, namely (and in brief)—

(a) low background noise level,
(b) good physical comfort,
(c) reasonable reverberation time,
(d) reasonable early echoes—no strong late echoes, good ensemble,

(e) time for audiences to become accustomed to the sound—

form the basis for the accepted acoustical procedures mentioned above.   Nevertheless, halls which have good early reflections, adequate reverberation, good diffusion and so on, often do not produce the premium quality of sound known variously through the years as "presence", "ambience", "singing tone", "reverberance", "enveloping sound", "acoustical intimacy", and "spatial responsiveness". In identification of this quality one notes that (a) as a property of the *room,* one feels that the hall responds to the music, allowing its nuances and contrasts to develop fully and acting as a spatial complement to the source: (b) for the listener it generates a sense of envelopment in the sound and of direct involvement with it, in the same way that an observer is aware of his involvement with the space of a room he has entered. One may therefore reasonably expect a "yes" or "no" answer to the question, "Am I experiencing this quality?"   In antithesis, rooms which do not produce this effect give one the impression of "looking *at* the music, instead of being *in* it".

Because rooms having this property usually have long reverberation times (e.g., Concertgebouw, Grosser Musikvereinsaal), the quality of sound they produce has frequently been attributed to some aspect of the reverberation characteristic, and

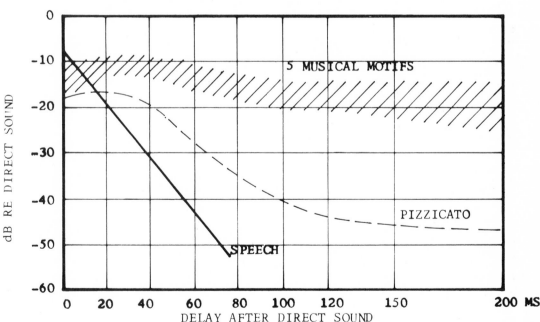

Fig. 1: Masking of a single reflection by direct sound for speech and a variety of motifs.

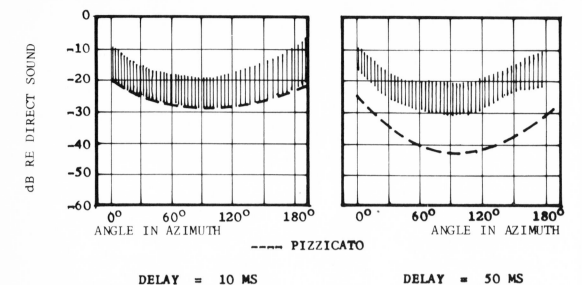

DELAY = 10 MS   DELAY = 50 MS

Fig. 2: Dependence of masking on delay and direction of incidence 0° = ahead, masking by direct sound.

the prediction of reverberation time has remained a continuing point of interest and controversy (refs. 10, 11). It is suspected that the invention of "assisted resonance", as used at the Royal Festival Hall, and similar systems to increase reverberation, were initiated by this belief. Nickson and Muncey (ref. 9), and Lochner and Burger (ref. 5), however, have shown that "optimum reverberation time" is not a critical value for either speech or music within considerable limits, and it is clear from recent experience that premium quality sound does not necessarily correlate with long reverberation times. It will be shown that rooms of the modern type which characteristically do not produce the quality, differ from the classical halls that do, not primarily, in reverberation time, but in the early reflections which are perceptible in the reflection sequences they produce.

Psychophysical Data Available

Burgtorf (ref. 12), Seraphim (ref. 13) and Schubert (ref. 2) in Germany, and Sommerville et al (ref. 14) in England, have studied the auditory masking produced by a variety of speech, noise, and music signals upon the perceptibility of delayed repetitions of the same signal (reflections). The data with immediate relevance here are those of Seraphim and Schubert, which give the *absolute threshold of perceptibility* of a reflection for speech signals and for a number of musical motifs, respectively. Figs. 1 and 2 summarise this information, after Schubert. Seraphim showed that the masking level is sustained when a number of speech reflections occur particularly when they follow from the same direction as the direct sounds. Fig. 2 gives the dependence of the masking upon direc-

tion, but only for masking by direct sound. We do not know if the sustaining effects shown by Seraphim with multiple reflections for speech occur with music, but it is probable that they do. In any case, the musical masking is evidently much greater than that for speech. As Seraphim (ref. 13) points out, these thresholds are measured by successive and immediate comparisons of two sounds fields, one of which contains the test "reflection", in anechoic conditions of minimum noise. He suggests that the threshold for effective contribution to acoustical percepts in a room would lie about 6 dB higher.

Application to Concert Halls

Fig. 3 shows two rectangular concert halls in cross-section together with the main orchestral image positions produced by walls and ceilings. One is based on the Boston Symphony Hall, the other has some of the cross-sectional characteristics of many modern halls (e.g., broad seating area, lower ceiling). It has, in fact, the mid-hall cross-section of the Royal Festival Hall. Both, for simplicity, are assumed to have horizontal floors and ceilings. These two hall shapes are chosen to provide a basis for discussion to which the effects of raked floors, scattering surfaces, stage reflectors and the like may later be added. Experience of listening in such rooms leads one to expect that the broad low-ceilinged hall "X" would lack premium quality sound while the other "Y" is known to have it.

Before the masking effects can be applied to the reflection patterns which will occur in these rooms, one needs to know the level of each reflection relative to the direct sound. Reductions in transit occur by: (a) spherical divergence (boundary absorption will be ignored), (b) interaction of the seated audience on waves at grazing incidence. This interaction has been studied in some detail by Meyer, Kuttruff and Schulte (ref. 15) and by Watters and Schulz (ref. 16). The main effect appears to be that from 100-500 Hz all sound propagated more than 45 feet across audiences suffers 10-20 dB attenuation independently of further distance (ref. 16). The onset of this attenuation as one moves away from the source was not investigated by Watters and Schulz, but for higher frequencies a similar effect was shown by Meyer, Kuttruff and Schulte to occur over the first few rows; in any case the effect is established at a distance of 45 ft. from the stage and remains substantially constant thereafter. The effects shown by Meyer et al (ref. 15) at higher frequencies appear to be due to complex scattering from audience heads, and it is doubtful whether they can be applied directly in the present context (ref. 15b). Watters and Schulz showed also that the middle and low frequency effects may be 10 dB more pronounced for sound propagating at 45° in azimuth relative to the audience rows

than for sound normal to the rows. Conservatively, therefore, in the following, 15 dB attenuation has been assumed for all sound propagated across the audience in each hall independent of distance travelled. In the case of the wide hall the situation is almost certainly worse than this value indicates. Further, even if the conclusions are limited to the 100-500 Hz frequency band, the effects to be shown must rank as having a gross influence on the perceived sound in the room, and almost certainly account for the renowned "warmth" of tone which occurs in halls with unmasked lateral reflections.

One further point which should be considered in a real situation, but which for simplicity is omitted here, is the dependence Schubert showed of the absolute threshold of perceptibility for a reflection upon its direction of incidence (Fig. 2). The masking is about 10 dB **less** for sound arriving normal to the direct sound than it is for sounds having nearly the same direction as the direct sound. This fact has important consequences in the design of concert halls, for a fan-shaped room in which the side wall image lies near the source, will produce

Fig. 3: Cross section of halls, showing image space and main orchestral images, looking toward stage.

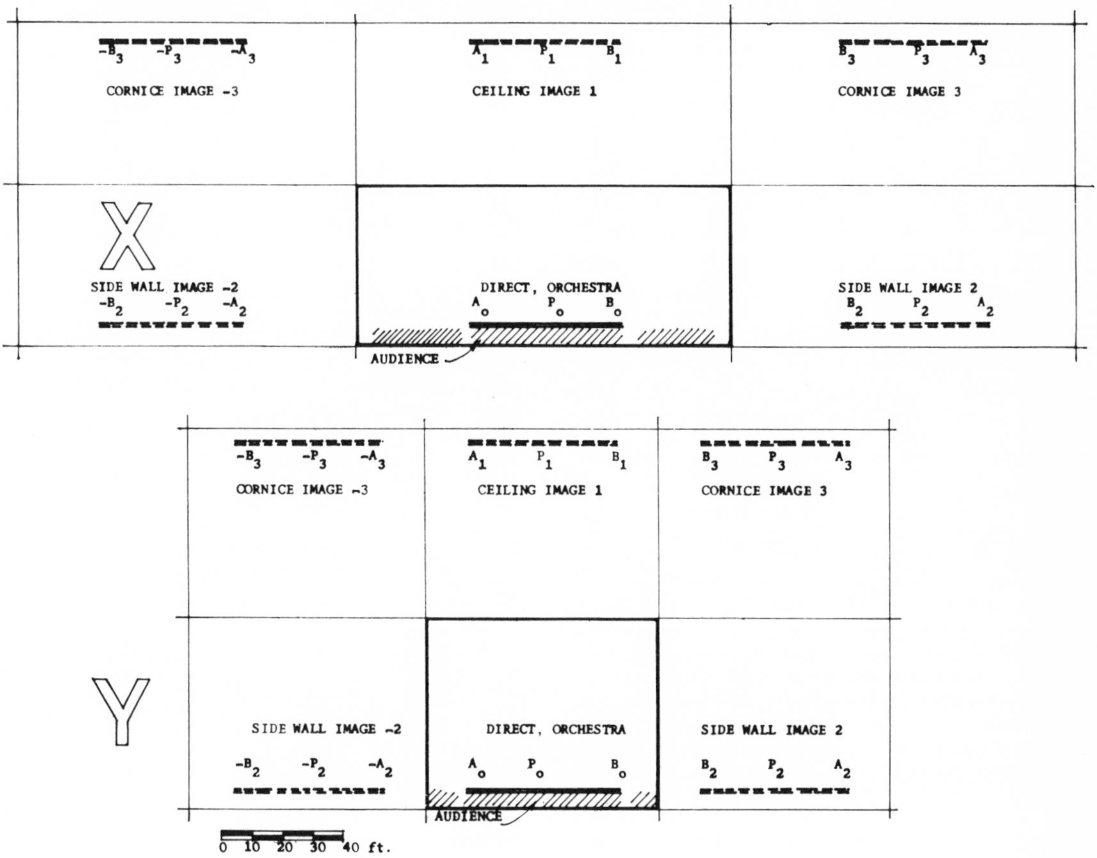

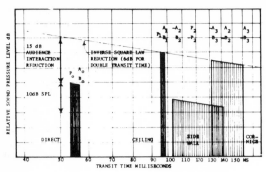

Fig. 4A: Reflection pattern for whole orchestra.

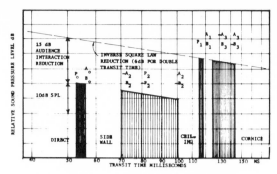

Fig. 5A: Reflection pattern for whole orchestra.

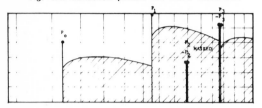

Fig. 4B: "P" reflections showing perception threshold (centre of orchestra) and masking.

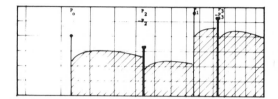

Fig. 5B: "P" reflections showing perception threshold (centre of orchestra) and masking.

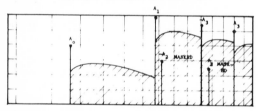

Fig. 4C: "A" reflections showing perception threshold and masking (extremity of orchestra).

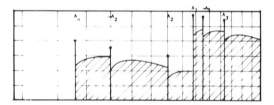

Fig. 5C: "A" reflections showing perception threshold and masking (extremity of orchestra).

Fig. 4: Hall **X** — Reflections and Masking.

Fig. 5: Hall **Y** — Reflections and Masking. See note on Fig. 4.

Fig. 4A: Shows the reflection pattern for the whole orchestra. Each block represents the sound from one entire reflection; in the case of overhead or direct sound, the first sound to arrive comes from the orchestral centre; while in the case of lateral reflections, the block is bounded by sound from the orchestral extremities. Figs. 4B and C show respectively the masking for these orchestral positions. Note that the time scale is logarithmic.

more masking of the lateral sound than a room with parallel or reverse splay walls.

Figures 4A and 5A show the resultant reflection patterns for the two halls, respectively, for a seat 60 ft. from the stage, and assuming an orchestral width of 40 ft. The inverse square law and audience reductions in sound pressure level are taken into account. Figures 4B and 5B show the corresponding masking for an instrument in the centre of the orchestra and figures 4C and 5C the masking for an instrument at one side of the orchestra. Levels are shown relative to the free field direct sound reduced by the 15 dB for its passage over the audience. The curved lines are

the maximum values given by Schubert, plus 5 dB, and it is suggested that these values might apply in a real room for a symphonic programme of romantic music. No sustaining effects are included, and actual masking is probably greater.

Even allowing for the dependence of masking upon motif, it is clear that in the "modern" hall the sound reflected from the side walls arrives well below the threshold of perceptibility, while in the other, every reflection tells. Here then is a gross difference in the spatial distribution of the most important part of the perceptible reflection sequences for the halls, which corresponds with the subjective impressions of "regarding" the sound in the one or of being "enveloped" by it in the other. Normally, too, the "cornice" reflections which arrive so much later than the direct sound that they cause incipient echo in a wide hall, are suppressed by cornice areas of absorption. The only significant sound which then remains effective during the time considered here comes from source, stage, or ceiling. With some confidence, then we may equate the occurrence of the premium quality of acoustical experience with the presence of unmasked lateral reflections in the early, integrable reflection sequence.

It is also clear that the strong masking is produced mainly by the ceiling reflection because it is not subject to audience absorption, and that the arrival time of this reflection just before, or just after that of the lateral sound, determines the audibility or not of the lateral reflections. Any other reflections occurring during the considered time interval, are similarly masked, or themselves produce masking.

In traditional halls of simple plan shape (fans or rectangles), the effective cross-section proportion is of primary importance in determining the position of the ceiling reflections, and thus in determining the masking. The following width to height ratios for halls renowned for possessing the premium quality sound illustrate this point: Vienna, Musikvereinsaal, 1.10:1, Boston Symphony Hall 1.9:1, St. Andrews, Glasgow, 1.31:1, Basel, Stadtcasino 1.35:1, Leipzig Neusgewandhaus 1.30:1, Amsterdam Concertgebouw 1.58:1. All these produce (or produced) ceiling reflections after the arrrival of lateral sound in the centre of the hall. Conclusions may also be drawn about maximum size of orchestra permitted in such halls without masking of some of the instruments, and similar details (ref. 17). Of much greater interest, however, are general conclusions arising from the requirement of unmasked lateral sound in the reflection sequence.

In a large hall of simple shape, the cross-section ratio requirement to avoid masking of the lateral reflections, gives a ceiling height unsuitable for speech intelligibility. Such halls are also limited in the clarity of music they produce. As Lochner and Burger have shown (ref. 5), speech intelligibility is determined by the amplitude and arrival time of the reflections, irrespective of direction, which occur during the time in which effective integration takes place, and that for moderately loud sounds, the echo threshold equals the direct level at about 35 millisecond delay. From inspection of the reflection sequences of figs. 4 and 5, with their delayed strong ceiling reflections, it is obvious why conditions for speech have been held to be irreconcilable with those which are excellent for romantic music. Even hall X would require major modification to be acceptable for speech. (Once again this has usually been seen as a conflict in Reverberation Time requirements, though Lochner and Burger show that even for speech there is no close connection between room volume, reverberation time and intelligibility.)

If, however, unmasked lateral sound could be provided in an audience by subdividing the seating into blocks with suitable lateral surfaces, the ceiling could be lower (i.e., the overall cross-section ratio could be increased), perhaps to a point where excellent speech intelligibility or at least

music of great clarity could be assured without loss of the enveloping sound. It might thus be possible to design a truly multi-purpose hall without having recourse to compromise for either speech or music. The application of this idea to opera is obvious. Before discussing in detail the forms such a concept would produce, let us consider the effect of modifications on the assumed shapes for the two concert halls discussed above.

(a) Floor rake: Increases effective level of direct sound and wall reflections, so that it would generally improve the range of music for which the lateral sound would be unmasked.

(b) Diffusing ceiling: Degrades the strength of overhead masking sound and so improve audibility of lateral sound.

(c) Fan-shaped hall, or reverse fan: See earlier discussion in Section 5.

(d) Lateral Diffusing Elements: Will decrease the effective strength of lateral reflections and so reduce their audibility.

(e) Over-stage reflector such as at Festival Hall: Upgrades masking level and reduces audibility of lateral sound.

For seats near the stage, including, of course, those of the musicians, the direct sound is so much louder than the early reflections that the most important audible feedback from the room is the field-reverberation. Hence it is suggested, the almost universal preference one finds among musicians for "live" halls with RT in excess of two seconds. The situation of a listener in the body of the audience is very different. This casts doubt upon the value of musicians' opinions on the merits of a hall, yet it is usually eminent musicians who are approached when a consensus is required. I believe that musicians' preference for these reverberant rooms has perpetuated the notion that "reverberation" as characterised by RT is the main factor in concert hall design. One notes that at the Festival Hall it was the musicians who first noticed the "improved" quality due to assisted resonance. It is probably by chance that the halls with "acoustical intimacy", call it what you will, were also highly reverberant, but in this connection one notes that the surfaces which produce sustained reverberation also produce the strong reflections. This is not to denigrate the importance of adequate, or as Nickson and Muncey put it, "reasonable", field-reverberation, which quite apart from the support it gives the performers, plays, as has been shown elsewhere (ref. 18), a most important role in the sense of place and occasions a room produces; field-reverberation is, of course, particularly telling on entry to the hall, during staccato music, at pauses in the music, or during applause.

Work has begun at the University of Western Australia School of Architecture, on a study of the forms for concert halls which might follow from the foregoing. It is emphasised that the design shown in plate 1 is but one of many possible solutions, is a preliminary study, and is still unre-

Plate 1: Two views of a model concert hall designed to embody the acoustical features proposed by the author.

solved in a number of architectural respects. It is considered, however, that the acoustical, sight line, and egress problems are well on the way to a solution. Further, the room shown is intended as a symphony concert hall, and does not fulfil the multi-purpose programme mentioned earlier as a possible application of the subdivided audience. It is designed to produce unmasked lateral sound at all seats further than 30 feet from the orchestra, together with clarity of sound. One notes the following features:

    (a) Subdivision of the peripheral audience into "boxes" seating about 300;

    (b) Steeply raked seats to guarantee minimum audience interaction with direct and lateral sound;

    (c) Lateral walls angled in reverse splay and sloping inwards to minimise masking.

The capacity of this hall is 2,500 seats, excluding orchestra and choir, but it is apparent that a similar scheme could include an audience in the order of 4,000, with similar reflection patterns, and therefore, one believes, with similar acoustical quality.

Finally, one notes that it is only as the total architectural problems posed by such a programme moves towards a solution that the acoustical determinants outlined in this paper can contribute to the design of a room for music. They do not, by themselves, form the room.

The premium quality of musical sound in concert halls known variously as "presence", "singing tone", "acoustical intimacy", "reverberance", "enveloping sound" and "spatial responsiveness", is tentatively equated with the presence of unmasked lateral reflections in the early reflection sequence. The effect will be particularly pronounced between 100 and 500 Hz. Gross differences in the early perceptible sound arriving at a listener's position are shown to depend upon the room shape, and a suggested new form for a concert hall is suggested on the basis of the hypothesis.

Acknowledgements:

The encouragement and advice of Mr. P. E. Doak, Hawker Siddley Reader in Noise Research at Southampton University, during the initial work on this topic, is gratefully acknowledged, as is the travel assistance from the German Academic Exchange Service in support of a visit to Germany.

The design for a concert hall is being carried out by Mr. R. P. J. Marshall, fourth year student in the School of Architecture, University of Western Australia.

REFERENCES

(1) BURGTORF, W. and SERAPHIM, H-P., *Acustica,* Vol. 11, 1961, p. 92. "Eine Apparatus zur Electroakustischen Herstellung einfacher Schallfelder."

(2) SCHUBERT, P., *Techn. Mitt. RFZ,* Vol. 3, 1966, p. 124 (East German Radio, Aldershof Berlin). "Untersuchungen über die Wahrnehmbarkeit . . ."

(3) MEYER E. and RICHARDSON E. G., *Technical Aspects of Sound,* Vol. III, 1962, Chapter 5, p. 292.

(4) HAAS, H., *Acustica,* Vol. 1, 1951, p. 49. "Uber den Einfluss eines Einfachechos auf die Hörsamkeit von Sprache".

(5) LOCHNER, J. P. A. and BURGER, J. F., *J. Sound Vib.,* Vol. 1, No. 4, 1964, p. 426. "The influence of reflections on auditorium acoustics."

6) MUNCEY, R. W. NICKSON, A. F. P. and DUBOUT, P., *Acustica,* Vol. 3, 1953, p. 168. "The acceptability of speech and music with a single artificial echo."

(7) MUNCEY, R. W., NICKSON, A. F. B., and DUBOUT, P., *Acustica,* Vol. 4, 1954, p. 515. "The acceptability of artificial echoes with reverberant speech and music."

(8) BOLT, R. H. and DOAK, P. E., *J. Acoust. Soc. Am.,* Vol. 22, p. 507. "A tentative criterion for the short term transient response of an auditorium."

(9) MUNCEY, R. W. and NICKSON, A. F. B., *C.I.B. Bulletin,* Vol. 2, June, 1960. "A lesson from Australian studies on room acoustics."

(10) KOSTEN, C. W., *Acustica,* Vol. 16, 1965, pp. 325-350. "New method for the calculation of the reverberation time of halls for public assembly."

(11) GOMPERTS, M. C., *Acustica,* Vol 16, 1966, pp. 256-268. "Classical reverberation formulae."

(12a) BURGTORF, W., *Acustica,* Vol. 11, 1961, p. 97.

(13) SERAPHIM, H-P., *Acustica,* Vol. 11, 1961, p. 81. "Uber die Wahrnehmbarkeit mehrerer Rückwürfe von Spachschall."

(14) SOMERVILLE, T., GILFORD, C. L. S., SPRING, N. F. and NEGUS, R. D. M., *J. Sound Vib.,* Vol. 3, (2), 1966, p. 127. "Recent work on the effects of reflections in concert halls and music studios."

(15) MEYER, E. et al, *Acustica,* Vol. 15, 1965, p. 178. "Schallausbreitung über Publikum."

(15b) KUTTRUFF, H., 1966. "Personal communication."

(16) WATTERS, B. C. and SCHULZ, T., *J. Acoust. Soc. Amer.,* Vol. 36, 1964, p. 885. "Propagation of sound across audience seating."

(17) MARSHALL, A. H., *J. Sound Vib.,* Jan., 1967. "On the importance of room cross-section in concert halls."

(18) MARSHALL, A. H., *Ph.D. Dissertation,* The Institute of Sound & Vibration Research, Uni. of Southamptcn, 1967. "The architectural significance of reverberation."

Part II

ROOM ACOUSTICS—TECHNIQUES AND THEORIES

Editor's Comments
on Papers 11 Through 17

The sounds of speech and music, which are the primary concern in room acoustics, constitute a sequence of individual transient sounds varying widely in frequency content, loudness, and shape. When a sequence of sounds is produced in a room, what reaches a given listening point is the original bare sound plus a multiplicity of reflected versions that may reinforce and embellish the original or may distort and blur it beyond recognition. Room acoustics is a study of the influence of the room on the end product reaching each listening position.

114

The varied character of the sounds involved makes it impossible to formulate a single universal statement about the propagation process. Instead one learns what one can from several different idealized models of the real situation, which are often categorized under the headings geometrical acoustics, statistical acoustics, and wave acoustics.[1] This approach will be followed here.

GEOMETRICAL ACOUSTICS

Geometrical acoustics is the application of the principles of geometrical optics to acoustical phenomena. The main application is in a graphical ray-tracing process from a source, through reflections from various surfaces of a hall, to their ultimate arrival at a listening point. As in geometrical optics, there are limitations: for specular reflection a surface must be large in lateral dimensions compared to a wavelength, and smooth compared to a wavelength. Smaller surfaces and "rough" surfaces tend to scatter the incident sound, and may not follow the simple specular reflection laws. These distinctions are especially important in the acoustical case, where the sounds of interest have wavelengths ranging from a fraction of an inch to, say, 50 feet, and thus encompass the range of dimensions of most room surfaces.

Another complication of the acoustical analogy is that the various direct and reflected versions of a brief sound may arrive at a listening point at noticeably different times, with a resultant garbling of the original. The acoustician must therefore consider both the spatial and temporal distributions of sounds.

The application of geometrical acoustics to design is well exemplified by W. C. Sabine's paper "Theatre Acoustics," a portion of which is presented here (Paper 11). It illustrates his very successful use of an optical–acoustical pulse technique for studying the progression of wave fronts in two-dimensional model sections of auditoria. Variations of Sabine's approach have been developed by A. H. Davis,[2] Takeo Satow,[3] and R. Vermeulen and J. de Boer.[4]

STATISTICAL ACOUSTICS—REVERBERATION

In contrast to geometrical acoustics, which provides a way of examining the direct sound plus the first few reflections, statistical acoustics deals with the "reverberant" sound resulting from many reflections from the room boundaries. The multiplicity of reflected sounds is treated statistically to obtain an expression relating the average level of reverberant sound in the room and the rate at which it decays to the absorption properties of the room surfaces. The first such relation was Sabine's famous formula $T = 0.161V/A$, where T is the reverberation time in seconds, V is the room volume in cubic metres, and A is the room absorption expressed in metric sabins. An early paper giving a play-by-play account of Sabine's discovery of this formula appears also in a previous Benchmark volume.[5] Only his final theoretical derivation is included in the present volume (Paper 12).

The assumption underlying Sabine's theory is that the growth steady-state, and decay of sound in a room may be treated as continuous processes, with equilibrium at all times between the energy density in the room, the power being added, to the room, and the power being lost by transmission or absorption. Implicit in this is the assumption that the sound field is diffuse, i.e., that on the average it looks everywhere the same, with equal probability of waves traveling in all directions. Sabine recognized these limitations, but he found that in many rooms his theory described the reverberation processes with adequate accuracy. Indeed it still does.

Sabine's formula has the formal defect that it yields a finite reverberation time in the limit when all surfaces are perfectly absorptive. A number of attempts were subsequently made to eliminate this defect and thus provide a formula that gives more plausible results in relatively absorptive spaces. The most successful of these is the formula developed independently by K. Schuster and E. Waetzmann,[6] by R. F. Norris (Paper 13) and by Carl Eyring (Paper 14). Norris's paper was presented to a meeting of rhe Acoustical Society on 11 May 1929, but never published as such; the version reproduced here appeared as an Appendix in V. O. Knudsen's book, *Architectural Acoustics*. The two presentations form an interesting contrast, the Norris version having the elegance of simplicity and brevity, Eyring's being an exhaustive analysis employing image theory. Both lead to the same formula, quite widely used today, which agrees with Sabine's formula when room absorption is low.

One question that was controversial for a time, and indeed still is, was how to combine the effects of surfaces having different absorption coefficients. This led, for example, to still another theory, by Millington[7] and by Sette,[8] but their derivations can be disputed on several counts, as was most lucidly done by Eyring.[9]

Of the various reverberation theories, the most satisfactory, within the restriction of highly reverberant rooms, is still Sabine's theory. The other approaches involve certain assumptions about transit times between successive reflections and the use of ray-tracing concepts that cannot readily be pushed to the statistical limit. These difficulties have been examined in some detail in more recent papers by F. V. Hunt,[10] L. Batchelder,[11] and W. B. Joyce.[12]

The particular application to the measurement of absorption coefficients of materials was examined experimentally in a joint project organized by C. W. Kosten (Paper 15). Other aspects of the measurement problem have been examined by T. D. Northwood[13] and by C. G. Balachandran and D. W. Robinson.[14] For the accuracy needed in design, however, the Sabine formula provides the basis of a method of measuring absorption coefficients of materials; and either the Sabine or Eyring formula provides a method of calculating the reverberation properties of a hall.

WAVE ACOUSTICS

The application of wave theory to room acoustics goes back at least to Lord Rayleigh, who wrote down the appropriate wave equations and the expressions for normal modes in rooms and discussed, at least in a qualitative way, methods of controlling "resonance" in rooms.[15] The concept of acoustic impedance and its relation to sound absorption was further developed in some detail by E. T. Paris, who invented the "stationary-wave" tube for measuring normal-incidence absorption and impedance of materials.[16]

The major application of wave theory to room acoustics began, however, with the work of P. M. Morse and his coworkers in the mid-thirties. Morse stimulated so much activity and enthusiasm that people predicted that all the phenomena of room acoustics would soon be explained by wave theory. It turns out, however, that sound fields in most rooms are far too complicated to be handled by wave acoustics. Nevertheless, the studies thus begun have provided a great deal of insight into the behavior of sound in rooms and into the physics of the absorption process. A concise survey of the application of wave acoustics to rooms is contained in a review paper by P. M. Morse and R. H. Bolt.[17]

EXPERIMENTAL STUDIES OF AUDITORIA

Reverberation time has, since Sabine's day, provided a measure of what might be called the long-term transient behavior of sounds in rooms. Equally important is the short-term transient process. This has not yet been successfully treated, although Sabine began the use of qualitative two-dimensional techniques (Paper 11). Among other small-scale techniques, a notable one, initiated by F. Spandock,[18] utilized a complete three-dimensional acoustic model, including surface absorptions of proper values for the range of scaled frequencies. Taped music, appropriately speeded up, is played in the model hall, picked up at various listening points, and then scaled down again to the original frequency range. Thus, in principle, one can listen to the sound as it would exist in the full-sized hall.

In full-scale halls, the period of analysis following the Royal Festival Hall included a number of experimental studies in search of new objective measures. Erwin Meyer and his colleagues were particularly active in this field, and Paper 17, by Meyer and Thiele, is an early product of this activity. Here and in related papers, for example by R. Thiele[19] and G. R. Schodder,[20] the aim was to find some measure of the short-term transient performance of a hall that would correlate with subjective assessments of hall quality. Some investigators (Paper 17) studied the directional distribution of arriving sounds; others (Papers 7 and 9) have emphasized the timing and relative magnitudes of successive arrivals.

Techniques for investigating room acoustics have come a long way since Sabine's use of a few organ pipes, a stopwatch and his two ears. In particular, the application of the computer to room acoustics, pioneered by M. R. Schroeder and colleagues at Bell Telephone Laboratories, has facilitated great leaps forward in all aspects of the subject. On the experimental side, studies of the transient properties of a room, requiring the collection and digestion of masses of data, utilize the computer for control of the measurements and analysis of results, as exemplified in Paper 16 by Atal, Schroeder, Sessler, and West. For subjective studies of halls in general, the whole process of sound propagation in rooms can be synthesized.[21] On the theoretical side, the use of the computer simply as a mathematical tool is narrowing the gap between statistical acoustics and wave acoustics.[22]

REFERENCE NOTES

1. Lothar Cremer, *Die Wissenschaftlichen Grundlagen der Raumakustik.* Band I. *Geometrische Raumakustik,* S. Hirzel, Stuttgart (1948). Band II, *Statische Raumakustik,* S. Hirzel, Stuttgart (1961). Band III, *Wellentheoretische Raumakustik,* S. Hirzel, Leipzig (1960, rev. 1976).
2. A. H. Davis, "The Analogy Between Ripples and Acoustical Wave Phenomena," *Proc. Phys. Soc.* (London) **38**:234–246 (1926).
3. Takeo Satow, *Acoustics of Auditorium Ascertained by Optical Treatment in Models,* World Engineering Congress, Paper 118 (1929).
4. R. Vermeulen and J. de Boer, "Optische Modellversuche zur Untersuchung der Horsamkeit von Schauspielhausern," *Philips Tech. Rundshau* **5**:329 (1940).
5. W. C. Sabine, "Reverberation," in *Acoustics: Historical and Philosophical Development,* ed. R. Bruce Lindsay, pp. 418–457. Stroudsburg, Pa.: Dowden, Hutchinson & Ross, Inc., 1973.
6. K. Schuster and E. Waetzmann, "Über den Nachhall in geschlossenen Räumen," *Ann. Phys.* (Leipzig) **1**:671–695 (1929).
7. G. Millington, "Modified Formula for Reverberation," *J. Acoust. Soc. Am.* **4**:69–82 (1933).
8. W. H. Sette, "A New Reverberation Time Formula," *J. Acoust. Soc. Am.* **4**:193–210 (1933).

9. C. F. Eyring, "Methods of Calculating the Average Coefficient of Sound Absorption," *J. Acoust. Soc. Am.* **4**:178–192 (1933).

10. R. V. Hunt, "Remarks on the Mean Free Path Problem," *J. Acoust. Soc. Am.* **36**:556–564 (1964).

11. L. Batchelder, "Reciprocal of the Mean Free Path," *J. Acoust. Soc. Am.* **36**:551–555 (1964).

12. W. B. Joyce, "Sabine's Reverberation Time and Ergodic Auditoriums," *J. Acoust. Soc. Am.* **58**:643–655 (1975).

13. T. D. Northwood, "Absorption of Diffuse Sound by a Strip or Rectangular Patch of Absorptive Material," *J. Acoust. Soc. Am.* **35**:1173–1177 (1963).

14. C. G. Balanchandran and D. W. Robinson, "Diffusion of the Decaying Sound Field," *Acustica* **19**:245–257 (1967/68).

15. Lord Rayleigh, *The Theory of Sound* (Dover, New York, 1945) Vol. II, pp. 69–72, 128–129.

16. E. T. Paris, "On the Stationary Wave Method of Measuring Sound Absorption at Normal Incidence," *Proc. Phys. Soc.* (London) **39**:269–295 (1927).

17. P. H. Morse and R. H. Bolt, "Sound Waves in Rooms," *Rev. Mod. Phys.* **16**:69–150 (April 1944).

18. F. Spandock, "Akustiche Modellversuche," *Ann. Phys.* (Leipzig) **20**:345 (1934).

19. R. Thiele, "Richtungsverteilung und Zeitfolge der Schallrückwürfe in Räumen," *Akust. Beih.* No. 2 (*Acustica* 3) 291–302 (1953).

20. G. R. Schodder, "Uber die Verteilung der Energiereicheren Schallrückwürfe in Sälen," *Acustica* **6**:445–465 (1956).

21. B. S. Atal, M. R. Schroeder, G. M. Sessler, "Subjective Reverberation Time and Its Relation to Sound Decay," *Proc. 5th Int. Cong. on Acoustics,* v. 1b, Paper G32.

22. M. R. Schroeder, "Digital Simulation of Sound Transmission in Reverberant Spaces," *J. Acoust. Soc. Am.* **47**:424–431 (1970).

11

Reprinted from pp. 163–164, 176, and 177–187 of *Collected Papers on Acoustics*, Dover Publications, Inc., New York, 1964, 299 pp.

THEATRE ACOUSTICS[1]

Wallace Clement Sabine

Vitruvius, De Architectura, Liber V, Cap. VIII. (*De locis consonantibus ad theatra eligendis.*)

"All this being arranged, we must see with even greater care that a position has been taken where the voice falls softly and is not so reflected as to produce a confused effect on the ear. There are some positions offering natural obstructions to the projection of the voice, as for instance the dissonant, which in Greek are termed κατηχοῦντες; the circumsonant, which with them are named περιηχοῦντες; and again the resonant, which are termed ἀντηχοῦντες. The consonant positions are called by them συνηχοῦντες.

The dissonant are those places in which the sound first uttered is carried up, strikes against solid bodies above, and, reflected, checks as it falls the rise of the succeeding sound.

The circumsonant are those in which the voice spreading in all directions is reflected into the middle, where it dissolves, confusing the case endings, and dies away in sounds of indistinct meaning.

The resonant are those in which the voice comes in contact with some solid substance and is reflected, producing an echo and making the case terminations double.

The consonant are those in which the voice is supported and strengthened, and reaches the ear in words which are clear and distinct."

This is an admirable analysis of the problem of theatre acoustics. But to adapt it to modern nomenclature, we must substitute for the word dissonance, interference; for the word circumsonance, reverberation; for the word resonance, echo. For consonance, we have unfortunately no single term, but the conception is one which is fundamental.

It is possible that in the above translation and in the following interpretation I have read into the text of Vitruvius a definiteness of conception and an accord with modern science which his language only fortuitously permits. If so, it is erring on the better side, and is but a reasonable latitude to take under the circumstances. The only passage whose interpretation is open to serious question is that re-

---

[1] The American Architect, vol. civ, p. 257.

lating to dissonant places. If Vitruvius knew that the superposition
of two sounds could produce silence, and the expression "*opprimit
insequentis vocis elationem*" permits of such interpretation, it must
stand as an observation isolated by many centuries from the modern
knowledge of the now familiar phenomenon of interference.

[*Editor's Note:* Material has been omitted at this point.]

Echo

When a source of sound is maintained constant for a sufficiently
long time — a few seconds will ordinarily suffice — the sound becomes steady at every point in the room. The distribution of the
intensity of sound under these conditions is called the interference
system, for that particular note, of the room or space in question.
If the source of sound is suddenly stopped, it requires some time for
the sound in the room to be absorbed. This prolongation of sound
after the source has ceased is called reverberation. If the source of
sound, instead of being maintained, is short and sharp, it travels as
a discrete wave or group of waves about the room, reflected from
wall to wall, producing echoes. In the Greek theatre there was ordi-

[*Editor's Note:* Figure 10, a photograph of the interior of the New Theatre, New
York, has been omitted due to limitations of space.]

narily but one echo, "doubling the case ending," while in the modern theatre there are many, generally arriving at a less interval of time after the direct sound and therefore less distinguishable, but stronger and therefore more disturbing.

This phase of the acoustical problem will be illustrated by two examples, the New Theatre, the most important structure of the

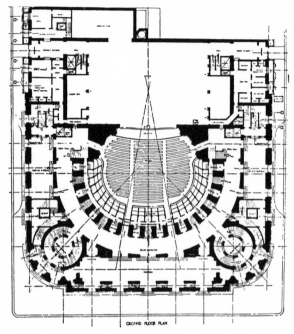

Fig. 11

kind in New York, and the plans of the theatre now building for the Scollay Square Realty Company in Boston.

Notwithstanding the fact that there was at one time criticism of the acoustical quality of the New Theatre, the memory of which still lingers and still colors the casual comment, it was not worse in proportion to its size than several other theatres in the city. It is, therefore, not taken as an example because it showed acoustical defects in remarkable degree, but rather because there is much that can be learned from the conditions under which it was built, because such defects as existed have been corrected in large measure, and above all in the hope of aiding in some small way in the restoration of a magnificent building to a dignified use for which it is in so many ways eminently suited. The generous purpose of its Founders, the high ideals of its manager in regard to the plays to be produced, and the perfection otherwise of the building directed an exaggerated and morbid attention to this feature. Aside from the close scrutiny which

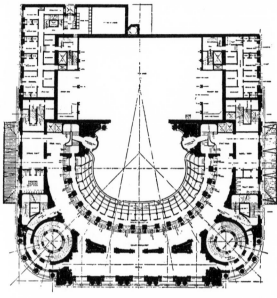

Fig. 12

always centers on a semi-public undertaking, the architects, Messrs. Carrère and Hastings, suffered from that which probably every architect can appreciate from some similar experience of his own, — an impossible program. They were called on to make a large "little theatre," as a particular type of institution is called in England; and, through a division of purpose on the part of the Founders and Advisers, for the Director of the Metropolitan Opera was a powerful factor, they were called on to make a building adapted to both the opera and the drama. There were also financial difficulties, although very different from those usually encountered, a plethora of riches. This necessitated the provision of two rows of boxes, forty-eight originally, equally commodious, and none so near the stage as to thereby suffer in comparison with the others. Finally, there was a change of program when the building was almost complete. The upper row of boxes was abandoned and the shallow balcony thus created was devoted to foyer chairs which were reserved for the

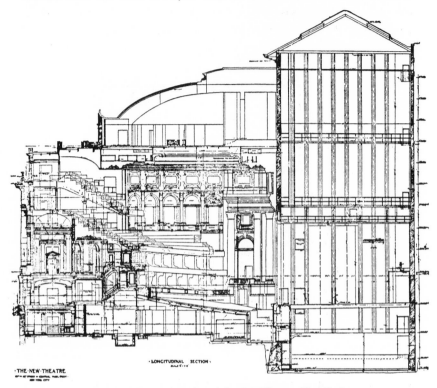

·THE·NEW·THEATRE·

·LONGITUDINAL SECTION·

Fig. 13. Plans and Section of the New Theatre, New York.
Carrère and Hastings, Architects.

annual subscribers. As will be shown later these seats were acoustically the poorest in the house.

Encircling boxes are a familiar arrangement, but most of the precedents, especially those in good repute, are opera houses and not theatres, the opera and the drama being different in their acoustical requirements. In the New Theatre this arrangement exerted a three-fold pressure on the design. It raised the balcony and gallery 12 feet. It increased both the breadth and the depth of the house. And, together with the requirement that these boxes should not extend near the stage, it led to side walls whose most natural architectural treatment was such as to create sources of not inconsiderable echo.

The immediate problem is the discussion of the reflections from the ceiling, from the side walls near the stage, from the screen and parapet in front of the first row of boxes and from the wall at the rear of these boxes. To illustrate this I have taken photographs of the actual sound and its echoes passing through a model of the

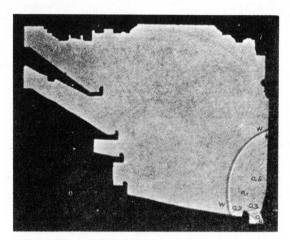

Fig. 14. Photograph of a sound-wave, WW, entering a model of the New Theatre, and of the echoes a_1, produced by the orchestra screen, a_2 from the main floor, a_3, from the floor of the orchestra pit, a_4, the reflection from the orchestra screen of the wave a_3, a_5 the wave originating at the edge of the stage.

theatre by a modification of what may be called the Toeppler-Boys-Foley method of photographing air disturbances. The details of the adaptation of the method to the present investigation will be explained in another paper. It is sufficient here to say that the method consists essentially of taking off the sides of the model, and, as the sound is passing through it, illuminating it instantaneously by the light from a very fine and somewhat distant electric spark. After passing through the model the light falls on a photographic plate placed at a little distance on the other side. The light is refracted by the sound-waves, which thus act practically as their own lens in producing the photograph.

In the accompanying illustrations reduced from the photographs the enframing silhouettes are shadows cast by the model, and all

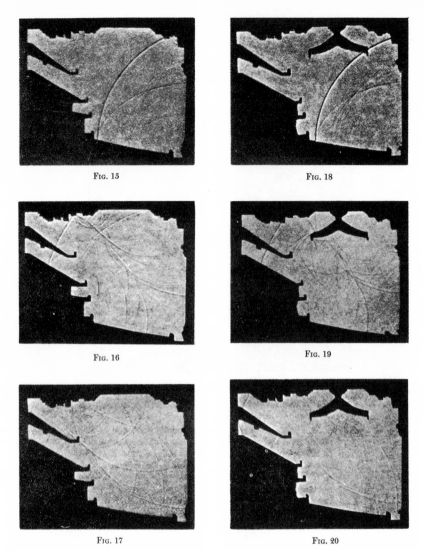

FIG. 15   FIG. 18

FIG. 16   FIG. 19

FIG. 17   FIG. 20

Two series of photographs of the sound and its reflections in the New Theatre, — 15 to 17 before, 18 to 20 after the installation of the canopy in the ceiling. The effect of the canopy in protecting the balcony, foyer chairs, boxes, and the orchestra chairs back of row L is shown by comparing Figs. 19 and 20 with Figs. 16 and 17.

within are direct photographs of the actual sound-wave and its echoes. For example, Fig. 14 shows in silhouette the principal longitudinal section of the main auditorium of the New Theatre. WW is a photograph of a sound-wave which has entered the main auditorium from a point on the stage at an ordinary distance back of the proscenium arch; a_1, is the reflection from the solid rail in front of the orchestra pit, and a_2, the reflection from the floor of the sound which has passed over the top of the rail; a_3 is the reflection from the floor

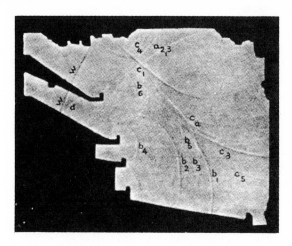

FIG. 21. Photograph of the direct sound, WW, and of the echoes from the various surfaces; $a_{2,3}$, a wave, or echo, due to the combination of two waves which originated at the orchestra pit; c_1 from the oval panel in the ceiling; c_a and c_3, from the ceiling mouldings and cornice over the proscenium arch; c_4, a group from the moulding surrounding the panel; c_5, from the proscenium arch; b_1, b_2, b_6 from the screens in front, and the walls in the rear of the boxes, balcony and gallery.

of the pit, and a_4 the reflection of this reflected wave from the rail; while a_5 originated at the edge of the stage. None of these reflections are important factors in determining the acoustical quality of the theatre, but the photograph affords excellent opportunity for showing the manner in which reflections are formed, and to introduce the series of more significant photographs on page 181.

Figures 15, 16, and 17 show the advance of the sound through the auditorium at .07, .10, and .14 second intervals after its departure from the source. In Fig. 15, the waves which originated at the orchestra pit can be readily distinguished, as well as the nascent waves where the primary sound is striking the ceiling cornice immediately over the proscenium arch. The proscenium arch itself was very well designed, for the sound passed parallel to its surface. Otherwise reflections from the proscenium arch would also have shown in the photograph. These would have been directed toward the audience and might have been very perceptible factors in determining the ultimate acoustical quality.

The system of reflected waves in the succeeding photograph in the series is so complicated that it is difficult to identify the several reflections by verbal description. The photograph is, therefore, reproduced in Fig. 21, lettered and with accompanying legends. It is interesting to observe that all the reflected waves which originated at the orchestra pit have disappeared with the exception of waves a_2 and a_3. These have combined to form practically a single wave. Even this combined wave is almost negligible.

The acoustically important reflections in the vertical section are the waves c_1, c_2, and c_3. The waves b_1 and b_2 from the screen in front of the boxes and from the back of the boxes are also of great importance, but the peculiarities of these waves are better shown by photographs taken vertically through a horizontal section.

The waves c_1, c_2, c_3, and b_1 and b_2 show in a striking manner the fallacy of the not uncommon representation of the propagation of sound by straight lines. For example, the wave c_1 is a reflection from the oval panel in the ceiling. The curvature of this panel is such that the ray construction would give practically parallel rays after reflection. Were the geometrical representation by rays an adequate one the reflected wave would thus be a flat disc equal in area to the oblique projection of the panel. As a matter of fact, however, the wave spreads far into the geometrical shadow, as is shown by the curved portion reaching well out toward the proscenium arch. Again, waves c_a and c_3 are reflections from a cornice whose irregularities are not so oriented as to suggest by the simple geometrical representation of rays the formation of such waves as are here clearly shown. But each small cornice moulding originates an almost hemispherical wave, and the mouldings are in two groups, the position of each being such that the spherical waves conspire to form these two master waves. The inadequacy of the discussion of the subject of architectural acoustics by the construction of straight lines is still further shown by the waves reflected from the screens in front of the boxes, of the balcony, and of the gallery. These reflecting surfaces are narrow, but give, as is clearly seen in the photograph, highly divergent waves. This spreading of the wave beyond the geometrical projection is more pronounced the smaller the opening or the reflecting obstacle and the greater the length of the wave. The phenomenon is called diffraction and is, of course, one of the well-known phenomena of physics. It is more pronounced in the long waves of sound than in the short waves of light, and on the small areas of an auditorium than in the large dimensions of out-of-door space. It cannot be ignored, as it has been heretofore ignored in all discussion of this phase of the problem of architectural acoustics, with impunity. The method of rays, although a fairly correct approximation with large areas, is misleading under most conditions. For example, in the present case it would have predicted almost perfect acoustics in the boxes and on the main floor.

Figures 17 and 20 show the condition in the room when the main sound-wave has reached the last seat in the top gallery. The wave c_1 has advanced and is reaching the front row of seats in the gallery, producing the effect of an echo. A little later it will enter the balcony, producing there an echo greater in intensity, more delayed, and affecting more than half the seats in the balcony, for it will curve under the gallery, in the manner just explained, and disturb seats which geometrically would be protected. Still later it will enter the foyer seats and the boxes. But the main disturbance in these seats and the boxes, as is well shown by the photograph, arises from the wave c_2, and in the orchestra seats on the floor from the wave c_3.

In the summer following the opening of the theatre, a canopy, oval in plan and slightly larger than the ceiling oval, was hung from the ceiling surrounding a central chandelier. The effect of this in preventing these disturbing reflections is shown by a comparison, pair by pair, of the two series of photographs, Figs. 15 to 17 and Figs. 18 to 20. It is safe to say that there are few, possibly no modern theatres, or opera houses, equal in size and seating capacity,

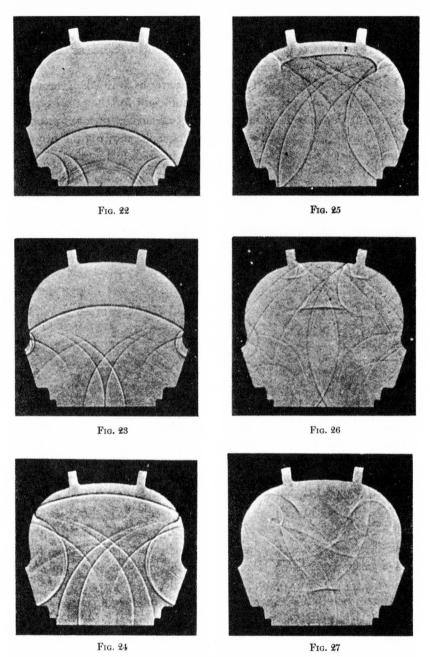

FIG. 22

FIG. 25

FIG. 23

FIG. 26

FIG. 24

FIG. 27

Photographs showing the reflections, in a vertical plane, from the sides of the proscenium arch, the plain wall below the actors' box, and the rail or screen in front of the boxes. The photographs taken in numerical sequence show the progress of a single sound-wave and its reflections.

130

which are so free from this particular type of disturbance as the New Theatre at the present time.

In the study of the New Theatre, photographs were taken through several horizontal sections. It will be sufficient for the purposes of the present paper to illustrate the effect of curved surfaces in producing converging waves by a few photographs showing the propagation of sound through a single section in a plane passing through the parapet in front of the boxes. The reflected waves shown in

Fig. 28. A photograph, one of many taken, showing in vertical section one stage of the reflection b_2, Fig. 21. These reflections were eliminated by the architects in the summer following the opening of the theatre, but have been in part restored by subsequent changes.

Fig. 22 originating from the edge of the proscenium arch and from the base of the column can be followed throughout all the succeeding photographs. In Fig. 23 are shown waves originating from the plain wall beneath the actor's box and the beginning of some small waves from the curved parapet. It is easily possible, as it is also interesting and instructive, to follow these waves through the succeeding photographs. In Fig. 25 the sound has been reflected from the rear of the parapet; while in Fig. 26 it has advanced further down the main floor of the auditorium, narrowing as it proceeds and gaining in intensity. The waves reflected from the parapet outside of the aisles are here shown approaching each other behind the wave which has been reflected from the parapet between the aisles. Waves are also shown in Fig. 26 emerging from the passages between the boxes.

Indeed, it is possible to trace the waves arising from a second reflection from the proscenium arch of the sound which, first reflected from the corresponding surfaces on the other side, has crossed directly in front of the stage. With a little care, it is possible also to identify these waves in the last photograph.

Although many were taken, it will suffice to show a single photograph, Fig. 28, of the reflections in the plane passing through the back of the boxes. These disturbing reflections were almost entirely eliminated in the revision of the theatre by the removal of the boxes from the first to the second row and by utilizing the space vacated together with the anterooms as a single balcony filled with seats.

An excellent illustration of the use of such photographs in planning, before construction and while all the forms are still fluid, is to be found in one of the theatres now being built in Boston by Mr. C. H. Blackall, who has had an exceptionally large and successful experience in theatre design. The initial pencil sketch, Fig. 29, gave in the model test the waves shown in the progressive series of photographs, Figs. 31 to 33. The ceiling of interpenetrating cylinders was then changed to the form shown in finished section in Fig. 30, with the results strikingly indicated in the parallel series of photographs, Figs. 34 to 36. It is, of course, easy to identify all the reflections in each of these photographs,—the reflections from the ceiling and the balcony front in the first; from the ceiling and from both the balcony and gallery front in the second; and in the third photograph of the series, the reflections of the ceiling reflection from the balcony and gallery fronts and from the floor. But the essential point to be observed, in comparing the two series pair by pair, is the almost total absence in the second series of the ceiling echo and the relatively clear condition back of the advancing sound-wave.

[*Editor's Note:* Material has been omitted at this point.]

12

Reprinted from pp. 43–51 of *Collected Papers on Acoustics,*
Dover Publications, Inc., New York, 1964, 299 pp.

REVERBERATION

Wallace Clement Sabine

EXACT SOLUTION

THE present paper will carry forward the more exact analysis proposed in the last paper.

For the sake of reference the nomenclature so far introduced is here tabulated.

t = time after the source has ceased up to any instant whatever during the decay of the sound.

t', t'', t''' = duration of the residual sound, the accents indicating a changed condition in the room such as the introduction or removal of some absorbent, the presence of an audience, or the opening of a window.

t_1, t_2, . . . t_n = whole duration of the residual sound, the subscripts indicating the number of organ pipes used.

T = duration of the residual sound in a room when the initial intensity has been standard.

i = intensity of the residual sound at any instant.

i' = intensity of minimum audibility.

I_1, I_2, . . . I_n = intensity of sound in the room just as the organ pipe or pipes stop, the subscripts indicating number of pipes.

I = standard initial intensity arbitrarily adopted, $I = 1,000,000\ i'$.

w = absorbing power of the open windows, minus their absorbing power when closed = area $(1 - .024)$.

a = absorbing power of the room.

a_1, a_2, . . . a_n = coefficients of absorption of the various components of the wall-surface.

s = area of wall (and floor) surface in square meters.

s_1, s_2, . . . s_n = area of the various components of the wall-surface.

V = volume of the room in cubic meters.

k = hyperbolic parameter of any room.

K = ratio of the parameter to the volume, $aT = k = KV$.

A = rate of decay of the sound.

p = length of mean free path between reflections.

v = velocity of sound, 342 m. per second at 20° C.

Let E denote the rate of emission of energy from the single organ pipe.

$$\frac{p}{v} = \text{the average interval of time between reflections.}$$

$$\frac{p}{v} E = \text{amount of energy emitted during this interval.}$$

$$\frac{p}{v} E \left(1 - \frac{a}{s}\right) = \text{amount of energy left after the first reflection.}$$

$$\frac{p}{v} E \left(1 - \frac{a}{s}\right)^2 = \text{amount of energy left after the second reflection, etc.}$$

If the organ pipe continues to sound, the energy in the room continues to accumulate, at first rapidly, afterwards more and more slowly, and finally reaches a practically steady condition. Two points are here interesting, — the time required for the sound to reach a practically steady condition (for in the experiments the organ pipes ought to be sounded at least this long), and second, the intensity of the sound in the steady and final condition. At any instant, the total energy in the room is that of the sound just issuing from the pipe, not having suffered any reflection, plus the energy of that which has suffered one reflection, that which has suffered two, that which has suffered three, and so on back to that which first issued from the pipe, as:

$$\frac{p}{v} E \left[1 + \left(1 - \frac{a}{s} \right) + \left(1 - \frac{a}{s} \right)^2 + \left(1 - \frac{a}{s} \right)^3 + \ldots \left(1 - \frac{a}{s} \right)^n \right],$$

where n is the number of reflections suffered by the sound that first issued from the pipe, and is equal to the length of time the pipe was blown divided by the average interval of time between reflections. The above series, which is an ordinary geometric progression, may be written

$$\frac{p}{v} E \frac{1 - \left(1 - \frac{a}{s} \right)^n}{1 - \left(1 - \frac{a}{s} \right)} ; \tag{1}$$

$\frac{a}{s}$ is by nature positive and less than unity. If n is very large or if $\left(1 - \frac{a}{s} \right)$ is small this may be written

$$\frac{pEs}{va} = \text{the total energy in the room in the steady condition.} \tag{2}$$

$$I_1 = \frac{pEs}{avV} \tag{3}$$

is the average intensity of sound in the room as the organ pipe stops. Substituting in this equation the values of a and p already found,

$$a = \frac{KV}{T}, \tag{4}$$

and $$p = \frac{va}{sA} = \frac{vKV}{sAT}, \tag{5}$$

we have
$$I_1 = \frac{vKV}{sAT} \cdot \frac{T}{KV} \cdot \frac{Es}{vV} = \frac{E}{AV} . \tag{6}$$

Also
$$I_1 = log_e^{-1} At_1, \tag{7}$$

whence
$$E = VA \, log_e^{-1} At_1, \tag{8}$$

where the unit of energy is the energy of minimum audibility in a cubic meter of air.

It remains to determine K and a. To this end the four organ pipe experiments must be made in a room with the windows closed and with them open, and the values of A' and A'' determined. The following analysis then becomes available:
$$a = \frac{KV}{T'}, \text{ and } a + w = \frac{KV}{T''},$$

whence
$$\frac{a}{a+w} = \frac{T''}{T'} .$$

For standard conditions in regard to initial intensity
$$A' \, T' = A'' \, T'' = log_e I = log_e (10^6) = 13.8,$$
$$\frac{T''}{T'} = \frac{A'}{A''}, \text{ and } T' = \frac{13.8}{A'} .$$

Substituting these values,
$$\frac{a}{a+w} = \frac{A'}{A''}, \, K = \frac{aT'}{V} = \frac{a \, 13.8}{A' \, V};$$

whence
$$a = \frac{A' \, w}{A'' - A'}, \tag{9}$$

and
$$K = \frac{13.8w}{V (A'' - A')} . \tag{10}$$

Or if K has been determined (9) may be written
$$a = \frac{A' \, KV}{13.8}, \tag{11}$$

a useful form of the equation.

From equation (1) and (2) we may calculate the rate of growth of sound in the room as it approaches the final steady condition.

Thus, dividing (1) by (2), the result, $1 - \left(1 - \frac{a}{s}\right)^n$, gives the intensity at any instant $n\frac{p}{v}$ seconds after the sound has started, in terms of the final steady intensity. Of all the rooms so far experimented on, the growth of the sound was slowest in the lecture-room of the Boston Public Library in its unfurnished condition. For this room $\frac{a}{s} = .037$, and $p = 8.0$ meters. The following table shows the growth of the sound in this room, and the corresponding number of reflections which the sound that first issued from the pipe had undergone.

LECTURE-ROOM, BOSTON PUBLIC LIBRARY

n	Time	Average Intensity	n	Time	Average Intensity
1	.02	.04	30	.69	.68
5	.11	.17	40	.92	.78
10	.23	.31	50	1.15	.85
15	.34	.43	100	2.30	.98
20	.46	.53	150	3.45	.997
			∞	∞	1.00

It thus appears that in this particular room the organ pipe must sound for about three seconds in order that the average intensity of the sound may get within ninety-nine per cent of its final steady value. As throughout this work we are concerned only with the logarithm of the initial intensity, ninety-nine per cent of the steady condition is abundantly near. This consideration — the necessary length of time the organ pipe should sound — is carefully regarded throughout these experiments. It varies from room to room, being greater in large rooms, and less in rooms of great absorbing power.

To determine the value of E, the rate of emission of sound by the pipe, formula (8), $E = VA \ log_e^{-1} At_1$, is available. It is here to be observed that as this involves the antilogarithm of At_1 these quantities must be determined with the greatest possible accuracy. The first essential to this end is the choice of an appropriate room. Without giving the argument in detail here, it leads to this, that the best rooms in which to experiment are those that are large in volume and have little absorbing power. In fact, for this purpose, small rooms are almost useless, but the accuracy of the result increases rapidly with an increase in size or a decrease in absorbing power. On this account the lecture-room of the Boston Public Library in its unfurnished condition was by far the best for this determination of all the available rooms. Inserting the numerical magnitudes obtained in this room in the equation,

$$E = VA \, log_e^{-1} \, At_1 = 2{,}140 \times 1.59 \, log_e^{-1} \, (1.59 \times 8.69) = 3{,}400{,}000{,}000.$$

If the observations in the same room after the introduction of the felt, already referred to, are used in the equation the resulting value of E is 3,200,000,000. The agreement between the two is merely fortunate, for the second conditions were very inferior to the first, and but little reliance should be placed on it. In fact, in both results the second figures, 4 and 2, are doubtful, and the round number, 3,000,000,000, will be used. It is sufficiently accurate.

The next equation of interest is that giving the value of K, number (10). It contains the expression, $A'' - A'$, the difference between the rates of decay with the windows open and with them closed; A'' and A' depend linearly on the difference in duration of the residual sound with four organ pipes and with one, and as both sets of differences are at best small, it is evident that these experiments also must be conducted with the utmost care and under the best conditions. The best conditions would be in rooms that are large, that have small absorbing power, and that afford window area sufficient to about double the absorbing power of the room. Practically this would be in large rooms that are of tile, brick, or cement walls, ceiling and floor, and have an available window area equal to about one-thirtieth of the total area.

The lobby of the Fogg Art Museum, although rather small, best satisfied the desired conditions. Sixteen organ pipes were used, arranged four on each air tank and, therefore, near together. Thus arranged, the sixteen pipes had 7.6 times the intensity of one, as determined by a subsequent experiment in the Physical Laboratory. The following results were obtained:

$$A' = \frac{log_e \, 7.6}{t'_{16} - t'_1} = \frac{log_e \, 7.6}{5.26 - 4.59} = 3.0,$$

and

$$A'' = \frac{log_e \, 7.6}{3.43 - 3.00} = 4.7.$$

$$K = \frac{13.8w}{V(A'' - A')} = \frac{13.8 \times 1.85}{96 \times 1.7} = .156.$$

Here, however, it is easy to show by trial that errors of only one-hundredth of a second in the four determinations of the duration of the residual sound would, if additive, give a total error of twenty per cent in the result.

It is impossible, especially with open windows, to time with an accuracy of more than one-hundredth of a second, and, therefore, this formula,

$$K = \frac{13.8w}{V(A'' - A')},$$

while analytically exact and attractive in its simplicity, is practically unserviceable on account of the sensitive manner in which the observations enter into the calculations.

The following analysis, however, results in an equation much more forbidding in appearance, it is true, but vastly better practically, for it involves the data of difficult determination only logarithmically, and then only as part of a comparatively small correcting term. For the room with the windows closed:

$$A' t'_1 = log_e I'_1;$$

and for standard conditions in regard to initial intensity

$$A' T' = log_e I,$$

whence

$$T' = t'_1 - \frac{1}{A'} log_e \frac{I'_1}{I};$$

$$T' a = KV,$$

hence

$$KV = t'_1 a - \frac{a}{A'} log_e \frac{I'_1}{I};$$

and similar steps for the same room with the windows open give

$$KV = t''_1 (a + w) - \frac{(a + w)}{A''} log_e \frac{I''_1}{I}.$$

Multiplying the first of the last two equations by t''_1, and the second by t'_1,

$$K = \frac{1}{(t'_1 - t''_1) V} \left[wt'_1 t''_1 + \left(\frac{at''_1}{A'} log_e \frac{I'_1}{I} - \frac{(a + w)t'_1}{A''} log_e \frac{I''_1}{I} \right) \right].$$

By equation (5)

$$\frac{a}{A'} = \frac{sp}{v};$$

and similarly

$$\frac{a+w}{A''} = \frac{sp}{v}.$$

Substituting these values in the above equation,

$$K = \frac{1}{(t'_1 - t''_1)V}\left[wt'_1t''_1 + \frac{sp}{v}\left(t''_1 \log_e \frac{I'_1}{I} - t'_1 \log_e \frac{I''_1}{I} \right) \right]. \quad (12)$$

As an illustration of the application of the last equation, the case of the lobby of the Fogg Art Museum is here worked out at length.

$$t'_1 = 4.59$$
$$t''_1 = 3.00$$
$$V = 96 \text{ cu. m.}$$
$$s = 125 \text{ sq. m.}$$
$$w = 1.86$$
$$a = \frac{.171\,V}{t'_1} = 3.58 \text{ as a first approximation}$$
$$p = 2.8$$
$$I'_1 = \frac{pEs}{vaV} = 8.8 \times 10^6\, i'$$
$$I''_1 = \frac{pEs}{v(a+w)\,V} = 5.8 \times 10^6\, i'$$

Substituting these values in the above equation,

$$K = \frac{1}{152}[25.7 + 1.02\,(6.53 - 8.1)] = .169 - .010 = .159,$$

where the term .169 is the value of K that would be deduced disregarding the initial intensity of the sound, $-$ .010 is the correction for this, and .159 is the corrected value of K. The magnitude as well as the sign of this correction depends on the intensity of the source of sound, the size of the room and the material of which it is constructed, and the area of the windows opened. This is illustrated in the following table, which is derived from a recalculation of all the rooms in which the open-window experiment has been tried, and which exhibits a fairly large range in these respects:

Room	V	I'_1	w	Uncorrected	Correction	K
Lobby Fogg Museum.	96	8,800,000	1.86	.169	$-.010$	.159
Lobby Fogg Museum, 16 pipes. . .	96	67,000,000	1.86	.191	$-.027$	.164
Jefferson Physical Laboratory 15 .	202	1,700,000	5.10	.164	$+.005$	.169
Jefferson Physical Laboratory 1 . .	1,630	390,000	12.0	.150	$+.017$	.167
Jefferson Physical Laboratory 41 .	1,960	300,000	14.6	.137	$+.024$	.161

Average value of K = .164

The value, K = .164, having been adopted, interest next turns to the determination of the absorbing power, a, of a room. For this purpose we have choice of three equations, two of which have already been deduced, (9) and (11),

$$a = \frac{A' w}{A'' - A'},$$

and

$$a = \frac{A' KV}{13.8},$$

and a third equation may be obtained as follows:
It has been shown that

$$T' = t'_1 - \frac{sp}{va} \log_e \frac{I'_1}{I},$$

and

$$T'a = KV.$$

Therefore

$$at'_1 - \frac{sp}{v} \log_e \frac{I'_1}{I} = KV,$$

and

$$a = \frac{1}{t'_1} \left(KV + \frac{sp}{v} \log_e \frac{I'_1}{I} \right). \tag{13}$$

Of these three equations the first, (9), for reasons already pointed out in regard to a similar equation for K, while rigorously correct, yields a result of great uncertainty on account of its sensitiveness to slight errors in the several determinations of the duration of the residual sound. The second, (11), is very much better than the first, but still not satisfactory in this respect. The third, (13), is wholly satisfactory. It has the same percentage accuracy as t'_1, and the only elements of difficult determination enter logarithmically in a small correcting term.

As an illustration of the application of these equations we may again cite the case of the lobby of the Fogg Art Museum:

by equation (9), $a = \dfrac{3.0 \times 1.86}{4.7 - 3.0} = 3.3$;

by equation (11), $a = \dfrac{3.0 \times .164 \times 96}{13.8} = 3.4$;

by equation (13), $a = \dfrac{1}{4.59} (.164 \times 96 + 1.02 \times log_e\, 8.8) = 3.8.$

The first two are approximate only, the last, 3.8, is correct, with certainty in regard to the last figure.

[*Editor's Note:* Equation 13 reverts to the familiar modern form of the Sabine formula if the decay time is taken to be for a decay of 60 dB, i.e. if $I' = I = 10^6 i'$. Then $T = KV/a$.]

13

Reprinted from pp. 603–605 of V. O. Knudsen, *Architectural Acoustics,*
John Wiley & Sons, Inc., New York, 1932, 617 pp.

A DISCUSSION OF THE TRUE COEFFICIENT OF SOUND-ABSORPTION — A DERIVATION OF THE REVERBERATION FORMULA

BY R. F. NORRIS

" In going over some experimental results which persisted in showing the sound-absorption value of certain materials as more than 100 per cent it seemed advisable to start with the standard assumptions adopted by Wallace C. Sabine and to check through to determine the maximum value of his absorption coefficient (a). If it should be possible to show that this coefficient could reach values greater than unity the experimental results would be justified. A certain amount of doubt was cast on the validity of Sabine's coefficient since his reverberation formula $0.05 \ V/as = t$ did not fit the facts at both limits. If a becomes 0 then t becomes infinite as it should, but if a becomes unity t does not become 0 as it should. For this reason it appeared that the equation did not correctly express the relation between volume, surface, absorption, and period of reverberation.

" Let the following assumptions be made:

E = the average sound intensity in a room at any moment.

E' = the maximum average sound intensity in a room when the source has just ceased.

t_p = the period of reverberation.

$\dfrac{4V}{S}$ = mean free path (Jaeger).

V = volume of room in cubic feet.

S = surface of room in square feet.

t' = time sound takes to travel the mean free path in seconds.

t = total time expressed in units of t'.

α = true coefficient of absorption.

v = velocity of sound in air.

" Assuming the sound source to have just stopped and that the sound is decaying in the room, E and t may be tabulated as follows:

t	E
0	E'
t'	$E'(1 - \alpha)$
$2t'$	$E'(1 - \alpha)^2$
$3t'$	$E'(1 - \alpha)^3$
....	......
$(n - 1)t'$	$E'(1 - \alpha)^{(n-1)}$
nt'	$E'(1 - \alpha)^n$

"The value of E at any moment may be written as:

$$E = E'(1 + \alpha)^{\frac{t}{t'}}. \tag{1}$$

Since t' is the time it takes sound to travel the mean free path it may be expressed as the mean free path divided by the speed of sound, or

$$t' = \frac{4V}{S} \times \frac{1}{v} = \frac{4V}{Sv}.$$

Substituting this value of t' in Eq. (1)

$$E = E'(1 - \alpha)^{\frac{tSv}{4V}}. \tag{2}$$

Now let $(1 - \alpha) = e^x$ from which $x = \log_e (1 - \alpha)$.
Eq. (2) now becomes

$$E = E'e^{\frac{xtSv}{4V}}. \tag{3}$$

Or substituting $\log_e (1 - \alpha)$ for x

$$E = E'e^{\log_e \frac{(1-\alpha)vSt}{4V}}, \tag{4}$$

Eq. (4) is the equation for the decay of sound.
"Compare (4) with the equation developed by Sabine,

$$E = E'e^{\frac{-avSt}{4V}},$$

and it is evident that Sabine's coefficient (a) has been replaced by $-\log_e (1 - \alpha)$; otherwise the equations are identical. From this it is evident that Sabine's coefficient (a) is related to the true absorption (α) by the equation $a = -\log_e (1 - \alpha)$.

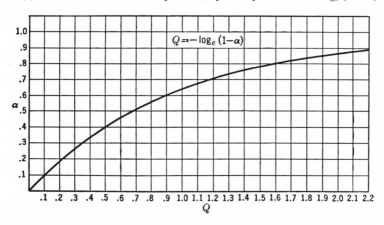

"Substituting this value for a in Sabine's reverberation formula it becomes

$$t_p = \frac{0.05V}{-\log_e (1 - \alpha)S}.$$

"This expression fits at both limits, for when α becomes 0, t_p becomes infinite; and when α becomes unity, t_p becomes 0, as it patently should.

" In the accompanying curve Sabine's (a) has been plotted against the corresponding values of α in order to allow the true values of absorption to be read off from the Sabine coefficient.

"Sabine in assuming that $a = $ unity for a perfect absorber has actually assumed a value of approximately 63 per cent as perfect absorption. From the curve it is evident that values of (a) may be obtained which range between 0 and α, whereas the true absorption α will vary only between 0 and unity."

14

Reprinted from pp. 217–234 of *J. Acoust. Soc. Am.* 1:217–235 (Jan. 1930)

REVERBERATION TIME IN "DEAD" ROOMS

By Carl F. Eyring

Bell Telephone Laboratories

Introduction

With the advent of radio broadcasting and sound pictures very "dead" rooms have been built, and the significant problem of just how much reverberation should be used in broadcasting and recording presents itself. The direct measurement of reverberation time or its calculation by the aid of a reliable formula, then, is an important aspect of applied acoustics. A reverberation time formula enables one to calculate the reverberation time once the volume, surface area and average absorption coefficient of the surface of the room are known; or if the reverberation time is measured it enables one to calculate the average coefficient of absorption of the surface treatment. A correct reverberation time formula is, therefore, much to be desired.

Theories of reverberation leading to Sabine's reverberation time equation have been given by W. C. Sabine (1900),[1] Franklin (1903),[2] Jaeger (1911),[3] Buckingham (1925).[4] Recently Schuster and Waetzmann (1929)[5] have pointed out that Sabine's formula is essentially a "live" room formula and they have shown as we also show that the reverberation time equation varies somewhat with the shape of the room. The present paper presents an analysis based on the assumption that image sources may replace the walls of a room in calculating the rate of decay of sound intensity after the sound source is cut off, which gives a form of reverberation time equation more general than Sabine's; it points out the difference between the basic assumptions leading to the two types of formulae; it adds experimental data which support the more general type; and it ends with the conclusion that no one formula without modification is essentially all inclusive.

Reverberation Time Formulae

Sabine's Formula. If sound is emitted at a constant rate in a room, the sound energy density will build up till an equilibrium is reached

[1] W. C. Sabine, Collected Papers on Acoustics.

[2] Franklin, Phys. Rev. 16, 372 1903.

[3] Jaeger, Wiener Akad, Ber., Math.-Naturw, Klasse, Bd. 120 Abt. IIa. 1911.

[4] Buckingham, Bur. Standards, Sci. Paper, No. 506, 1925.

[5] Schuster and Waetzmann, Ann. d. Phys. March 1929; also Textbook by Muller-Pouillet, Vol. on Acoustics Chapter VII, pp. 456–460.

between the energy emitted and the energy absorbed. When the source is turned off, this energy density will drop off at a rate depending on the absorbing power of the walls and fixtures. Early in his researches Sabine found "the general applicability of the hyperbolic law of inverse proportionality" between reverberation time and absorbing power as given by the relation

$$T = \frac{KV}{a} \tag{1}$$

where T is the duration of residual sound, V the volume of the room, and a the absorbing power of the walls. Standardizing reverberation time as the time required for the intensity of sound to drop to one millionth of its value, Sabine determined the constant K, and obtained in English units,

$$T = \frac{0.05V}{a} = \frac{0.05V}{S\alpha_a} \tag{2}$$

where S is the surface of the room and α_a is the average coefficient of absorption defined by the relation

$$\alpha_a = \frac{s_1\alpha_1 + s_2\alpha_2 + s_3\alpha_3 + \cdots}{s_1 + s_2 + s_3 + \cdots} = \frac{\sum s\alpha}{S} \tag{3}$$

where $s_1, \alpha_1, s_2, \alpha_2$, etc., are the elements of surface and the corresponding absorption coefficients, and S is the total surface. This method of averaging assigns equal weights to the elements of surface, which means that it tacitly assumes a perfectly diffuse condition of the energy density. If, on the other hand, an ordered condition exists, proper weights will need to be assigned to the elements of surface.

Although Sabine tested his reverberation time formula for rooms of various shapes and volumes, ranging from a "small committee room to a theatre having a seating capacity of nearly fifteen hundred," he did not increase the absorption power to the extent that the rooms became "dead," the majority of reverberation times ranging from 4 sec. to 1.5 sec. Thus Sabine did not test his formula in "dead" rooms and no doubt he did not expect it to have a meaning for the extreme case when the average coefficient is unity, for under this condition there can be no hang over of sound and hence no reverberation time, except as one wrongly calls the time for sound to travel from the source to the observer a reverberation time. Yet formula (2) does not become zero but simply reduces to $T = 0.05V/S$ for this extreme case of absorption, and the reverberation time, strange as it may seem, becomes a function of the shape of the room, and reduces to zero only when S becomes very much greater than V. What meaning the equation has for the out-of-doors is certainly very vague. It is evident that the formula fails for this extreme case, and it seems natural to test its validity also for very "dead" rooms.

Recent experiments in the Sound Stage, Sound Picture Laboratory, Bell Telephone Laboratories, indicate the failure of Sabine's formula when the average absorption coefficient for a room is rather high—above 0.5. The results are recorded in another part of this paper.

The More General Formula. Although Sabine's formula had its beginning in the experimental study of "live" rooms, it is also derivable from a theorectical study. A careful analysis of the theoretical basis of reverberation theory is, therefore, important in the development of a new formula.

This necessary analysis is aided by the method of images. Just as a plane mirror produces an image of a source of light, so also will a reflecting wall with dimensions large as compared with the wave length of the sound wave produce the image of a source of sound. An image will be produced at each reflection. In a rectangular room, the source images will be discretely located through space. This infinity of image sources may replace the walls of the room, for they will produce an energy density at a point in the room just as if they were absent and the walls were present.

One may picture the building up of the sound as follows. As soon as the source is turned on the infinity of image sources are at that instant all turned on. The walls are imagined removed, and hence at a given place in the former enclosure sound energy will begin to arrive first from the source, then from the first reflection image sources, then from the second reflection image sources, and so on till the energy arrives from the most distant sources.

The decay of sound may be pictured thus. When the source is stopped one may imagine that all the image sources are simultaneously stopped. The first drop will be heralded by the direct wave, then a series of drops by the waves from the first reflection image sources, then drops from the second reflection image sources and so on, and so on, till the contribution of all the remaining image sources is not sufficient to affect the hearing. Thus the decay ends. The effect of all the image sources is a million times greater than the effect of all those located *beyond* a distance cT; this is the meaning of reverberation time T in the new picture, c being the speed of sound in air.

We wish to emphasize that this picture which gives the details of the decay, involving as it does the geometry of the room, the distribution of the absorbing material, the discontinuous nature of the energy decay, and interference phenomena, is fundamental in developing reverberation theory. Just how many details we shall represent by averages depends upon the degree of simplification desired. We shall ignore interference phenomena in the following considerations, and accordingly we shall attempt to make reverberation time meter measurements independent of this effect. We shall, following Sabine's experimental results, assume that α_a defined by equation (3) may be considered as the uniform absorption coefficient of the walls; yet we realize that for an ordered condition of the sound waves this procedure may need to be modified. We shall apply the method of images first to a few special cases involving rooms of simple geometrical form in order that later we may with better understanding approach the general case.

Consider a spherical room of diameter D and volume V, with uniform absorption over the surface and with a sound source emitting $\overline{E}$ energy units per second located at its center. In a time $D/2c$ sec. after the source is turned on, the sphere will be filled with energy directly from the source of average density $\overline{E}D/2cV$, and then a first reflection will begin. The reflected wave will travel back to the center, then spread out and in a time D/c sec. after the reflection first started a second reflection will begin. The average density due to the first reflected energy then is $\overline{E}DR/Vc$ where R is the coefficient of reflection defined by the relation

$$R = (1 - \alpha_n). \tag{4}$$

After another D/c sec. interval, the third reflection will begin and the average energy density of the second reflected energy is, $\overline{E}DR^2/Vc$. As time proceeds, at the end of each D/c interval, a new reflection begins and a new increment will have been added to the total average density of the enclosure. Finally equilibrium is established and the total average energy density is given by the infinite sum

$$S_\infty = \frac{\overline{E}L}{Vc}\left(\frac{1}{2} + R + R^2 + R^3 + \cdots R^n + \cdots\right). \tag{5}$$

When the source is turned off, the decay begins. At a point of observation near the center the energy directly from the source stops, then in D/c seconds the average energy represented by the second term of equation (5) vanishes; in D/c seconds more the third term vanishes, etc., etc. The total average energy at any D/c interval during the decay will be given by the equation

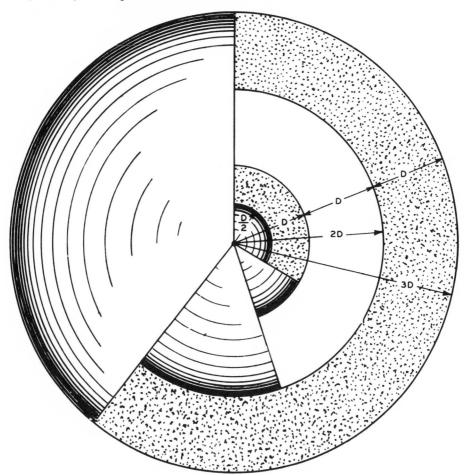

Fig. 1. *Illustrating a source at the center of a spherical room and a series of spherical concentric image sheets.*

$$S_\infty - S_n = \frac{\overline{E}DR^n}{Vc(1-R)} = \frac{6\overline{E}}{S\alpha_a c} e^{\dfrac{cS \log_e (1-\alpha_a)t}{6V}} \tag{6}$$

remembering that $n = ct/D$, that $R = (1-\alpha_a)$, and that for a sphere $D = 6V/S$. We must keep in mind that t is not a continuous function but given by, $t = nD/c$ where n is an integer, and also that the absorbing material is assumed uniformly distributed over the surface.

One may interpret the process just described by the aid of the method of images. The walls are imagined removed and image sources are substituted. These images are imagined to be spherical concentric sheets. (See Figure 1). They are separated by a distance D, the distance between two successive reflections. The first image sheet contributes to the enclosure the average energy density given by the second term of

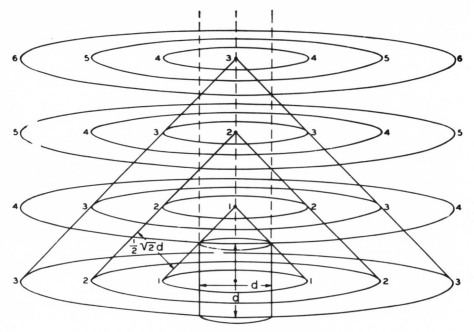

FIG. 2. *Illustrating a source at the center of a cylindrical room and a series of concentric circular line image sources.*

equation (5), the second one the amount given by the third term, etc., etc. When the source stops emitting energy all the image sheets do likewise. The persistence of sound is due to the fact that it takes time for the end of a sound wave which marks the termination of energy emission from a particular image sheet to reach the enclosure. The decay goes down in steps; due to the definition of reverberation time the end of the wave from a sheet located at a distance cT from the source passes thru the enclosure at the instant the intensity reaches one millionth of its original value.

Next consider a cylindrical room with diameter d equal to length, with uniform absorption over the surface and with a source at its center. Using the method of images we note that the source may be imagined imaged in the cylindrical walls as a series of concentric circular line sources, and in the plane ends as a series of point sources, which in turn are imaged in the projected cylinder as concentric line image sources. (See Figure 2.) The numbers indicate the order of the reflection and if images of the same order are included in a surface, we shall have a series of concentric cones separated by a distance $\frac{1}{2}\sqrt{2}d$. If we identify this as the average distance between reflections[6] we may replace D

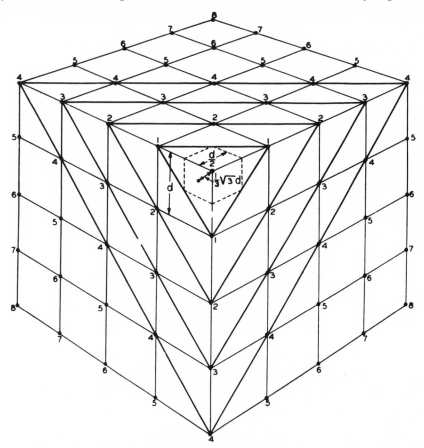

Fig. 3. *Illustrating a source at the center of a cubical room and a series of point image sources.*

of equation (6) by this value, and remembering that $d = 6V/S$ for an enclosure of this sort we rewrite (6) and obtain the following decay equation for a point near the center of this cylindrical room,

$$S_x - S_n = \frac{3\sqrt{2E}}{S\alpha_a c} e^{\dfrac{cS \log_e (1-\alpha_a)t}{3\sqrt{2}V}}. \tag{7}$$

The same method may be applied to a cubical room with the source

[6] Schuster and Waetzman loc. cit. have obtained these values by a more rigorous method.

at its center. The images are point sources and are located discretely throughout space as shown in Figure 3. As before the numbers indicate the order of reflection, and if the image sources of the same order are included in a given surface we get a series of plane surfaces separated by the distance $\frac{1}{3}\sqrt{3}\,d$, where d is the length of the cubical room. Identifying this distance as the average distance between reflections[6] and remembering that for a cube $d = 6V/S$, we have from equation (6) the

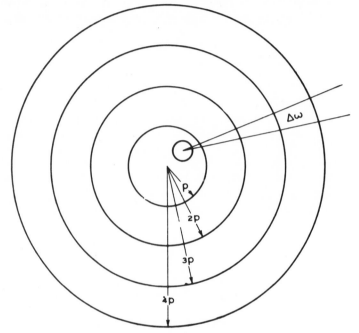

Fig. 4. *Illustrating zones of image sources.*

following decay equation for a point near the center of the cubical room,

$$S_\infty - S_n = \frac{2\sqrt{3}\,\overline{E}}{S\alpha_a c}\, e^{\frac{cS \log_e (1-\alpha_a)t}{2\sqrt{3}\,V}}. \qquad (8)$$

Formula (7) is valid for a cubical room with a source at its center provided two opposite walls are completely absorbing. The α_a in the coefficient of the exponential term is the average value for the whole room, but the α_a in the exponent of this term is the average value for the four reflecting walls. Again if the cubical room has two pairs of opposite walls completely absorbing equation (6) is valid, but just as before the proper values for α_a must be used.

So far we have considered rooms of very simple geometrical shape and we have located the source at the center in each case. We now consider the more general case, make the shape more complex, move the source from the center and thus approach a diffuse condition of the waves. In the interest of simplicity we shall be content with average effects and assume that the image sources are discretely located in zones—surfaces of concentric spheres with radii, p, $2p$, $3p$, $\cdots np$, $\cdots$ etc., where p is the mean free path between reflections (See Figure 4). The mean free path is the average distance between reflections which the sound establishes on its many trips across the room from wall to wall, from wall to ceiling, from ceiling to floor, etc., as it travels in all conceivable directions. At present, the centers of these spheres are assumed located at the source of sound.

This means that we have replaced the source of sound and the walls of the room, by the source of sound surrounded by image sources located in evenly spaced discrete zones. To determine the energy density produced in the enclosure by any zone, we proceed as in the case of the spherical room and follow through the growth of sound in the room as follows.[7] The rate of energy emission is denoted by $\overline{E}$, c is the speed of sound, p/c is the average interval of time between reflections, and R is the coefficient of reflection already defined. The amount of energy emitted into the room directly from the source during this interval is $p\overline{E}/c$; the amount left over after the first reflection, or the amount emitted into the enclosure by the first zone of image sources during the same interval of time, is $p\overline{E}R/c$; and that emitted in the same time into the enclosure by the nth zone is $p\overline{E}R^n/c$. Because of the meaning of the mean free path p, it takes just this interval of time for the energy flow to fill the enclosure once it starts to enter it. At once the energy density in the enclosure due to the nth zone is

$$E_n = \frac{p\overline{E}}{cV}R^n. \tag{9}$$

It now becomes interesting to determine whether E_n is an average value or is constant for all parts of the enclosure. The sound intensity varies inversely with the square of the distance from the source or its images. The intensity of the direct sound, then simply obeys the inverse square law. To investigate the point to point variation within the enclosure of the sound energy density due to the various zones we proceed as follows. For simplicity neglect differences in phase and assume that

[7] Collected papers, Sabine, p. 43. Vibrating Systems and Sound, Crandall, p. 201.

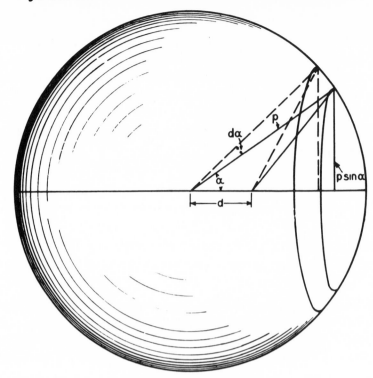

the image sources are evenly distributed over the zones. Call the surface density of image sources i. Then for the first zone, the radius of the sphere is p, and if we let d be the distance of the point of observation from the center of the sphere, then the energy density at the point of observation is given by the relation (See Figure 5),

$$E_1 = M i \int_0^\pi \frac{2\pi p^2 \sin \alpha d\alpha}{p^2 - 2pd \cos \alpha + d^2} = \frac{Kp}{d} \log_e \frac{p+d}{p-d} \qquad (10)$$

where M and K are constants.

Now expanding $\log_e (p+d) (p-d)$ we get

$$E_1 = 2K \left[1 + \frac{1}{3} \left(\frac{d}{p} \right)^2 + \cdots \right]$$

and for any zone

$$E_n = 2K_n \left[1 + \frac{1}{3} \left(\frac{d}{n_i'} \right)^2 + \cdots \right]. \qquad (11)$$

For all but the first zone, $(d/np)^2$ is small and E_n is essentially independent of d, the distance of the point of observation from the center of the sphere. This means that neglecting interference any zone beyond the first, viz. the rth zone, contributes approximately the same energy to all points in the enclosure, and therefore, that the energy density produced by this zone is constant throughout the room, and of a magnitude E_r as given by equation (9). However, E_0 and E_1 are average values, the actual point to point variation of the energy density being obtained for each case by the use of the inverse square law and the method of images. The actual energy density at a point in the enclosure, then, may be divided into three parts: (1) that which is produced by the source only and this varies inversely as the square of the distance from the source, (2) that which is produced by the first reflection image sources, and the point to point variation of intensity may be calculated for each particular enclosure; (3) that which is produced by all the remaining image sources, a term with a magnitude $\sum_2^\infty E_n$, which shows no point to point variation.

In the following development of the sound intensity decay equation, we assume that the average densities E_0 and E_0R may be substituted for the actual densities produced at the point of observation by the source and the first reflection image sources respectively. Because of this the equation developed will not give an exact statement of the decay during the first two (p/c) sec. intervals, but this is not important because from then on it will give the correct statement and in good experimental practice the rate of decay obtained need never depend upon the changes during these first two intervals.

Hence when sound is established in the enclosure, the average density due to the source and all the zones of image sources is $S_\infty = \sum_0^\infty E_n$. Remembering equation (9), we have $E_0 = p\bar{E}/cV$ and at once

$$S_\infty = E_0(1 + R + R^2 + R^3 + \cdots R^n + \cdots).   \tag{12}$$

When the source is stopped all the image sources stop, and, of course, the energy supply also. At the source, or very near it, the portion E_0 of the total energy at once vanishes. The intensity then remains constant for an interval p/c seconds, the time it takes sound to travel from the first zone to the point of observation announcing the stopping of the energy supply of this zone. At the close of this interval the portion E_0R abruptly vanishes. This is followed by another interval of constant intensity. Then a sudden disappearance of the portion E_0R^2 occurs. By this time the sound has traveled from the second zone to the point of observation and announces the stopping of the energy supply of that zone. Thus the decay continues and remembering that equation (12) is a geometric series, the energy density during the nth interval is

$$S_\infty - S_n = \frac{E_0}{1-R} - \frac{E_0(1-R^n)}{1-R} = \frac{E_0 R^n}{1-R}.$$

(13)

We developed this equation on the assumption that the point of observation was very near the source, which was also the center of the zones. There seems to be no *a priori* reason why the center should be chosen at the source rather than at some other point in the enclosure, but once the center is located it does not seem legitimate to move it about without a careful investigation of what would happen if it were not moved to each point of observation.

Hence we select for study some point of observation not at the center of the zones. Waves leaving a given zone at the same instant will not arrive simultaneously at this new location. This means that unless one judges the energy density at a point by the flow of energy out in a *given direction*, equations (12) and (13) will not give a correct picture. But the decay history at a point as witnessed in a *given direction* is given exactly by equation (13). We propose to determine the decay history of the energy density by *adding up the complete decay histories of all the directions*; we shall not obtain the decay history by adding the instantaneous values of the directional histories, and then tracing the history of this instantaneous sum. Therefore, let e_0, e_1, e_2, etc. be the energy density due to the image sources, included within any solid angle $\Delta\omega$. (See Figure 4). Then

$$e_0 = \frac{\Delta\omega}{4\pi} E_0, \quad e_1 = \frac{\Delta\omega}{4\pi} E_1, \text{ etc.}$$

By equation (13)

$$S_\infty = \sum_0^\infty e_n = \frac{e_0}{1-R}$$

$$S_n = \sum_0^n e_n = \frac{e_0(1-R^n)}{1-R}$$

and

$$S_\infty - S_n = \frac{e_0 R^n}{1-R} = \frac{\Delta\omega}{4\pi} \frac{E_0 R^n}{1-R}.$$

(14)

Now "n" is an integer and simply denotes during which p/c sec. interval the intensity is measured. Except in extreme cases, waves which left simultaneously from a given zone will all arrive within the interval p/c sec. after the first arrival takes place. Hence we are justified in adding up equations (14) for all directions keeping "n" constant. At once

$$\sum(s_\infty - s_n) = \frac{E_0 R^n}{4\pi} \frac{\sum \Delta\omega}{(1-R)} = \frac{E_0 R^n}{1-R}. \tag{15}$$

That this summation is legitimate, even for the extreme cases where a slightly greater interval than p/c sec. is needed to have all the waves arrive, is illustrated by Fig. 6, in which it is clear that the staggering effect simply smooths out the distinct drops of the *total* energy but does not change the slope of the decay. From this it becomes clear that so far as the energy density at a point is concerned, "n" which was introduced

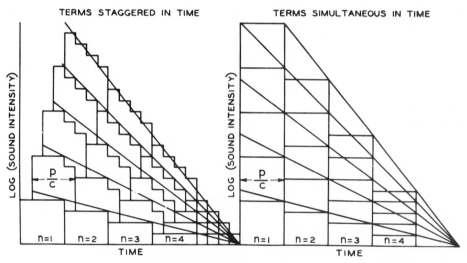

Fig. 6. *Illustrating that the staggering effect simply smooths out the distinct drops but does not change the slope.*

as an integer may after the summation is over, be considered, approximately at least, a continuous function. This method of tracing the "life history" (equation 14) for each particular direction and then adding the "life histories" is essentially the uniqueness of the new analysis.

In the old theory during the decay the energy density is assumed strictly continuous, but we have found this to be approximately true. The difference does not lie here. The old theory also assumes that the energy density at a point appears to have a continuous change as viewed from an element of surface. It tacitly assumes that an absorbing surface has the ability to register back into the oncoming wave the decrease in intensity which it is causing. This is obviously not true; this particular absorption is registered in the outgoing wave, not in the oncoming wave. During decay the surface will receive *from a given direction* a constant energy flow for a time equal to the time it takes sound to travel the mean free path between reflections; then there will be an abrupt change. *This constant energy flow followed by an abrupt drop, rather than a continuous drop to this same level, means a greater absorption during the same interval of time and hence a more rapid decay of the sound.* (See Figures 7 and 8). This is in essence the physical difference between the foundation on which Sabine's formula has been erected, and the foundation on which we propose to erect a more general type of formula. The new type of equation must give Sabine's formula as a special case for "live" rooms and must do so simply because under these circumstances the two foundations become equivalent but not because the new picture ceases to hold for "live" rooms.

Rewriting equation (15) and remembering that

$$n = \frac{ct}{p} \tag{16}$$

we have

$$\sum (s_x - s_n) = \frac{E_0 R^{ct/p}}{1 - R}$$

and remembering that

$$E_0 = \frac{p\overline{E}}{cV}$$

we have

$$\sum (s_x - s_n) = \frac{p\overline{E}}{cV(1-R)} e^{\frac{c(\log_e R)t}{p}} \tag{17}$$

and from what we have said above we are justified in assuming that t is a continuous function.

We now have to determine p, the mean free path between reflections. Sabine[1] obtained the experimental value, $p = 0.62(V)^{1/3}$, which reduces to $p = 3.7V/S$ for enclosures of reasonably compact proportions. Jaeger[3] using the theory of probability and assuming a perfect diffuse condition of the sound waves calculated the number of reflections per second in the same manner that the impacts of molecules are estimated in the kinetic theory of gases, and found $p = 4V/S$. Recently Schuster and Waetzmann[5] have shown as we have also shown in the first part of this paper that for rooms of special shapes where an ordered not a diffuse condition of the sound waves is attained the mean free paths are as follows: cubical room, $p = 2\sqrt{3}V/S$; cylindrical room with length equal diameter, $p = 3\sqrt{2}V/S$; spherical room, $p = 6V/S$. Probably the formula which will be used most is one based on a perfect diffuse condition of the waves, a special formula being written down for each enclosure which cannot fulfill this condition. To obtain the formula for the diffuse condition we shall use the value obtained by Jaeger for the mean free path. That this value may be used for "dead" rooms follows from the fact that Jaeger's analysis does not in any way involve the absorption coefficients of the surfaces. Of course, we must demand that the "dead" room shall produce a diffuse condition among the sound waves; the assumption that the image sources are located on the surfaces of concentric spheres tacitly implies this diffuse state.

Substituting the value

$$p = \frac{4V}{S} \tag{18}$$

in equation (17) we get the decay equation

$$\rho = \rho_0 e^{\frac{cS \log_e (1 - \alpha_a)t}{4V}} \tag{19}$$

where

$$\rho_0 = \frac{4E}{cS\alpha_a} \tag{20}$$

As we have already pointed out, for practical and experimental purposes, t may be considered continuous, and hence (19) may be thought of as derivable from a differential equation.[*] In a very "live" room (and in this analysis we may make the room as "live" as we please) the absorption per reflection is so very small that the energy density as viewed from an element of surface can be considered continuous. This means that we may follow the old theory[4] and write down the differential equation for the sound intensity in a closed room as follows:

$$V\frac{\partial \rho}{\partial t} + \frac{1}{4}c\alpha_a S\rho = E \tag{21}$$

[*] Schuster and Waelzmann, loc. cit., have made use of a differential equation involving velocity potentials.

where V is the volume of the room, S its surface, α_a the average coefficient of absorption, c the speed of sound in air, ρ the energy density and $\bar{E}$ the rate of sound energy emission, which is considered constant.[9]

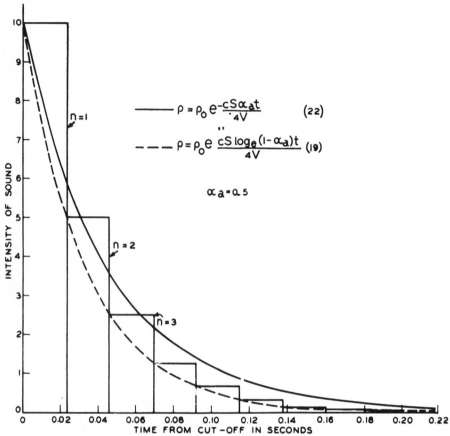

FIG. 7. *Illustrating rate of sound decay in a room with volume 80400 cu. ft., surface 12180 sq. ft. and walls of average coefficient of absorption 0.5.*

The solution of this equation under proper boundary conditions gives the following for the equation of decay

$$\rho = \rho_0 e - \frac{c\alpha_a S}{4V} t \qquad (22)$$

and from this Sabine's reverberation time equation is obtained. Now

$$\log_e R = \log_e (1 - \alpha_a) = -[\alpha_a + \frac{\alpha_a^2}{2} + \frac{\alpha_a^3}{3} + \cdots] \qquad (23)$$

and when α_a is very small as is true for *very live* rooms,

$$\log_e (1 - \alpha_a) = -\alpha_a.$$

[9] Vibrating Systems and Sound, Crandall, p. 207.

This means that (22) is a special case of (19). The value of ρ_0 is the same for both equations and may be derived from differential equation (21) as the maximum value reached by the energy density as a steady state is established. This agreement is natural since the steady state value depends simply on the conservation of energy principle, and not upon the rate of growth or decay when the sound source is turned on or cut off. This differential equation cannot give the correct damping factor,

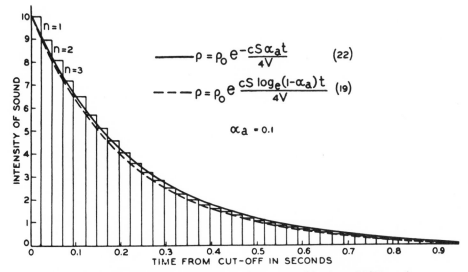

$$\rho = \rho_0\, e^{-\frac{cS\alpha_a t}{4V}} \qquad (22)$$

$$\rho = \rho_0\, e^{\frac{cS\log_e(1-\alpha_a)t}{4V}} \qquad (19)$$

$$\alpha_a = 0.1$$

FIG. 8. *Illustrating rate of sound decay in a room with volume 80400 cu. ft., surface 12180 sq. ft. and walls of average coefficient of absorption 0.1.*

because as we have already pointed out, it does not describe the true nature of the absorption. However, if we put $\overline{E}=0$, substitute $-\log_e(1-\alpha_a)$ for α_a and set down the boundary condition, $\rho=\rho_0$ when $t=0$, we get a differential equation which gives (19) as its solution.

In Figures 7 and 8, we have plotted equation (19) first with t considered continuous, then with $n=ct/p$ considered as an integer. The decay is shown to be greater than that for equation (22), the old decay formula, the difference becoming very marked for "dead" rooms.

To obtain the reverberation time formula involving the true nature of the absorption we make use of equation (19) and write

$$\frac{\rho_1}{\rho_2} = \frac{e^{\dfrac{cS\log_e(1-\alpha_a)t_1}{4V}}}{e^{\dfrac{cS\log_e(1-\alpha_a)t_2}{4V}}} = 10^6$$

and put $(t_2 - t_1) = T$, the reverberation time.

At once

$$\log_e 10^6 = \frac{-cS \log_e (1 - \alpha_a)}{4V} T$$

$$T = \frac{4 \log_e 10^6 V}{cS \log_e (1 - \alpha_a)} = \frac{0.05V}{-S \log_e (1 - \alpha_a)} . \tag{24}$$

From equation (23) it follows at once that for very small values of α_a the reverberation time formula (24) becomes

$$T = \frac{0.05V}{S \alpha_a} = \frac{0.05V}{a} \tag{25}$$

which is Sabine's formula—*a special case of the more general formula* which we have developed.

[*Editor's Note:* Material has been omitted at this point.]

ERRATUM

Page 148, lines 2 and 3 from the bottom should read, "energy density is given by the infinite sum

$$S_\infty = \frac{\bar{E}D}{Vo} \left(\frac{1}{2} + \dots \right)"$$

15

Reprinted from *Acustica* **10**:400–411 (1960)

INTERNATIONAL COMPARISON MEASUREMENTS IN THE REVERBERATION ROOM

by C. W. Kosten

Technological University, Delft, Netherlands

Sommaire

ISO/TC 43 est en train d'établir une norme internationale pour la mesure de l'absorption du son à l'aide de salles réverbérantes. La publication qui suit est le compte-rendu des nombreuses mesures de comparaison internationale entreprises afin de fournir les éléments d'information nécessaires pour l'établissement d'une telle norme.

Les résultats sont les suivants:

1) Il faut un échantillon de 10 m².

2) Un tel échantillon exige une salle plus grande que 180 m³ dans laquelle sont installés de nombreux éléments diffusants de grande dimension permettant d'assurer la parfaite diffusion du champ sonore réverbérant.

3) Probablement les autres différences entre les différents instituts sont dues essentiellement aux effets de bord. Ce dernier point réclame une étude ultérieure.

Zusammenfassung

ISO/TC 43 bereitet eine internationale Normvorschrift hinsichtlich der Messung der Schallabsorption nach dem Nachhallverfahren vor. Diese Veröffentlichung ist ein Bericht über umfangreiche internationale Vergleichsmessungen, ausgeführt mit dem Zweck, eine solide Grundlage für eine derartige Norm zu schaffen. Die Ergebnisse sind:

1. eine Prüfstofffläche von etwa 10 m² ist erwünscht;

2. der Hallraum sollte sodann nicht kleiner sein als etwa 180 m³ und ausgestattet sein mit einer großen Zahl von großen Streuelementen, um zu gewährleisten, daß das abklingende Schallfeld genügend diffus ist;

3. die restlichen Abweichungen zwischen den Laboratorien rühren wahrscheinlich zum größten Teil her von dem Kantenbeugungseffekt. Dieser Punkt bedarf weiterer Untersuchung.

1. Introduction

During the last decade many efforts have been made to arrive at an agreement with respect to an international standard for the measurement of sound absorption in the reverberation room. At the occasion of a meeting of Technical Committee 43 of the International Organization for Standardization, under whose aegis such a standard comes, it was decided that comparison measurements would be carried out in many institutes, all measuring a sample of the same material, Sillan, in order to check the effectiveness of the proposed standard at that time, 1958. The chairman of the working group, Prof. Dr. L. Cremer, organized the test, that we shall call the First Round Robin. The material was put at

the disposal of all participants by Grünzweig and Hartmann, Ludwigshafen. The author undertook the evaluation of the reports of the participating institutes. A first verbal report was brought before the members of the Third ICA-congress, Stuttgart 1959 by the author [1].

The First Round Robin was successful in that respect that it became clear that results could be obtained with reasonable accuracy, results that compare reasonably with the values that were expected from tube measurements, provided extensive measures are taken to guarantee a diffuse reverberant sound field in the room. Many problems were only partly solved, however, such as the dependence of the absorption coefficient upon the surface of the sample, the volume of the room, the degree of diffusivity of the field, etc. It was, therefore, decided to continue and extend the international measurements. This work will be called the Second Round Robin. Participants were asked to measure again, this time as a function of the area of the sample (4, 8 and 12 m²) and as a function of the degree of diffusivity (absent, probably insufficient, probably sufficient, certainly sufficient).

The author, again, undertook the evaluation of the reports. The result of that evaluation was brought to the attention of those present at the meeting of ISO/TC 43/working group 3 at Rapallo in March 1960 and of the participating institutes. In our opinion – and that of many others we daresay – the result was very satisfactory and decisive, so that a reasonable agreement as to a draft-recommendation could be reached.

There are others, however, who are of opinion that we did in Rapallo a very decisive step in an entirely wrong direction, so that the reverberation method finally lost its raison d'être completely.

Since the problem is, in our opinion, of great importance and since also those who did not participate in the international discussions of ISO/TC 43/ WG 3 may be interested in the work, we thought it to be desirable to publish the result of the Second Round Robin. It should be made clear, however, that this publication is of a personal character only. Statements made need not necessarily be supported by ISO/TC 43. The results of the measurements are reproduced here with the kind permission of the participating institutes.

To begin with, we shall discuss in general what we should aim at while standardizing the reverberation method. It is obvious that this point is perhaps liable to the largest divergence of opinion, and is – more than the facts that follow – to be considered as the personal feeling of the author only.

2. What we should aim at with the reverberation method

The sound absorption coefficient of a material can be measured easily and with relatively great accuracy in a standing wave tube. It is considered a disadvantage of this method, however, that it is only suited to measure at normal incidence of sound. In practice normal incidence of sound is usually an exception. Problems like the computation of the reverberation time T of a hall, a classroom, a studio, or the reduction of the noise level in a factory, an office, are frequently dealt with in a statistical way. The sound field is supposed to be diffuse. E. g., for the computation of the reverberation time T a formula is used that is derived on the basis of a diffuse field. The sound absorption coefficient wanted for

these calculations is then necessarily that for random incidence. The situation in cases of noise reduction is frequently similar.

It is considered the great advantage of the reverberation method that it yields this absorption coefficient at random incidence. Let us assume that this is the case indeed; the results that follow corroborate this statement, provided special precautions are taken to assure that the reverberant sound field is diffuse indeed.

Whether this is really an advantage or not is a source of divergence of opinion. Many acousticians, the author included, are of opinion that we should try to measure as precisely as we can the absorption coefficient at random incidence indeed. Diffuseness of the sound field is the basis for the reverberation formula. Moreover it seems feasible to specify how to make a field sufficiently diffuse, whereas a semi-diffuse field is difficult to define.

Others are not unconditionally in favour of the diffuse field as being our ideal. They state that practical sound fields are seldom approximately diffuse, e. g. in a classroom or an office where only the ceiling is highly absorbent, so that α for random incidence would predict a too short reverberation time. They prefer therefore the measurement of α under conditions of diffusivity that are comparable with practice. In our opinion, however, this is like pleading for a reduction of the unit of mass since there is a general tendency to sell less than a unit when a unit is asked for.

Personally I am strongly in favour of maintaining our high standard. We ought to measure a constant of the material, the random incidence coefficient. If sound fields in practice are not sufficiently diffuse to justify the use of this coefficient this means that an unsolved problem faces us. It seems wiser to separate the problems clearly and to solve them individually rather than to mix the problems to an unsolvable aggregate that will keep us busy indefinitely.

Another controversy has arisen with respect to the edge effect. It is well-known that the measured coefficient α increases with decreasing area F of the sample. The free edges of the sample invoke diffraction of the waves, resulting in greater absorption coefficients. When the dimensions of the sample exceed a few wavelengths the increase of the equivalent absorption area A will be approximately proportional to the total length of the free edges, so that

$$\alpha = \alpha_\infty + \beta E \qquad (1)$$

where α_∞ is the coefficient for an infinite sample, E the length of free edge per unit area of the sample, and β a constant, viz. the incremental absorption coefficient per unit of free edge per m² area, viz. per unit of E. Let us call E the relative edge length.

The best scientific approach of the entire problem would undoubtedly be:

i) to measure α_{∞},

ii) to measure β,

iii) to develop methods of calculation for the use of α_{∞} and β in practice, suitable even in cases where the sound field is only semi-diffuse.

It has been stated (MEYER and KUTTRUFF [2], INGERSLEV [3]) that it is possible to measure α_{∞} in the reverberation room by covering the whole floor or one wall with the material under test. This would solve only part of the problems, since β is needed too. Another possible approach is of course to measure α for several values of E and to separate α_{∞} and β with the approximate equation (1). This method has several advantages: it is not necessary to cover the whole floor, it yields α_{∞} and β, and it enables one to measure β for various conditions of the edges, on which β is undoubtedly dependent.

It was accepted by the working group in Rapallo that a standard should aim at the measurement of α in a perfectly diffuse field, and that a reasonable amount of edge effect should be included.

To give an idea of the order of magnitude of β: in the Second Round Robin β was found to be between 0.04 and 0.14. When 12 m² of the material is used the relative edge length is approx. $1.2\ \mathrm{m^{-1}}$, so the increment βE would be between 0.05 and 0.17. It goes almost without saying that βE in practice will depend upon the material in question, the dimensions of the area and the way of application.

3. Material and instructions

The material distributed amongst the laboratories was a kind of rockwool, Sillan, 5 cm thick, $100\ \mathrm{kg/m^3}$ (see [1] for more information). In order not to increase E too much, the material was to be applied in one single area near the middle of the floor or one wall. Measurements were requested with the areas 4, 8 and 12 m².

The results of the First Round Robin [1] had shown clearly that the discrepancies between the institutes were small only when many diffusing elements were present in the room, and that the results, then obtained, were trustworthy since they compared favourably with the results of the tube measurements (after conversion for random incidence and some addition for edge effect). The following instructions for obtaining a diffuse reverberant field were, therefore, given:

"Rooms with flat walls, oblique or not, and rooms with only a few m² of vanes, rotating or not, and arrangements with many loudspeakers and/or microphones

should be dealt with as rooms without any diffusivity at all. It is not at all easy to make a reverberant sound field sufficiently diffuse. Much more is needed than is generally thought. Sufficient diffusivity is probably almost reached when three non-parallel boundaries (floor, walls and/or ceiling) are polycylindric. If you have a room without sufficient diffusivity it is impossible to make three walls polycylindric for the mere benefit of this Round Robin only. Moreover it is not at all certain that 3 polycylindric walls are entirely satisfactory. There is, however, a very simple way of introducing diffusing elements. An adequate diffusing element is a plate of plywood of only a few millimeters thickness and an area between 0.8 and 2 m² (both sides 1.6 to 4 m²). The plate might be slightly curved. Sufficient diffusivity is probably reached by hanging in a thoroughly random way a large number of these plates in your room. *The total area should probably be approximately equal to the surface of the floor* (both sides of the plates together twice the floor surface). In order to measure *as a function of diffusivity* the total surface may be increased, say, from 0.4 to 0.8 and further to 1.2 times the floor surface. Instead of plates (that may perhaps not be available immediately) you can use all other obstacles or elements that are sufficiently large and reflecting, such as tables, wooden chairs, light panels, wooden doors etc. etc. It is, however, certainly insufficient to put a number of these objects on the floor thus 'filling' only the lower half of the room. In doing so you do not diffuse sufficiently *at random*. A way to verify whether or not the elements are hanging at random is the following. The projections of the elements on each boundary (wall, floor or ceiling) should cover almost the same percentage of that boundary. So, if all plates are horizontal, the projection on a vertical wall is small; the plates are not arranged at random.

Fig. 0. A typical reverberation room with diffusing elements as requested for the Second Round Robin.

If they all are hanging vertically, the projection on floor and ceiling is too small. So, in order to get randomness all angles of inclination must occur equally frequently."

A typical room in the "sufficiently diffuse" state is shown in the photograph.

The absorption coefficient was to be computed with the aid of SABINE's formula:

$$T = 0.163 \, V/A \, .$$

4. The participating laboratories

Of the laboratories that received material the following ones sent a report up till now (1st June 1960):

1. Stockholm, Kung. Tekniska Högskolan, Institut för Buggnadsakustik,
2. Dresden, Institut für Elektro- und Bauakustik,
3. Copenhagen, Lydteknisk Laboratorium,
4. Göteborg, Chalmers Tekniska Högskola, Akustiklaboratoriet,
5. Palaiseau, Départment Acoustique du C.N.E.T.,
6. Helsinki, The Finnish Broadcasting Corporation Ltd.,
7. Meudon, Radiodiffusion-Télévision Française,
8. Braunschweig, Physik.-Technische Bundes-anstalt,
9. Beograd, Physical department of the Institute for Testing Materials,
10. Stockholm, The Swedish Institute for Materials Testing,
11. Ottawa, Building Physics Section, National Research Council Canada,
12. Delft, Technisch Physische Dienst TNO-TH.,
13. Berlin, Institut für Technische Akustik der TU,
14. London, British Broadcasting Corporation,
15. Prague (parallel), Research Institute of Sound Picture and Reproduction in collaboration with the Institute of Physics, Electr. Faculty of the Techn. University,
16. Prague (non-par.), ditto,
17. Moscow, Acoustical Laboratory Architectural Design Bureau,
18. Moscow, Acoustical Institute Academy of Sciences USSR,
19. Teddington, National Physical Laboratory, Applied Physics Division.

The reports will be quoted by the reference number.

5. The data (Figures 1 ... 19)

The reports contain so much information that for this concise paper a choice had to be made. The result of this choice is to be found in figures 1 ... 19, each figure bearing the same number as the reference number of the institute. Moreover each figure bears

the name of the city of the institute. The following legend applies to all figures 1 ... 19.

i) The curves represent SABINE's coefficient a_S in percents as a function of frequency for a few areas F and a few states of diffusivity $a \ldots d$;

ii) the curves are labelled and drawn in the following way:

number of curve		F approx.	diffusivity
12a	□━━━━□	12 m²	maximum
8a	□━━━━□	8 m²	maximum
4a	■━━━■	4 m²	maximum
12b	△—·—·—△	12 m²	medium
8b	△—·—·—△	8 m²	medium
4b	▲—·—·—▲	4 m²	medium
12c	○————○	12 m²	poor or absent
8c	⊙————⊙	8 m²	poor or absent
4c	●————●	4 m²	poor or absent
12d	··········	12 m²	absent
8d	··········	8 m²	absent
4d	··········	4 m²	absent

When the area F is not exactly 4 or 8 or 12 m² the number of the curve indicates the exact value, e. g. 8.25 a (see Fig. 2);

iii) the right hand corner at the bottom of each figure contains information about the dimensions and the diffusivity, viz. V = the volume of the room; S = the surface of floor + walls + ceiling; B = the surface of the boundary to which the sample was applied; the drawing represents B and the position of the samples: the numbers near the middle of the boundaries indicate their lengths in m, the numbers at the corners the height at that corner;

iv) the reverberation time T_0 of the empty room is indicated above the frequency scale for the various states of diffusivity (see indication near 1000 Hz);

v) the curves a apply to the state of maximum diffusivity. Since this is in our opinion the most desirable state these curves are drawn heavily.

6. General remarks concerning figures 1 ... 19

The aim of the measurements was not to get acceptable results, but to contribute to a concerted effort to understand the behaviour of reverberation. Many curves were obtained knowing that the results would be "bad". Several institutes reported not to have entirely satisfactory facilities, but participated in order to supply data for a better understanding. So one should be very careful when criticizing a specific result.

The conclusion from a first glance at the figures 1 ... 19 is obviously that we can obtain any result wanted by properly choosing the conditions of measurements.

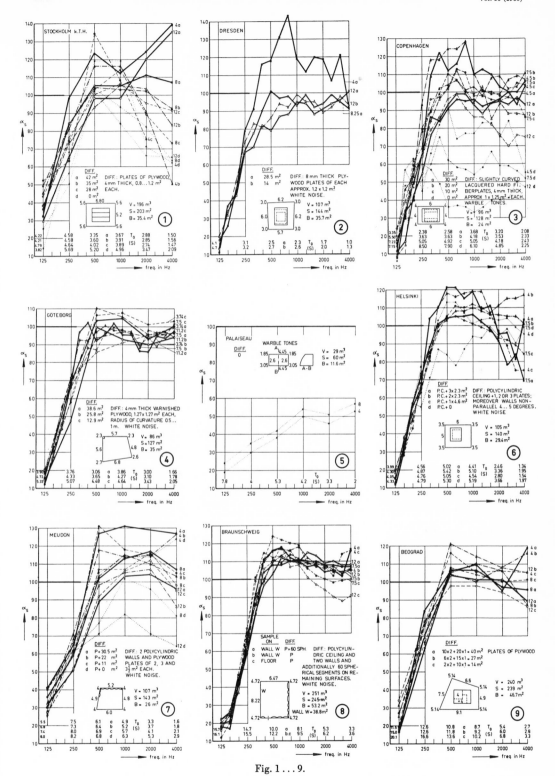

Fig. 1 ... 9.

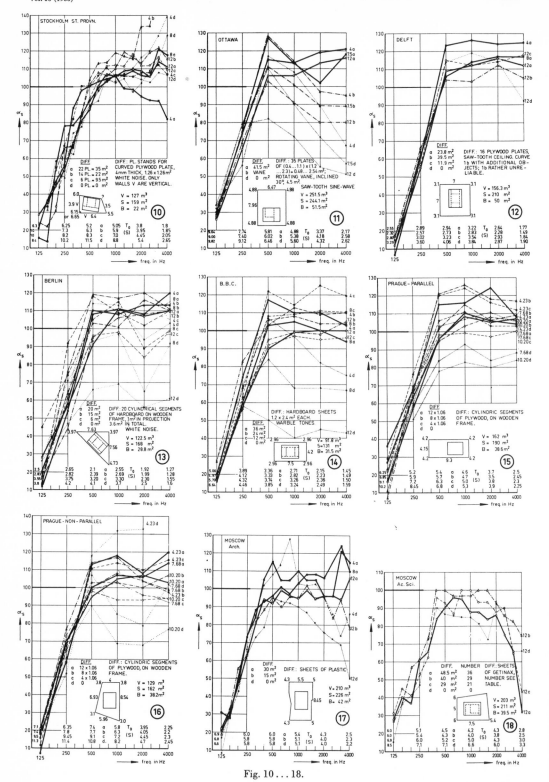

Fig. 10...18.

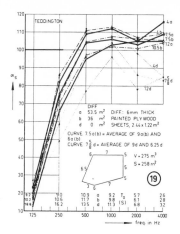

Fig. 19.

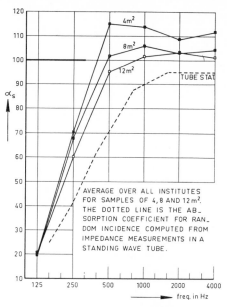

Fig. 20. Influence of area F; diffuse rooms.

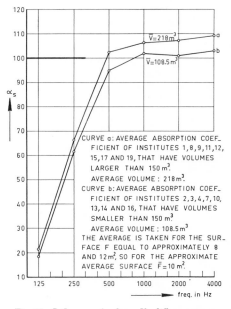

Fig. 21. Influence of volume V; diffuse rooms.

Another conclusion that is readily obtained is that the curves corrsponding to poor or absent diffusivity are worthless and entirely unreliable. The dotted line in Fig. 20 represents the absorption coefficient for random incidence as calculated from the normal incidence coefficient as measured in a standing wave tube. It was tacitly assumed, in doing so, that the material is locally reacting. Assuming furthermore that α_S in the reverberation room must be even greater due to edge effect we expect approximately $\alpha_S = 100$ at high frequencies. From Figs. 1, 3, 5, 7, 11, 13, 14, 15, 16, 17 and 19 it is clear that poor diffusivity is entirely unacceptable.

We will, therefore, concentrate in this section on the results obtained in the most diffuse state, i. e. on the heavy curves (a).

From a historical point of view it seems interesting to study the influence of the area F and the volume V (Fig. 20 and 21 respectively). It is clear that there is — as mentioned many a time in literature — a rather strong dependence on F. The dotted line in Fig. 20 is given for comparison, which shows that on the average α_S is larger than predicted from tube measurements after conversion by computation to random incidence. Bearing in mind that the relative edge length E increases with decreasing area F, the conclusion forces itself upon us that the edge effect is responsible for the dependence on F (as known from literature).

The dependence on the volume V (Fig. 21) is less pronounced, but in our opinion significant.

Since α_S decreases with F and increases with V, it seems logic trying to combine these effects by supposing, e.g., that α_S depends only upon the parameter F/S or $F/V^{2/3}$. If a correlation would exist between α_S and F/S or $F/V^{2/3}$ this would provide us of a possibility to arrive at the same result in rooms of different size by requiring a larger sample in a larger room.

The results of the First Round Robin (mis)led the author, too, to give credence to such a reasoning. He proposed $F/V^{2/3}$ as a possible parameter [1]. We made a plot, therefore, of α_S against $F/V^{2/3}$ (Fig. 22). In order to get average results we used the average value of α_S at 500, 1000, 2000 and 4000 Hz (abbreviated symbol $\bar{\alpha}_{500}\ldots$). The dotted line is the straight line that gave the best fit in the First Round Robin.

The tremendous spread in Fig. 22, especially at low values of $F/V^{2/3}$, dashes all expectations of giving an explanation, at least at first sight. There is a general tendency towards a high value of α for small areas F (which is well-known). At values $F/V^{2/3} > 0.4$ the absorption coefficient decreases less rapidly. This seems to corroborate, or at least not to violate, the statement of MEYER, KUTTRUFF [2], that α_S reaches the "true" value when the entire (floor) surface is covered with the material, which is the case approximately when $F/V^{2/3} = 1$. The "true" value is about $\bar{\alpha}_\infty = 88$, this being the value of α from tube measurements converted to random incidence and averaged over the frequencies 500, 1000, 2000 and 4000 Hz.

Also a plot was made of $\bar{\alpha}_{500}\ldots$ against F/S, a parameter that was suggested in literature several times. The figure (not reproduced here) is almost identical to Fig. 22, as was to be expected since approximately $S = 6\,V^{2/3}$.

Notwithstanding Figs. 20, 21, 22, that seem to show that there is some dependence on area, volume or a combined parameter, we maintain that this way of representation is misleading. It explains nothing, it only states a fact. The true explanation can be given only *by taking into account simultaneously the edge effect and the diffusivity.*

In the following sections we shall mainly use the quantity $\bar{\alpha}_{500}\ldots$ to discuss the various effects, i.e. we concentrate on the behaviour at medium and high frequencies. This is justified since neither the area F, nor the volume V, nor the degree of diffusivity show any measurable influence on the result at 125 Hz. The average value of α_{125} is approximately 20.5 with a standard deviation of approximately 7. The difference between "with diffusivity" and "without diffusivity" is less than 1. So is the difference between the group of rooms smaller and larger than 150 m³. The differences between the results with $F = 12$ m² and with 4 m² is of the order of 2.

So, far within the accuracy of measurement no influence on α_{125} can be detected. We therefore shall continue to concentrate on $\bar{\alpha}_{500}\ldots$, i.e. on medium and high frequencies.

7. Edge effect in diffuse rooms

If eq. (1), $\alpha = \alpha_\infty + \beta\,E$, would hold, we ought to find a linear relation between α and E. If β would be different for the various laboratories – which is very likely since there have not been given strict instructions what to do with the edges, so that there is much variation in this respect between the laboratories – the linear relation should hold for each laboratory separately.

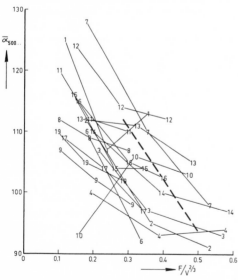

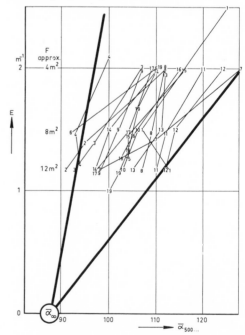

Fig. 22. $\bar{\alpha}_{500}\ldots$ as function of $F/V^{2/3}$ for all institutes. The dotted line is the straight line that gave the best fit for diffuse rooms in the First Round Robin [1].

Fig. 23. There seems to be a linear relation per institute between $\bar{\alpha}_{500}\ldots$ and the relative edge length E; diffuse rooms.

We, therefore, made a plot (Fig. 23) of $\bar{a}_{500}\cdots$ for the most diffuse state against E for all laboratories. Instead of points, crosses, etc. we used the reference numbers of the laboratories. Since most institutes measured with three areas, we have as a rule three points per institute, which were connected by straight lines. We expect, then, a straight line per institute, which, extrapolated to $E=0$, would intersect the α-axis at $\bar{a}_\infty$, the true value at random incidence. As said before we expect about $\bar{a}_\infty=88$ from tube measurements. The heavy lines in Fig. 23 are drawn with this in mind. The individual curves should be straight lines between these boundaries, all pointing to the apex $\bar{a}_\infty$. With a view to the limited accuracy, say 2 or 3% at least, this is the case indeed. After Figs. 20 and 21, that explain nothing, and Fig. 22, that consists of many awkwardly curved lines giving no insight at all, it is a relief to have found a way of representation, Fig. 23, that corroborates, or at least does not violate, a reasonable hypothesis, namely that the edge effect is responsible for what is generally called the area effect. More evidence is available, however, to prove the soundness of the hypothesis (see later).

8. Effect of diffusivity

Fig. 24 ist identical with Fig. 23 but for a change in the absorption-scale. Although all details are now

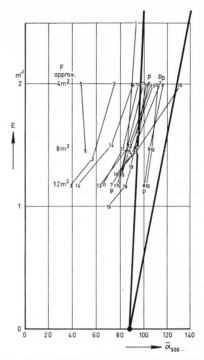

Fig. 25. As Fig. 24 but for rooms without diffusivity. Obviously lack of diffusivity is detrimental.

lost, it is even clearer from this figure than from Fig. 23 that eq. (1) holds approximately. Fig. 25 represents the same results, but this time for rooms without any diffusing elements. The heavy lines are copied from Fig. 24. Without diffusing elements α is, on the average, smaller than with such elements, the more so the larger the area F is (large F corresponds with small E). Moreover the individual lines for each institute do no longer point to $\bar{a}_\infty=88$ for $E\to0$.

We may conclude: lack of diffusivity yields too low values of α, especially at large areas F of the sample. A few institutes remain, even without diffusing elements, more or less between the heavy boundaries. But they are the institutes with polycylindric surfaces. So even without intentionally added diffusing elements they are more or less diffuse. The corresponding lines (6, 7 and 8) are marked with p. However, even for these institutes 12 m² area of a highly absorbing sample is almost more than they can cope with.

These facts are commonly known, although not quantitatively. In a very reverberant room it is rather easy to achieve a diffuse sound field, even when the boundaries are flat and when no diffusing elements have been added intentionally. A patch of highly absorbing material applied to one of the

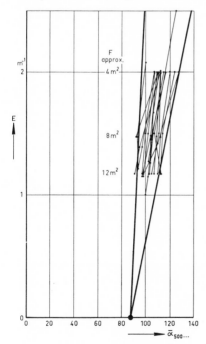

Fig. 24. As Fig. 23 but for a change in the α-scale.

boundaries of such a room tends to distort the diffuse field, absorbing those waves best to which it is exposed more intensely. A sufficient diffusivity can be maintained in that case, but only by introducing diffusing elements. The larger the absorbing area is and the higher the absorption coefficient is, the more it becomes difficult to maintain a high degree of diffusivity of the field. If the problem is to achieve a diffuse field in the steady state it may be helpful to use a number of sound sources, skilfully distributed in space and direction.

For a decaying reverberant field, as in a reverberation room after the sound source has stopped, the way in which the sound field has been generated is probably entirely indifferent, contrary to what is frequently hoped.

9. The combined influence of edge effect and diffusivity

If our reasoning – that is still in need of further evidence – is correct, the situation of the reverberation method is the following. When the sample is taken small in order to warrant a sufficiently diffuse field the absorption coefficient may be intolerably increased by edge effect. When, on the contrary, a large sample is used in order to reduce the edge effect to acceptable proportions it may turn out to be impossible to achieve a sufficiently diffuse reverberant field, especially when the sample is highly absorbing. So, the problem really is to find the best compromise, if such a compromise does exist, between the two dangers mentioned, intolerable edge effect and lack of diffusivity.

In order to reduce the edge effect to reasonable proportions ISO/TC 43/WG 3 thought an area of 10 to 12 m² of material necessary. A room of 180 m³ or more can be maintained sufficiently diffuse with such a sample. At the moment it is doubtful whether rooms of about 100 m³ can be made acceptable (see section 10). The floor surface will be of the order of 5 m × 5 m, which is too small for a sample of 3 m × 4 m. On the one hand there must remain sufficient room between sample and wall to call the edge really free. On the other hand it is doubtful whether it is feasible, or at least it will be very difficult, to maintain a diffuse field in such cases. For an international standard this seems sufficient reason not to allow rooms smaller than 180 m³ up to the moment that it has been shown clearly that smaller rooms can be used and how this should be done. ISO/TC 43 has, therefore, prepared a draft-recommendation with 180 m³ as the smallest allowed volume.

10. Further proof of the validity of the hypothesis

The final proof of our reasoning is that the results of both Round Robins can be explained along this line of thought without any difficulty. None of the results is in contradiction with it.

Fig. 26 represents $\bar{a}_\infty$, from measurements in diffuse and non-diffuse rooms respectively, as read from Figs. 24 and 25 by extrapolation to $E = 0$ of the lines for all individual institutes, as a function

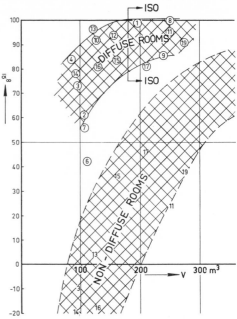

Fig. 26. $\bar{a}_\infty$ is a function of the volume for diffuse and nondiffuse rooms (derived from Fig. 24 and 25 for each institute). Values with and without diffusivity resp. are represented with and without a circle around the reference number.

of the volume V of the rooms. Even in the very diffuse rooms $\bar{a}_\infty$ is measured considerably too low in rooms of less than 120 m³. The ISO-proposal for the smallest volume, 180 m³, looks to be on the safe side at least when $\bar{a}_\infty$ is taken as a criterion. The values of $\bar{a}_\infty$ found in rooms with no diffusivity at all, is undoubtedly much too low. Here again, the smallest volumes are the worst.

Finally we will discuss Fig. 27 that is a representation of the results of the first Round Robin in the $E\text{-}\bar{a}_{500}\cdots$-plane. The heavy boundaries are the same as those of Figs. 23 ... 25. The reference numbers of the 1st Round Robin differ from those of the present one (see legend in the left hand corner). Upright reference numbers refer to rooms with parallel walls, obliquely written reference numbers

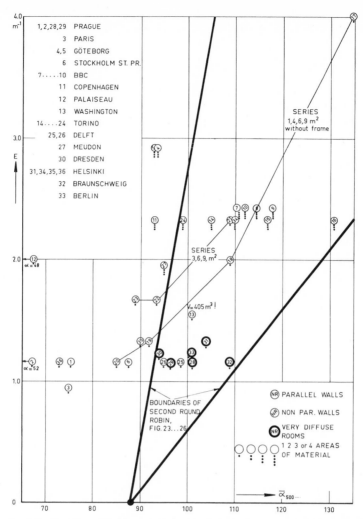

Fig. 27. The results of the First Round Robin explained with the aid of a representation in the E-$\bar{\alpha}_{500}\cdots$-plane.

to rooms with non-parallel walls. The number of patches of material used, n, is indicated under the corresponding reference number by 1, 2, 3 or 4 dots. Six circles (28...33) are drawn heavily. They correspond to rooms with much diffusivity; they constitute the group that I called the ISO-group in the report of the 1st Round Robin [1]. It should be borne in mind that most of the rooms as used during the First Round Robin would be called "without any diffusivity at all" in the Second Round Robin.

All results of Fig. 27 can be explained easily as follows:

i. The ISO-group (28...33) falls in between the heavy straight lines, as was to be expected, since the rooms were very diffuse.

ii. The other points show up, as a rule, too low values for $\bar{\alpha}_{500}\cdots$ when only one area of material ($n=1$) was used. This is due to lack of diffusivity (compare Figs. 24 and 25). The reference numbers 2 and 12 even fall outside the graph.

iii. The points with more areas ($n=3$ or 4) *seem* satisfactory, at least most of them lie between the heavy straight lines. Application in more than one area increases the diffusivity (greater $\bar{a}$), but increases the edge effect also (again greater $\bar{a}$) resulting in a value of $\bar{a}$ that is too large. Moreover the spread is tremendous, viz. from $\bar{a}=92\ldots131$. Probably the diffusivity in this group is not entirely sufficient. If it would have been, the results would have shown even larger values of $\bar{a}$, thus bringing the group between the heavy lines. The centre of

173

gravity of this group would then lie some 10% above that of the ISO-group, due to excessive edge effect.

iv. The idea of dividing the material in more areas, borrowed from the German standard, is not bad. Since the phenomenon of diffusivity has been studied, however, we must say that using more than one area, is an awkward means of arriving at a more or less diffuse sound field. The disagreeable, and in fact unacceptable, consequence is a large increase in edge effect and a large decrease of accuracy or reliability as a consequence of the variance in the edge effect from institute to institute.

v. The group with $n = 3$ or 4 compares very well with the group $n = 1$, $F = 4$ m^2 in diffuse rooms, as is evident from a comparison with Fig. 24. Since the diffusivity was probably rather insufficient a comparison with Fig. 25 is reasonable too. The resemblance is very good.

vi. Nr. 13 is interesting. $F = 6.5$ m^2, $V = 405$ m^3. For achieving diffusivity, only a rotating vane was used. Although these circumstances can be improved the situation is not bad. The edge effect is rather large; the field must have been more or less diffuse due to the small value of F and the large value of V.

vii. Nr. 25 and 26 (our own room) are not bad. Volume rather large, 156 m^3; walls and ceiling rather diffuse. The situation can be improved certainly, as is likely from the absorption characteristics [1] and has been shown by the measurements for the second Round Robin (Nr. 12 in Fig. 23).

viii. Torino carried out two series (17...20 with one sample of 1, 4, 6, 9 m^2 respectively and 21...23 with one sample of 3, 6, 9 m^2 respectively). The results have been connected by thin lines. The first series was without frame at all (large edge effect), the second with a frame. For the larger areas (small E) lack of diffusivity explains the bend of the curve to the left, outside the heavy lines.

11. Conclusions and desirability of future work

In order not to get too much edge effect one area of 10...12 m^2 should be used. A diffuse reverberant field can be maintained in rooms larger than 180 m^3, but a number of diffusing elements, e.g. panels with a total surface sufficient to cover the floor, should be installed; this should be done with much care, as described in section 3.

Although these conclusions seem to warrant that reasonable results will be obtained, there remain various problems that are in need of further study.

i. Up till now rooms much smaller than 180 m^3, say around 100 m^3, seem unacceptably small, since with a sample of 10...12 m^2 a diffuse reverberant field can generally not be maintained in such small rooms. It has not been proved, however, that this should necessarily be so. Since many laboratories have such small rooms it is worthwhile trying to find methods for obtaining sufficient diffusivity in such rooms with a highly absorbing sample of at least 10 m^2. The requirement that the smallest distance between the sample and the walls should exceed 1 m, should be kept in mind since otherwise the edges are not sufficiently free.

ii. Hanging numerous plates in a random way all over the volume of the room makes the room hardly accessible. It seems logic to suppose that the plates can be moved somewhat sideways and to the ceiling, keeping them at distances not smaller than say 0.5...1 m from the boundaries, without decreasing the diffusivity. How and to which extent this can be done is not yet known.

iii. The edge effect needs further study, i.e. the value of β for various ways of finishing the edges for various materials and absorbing constructions is wanted. Since absorbing material is often used as patches distributed in a regular or irregular way, e.g. by mounting patches in a checker board pattern, the behaviour of such arrangements should be studied.

iv. As said before, absorbing materials are frequently used in a semi-diffuse field. Although it is difficult to define the problem, it is obvious that the consultant, who has to deal with such cases in practice, is badly in need of information to this effect.

(Received 27th June 1960.)

References

[1] Kosten, C. W., Die Messung der Schallabsorption von Materialien, einschließlich die Ergebnisse der internationalen Vergleichsmessung 1959, Proc. 3rd ICA-Congress, in press.

[2] Meyer, E. and Kuttruff, H., Akad. Wiss. Göttingen, Math.-physik. Kl., II a, Mathem.-physik.-chem. Abt., (1958) 6.

[3] Ingerslev, F., Unpublished communication at the meeting at Rapallo of ISO/TC 43/WG 3, March 1960.

16

Reprinted from *J. Acoust. Soc. Am.* 40(2):428–433 (1966)

Evaluation of Acoustic Properties of Enclosures by Means of Digital Computers

B. S. Atal, M. R. Schroeder, G. M. Sessler, and J. E. West

Bell Telephone Laboratories, Inc., Murray Hill, New Jersey

A method employing a digital computer for evaluating the acoustic properties of enclosures is described. Specially shaped tone bursts, generated on the computer, are radiated into the enclosure under study. The sound-pressure responses at different locations in the enclosure are recorded on a magnetic tape. The data are converted into digital form by an analog-to-digital converter and are processed by the digital computer. The processing by the computer includes filtering (to improve signal-to-noise ratio), envelope detection, and evaluation of different quantities having subjective or physical significance. A microfilm plotter attached to the computer is used to plot the results. Among the quantities evaluated are reverberation times based on different portions of the decay; direct, early, and reverberant energies; and directional distribution of sound-energy flux (diffusion). The different quantities are evaluated both as a function of frequency and location in the enclosure. Spatial and frequency averages of the different quantities are also evaluated,

INTRODUCTION

THE study of sound transmission in enclosures is often a laborious task. It requires evaluation of numerous quantities at different frequencies and at

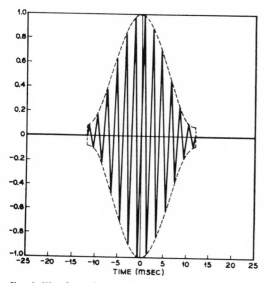

FIG. 1. Waveform of a tone burst with a center frequency of 500 Hz and a duration of 24 msec. The dashed curve is the envelope of the tone burst.

many locations in the enclosure. In addition, for convenient interpretation, the measured quantities must be averaged over frequency and location. Considerable time can be saved by employing digital computers for such studies. A general method of evaluating the acoustic properties of enclosures, using a digital computer, is described in this paper. The computer, together with its peripheral equipment, is used for (1) generating the signals for acoustical tests in the enclosure, (2) processing and evaluating the sound-pressure responses in the enclosure, and (3) tabulating and plotting the final results.

Among the quantities evaluated are reverberation times based on different portions of the sound decay; energy of direct sound, early- and late-arriving reflections; and directional distribution of early sound. The different quantities are evaluated as functions of both frequency and location in the enclosure. Spatial and frequency averages of the data are also evaluated. Functions of different parts of the computer program as well as the measurement procedure in the enclosure are described in the next five sections.

I. SIGNALS FOR ACOUSTICAL TESTS

A. Choice of Text Signal

In order to conserve time spent on measurements in the actual enclosure, it is desirable to use only one type

175

of test signal for all measurements in the enclosure. Since the *impulse response* between any two points in an enclosure contains all the information about the sound-transmission characteristics between these points, short pulses can be used as a general-purpose test signal for acoustic measurements in an enclosure.

There are, however, practical difficulties in measuring the response of an enclosure to pulses of very short duration. Because of peak power limitations, a loudspeaker can radiate only little energy during short pulses. In addition, a short pulse has a large bandwidth, and narrow-bandpass filters cannot be used to reduce the ambient-noise level. These difficulties can be overcome by dividing the frequency band over which measurements are to be taken into a number of relatively narrow frequency bands.[1] This is accomplished by measuring the response of the enclosure to tone bursts with contiguous, narrow-band spectra. This procedure has another advantage. Generally, it is necessary to determine the different acoustical properties of the enclosure as a function of frequency. With tone bursts having narrow-band spectra, this is achieved automatically.

A tone burst can be represented by a time function $s(t)$ given by

$$s(t) = e(t)\sin(2\pi f_n t + \varphi), \qquad (1)$$

where $e(t)$ is the envelope, f_n is the center frequency, and φ is the "carrier" phase. The envelope $e(t)$ is chosen so that the spectrum of $s(t)$ has the desired bandpass characteristics. A preferred shape of the envelope is

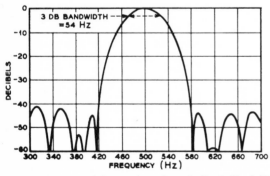

FIG. 2. Spectrum of the tone burst shown in Fig. 1. The 3-dB bandwidth of the spectrum is 54 Hz. The amplitude of the largest side lobe is more than 41 dB below the amplitude of the main lobe.

---

[1] Another method of improving signal-to-noise ratio is to radiate time-stretched pulses from the loudspeaker. The pulse can be stretched by passing it through an all-pass network with an impulse response that is long as compared to the duration of the pulse. To obtain the response of the room to the original unstretched pulse, the response of the room to the stretched pulse is passed through another all-pass network whose transfer function is the inverse of the all-pass network employed for stretching ("collapsing filter"). For a "stretch factor" of 1000, the signal-to-noise ratio is improved by about 30 dB.

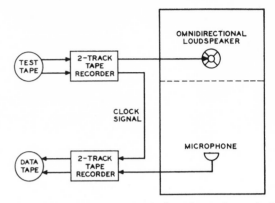

FIG. 3. Measurement procedure in the enclosure. One track of the tape is used for the signal. The other track is used for recording the "clock signal" from the test tape on to the data tape.

given by[2]

$$e(t) = 0.54 - 0.46\cos(2\pi t/d), \quad 0 \le t \le d, \qquad (2)$$
$$= 0, \qquad\qquad\qquad\qquad\qquad t > d,$$

where d is the total duration of the envelope. Thus, the exact time function describing the tone bursts is

$$s_n(t) = [0.54 - 0.46\cos(2\pi t/d_n)]\sin(2\pi f_n t + \varphi),$$
$$0 \le t \le d, \qquad (3)$$
$$= 0, \qquad\qquad\qquad\qquad t > d,$$

where d_n is the duration and f_n is the center frequency of the nth tone burst.

The choice of center frequencies f_n depends upon the particular application. For example, the frequency difference $f_n - f_{n-1}$ between adjacent tone bursts could be the same for all tone bursts. Alternatively, the difference $f_n - f_{n-1}$ could be made proportional to f_n. The latter choice will result in center frequencies forming a geometric series and is often desirable for analysis over wide frequency ranges. The duration d_n of the tone bursts is selected so that spectra of adjoining tone bursts cross over at approximately -3 dB.

Another useful frequency spacing follows the "Koenig scale," a modified mel scale.[3] Here, the center frequencies f_n of the tone bursts are spaced along the frequency axis so that the differences $f_n - f_{n-1}$ are constant up to 1000 Hz and thereafter increase proportional to f_n. It may be pointed out that the choice of f_n and d_n need not be restricted to alternatives described above. They can be selected to cover the appropriate frequency range in any desired manner.

As an example, a tone burst having a center frequency of 500 Hz and a duration of 24 msec is shown in

---

[2] R. B. Blackman and J. W. Tukey, *The Measurement of Power Spectra* (Dover Publications, Inc., New York, 1958), pp. 14–15.
[3] W. Koenig, "A New Frequency Scale for Acoustic Measurements," Bell Labs. Record **27**, 299–301 (1949).

$\cos(2\pi f_n t)$

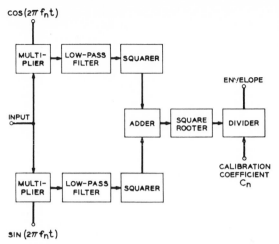

FIG. 4. Block diagram of the heterodyne method of filtering and envelope detection used in the digital computer.

$\sin(2\pi f_n t)$

Fig. 1. The phase φ of the carrier is made zero since this minimizes the amplitude of side lobes in the spectrum of $s_n(t)$. The spectrum of $s_n(t)$ is shown in Fig. 2. The amplitude of the largest side lobe is 41 dB below the amplitude of the main lobe. For comparison, a rectangular envelope would give a largest side lobe that is only 13.5 dB below the main lobe. The 3-dB bandwidth of the tone burst is 54 Hz. A time interval of 10 msec about the center of the tone burst contains 80% of the total energy of the burst.

B. Generation of Test Signal

The digital computer is programmed to compute and write, on a digital tape, sample values of the tone bursts $s_n(t)$. The sampling frequency must be greater than twice the upper cutoff frequency of the tone burst with the highest center frequency. For example, if the highest center frequency is 1000 Hz and the duration of the tone burst is 24 msec, the spectral components are negligible above 1100 Hz. A sampling frequency of 2500 Hz is suitable in this case. A digital-to-analog converter transforms the binary-coded numbers on the digital tape into a continuous voltage signal representing the tone bursts. The digital tape is written so that a time gap of 3 sec appears between adjacent tone bursts to allow the sound to decay in the enclosure. The output of the digital-to-analog converter is recorded on one track of a magnetic tape by a two-track tape recorder. The other track is used to record the "clock signal" of the digital-to-analog converter. This is done to reduce timing errors due to tape-recorder speed variations and tape stretching. Speed variations in the tape recorder cause the time gap between adjacent tone bursts to vary; however, the time interval between adjacent tone bursts still contains the same number of periods of the sampling frequency every time that the tape is played

back. Correct timing between adjacent tone bursts is needed for later processing in the digital computer.

II. MEASUREMENTS IN THE ENCLOSURE

The measurement procedure is illustrated in Fig. 3. The test tape, consisting of tone bursts with different center frequencies, is played back in the enclosure under study using a special nearly omnidirectional sound source. The sound source, for low frequencies (below 1 kHz) consists of five large loudspeakers mounted on five sides of a box. For frequencies above 1 kHz, a driver unit with a small neck opening is used. The response of the enclosure to the tone bursts is picked up by omnidirectional and directional microphones and is recorded on one track of a magnetic tape by a two-track tape recorder. The clock signal is recorded on the other track of the tape. Measurements are repeated for several positions of microphones and sound source.

III. CONVERSION OF RECORDED DATA INTO DIGITAL FORM

The recordings made in the enclosure are converted to digital form and recorded on a digital tape by means of an analog-to-digital converter.[4] The clock signal on the second track of the analog tape is used for sampling, replacing the standard clock of the analog-to-digital converter. This procedure ensures that the adjacent tone-burst responses are separated by a constant number of samples on the digital tape.

IV. COMPUTER PROCESSING OF THE RECORDED DATA

The computer processing of the data is divided into three major steps: (1) filtering and envelope detection, (2) computation of various quantities having physical or subjective significance, and (3) preparing plots and Tables of the results.

A. Filtering and Envelope Detection

In Step 1 of the computer processing, the recorded signal is filtered and envelope is detected by a heterodyne method[5] (see Fig. 4). In this procedure, the input data are multiplied by $\sin(2\pi f_n t)$ and $\cos(2\pi f_n t)$, generating sum and difference frequencies. Only the difference frequencies are passed by the low-pass filters following each multiplier. The outputs of the low-pass filters are squared and added, and the square root of the sum is taken. The impulse response of the low-pass filters is chosen to be the same as the envelope of the tone bursts. This method of filtering corresponds to

[4] E. E. David, Jr., M. V. Mathews, and H. S. McDonald, "Digital Computer Simulation as a Tool in Speech Research," in *Proceedings of the Third International Congress on Acoustics, Stuttgart, 1959*, L. Cremer, Ed. (Elsevier Publ. Co., Amsterdam, 1961), Vol. 1, pp. 224–226.
[5] M. R. Schroeder and B. S. Atal, "Generalized Short-Time Power Spectra and Autocorrelation Functions," J. Acoust. Soc. Am. 34, 1679–1683 (1962).

"matched filtering,"[6] and maximizes the signal-to-noise ratio for signals contaminated by flat-spectrum noise.

The computer processing combines the operation of filtering and envelope detection. For an input signal consisting of $s(t)$ as defined in Eq. 1, the output $y(t)$ after filtering and envelope detection is approximately equal to $e(t)$ convolved with itself. The total duration of $y(t)$ is thus twice that of $e(t)$, but 80% of the total energy is still contained in an interval $\pm d/4$ around the center of the tone burst. The 3-dB bandwidth of the output signal is $0.92/d$ Hz.

Next, the data are adjusted in amplitude by applying a proper calibration coefficient for each frequency band. This is done to compensate for the "nonflat" frequency responses of the various transducers and amplifiers used in the measurements. This includes loudspeakers, microphones, tape recorders, amplifiers, and filters. (The procedure for determining these calibration coefficients is described in Sec. V.)

The sampling rate in the computer for all the operations described so far is based on the upper cutoff frequency of the tone burst with the highest center frequency. However, the squared envelopes of the tone-burst responses can be sampled at a lower rate. For example, for tone bursts with a duration of 24 msec, the squared envelopes $y^2(t)$ have negligible spectral components above 80 Hz. Thus, a sampling rate of 160 Hz will be sufficient for squared envelopes.

However, to facilitate the plotting of the envelope directly from the sample values, a sampling rate higher than one based entirely on bandwidth considerations was selected. For tone bursts with a duration of 24 msec, a sampling rate of 500 Hz has been used. For further processing of the data in subsequent computer programs, the envelopes of the tone-burst responses are stored on digital tapes with the reduced sampling rate.

The digital tape obtained is fed into the computer for plotting of the envelopes both on a linear and a logarithmic scale. With the use of a microfilm plotter attached to the computer, one can obtain such plots, sometimes called "echograms," with any desired labels, grids, and scales. An example of a logarithmic echogram is shown in the upper part of Fig. 5. The ordinate is the envelope, in decibels, of the response of an enclosure to a tone burst. Such echograms are used for detailed analyses of the transmission of the direct sound and early reflections as well as later portions of the reverberation process.

B. Computation of Frequency Responses

The digital tape containing the envelopes of the tone-burst responses is fed into the computer for further processing. Among the parameters calculated by the computer are the following—

• *Reverberation times.* These are derived from an integrated tone-burst decay $b(t)$ defined by the following

[6] G. L. Turin, "An Introduction to Matched Filters," IRE Trans. Inform. Theory **6**, 311–329 (1960).

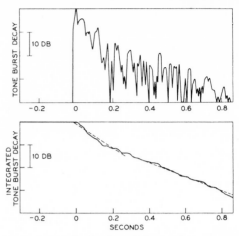

FIG. 5. *Upper half:* Envelope of the tone-burst response in decibels. *Lower half:* Integrated tone-burst decay obtained according to Eq. 4. Reverberation times T_{15} and T_{15-30} are obtained by best straight-line fits to first 15 and the next 15 dB of decay, respectively. $T_{15} = 1.11$ sec. $T_{15-30} = 2.09$ sec.

equation:

$$b(t) = \int_{t}^{1.5 \text{ sec}} y^2(\tau) d\tau, \tag{4}$$

where $y(t)$ is the envelope of the tone-burst response. Typically, three different reverberation times T_{15}, T_{15-30}, and T_{5-35} are derived from different portions of the integrated tone-burst decay. T_{15} is based on decay over the first 15 dB, T_{15-30} is based on decay from -15 to -30 dB, and T_{5-35} is based on decay from -5 to -35 dB. It has been shown in Ref. 7 that the integrated tone-burst decay $b(t)$ is equal to an ensemble average of the squared noise decay. The reverberation times are, thus, based on average noise decays. Straight lines yielding minimum mean-square errors are fitted to the logarithm of the integrated tone-burst decay over different portions as needed. T_{15}, T_{15-30}, and T_{5-35} are then obtained from the slopes of the straight lines by computing the time that would be required for 60-dB decay. This is illustrated in the lower half of Fig. 5. The solid curve is the integrated tone-burst decay, and the dashed lines are the best-fitted straight lines over 0 to -15 and -15 to -30 dB, respectively.

• *Energy of the direct sound $D(f_n)$.* The direct energy is computed by integrating the squared envelope over a time interval $d/2$ centered on the arrival time of the direct sound.

• *"Early energy" $E(f_n)$.* The early energy (arriving within 50 msec of the direct sound) is obtained by integrating the squared envelope from $t_0 - d/4$ to $t_0 + d/4 + 50$ msec, where t_0 is the arrival time of the direct sound.

[7] M. R. Schroeder, "New Method of Measuring Reverberation Time," J. Acoust. Soc. Am. **37**, 409–412 (1965).

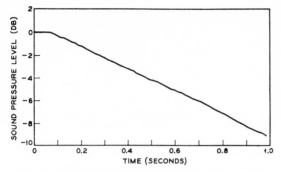

<figure>FIG. 6. Average integrated tone-burst decay in an empty re-
verberation chamber ($T_{15}=6.0$ sec) for 1 oct band from 200 to
400 Hz. The averaging was done over 12 positions in the reverberation chamber.</figure>

- *"Total energy"* $A(f_n)$. The total energy is computed
by integrating the squared envelope of the tone-burst
response over a 1-sec time interval (which includes at
least 99% of the total energy for reverberation times
up to 3 sec). $A(f_n)$ corresponds, approximately, to the
steady-state frequency response of the enclosure between loudspeaker and microphone averaged over the
3-dB bandwidth of the filtered tone-burst envelope.[8]

- *Ratio of early energy to reverberant energy* $S(f_n)$.
This ratio is obtained as the ratio of $E(f_n)$ and $[A(f_n)
-E(f_n)]$.

- *Energies of early reflections from various surfaces*
$C(f_n)$. These are computed by centering the integration
interval $d/2$ on the arrival times of these reflections.

- *Directional distribution factor* $F(f_n)$ *of early sound.* It
is defined as S_1/S_{n1}, where S_1 is the ratio of early energy
to reverberant energy for a laterally oriented directional
microphone, and S_{n1} the same ratio for a microphone
covering the complementary directions.

1. Space and Frequency Averages

The computer program has provisions for computing
any specified space and frequency averages of the envelopes or of any of the parameters mentioned earlier
over any set of selected locations measured in the enclosure. Any combination of frequency and space averaging may also be performed. An example of such an
averaging is shown in Fig. 6, where integrated tone-
burst decays in a reverberation chamber at 12 different
positions were combined to cover an octave band from
200 to 400 Hz. The individual tone bursts used for
measurement in the reverberation chamber had a 3-dB
bandwidth of only 54 Hz.

[8] M. R. Schroeder, B. S. Atal, G. M. Sessler, and J. E. West,
"Acoustical Measurements in Philharmonic Hall (New York),"
in *Proceedings of the Third Acoustical Conference, Budapest, 1964*
(Hungarian Society for Optics, Acoustics, and Film Techniques,
Budapest, 1964), pp. 380–385.

C. Plotting of Results

The different energies $A(f_n)$, $E(f_n)$, and the ratio
$S(f_n)$ are converted to decibels and plotted as functions
of frequency for each location (see Fig. 7). Reverberation times and directional distribution factors are likewise plotted as functions of frequency. The frequency-
averaged energies and reverberation times are, however,
plotted as functions of microphone position.

V. CALIBRATION PROCEDURE

The calibrating procedure resembles closely the actual
measuring process and involves the entire chain from
test-signal generation to information processing and
plotting.

For the calibration, tone bursts are played in a reverberation chamber through the same amplifiers and
sound sources that are used in the actual measurements.
The response of the chamber to these tone bursts is recorded by the various microphones used in the measurements at a dozen or more well-separated positions. The
recordings are reconverted into digital form and fed into
the computer. There they are filtered and envelope-
detected. The envelopes are squared and averaged over
all the measured positions for each tone burst. The total
energy $A(f_n)$ and an initial reverberation time $T_i(f_n)$
are then determined for the space-averaged square envelope. $T_i(f_n)$ is determined by fitting a straight line to
the first 5 dB of the integrated tone-burst decay. The
total energy radiated into a room for the tone burst at
frequency f_n is proportional to the ratio $A(f_n)/T_i(f_n)$.
The square root of this ratio is printed out as the calibration coefficient for each frequency band and microphone.
These calibration coefficients are applied in the computer program as described above.

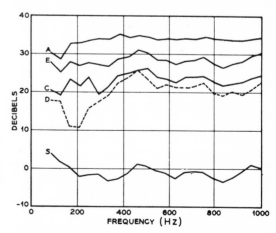

<figure>FIG. 7. Frequency responses obtained for one location in a con-
cert hall. A: "Total" energy (integrated over 1 sec). E: "Early"
energy (integrated over first 50 msec). C: Energy of early reflections from ceiling. D: Energy of the direct sound. S: Ratio of
early-to-reverberant energy.</figure>

179

VI. CAPABILITIES OF THE REQUIRED COMPUTER

An IBM-7094 computer has been used for the measurements to date. It has a core storage with a capacity of 32 768 words, each comprising of 36 data bits. The program makes use of most of this storage. If the processing is divided into smaller packages, a computer with a much smaller storage can be used. The first part of the computer processing that includes filtering and envelope detection uses most of the computer storage and running time. For a sampling frequency of 10 kHz, the total computation time is about 4 sec per tone burst. Approximately 80% of this time is spent for filtering and envelope detection.

The analog-to-digital converter uses a quantization of 12 bits/sample, which is considered adequate for room-acoustical evaluation.

VII. APPLICATION OF THE METHOD

This method has been used for evaluating the acoustics of several concert halls. The results of measurements made in Philharmonic Hall of Lincoln Center for the Performing Arts, in the City of New York, are described in Ref. 9. By employing this method, it has been possible to determine accurately the original state of Philharmonic Hall and to monitor the effects of various alterations.

[9] M. R. Schroeder, B. S. Atal, G. M. Sessler, and J. E. West, "Acoustical Measurements in Philharmonic Hall (New York)," J. Acoust. Soc. Am. **40**, 434–440 (1966).

VIII. CONCLUSIONS

By using the digital computer for acoustical analysis of enclosures, it has been possible to combine signal generation, evaluation of acoustical data, and plotting of the results in one general method. The computer method, in conjunction with appropriate recording and reproducing equipment, permits accurate and reproducible measurement of sound-transmission characteristics of enclosures. Results are presented in forms in which they can be conveniently interpreted. Meaningful comparisons of acoustical characteristics of different enclosures can thus be made.

The use of digital computers allows a high degree of flexibility. Acoustical criteria, suggested by new insights into auditory perception, can be incorporated into the method by additions to the computer program. The acoustical data can be evaluated according to such new criteria without making new measurements in the enclosure, because the basic information regarding the acoustical responses of the enclosure is stored on the digital tape.

It is hoped, by using this method, that a better understanding of the correlation between physical measurements and subjective qualities of the enclosure can be achieved.

ACKNOWLEDGMENTS

The authors express their appreciation to Carol Bird and Susan Murphy for writing the computer programs.

17

Reprinted from *Acustica* 6:425–444 (1956)

RAUMAKUSTISCHE UNTERSUCHUNGEN IN ZAHLREICHEN KONZERTSÄLEN UND RUNDFUNKSTUDIOS UNTER ANWENDUNG NEUERER MESSVERFAHREN

von E. MEYER und R. THIELE

1. Einleitung

Bei der Beurteilung der akustischen Eigenschaften von Räumen hat sich die Nachhallzeit als wichtigstes, objektives Kriterium erwiesen. Sie ist heute einfach zu messen und kann für Neubauten mit genügender Genauigkeit vorausberechnet werden. Dadurch besitzt sie für die praktischen Belange eine besonders große Bedeutung. Aus der Erkenntnis, daß die Nachhallzeit allein jedoch nicht ausreicht, um die akustischen Verhältnisse eines Raumes zu bewerten — Räume verschiedener Größe, aber gleicher Nachhallzeit werden akustisch oft verschieden beurteilt —, sind in mehreren Arbeiten Vorschläge für weitere Meßmethoden gemacht worden, deren Bedeutung vielfach diskutiert wurde.

Im Rahmen der raumakustischen Untersuchungen des III. Physikalischen Instituts der Universität Göttingen wurde nun geplant, diese neuen Meßmethoden systematisch bei einer großen Anzahl von Räumen anzuwenden, um ihre Brauchbarkeit zu erproben. Bei der hierfür notwendigen Aufstellung des Arbeitsprogrammes wurde zu Beginn erwogen, durch eine ausgedehnte Befragung derjenigen Personen, die die akustischen Verhältnisse der zu messenden Räume kennen sollten (Musiker, Dirigenten, Kritiker, Tonmeister, Konzertbesucher usw.), auch eine gewisse subjektive Bewertungsskala zu gewinnen. Wegen der vielfachen Schwierigkeiten einer solchen „Fragebogenaktion" — nur wenige Menschen kennen eine größere Zahl von Räumen akustisch sehr genau — und auch aus prinzipiellen Gründen wurde davon abgesehen. Es ist natürlich erforderlich, zu versuchen, eine Korrelation der subjektiven Beurteilung der Hörsamkeit eines Raumes mit den darin gewonnenen Ergebnissen der objektiven Messungen zu gewinnen. Diese Forderung, deren Lösung man als das Fernziel der Raumakustik bezeichnen muß, läßt sich jedoch nach unserer Meinung zunächst nur in der folgenden Weise angreifen: Man hat durch experimentelle und theoretische Untersuchungen diejenigen rein physikalischen Gesetzmäßigkeiten der Schallausbreitung in Räumen festzustellen, deren Kenntnis noch ungenügend ist. Auf Grund der erhaltenen Ergebnisse wird man dann überlegen, welche der gewonnenen Beobachtungen von Einfluß auf die Hörsamkeit eines Raumes sein können, und versuchen, durch ganz spezielle Tests mit wohl-definierter Fragestellung, wahrscheinlich nicht in Räumen, sondern in elektroakustischen Nachbildungsversuchen, auch über den subjektiven Einfluß einzelner Erscheinungen Aussagen zu gewinnen. Als Beispiel solcher Versuche sei auf die Methoden von HAAS [1] hingewiesen. Erst die Ergebnisse dieser Testversuche kann man dann benutzen, um Richtlinien aufzustellen, nach denen die in Frage kommenden Parameter des Schallfeldes z. B. durch architektonische Maßnahmen optimal gestaltet werden können.

Im Sinne dieser Überlegungen sollten sich die geplanten Untersuchungen im wesentlichen nur mit den physikalischen Eigenschaften des Schallfeldes in Räumen befassen, selbstverständlich jedoch in Hinsicht auf die Möglichkeit, Anhaltspunkte für die subjektive Beurteilung zu gewinnen.

Folgende Meßverfahren wurden verwendet:

1. Bestimmung der Nachhallzeit in Abhängigkeit von der Frequenz.
2. Aufnahme der Nachhallkurven bei gleitender Frequenz [2],
3. Aufnahme der Frequenzkurve des Schalldrucks und Ermittlung der „Frequenzkurvenschwankung" (Frequency irregularity) [3],
4. Messung der Richtungsverteilung der Schallrückwürfe,
5. Registrierung der Rückwürfe in Abhängigkeit von der Zeit und Bestimmung des „50-ms-Energie-Anteils" [4].

In den folgenden Abschnitten 2 bis 6 wird über die dabei verwendeten Apparaturen und die in den verschiedenen Räumen erhaltenen Ergebnisse berichtet.

2. Nachhallzeit in Abhängigkeit von der Frequenz

a) Meßverfahren

Zur Bestimmung der Nachhallzeit wurde der jeweilige Raum stationär mit Heultönen angeregt. Als Schallquelle für Frequenzen oberhalb von 400 Hz wurde der von HARZ und KÖSTERS [5] angegebene „Kugellautsprecher" verwendet. Für die Abstrahlung der tieferen Töne diente ein Einzellautsprecher. Der Schalldruckverlauf nach Abschalten der Schallquelle wurde von einem Kondensatormikrophon gemessen, dessen Ausgangsspannung nach entsprechender Verstärkung und unter Einschaltung eines Bandpasses mit einem Pegelschreiber aufgezeichnet wurde. Als Bandpaß diente in der Regel ein Terzsieb; nur in einigen Räumen mußte zur Zeitersparnis ein Oktavsieb verwendet werden. Die Messungen umfaßten den Frequenzbereich von 0,1 bis 10 kHz.

[*Editor's Note:* An English summary of this article prepared by T. D. Northwood follows.]

Bei der Auswertung der im logarithmischen Maßstab aufgezeichneten Abklingkurven ergibt sich in manchen Fällen die wohlbekannte Schwierigkeit, daß einzelne Kurven nicht gerade verlaufen, sondern mehr oder weniger stark „durchhängen", eine geknickte Gerade bilden oder bei starkem direktem Schall erst für mehr oder weniger tiefe Pegel einen Abklingvorgang zeigen (siehe Abb. 1). Als Regel

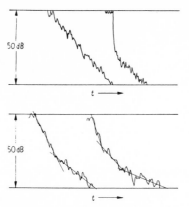

Abb. 1. Beispiele für verschiedene Formen von Nachhall-kurven.

für die Bestimmung der Nachhallzeit wurde in solchen Fällen festgelegt, daß der Verlauf der Abklingkurve im Bereich von −5 dB bis −35 dB, bezogen auf den Anfangs-pegel, möglichst gut durch eine Gerade angenähert wurde. Die Neigung dieser Geraden wurde als Maß für die Nach-hallzeit genommen [6].

b) Ergebnisse der Nachhallzeitmessungen

In Tabelle I sind 31 Räume zusammengestellt, deren akustische Eigenschaften in der vorliegenden Arbeit besprochen werden. Eine Anzahl Rundfunk-studios mit einem kleineren Volumen als 500 m³, in denen auch gemessen wurde, sollen hier unberück-sichtigt bleiben. Außer dem Raumvolumen ist die mittlere Nachhallzeit im Frequenzgebiet von 0,5 bis 1 kHz angegeben. In der Spalte „Bemerkungen" werden Angaben über das Gestühl im jeweiligen Raum gemacht, wenn solches vorhanden war, d. h. wenn der Raum auch für Aufführungen mit Publi-kum benutzt wird. Alle Messungen, die in dieser Arbeit angegeben werden, mußten aber wegen ihrer Langwierigkeit im unbesetzten Raum ausgeführt werden. Dies war jedoch nicht zu kritisch, da meist stark gepolstertes Gestühl vorhanden war. Die-jenigen Rundfunkstudios, bei denen keine Anmer-kung über das Gestühl gemacht ist, dienen in der Regel nur den Zwecken des reinen Rundfunkbetrie-bes, werden allerdings gegenüber dem Zustand bei den vorliegenden Messungen noch durch den Klang-körper verändert.

Die Abb. 2 zeigt die gemessenen Nachhallzeiten in Abhängigkeit von der Frequenz. Man bemerkt zwi-

Tabelle I

Nr.	Raum	Volumen m³	mittl. Nachhall-zeit (s) zwischen 500 und 1000 Hz	Bemerkungen
1	Studio i, Bremen	550	1,15 0,80	Raum mit va-riabler Nach-hallzeit
2	Studio 3, Frankfurt	560	0,75	—
3	Studio 12, Hamburg	810	0,80	—
4	Studio 7, Köln	850	0,95	—
5	Aula der schwedi-schen Handelshoch-schule, Helsinki	1 500	1,05	Polstergestühl
6	Studio 2, Frankfurt	1 600	1,25	—
7	Studio 2, Köln	1 800	1,05	—
8	Studio 7, Berlin	1 900	1,45	—
9	Unterhaltungsstudio Baden-Baden	2 000	1,65	—
10	Studio 1, München	2 100	1,45	—
11	Studio F, Bremen	2 900	1,45	Polstergestühl
12	Studio 1, Hamburg	3 200	1,85	—
13	Fernsehstudio B, Hamburg	3 200	0,80	mit Kulissen
14	Studio Berlin-Lankwitz	3 300	2,05	—
15	Villa Berg, Stuttgart	4 500	1,40	Polstergestühl
16	Stadttheater Kiel	5 000	1,60	Polstergestühl
17	Stadttheater Bremen	5 000	1,55	Polstergestühl
18	Studio 10, Hamburg	5 250	1,55	Polstergestühl
19	Schillertheater Berlin	5 800	1,95	Polstergestühl, sehr halliger Bühnenraum zur Zeit der Messung
20	Staatsoper Hamburg (Provisorium)	6 000	1,30	Polstergestühl
21	Musikstudio Baden-Baden	6 500	1,45	Polstergestühl
22	Studio 1, Köln	6 800	1,65	Polstergestühl
23	Konzertsaal der Musikhochschule Berlin-Charlottenbg.	9 600	1,85	Polstergestühl
24	Konzertsaal Turku (Finnland)	10 000	1,95	Polstergestühl
25	Konzertsaal „Die Glocke", Bremen	10 000	1,90	Holzgestühl mit Kunstleder-polsterung
26	Jesus-Christus-Kirche Berlin-Dahlem	10 000	2,75	Holzbänke
27	Musikhalle Hamburg	11 500	2,20	Gestühl mit Stoffbespan-nung
28	Kurhaussaal Wiesbaden	11 500	1,65	Polstergestühl
29	Studio 1, Frankfurt	12 000	1,80	Polstergestühl m. gelochter Kunststoff-bespannung
30	Herkulessaal München	14 000	2,40	Polstergestühl
31	Royal Festival Hall London	22 000	1,80	Polstergestühl

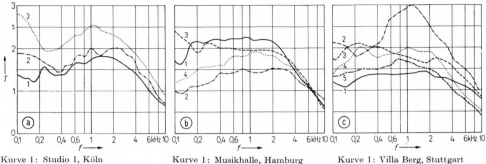

Kurve 1: Studio 1, Köln
Kurve 2: Studio 1, Frankfurt
Kurve 3: Herkulessaal, München

Kurve 1: Musikhalle, Hamburg
Kurve 2: Musikstudio, Baden-Baden
Kurve 3: Konzertsaal, Turku
Kurve 4: Royal Festival Hall,
London

Kurve 1: Villa Berg, Stuttgart
Kurve 2: J.-Christus-Kirche, Berlin
Kurve 3: Konzertsaal der Musikhoch-
schule Berlin-Charlottenbg.
Kurve 4: „Die Glocke", Bremen
Kurve 5: Kurhaussaal, Wiesbaden

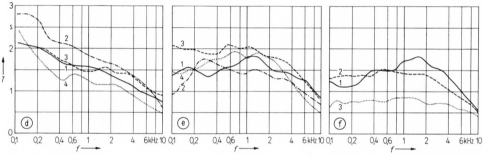

Kurve 1: Stadttheater, Kiel
Kurve 2: Schillertheater, Berlin
Kurve 3: Stadttheater, Bremen
Kurve 4: Staatsoper, Hamburg (Pro-
visorium)

Kurve 1: Unterhaltungsstudio,
Baden-Baden
Kurve 2: Studio 1, München
Kurve 3: Studio Lankwitz, Berlin
Kurve 4: Studio 1, Hamburg

Kurve 1: Studio 10, Hamburg
Kurve 2: Studio F, Bremen
Kurve 3: Fernsehstudio B, Hamburg

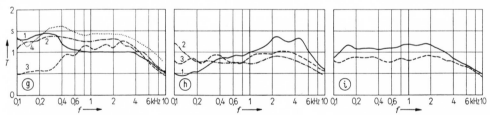

Kurve 1: Aula der schwedischen Han-
delshochschule Helsinki
Kurve 2: Studio 2, Frankfurt
Kurve 3: Studio 2, Köln
Kurve 4: Studio 7, Berlin

Kurve 1: Studio 7, Köln
Kurve 2: Studio 3, Frankfurt
Kurve 3: Studio 12, Hamburg

Studio i, Bremen; die Nachhallzeit ist
variabel durch bewegliche Wandteile

Abb. 2. Nachhallzeit in Abhängigkeit von der Frequenz.

schen verschiedenen Räumen starke Unterschiede im Frequenzgang. Das gilt besonders für die tiefen Frequenzen. Es gibt im Gebiet von 100 Hz sowohl größere als auch kleinere Nachhallzeiten in bezug auf den Wert bei 1 kHz. Einige Räume haben unterhalb 1 kHz eine verhältnismäßig gerade oder gleichmäßig steigende oder fallende Frequenzkurve, andere zeigen „Einbrüche" oder „Berge". Diese Ungleichheiten lassen sich auf zwei Ursachen zurückführen, zumindest für diejenigen Räume, bei deren Bau Raumakustiker beratend tätig waren:

1. Man besitzt in der Raumakustik noch keine einheitliche Meinung darüber, welcher Frequenzgang der Nachhallzeit unterhalb 1 kHz optimal ist.

2. Es ist sehr schwierig, bei der Vorausberechnung der Nachhallzeit die Gesamtabsorption für die tiefen Frequenzen genau anzugeben, selbst bei

vorherigen Schluckgradmessungen im Hallraum, da hier Einzelheiten der Montage eine erhebliche Rolle spielen können.

Ganz allgemein entnimmt man den Frequenz-kurven der Abb. 2, daß die Nachhallzeit für mittlere Frequenzen mit zunehmendem Raumvolumen ansteigt. Das bestätigen in gewissen Grenzen die Abb. 3a, b, in denen die mittlere Nachhallzeit für alle Räume in Abhängigkeit vom Volumen aufgetragen ist. Die eingezeichneten gestrichelten Geraden stellen die Funktion $T = a \sqrt[3]{V}$ dar, wobei a für den jeweiligen Frequenzbereich nach der Methode der kleinsten Quadrate berechnet wurde. Diese Abhängigkeit der optimalen Nachhallzeit $T_{opt} \sim \sqrt[3]{V}$ wird in der raumakustischen Literatur häufig empfohlen. Die Abweichungen, die hier auftreten, sind jedoch beträchtlich, besonders für den Frequenzbereich 2 bis 3 kHz bei kleineren Volumina. Man kann auch nicht erwarten, daß die Nachhallzeit über einen so großen Volumenbereich sehr genau einem einheitlichen Gesetz folgt. Als Faustformel könnte man immerhin auf Grund der vorliegenden Messungen für Musikdarbietungen $T \approx 0,09 \cdot \sqrt[3]{V}$ vorschlagen. Die neueren Untersuchungen von KUHL [7]

zur Bestimmung der optimalen Nachhallzeiten für Musik ergaben dagegen, allerdings nur für den Fall der einkanaligen Übertragung, daß die Nachhallzeit unabhängig vom Volumen sein soll und nur durch die Art des Musikstückes bestimmt ist. Danach wäre ein Raum je nach der Größe seiner Nachhallzeit nur für ganz bestimmte Kompositionen als optimal anzusehen.

Hingewiesen sei noch auf die besonders bei den großen Räumen auffällige Tatsache, daß die Nachhallzeit oberhalb von 2 kHz gleichmäßig abfällt. Wie durch die Kreuze in Abb. 3b ersichtlich, ist die Nachhallzeit bei der Frequenz von 8 kHz nur in zwei Räumen größer als 1 s. Der beobachtete Abfall beruht auf der Luftabsorption, die für höhere Frequenzen als 2 kHz immer stärker ins Gewicht fällt. Man korrigiert daher in bekannter Weise die Sabinesche Formel für die Nachhallzeit bei hohen Frequenzen, indem man zu der Schluckfläche $A = \sum a_i S_i$, die sich aus der Absorption an den Raumbegrenzungsflächen ergibt, die Größe $4\,mV$ addiert (m ist das Dämpfungsmaß der Luft pro Längeneinheit):

$$T = \frac{0,161\,V}{A + 4\,mV}.$$

Die Werte von m hängen außer von der Frequenz noch von der Temperatur und der Luftfeuchtigkeit ab. Sie sind zahlenmäßig bekannt durch die Messungen von V. O. KNUDSEN [8]. Man hat damit die Möglichkeit, bei bekanntem Volumen aus den gemessenen Nachhallzeiten die Schluckfläche A für hohe Frequenzen zu bestimmen. Das wurde für die Frequenz von 8 kHz ausgeführt. Die für die einzelnen Räume berechneten Werte sind in Abb. 4 durch Kreuze gekennzeichnet. Sie sind insofern ungenau, als für die Luftdämpfung in allen Fällen der Wert für 60% relative Luftfeuchtigkeit bei der Temperatur von 20⁰ C gewählt wurde ($m = 1,86 \cdot 10^{-2}$ Meter^{-1}). Ebenfalls angeführt sind die berechneten Schluckflächen für 1 kHz. Hier ist der Einfluß der Luftabsorption praktisch noch zu vernachlässigen. Generell liegen die Werte bei 8 kHz etwas höher, doch sind die Abweichungen oft nur gering. Für die Praxis, ergibt sich aus dieser Betrachtung die bekannte Forderung, die Schallschluckung durch poröse Absorber bei hohen Frequenzen möglichst niedrig zu halten, um dem Einfluß der Luftabsorption zu begegnen.

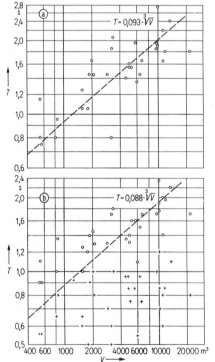

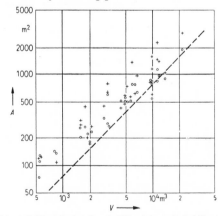

Abb. 3. (a) Mittlere Nachhallzeit zwischen 0,5 und 1 kHz in Abhängigkeit vom Raumvolumen;
(b) o mittlere Nachhallzeit zwischen 2 und 3 kHz, + Nachhallzeit bei 8 kHz.

Abb. 4. Schluckfläche für 1 kHz (o) und 8 kHz (+); ———— $4\,mV$ bei 20⁰ C und 60% relativer Luftfeuchtigkeit.

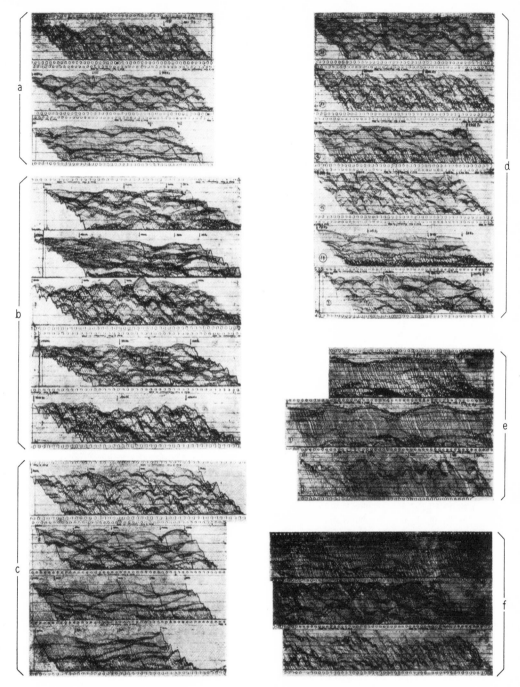

Abb. 5. Nachhallkurven bei gleitender Frequenz; (a) Musikstudio, Baden-Baden
(b) Herkulessaal, München
(c) Studio 1, München
(d) Villa Berg, Stuttgart
(e) Werbefunkstudio, Stuttgart ($V = 90 \, \text{m}^3$)
(f) Sprecherstudio 1, Baden-Baden ($V = 125 \, \text{m}^3$).

185

3. Nachhallkurven bei gleitender Frequenz

a) Meßverfahren

Gewissermaßen als Verfeinerung der Nachhallzeitmessung wurde durch T. SOMERVILLE die Methode zur Aufnahme der Nachhallkurven bei gleitender Frequenz eingeführt [2]. Er photographierte die Abklingkurven auf dem Schirm eines Oszillographen mit Hilfe eines Films, der langsam weitertransportiert wurde. In unserer Untersuchung wurde die von der Firma Brüel & Kjær in Anschluß an SOMERVILLES Arbeiten entwickelte Apparatur benutzt: Ein mechanisches Schaltwerk erzeugte mit Hilfe eines Schwebungssummers Einzelimpulse aus reinen Tönen von 200 ms Dauer, die von der Schallquelle in den zu untersuchenden Raum abgestrahlt wurden. Als Schallquelle dienten, je nach Frequenzbereich, die im Abschnitt 2a angegebenen Lautsprecher. Der durch jeden Einzelimpuls ausgelöste Schalldruckverlauf wurde von einem Kondensatormikrophon aufgenommen und mit einem Pegelschreiber (Brüel & Kjær) im logarithmischen Maßstab (50-dB-Potentiometer) auf einer bewegten Bandschleife aufgezeichnet. Das erwähnte Schaltwerk veränderte während des Meßvorganges auch kontinuierlich die Frequenz des Schwebungssummers, betätigte einen Abhebmagneten, so daß der Schreibstichel des Pegelschreibers nur während des Abklingvorganges die Bandschleife berührte, und regelte die Aufeinanderfolge der Impulse in der Weise, daß die Abklingkurven dicht nebeneinander geschrieben wurden. Zur Verbesserung der Dynamik war empfangsseitig ein Oktavsieb eingeschaltet. Nach Überwindung einiger Anfangsschwierigkeiten arbeitete die Apparatur sehr zufriedenstellend.

b) Ergebnisse

Von den aufgenommenen Abklingkurven sind in Abb. 5a—f einige Beispiele für verschiedene Frequenzbereiche aus sechs Räumen angegeben. (Die Zahlen bei den eingezeichneten Pfeilen geben die Frequenz in Hertz für die betreffende Abklingkurve an.) Auffallend sind die im allgemeinen in waagerechter Richtung verlaufenden „Musterungen", die durch die Interferenz der durch den jeweiligen Impuls angeregten Eigentöne des Raums beim Abklingvorgang entstehen. Sie hängen vom Ort des Mikrophons ab und sagen daher praktisch nichts aus.

Außer diesen normalerweise vorhandenen Musterungen wurden nur in zwei kleineren Rundfunkstudios Besonderheiten festgestellt. In Abb. 5e erkennt man im oberen Diagramm bei den Abklingkurven um 70 und 100 Hz eine starke Abknickung der Kurven im unteren Pegelbereich, ebenso im mittleren Diagramm bei 160 Hz. Die Ursache dieses Nachklingens konnte nicht festgestellt werden. Den gleichen Effekt beobachten man in noch stärkerem Maße in Abb. 5f im mittleren Diagramm bei 700 Hz. Dieser Abklingvorgang wurde durch den vorhandenen Studiogong hervorgerufen und konnte auch subjektiv bemerkt werden.

Allgemein läßt sich sagen, daß die von SOMERVILLE beobachteten Erscheinungen bestätigt werden. Nicht vertretbar erscheinen jedoch seine Überlegungen über die „komplexe" Diagrammstruktur bei hohen Frequenzen zu sein, die nach unserer Erfahrung nur entsteht, wenn die Frequenz zu schnell geändert wird. Außerdem ist sehr viel Erfahrung notwendig, wenn man versucht, aus den aufgenommenen Nachhallkurven wesentliche Aussagen über die akustischen Eigenschaften des betreffenden Raumes zu gewinnen. Es ist fraglich, ob dies überhaupt möglich ist. Daher wurde im Verlauf der Meßreisen von weiteren Aufnahmen der Nachhallkurven bei gleitender Frequenz abgesehen.

4. Die Frequenzabhängigkeit des Schalldrucks bei stationärer Anregung

Der unregelmäßige Verlauf der Frequenzkurve des Schalldrucks in Räumen ist schon lange bekannt, und man hat auch versucht, daraus Aussagen über die akustischen Eigenschaften des jeweiligen Raumes zu gewinnen, speziell in Hinblick auf die „Diffusität" des Schallfeldes. Hervorzuheben ist das von BOLT und ROOP [3] eingeführte Maß der „Frequenzkurvenschwankung" (Frequency irregularity)

$$F = \frac{\sum p_{\max} - \sum p_{\min}}{\Delta v} \text{ (dB} \cdot \text{s)} .$$

Dabei bedeuten $p_{\max}$ und $p_{\min}$ die Schalldruckpegel in den Maxima und Minima der Frequenzkurve im Frequenzintervall Δv. Um über die Brauchbarkeit der Größe F experimentelle Unterlagen zu gewinnen, wurde in 19 Räumen, darunter 14 aus Tabelle I, die Frequenzkurve des Schalldrucks aufgenommen und analysiert. Die Ergebnisse sind bereits veröffentlicht [9]. Sie zeigen, daß die Parameter der Frequenzkurve in den normalen Räumen im wesentlichen nur durch die Nachhallzeit bestimmt werden. So hängt die mittlere Anzahl $\overline{N}$ der Maxima pro Frequenzintervall nicht von der Zahl der Eigenfrequenzen ab, sondern ist direkt proportional der Nachhallzeit. Außerdem beträgt die mittlere Höhe $\overline{h}$ zwischen Maxima und Minima des Pegels, unabhängig vom untersuchten Raum, etwa 10 dB. Daraus ergibt sich, daß die Frequenzkurvenschwankung $F = \overline{N} \cdot \overline{h}$ proportional der Nachhallzeit ist. Dies soll noch einmal durch die Abb. 6 gezeigt werden. Dargestellt sind die Werte $F/\overline{T}$, die jeweils aus einer der in verschiedenen Räumen oder an verschiedenen Raumplätzen aufgenommenen Frequenzkurven für den angegebenen Frequenzbereich berechnet wurden. ($\overline{T}$ ist die mittlere Nachhallzeit im betreffenden Frequenzintervall.) Die Messung erstreckte sich von 70 bis 4000 Hz. Für diesen großen Bereich sind wie ersichtlich die Abweichungen vom Mittelwert $F/\overline{T} = 1{,}45$ am geringsten. Sie betragen im Mittel $\pm 15\%$. In diesen Fehlerbereich, der durch die gestrichelten Linien angegeben ist, fallen auch die meisten Werte für die schmaleren Frequenzbereiche. Die auf die Nachhallzeit bezogene Frequenzkurvenschwankung ist also praktisch konstant.

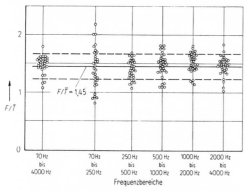

Abb. 6. Auf die mittlere Nachhallzeit bezogene Frequenzkurvenschwankung für einzelne Frequenzbereiche.

Die experimentell gefundenen Resultate stehen in bemerkenswert guter Übereinstimmung mit theoretischen Überlegungen [10]. Diese zeigen, daß die angeführten Ergebnisse immer dann erhalten werden, wenn die Eigenfrequenzen des Raumes genügend dicht liegen, genauer: wenn der Frequenzbereich der Halbwertsbreite einer einzelnen Eigenfrequenz mehr als 10 Eigenfrequenzen umfaßt. Durch diese Forderung wird eine Grenzfrequenz

$$\nu_g = 4000 \cdot \sqrt{V/T}$$

(T Nachhallzeit in s, V Raumvolumen in m³)

definiert, unterhalb von der die Theorie nicht mehr gültig ist. Die Verhältnisse in der Umgebung von ν_g sind ebenfalls experimentell untersucht [11]. Das wesentliche Ergebnis dieser Messungen wird durch Abb. 7a veranschaulicht. Aufgetragen ist das Verhältnis $\overline{N}/\overline{T}$ für Oktavbereiche in Abhängigkeit von der auf die Grenzfrequenz bezogenen Frequenz für einen quaderförmigen Raum von 24 m³. Die Berechnung erfolgte aus den Frequenzkurven bei drei verschiedenen Zuständen:

1. Der Raum ist leer und sehr hallig,

2. Der Raum ist so gedämpft, daß die Nachhallzeit praktisch frequenzunabhängig ist ($T \approx 0{,}8$ s),

3. Der Raum enthält acht halbzylindrische, geschlossene „Diffusoren" aus 25 mm starken lackierten Holzleisten (Maße dieser Körper: Höhe 1,8 m, Breite 1 m, Tiefe 0,25 m), die auf drei zueinander senkrechten Flächen verteilt waren.

Die erhaltenen Werte liegen im allgemeinen etwas über dem von der Theorie geforderten Wert $\overline{N}/\overline{T} = 0{,}14$, doch sind die Schwankungen für den jeweiligen Raumzustand nur gering. Erst unterhalb der halben Grenzfrequenz macht sich ein Abfall bemerkbar, und die Werte münden in den nach der Eigenfrequenztheorie zu erwartenden Verlauf von $\overline{N}/\overline{T}$, der ebenfalls angegeben ist. Ein ähnliches Bild ergibt sich für die auf die Nachhallzeit bezogene Frequenzkurvenschwankung $F/\overline{T}$, wie man der Abb. 7b entnimmt; insbesondere bemerkt man keinen besonders hervortretenden Einfluß der eingebrachten Strukturen.

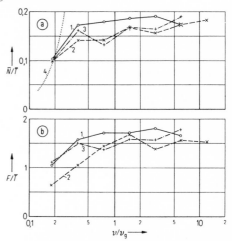

Abb. 7. (a) Mittlere Anzahl der Maxima pro Oktave, bezogen auf die Nachhallzeit in Abhängigkeit von ν/ν_g (Quaderraum von 24 m³)
Kurve 1: leerer Raum,
Kurve 2: gedämpfter Raum,
Kurve 3: Raum mit Strukturen,
Kurve 4: Verlauf nach der Eigenfrequenztheorie.

(b) Auf die Nachhallzeit bezogene Frequenzkurvenschwankung in Abhängigkeit von ν/ν_g in einem Quaderraum von 24 m³;
Kurve 1: leerer Raum,
Kurve 2: gedämpfter Raum,
Kurve 3: Raum mit Strukturen.

Die angeführten Ergebnisse zeigen, daß es nicht möglich ist, in normalen Räumen aus der Frequenzkurve des Schalldrucks oberhalb von ν_g ein raumakustisches Kriterium zu gewinnen, das mehr aussagt als die Nachhallzeit. Hinsichtlich der „Diffusität" des Schallfeldes in Räumen ist es daher günstiger, die geometrische Betrachtungsweise zu bevorzugen, d. h. die Schallrichtungsverteilung für einzelne Raumplätze zu bestimmen.

(Anmerkung bei der Korrektur: Inzwischen wurde ein weiteres Meßverfahren zur Bestimmung der Schalldiffusität in Räumen eingehend geprüft, nämlich die Verwendung zweier Mikrophone und die Kreuzkorrelation ihrer Ausgangsspannungen als Funktion ihres Abstandes. Es konnte gezeigt werden (P. DÄMMIG), daß auch diese Methode unbrauchbar ist.)

5. Richtungsverteilung der Schallrückwürfe

a) Meßapparatur

Als Schallquelle bei der Aufnahme der Richtungsverteilung der Schallrückwürfe diente zunächst der von HARZ und KÖSTERS entwickelte „Kugellautsprecher" [5], bei dem zwölf kleine dynamische Hochtonlautsprecher von 5,5 cm Durchmesser in den Oberflächen eines geschlossenen Pentagon-Dodekaeders angeordnet sind. Bei den Auswertungen der ersten Messungen ergab sich, daß diese Schallquelle für den hier interessierenden Zweck noch nicht genügend gleichmäßig in alle Raumrichtungen abstrahlt. Daher wurde für die weiteren Messungen die Öffnung eines 1,2 m langen zylindrischen Rohres von 1,5 cm Durchmesser zur Schallabstrahlung verwendet[1]. Zur Schallerzeugung diente dabei ein Druckkammersystem, das mit dem einen Rohrende verbunden war. Da die Ausdehnung der abstrahlenden Fläche in dem verwendeten Frequenzbereich klein zur Wellenlänge ist, erhält man eine praktisch konstante Abstrahlung in alle Raumrichtungen, was durch Aufnahme der Richtcharakteristik bestätigt wurde. Auch die Rohrresonanzen waren genügend gedämpft. Stationär abgestrahlt wurden Wobbeltöne ($f = 2$ kHz ± 200 Hz, Wobbelfrequenz 32 Hz).

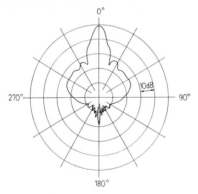

Abb. 8. Richtmikrophoncharakteristik ($f = 2000 \pm 200$ Hz).

Als Ort der Schallquelle wurde in Räumen für musikalische Darbietungen in der Regel ein mittlerer Orchesterplatz gewählt. Die Messung der aus den verschiedenen Raumrichtungen am Meßort ankommenden Schallwellen erfolgte mit dem schon früher benutzten Richtmikrophon, das aus einem metallischen Parabolspiegel von 1,2 m Durchmesser besteht, in dessen Brennpunkt ein dynamisches Mikrophon angebracht ist [4]. Die Richtcharakteristik im logarithmischen Maßstab zeigt Abb. 8. Bei der Messung der Schallrichtungsverteilung wurde in der Regel der Erhebungswinkel des Richtmikrophons in Schritten von 10^0 geändert. Für jeden dieser festen Erhebungswinkel α_i wurde der Azimutwinkel φ kontinuierlich um 360^0 ge

ändert. Die dem Schalldruck proportionale Mikrophonspannung wurde nach entsprechender Verstärkung und unter Einschaltung eines Oktavbandpasses von einem Pegelschreiber im logarithmischen Maßstab aufgezeichnet.

Zur anschaulichen Darstellung der gemessenen Richtungsverteilungen wird folgende Methode benutzt: In eine Metallhalbkugel von 4 cm Durchmesser sind in radialer Richtung Löcher gebohrt, die so verteilt sind, daß zu jedem Loch ein konstanter Raumwinkel gehört, dessen Größe etwa mit dem räumlichen Winkel übereinstimmt, der durch die Halbwertsbreite der Richtmikrophoncharakteristik gegeben ist. In jedes Loch ist ein Metallstab eingesetzt, dessen Länge dem Meßwert im linearen Energiemaßstab in der betreffenden Raumrichtung entspricht. Der Nullpunkt ist durch die Kugeloberfläche gegeben. Als Bezugsmaß dient der aus der Richtung der Schallquelle eintreffende Schall. Der ihm zugehörige Stab hat für alle dargestellten Schallrichtungsverteilungen, die photographisch festgehalten werden, die gleiche Länge von 25 cm.

Als quantitatives Maß wurde die Richtungsdiffusität d eingeführt [4]. Sie wird jetzt im Zusammenhang mit der angegebenen „Igel"-Darstellung der Schallrichtungsverteilung bestimmt. Bezeichnet man mit A_i die Länge der Metallstäbe und mit N ihre Anzahl im gemessenen Raumwinkelbereich, so ist der Mittelwert

$$M = \sum_{i=1}^{N} A_i / N.$$

Außer dem Mittelwert wird die mittlere absolute Abweichung

$$\Delta M = \sum_{i=1}^{N} |A_i - M| / N$$

berechnet. Setzt man zur Abkürzung $\Delta M / M = m$ und für den im reflexionsfreien Raum erhaltenen Wert $m = m_0$, so ist $d = 1 - (m/m_0)$ ein Maß für die „Richtungsdiffusität", das, im Gegensatz zur Nachhallzeit, sowohl vom Beobachtungsort als auch vom Ort der Schallquelle im Raum abhängt. Es ist so gewählt, daß der ideale Hallraum die Richtungsdiffusität $d = 100\%$ erhält. Denn da dort der Schall mit gleicher Intensität aus allen Raumrichtungen eintrifft, wird ΔM und damit $m = 0$. Im reflexionsfreien Raum kommt der Schall nur aus der Richtung der Schallquelle, und man erhält $d = 0$, da $m = m_0$ ist.

Es soll erwähnt werden, daß die Beschränkung der durchgeführten Messungen auf nur einen Frequenzbereich unbefriedigend bleibt. In engem Zusammenhang damit steht auch die noch ungeklärte Frage, welches Auflösungsvermögen des Richtmikrophons notwendig bzw. ausreichend für die Aufnahme der Schallrichtungsverteilung in Räumen ist.

---

[1] Eine Verbesserung besteht auch darin, den „Kugellautsprecher" um eine vertikale Achse genügend schnell rotieren zu lassen.

b) Meßergebnisse

Von den über 100 aufgenommenen Schallrichtungsverteilungen ist eine größere Anzahl in den Abb. 9 bis 18 dargestellt. Gezeigt werden Ansichten „von oben" oder auch „von der Seite". Im ersten Fall sind die Wandreflexionen, im zweiten die Rückwürfe von der Decke am besten zu erkennen. In der Regel tritt, wie zu erwarten, der Schall aus der Richtung der Schallquelle am stärksten hervor, doch besitzen häufig auch die Rückwürfe, besonders auf rückwärtigen Plätzen, Werte von der gleichen Größenordnung, wenn die reflektierenden Flächen genügend glatt und schallhart sind (siehe z. B. Abb. 9 d, 10 b und 10 c). In Abb. 11 c ist der Rückwurf von der Decke sogar erheblich größer als der aus der Richtung der Schallquelle eintreffende Anteil. Dies beruht auf einer im Längsschnitt konkav gekrümmten Decke.

Allgemein bemerkt man, daß die Rückwürfe aus allen Raumrichtungen in bezug auf den direkten

Schall meist deutlich größer werden, wenn man bei unverändertem Schallquellenort mit dem Mikro-

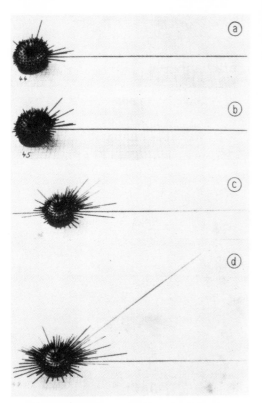

Abb. 9. Schallrichtungsverteilungen im Musikstudio in Baden-Baden;
    (a) Vorderkante Podium, $d = 46\%$,
    (b) Mittelgang neben Reihe 1, $d = 50\%$,
    (c) Mittelgang neben Reihe 8, $d = 57\%$,
    (d) Mittelgang neben Reihe 15, $d = 63\%$.

Abb. 10. Schallrichtungsverteilungen im Kurhaussaal in Wiesbaden, Parkett;
    (a) 4. Reihe, rechter Sitzblock, Mittelplatz, $d = 41\%$,
    (b) Mittelgang neben Reihe 12, $d = 56\%$,
    (c) Mittelgang neben Reihe 32, $d = 53\%$.

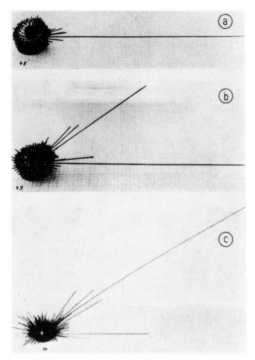

Abb. 11. Schallrichtungsverteilungen im Sendesaal der
Villa Berg in Stuttgart;
(a) 2. Reihe, Mittelplatz, $d = 43\%$,
(b) 7. Reihe, Mittelplatz, $d = 49\%$,
(c) 13. Reihe, Mittelplatz, $d = 63\%$,
(Rückwurf von der Decke größer als der direkte
Schall).

Abb. 12. Schallrichtungsverteilungen im Studio 1 in Frankfurt;
(a) Mittelgang neben Reihe 5, $d = 54\%$,
(b_1), (b_2) Mittelgang neben Reihe 15, $d = 55\%$,
(c) Mittelgang neben Reihe 29, $d = 59\%$.

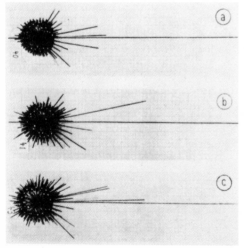

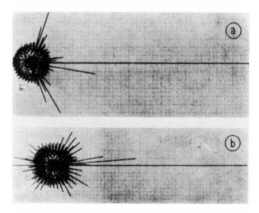

Abb. 13. Schallrichtungsverteilungen in der Royal Festival
Hall in London;
(a) Stalls, letzte Reihe, Mittelplatz, $d = 45\%$,
(b) Terrace Stalls, letzte Reihe, Mittelplatz, $d =$
54%,
(c) Grand Tier, 6. Reihe, Mittelplatz, $d = 57\%$.

Abb. 14. Schallrichtungsverteilungen im Herkulessaal in
München, Parkett;
(a) Reihe 7, Mittelplatz, $d = 38\%$,
(b) Reihe 28, Mittelplatz, $d = 63\%$.

Abb. 15. Schallrichtungsverteilungen im Konzertsaal in Turku;
(a) Reihe 7, Mittelplatz, $d = 53\%$,
(b) Reihe 19, Mittelplatz, $d = 54\%$.

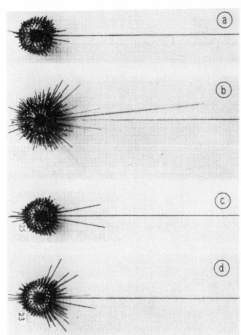

Abb. 16. Schallrichtungsverteilungen an mittleren Plätzen in verschiedenen Räumen;
(a) Musikhalle in Hamburg, $d = 44\%$,
(b) „Die Glocke" in Bremen, $d = 64\%$,
(c) Jesus-Christus-Kirche in Berlin, $d = 48\%$,
(d) Konzertsaal der Musikhochschule in Berlin, $d = 41\%$.

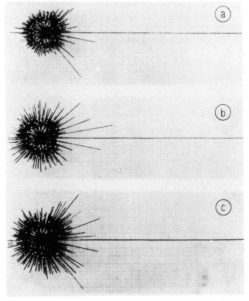

Abb. 17. Schallrichtungsverteilungen in Rundfunkstudios mit „aufgegliederten" Wänden;
(a) Studio F in Bremen, $d = 64\%$,
(b) Studio i in Bremen, $d = 74\%$,
(c) Studio 1 in Hamburg, $d = 73\%$.

phon von vorderen zu rückwärtigen Plätzen geht (siehe Abb. 9 bis 15). Wie ersichtlich, steigen in dieser Reihenfolge auch die Zahlenwerte der Richtungsdiffusität d, wenn die Richtungsverteilung nicht durch mehrere Rückwürfe besonders unregelmäßig wird (siehe Abb. 10c). In den Abb. 16 und 17 sind aus verschiedenen Räumen noch einige Schallrichtungsverteilungen gezeigt, die an mittleren Raumplätzen aufgenommen wurden. Bemerkenswert sind die recht hohen d-Werte in Abb. 17. Diese Richtungsverteilungen wurden in Rundfunkstudios mit „aufgelösten" Raumbegrenzungsflächen (Halbzylinderabschnitte in Abb. 17a und b, Galerien und Pfeiler in Abb. 17c) gewonnen. Abb. 18 zeigt zwei Meßergebnisse für verschiedene Plätze im antiken Theater in Orange, das in der Raumakustik durch die Untersuchungen von F. CANAC [12] bekannt geworden ist. Da der „Bühnenraum", im Gegensatz zu vielen anderen antiken Theatern, durch eine hohe Mauer abgeschlossen ist, kann man dies Theater als einen „Raum ohne Decke" bezeichnen ($V \approx 60000$ bis 70000 m³). Wie zu erwarten, sind die Richtungsdiffusitäten besonders niedrig, da nur noch wenige

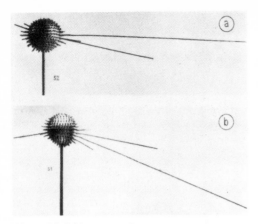

Abb. 18. Schallrichtungsverteilungen im antiken Theater in Orange, Mittelplätze;
(a) Reihe 1, $d = 32\%$, (b) Reihe 21, $d = 23\%$.

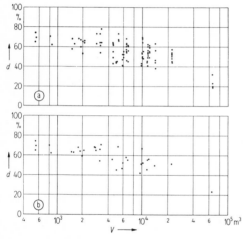

Abb. 19. (a) Meßwerte der Richtungsdiffusität d,
(b) Mittelwerte der Richtungsdiffusitäten.

Die angeführten Meßergebnisse aus normalen Räumen sollen noch durch die Resultate von Untersuchungen in einem Modellraum ergänzt werden[2]. Dieser Raum war quaderförmig und hatte die Abmessungen $2 \times 1,2 \times 0,8$ m³. Die Wände bestanden aus 2 cm dickem Sperrholz, das möglichst porendicht lackiert war. Als Schallquelle diente ein kugelförmiger elektrostatischer Lautsprecher von 5 cm Durchmesser. Die recht gleichmäßige Schallabstrahlung in die verschiedenen Raumrichtungen wurde

Abb. 20. Schallrichtungsverteilung im ungedämpften Modellraum. Schallquelle und Richtmikrophon auf der Längsachse des Raumes, Rauschfrequenzband 22,5 bis 28,5 kHz, Nachhallzeit $T \approx 100$ ms.
(a) Wände und Decke glatt, $d = 74\%$,
(b) Wände und Decke mit Rechteckgittern verkleidet, $d = 77\%$,
(c) Wände und Decke mit Halbzylindergittern verkleidet, $d = 80\%$.
Maße der Gitterstäbe: Breite 2,5 cm, Dicke 1,2 cm, Mittenabstand 5 cm.

Rückwürfe auftreten. Sie sind jedoch auf vorderen Plätzen größer als auf entfernteren.

Eine Zusammenstellung aller gemessenen Werte der Richtungsdiffusität wird in Abb. 19a gegeben. Die zugehörigen Räume sind nach der Größe ihres Volumens angeordnet. Man erkennt, daß z. B. d-Werte um 65% sowohl in großen als auch in sehr kleinen Räumen vorkommen. Generell bemerkt man jedoch ein Abfallen von d mit steigendem Volumen. Diese Tendenz zeigen auch die Mittelwerte im jeweiligen Raum, die etwa den Werten an mittleren Plätzen entsprechen (siehe Abb. 19b).

[2] Diese Messungen wurden von Dipl.-Phys. J. RUPPRECHT im III. Physikalischen Institut der Universität Göttingen ausgeführt.

noch durch genügend schnelle Rotation des Lautsprechers um eine vertikale Achse verbessert. Stationär abgestrahlt wurde Rauschen der Bandbreite
22,5 bis 28,5 kHz. Als Richtmikrophon diente ein
kreisförmiges elektrostatisches System von 12 cm
Durchmesser, dessen Halbwertsbreite etwa 10⁰ betrug.

Das besondere Interesse bei diesen Messungen
galt dem Einfluß von schallstreuenden Strukturen
auf die Schallrichtungsverteilung. Abb. 20 zeigt die
Ergebnisse für den ungedämpften Raum. Man erkennt, daß die Schallenergie durch die eingebrachten Strukturen gleichmäßiger auf die verschiedenen Raumrichtungen verteilt wird. Entsprechend steigt
auch die Richtungsdiffusität d. Der Abb. 21 ist zu
entnehmen, daß durch die Dämpfung der Bodenfläche mit Gesteinswolle (Annäherung an die Verhältnisse mit Publikum) die reflektierten Schall-
anteile verringert werden. Insgesamt verkleinern
sich dadurch auch die d-Werte. Es ist aber auch in
diesem Fall eine Verbesserung der Diffusität zu
beobachten, wenn Wände und Decke mit streuenden Elementen bekleidet sind. Wie nach den Ergebnissen früherer Untersuchungen [13] zu erwarten

war, ergeben halbzylindrische Strukturen eine
stärkere Verbesserung der Schallrichtungsverteilung
als rechteckförmige. Noch besser wirken wahrscheinlich Kugelabschnitte.

Eine Diskussion der vorstehend beschriebenen
Meßergebnisse führt naturgemäß zu der Frage, ob
aus der Kenntnis der Schallrichtungsverteilung bzw.
Richtungsdiffusität eine Beurteilung der Hörsamkeit eines Raumes möglich ist. Diese Frage ließe
sich bejahen, wenn man eine Richtungsverteilung
angeben könnte, die subjektiv als optimal anzusehen
ist. Nach der statistischen Nachhalltheorie hätte
man eine Verteilung zu erwarten, bei der der reflektierte Schall völlig gleichmäßig aus allen Richtungen am Beobachtungsort eintrifft, mit Ausnahme
des mehr oder weniger stark hervortretenden Anteils
aus der Richtung der Schallquelle. Die Richtungsdiffusität d hängt in diesem Fall nur noch von dem
Verhältnis der Schallintensitäten des direkten
Schalles und des Nachhallschalles ab und ist einfach
zu berechnen. In Wirklichkeit treten besonders
durch Einzelrückwürfe Ungleichmäßigkeiten in der
Richtungsverteilung auf, wodurch die Richtungsdiffusität kleiner wird. Wie oben gezeigt, können
diese Rückwürfe durch die Anwendung von schallstreuenden Strukturen „aufgesplittert" werden, so
daß die Schallrichtungsverteilung gleichmäßiger
wird und sich der nach der Nachhalltheorie zu erwartenden Gleichverteilung stärker nähert. Die von
den Raumakustikern in den letzten Jahren sehr
bevorzugte Verwendung von schallstreuenden
Strukturen und die damit gemachten guten Erfahrungen ließen danach indirekt den Schluß
zu, daß eine Gleichverteilung des reflektierten
Schalles auf alle Raumrichtungen das wünschenswerte Optimum sei. Diese Folgerung bedarf jedoch
noch der experimentellen Überprüfung, wobei besonders die Frage der Laufzeiten der ersten Rückwürfe in bezug auf den Primärschall genauer zu
untersuchen ist. Zweifellos spielen gerade diese
Rückwürfe für die Hörsamkeit eines Raumes eine
wesentliche Rolle. Über eine wünschenswerte Richtungsverteilung dieser ersten energiereicheren Rückwürfe liegen jedoch keine quantitativen Angaben
vor. Man weiß bislang nur, daß sie, zumindest bei
Sprache, in einer Zeitspanne von 30 ms sogar um
bestimmte Beträge größer als der Primärschall sein
dürfen, ohne die Lokalisation der Schallquelle zu
beeinträchtigen [14]. Das wiederum könnte bedeuten, daß man bei der Bewertung der Richtungsdiffusitäten diese Rückwürfe innerhalb einer gewissen noch festzulegenden Laufzeit ΔT als zum
Primärschall gehörend ansehen müßte, so daß man
nur die Schallrichtungsverteilung der Rückwürfe
mit Laufzeiten $> \Delta T$ überprüfen müßte. In diesem
Fall müßte man bei einer Messung der Schall-

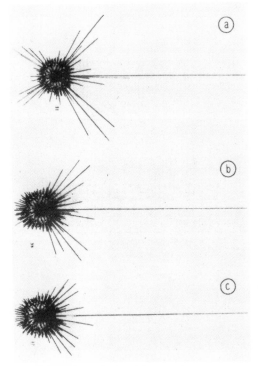

Abb. 21. Schallrichtungsverteilung im gedämpften Modellraum (Gesteinswolle auf der Bodenfläche); Nachhallzeit $T \approx 30$ ms;
(a) $d = 50\%$, (b) $d = 57\%$, (c) $d = 61\%$; sonst
wie in Abb. 20.

richtungsverteilung nicht mehr mit stationärer Anregung arbeiten, sondern mit kurzen Impulsen, und die Rückwürfe in bezug auf Raumrichtung, Intensität und Laufzeit registrieren. Versuche dieser Art sind bereits begonnen. Die endgültige Klärung der aufgeworfenen Frage nach der optimalen Schallrichtungsverteilung in einem Raum wird man erst durch systematische Untersuchungen und Testversuche mit elektroakustischen Nachbildungen der Rückwurffolgen gewinnen können.

6. Größe und zeitliche Folge der Schallrückwürfe

a) Meßverfahren

Zur Registrierung der Schallrückwürfe wurde das bereits früher benutzte Meßverfahren beibehalten [4]. Als Schallquelle diente eine Funkenstrecke, die durch Entladung eines Kondensators einen kurzen Einzelimpuls aussendet (über Einzelheiten wie Schalldruckverlauf und Spektrum siehe [15]). Der direkte Schall und die darauf folgenden Rückwürfe wurden am jeweiligen Beobachtungsort von einem Mikrophon mit kugelförmiger Richtcharakteristik aufgenommen und nach linearer Verstärkung auf dem Schirm eines Oszillographen sichtbar gemacht und photographiert. Auf den Sitzplätzen war das Mikrophon stets

in normaler Ohrhöhe angebracht. Um ein Bezugsmaß für die Amplituden an den verschiedenen Raumplätzen zu haben, wurde ein zweites Mikrophon, das sehr nahe bei der Schallquelle stand, dem Meßmikrophon parallel geschaltet. Auf den Oszillogrammen erscheint daher am linken Rand ein Einzelimpuls, der den Schalldruck der Schallquelle angibt. Dem Abstand der Rückwürfe von diesem Vergleichsimpuls entnimmt man deren Laufzeit bis zum Meßort.

Gleichzeitig mit der Aufnahme der Rückwurffolgen wurde wie schon früher [4] mit Hilfe von zwei Thermoumformern und Fluxmetern der „50-Millisekunden-Energie-Anteil" am jeweiligen Meßort bestimmt:

$$D = \int_0^{50\,\mathrm{ms}} p^2(t)\,\mathrm{d}t \Big/ \int_0^\infty p^2(t)\,\mathrm{d}t$$

($p(t)$ Schalldruck am Meßmikrophon in Abhängigkeit von der Zeit).

b) Meßergebnisse

Von den zahlreichen aufgenommenen Rückwurffolgen ist in den Abb. 22 bis 29 eine kleine Auswahl zusammengestellt. Als Meßorte sind hintereinanderliegende Plätze in der Längsrichtung des jeweiligen Raumes gewählt. Man bemerkt zunächst ganz allgemein auf vorderen Plätzen ein Überwiegen des direkten Schalles gegenüber den folgenden Rückwürfen. Auf entfernteren Plätzen sind dagegen die Rückwürfe von der gleichen Größenordnung wie der direkte Schall. Das gilt jedoch nicht für die Verhältnisse in dem antiken Theater in Orange (Abb. 29). Hier dominiert auch in großer Entfernung von der

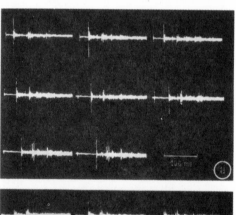
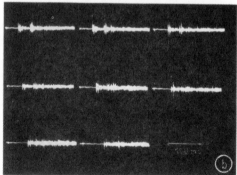

Abb. 22. Rückwurffolgen in der Royal Festival Hall, „Stalls"; von links nach rechts: Reihe B, D, F, H, L, N, S, U (siehe Abb. 30);
(a) Mittelplätze im mittleren Sitzblock,
(b) Plätze im rechten Sitzblock am Gang.

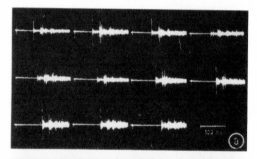
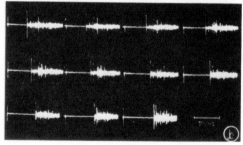

Abb. 23. Rückwurffolgen in der Royal Festival Hall, „Terrace Stalls"; von links nach rechts: Reihe A, C, E, G, K, M, O, R, T, W, Y (siehe Abb. 30);
(a) 2. Sitzblock von rechts, Plätze am Mittelgang,
(b) rechter Sitzblock, Mittelplätze.

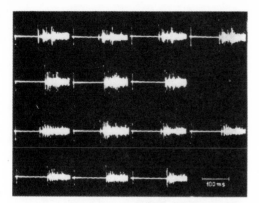

Abb. 24. Rückwurffolgen in der Royal Festival Hall, „Grand Tier"; von links nach rechts: Reihe B, C, E, G, K, M, O (siehe Abb. 30); in den oberen zwei Reihen: Mittlerer Sitzblock, Mittelplätze, in den unteren zwei Reihen: Plätze am rechten Gang.

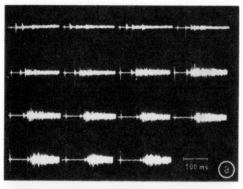

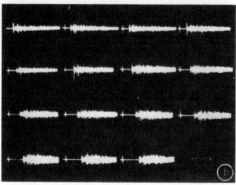

Abb. 25. Rückwurffolgen im Herkulessaal in München, Parkett, von links nach rechts: Reihe 1, 3, 5, 7, ⋯, 25, 27, 29 (siehe Abb. 31); (a) Mittelplätze, (b) rechte Außenplätze.

Schallquelle der direkte Schall, und alle Rückwurffolgen zeigen den Typus wie bei vorderen Plätzen innerhalb eines geschlossenen Raumes. Allerdings

nimmt die Lautstärke mit zunehmender Entfernung ab. In der Regel folgen die Rückwürfe mit so kleinem Zeitabstand, daß sie nicht mehr getrennt registriert werden und in der Umgebung der Nulllinie nur einen mehr oder weniger hohen kontinuierlichen „Untergrund" bilden. Es ist jedoch typisch für viele Raumplätze, daß sich Einzelrückwürfe oder mehrere, mit geringem zeitlichem Abstand aufeinanderfolgende Rückwürfe (Rückwurfgruppe) deutlich aus dem „Untergrund" herausheben. Die Amplitude kann dabei die des direkten Schalles überschreiten (siehe z. B. Kurhaussaal Wiesbaden).

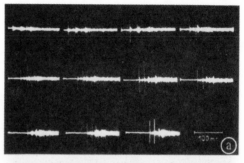

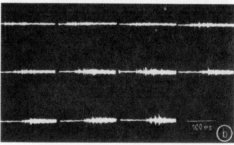

Abb. 26. Rückwurffolgen im Kurhaussaal in Wiesbaden, Parkett, von links nach rechts: Reihe 1, 4, 7, 10, ⋯, 25, 28, 31 (siehe Abb. 33);
(a) rechter Sitzblock, Plätze am Mittelgang,
(b) linker Sitzblock, Mittelplätze.

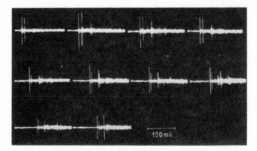

Abb. 27. Rückwurffolgen im Kurhaussaal in Wiesbaden, Parkett, von links nach rechts: Reihe 2, 5, 8, 11, ⋯, 23, 26, 29 (siehe Abb. 33).

195

In Abb. 28 ist eine Gruppe von besonders unterschiedlichen Rückwurffolgen aus verschiedenen Räumen zusammengestellt. Bei der ersten folgen auf den herausragenden direkten Schall sehr kontinuierlich zahlreiche Rückwürfe in dichter Folge, deren Amplituden von gleicher Größenordnung sind. Bei der zweiten Folge leiten gewissermaßen Einzelrückwürfe von der Amplitude des direkten Schalles zu der des Untergrundes über. Die dritte Folge beginnt mit einer Gruppe von Einzelrückwürfen in der Größenordnung des direkten Schalles. Die drei nächsten Oszillogramme zeigen besonders große Rückwürfe in verschiedenen Abständen vom direkten Schall, während die letzte Rückwurffolge an einem Platz aufgenommen wurde, an dem zur Schallquelle keine Sichtverbindung bestand.

Diese qualitative Übersicht der aufgenommenen Rückwurffolgen wird durch die umfangreichen statistischen Untersuchungen von G. R. SCHODDER in

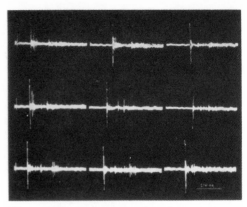

Abb. 29. Rückwurffolgen im antiken Theater in Orange; von links nach rechts: Reihe 1, 11, 21, von oben nach unten: Mitte, halblinks, links.

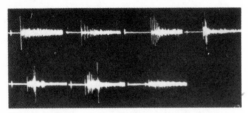

Abb. 28. Besonders unterschiedliche Rückwurffolgen aus verschiedenen Räumen; von links nach rechts: Herkulessaal (1), Royal Festival Hall, London (2, 3), Schillertheater, Berlin (4), Konzertsaal, Turku (5, 6), Kurhaussaal, Wiesbaden (7).

quantitativer Weise ergänzt [15]. Bislang fehlen aber noch ausreichende Unterlagen, um aus den Rückwurffolgen eine Beurteilung der Hörsamkeit des betreffenden Raumplatzes zu gewinnen. Das vorliegende Meßmaterial gestattet es jedoch, die verschiedensten Rückwurffolgen im reflexionsfreien Raum mit Hilfe von Lautsprechern und geeigneten Verzögerungseinrichtungen nachzubilden. Durch definierte Änderungen der Laufzeiten, der Amplituden und der räumlichen Richtung der Rückwürfe wird man dann versuchen, den Einfluß dieser Größen subjektiv zu bewerten. Von Interesse ist z. B. der schon bei der Richtungsdiffusität erörterte Einfluß

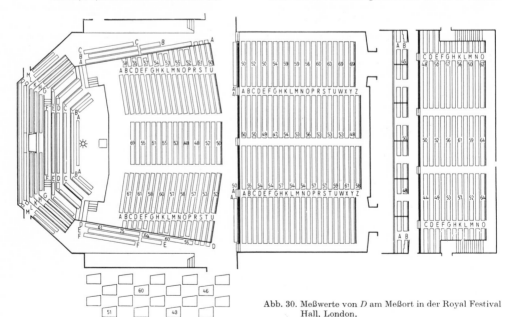

Abb. 30. Meßwerte von D am Meßort in der Royal Festival Hall, London.

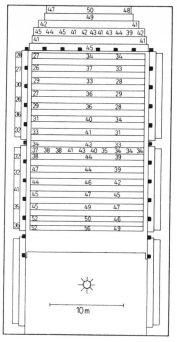

Abb. 31. Meßwerte von D am Meßort im Herkulessaal, München.

schallstreuender Strukturen, der sich in den Rückwurffolgen dadurch bemerkbar macht, daß besonders herausragende erste Rückwürfe nicht mehr auftreten. Daß solche systematischen Untersuchungen aussichtsreich erscheinen, wird durch die Erfahrungen von R. VERMEULEN mit „Stereo-Nachhall"-Anlagen bestätigt [16], der feststellte, daß nicht die Nachhallzeit, sondern die Laufzeitdifferenz der ersten Rückwürfe für die gehörmäßige Beurteilung der Raumgröße besonders wichtig ist. Genauere Bewertungen quantitativer Art, wie sie schon durch die Untersuchungen von HAAS [1] und E. MEYER und SCHODDER [14] für Einzelfälle aufgefunden wurden, fehlen jedoch noch bis heute. Daher wurden, gewissermaßen unter Vernachlässigung der „Feinstruktur" der Rückwurffolgen, die bereits vor längerer Zeit begonnenen Messungen des „50-ms-Energie-Anteils" D in größerem Umfang fortgesetzt.

Ähnlich wie die Richtungsdiffusität ist auch

$$D = \int_0^{50\,\text{ms}} p^2(t)\,\mathrm{d}t \left/ \int_0^\infty p^2(t)\,\mathrm{d}t \right.$$

eine Größe, die bei bestimmter fester Stellung der Schallquelle vom jeweiligen Beobachtungsort im Raum abhängt. Die Einzelmessung ist sehr schnell durchzuführen, und man kann daher die D-Werte an zahlreichen Raumplätzen messen. In den Abb. 30 bis 34 sind in den Grundrissen mehrerer großer Räume die Zahlenwerte D in Prozenten am jeweili-

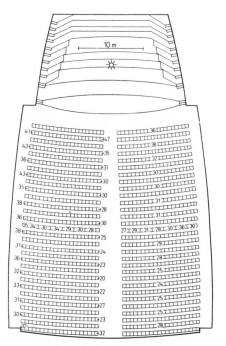

Abb. 32. Meßwerte von D am Meßort im Studio 1, Frankfurt.

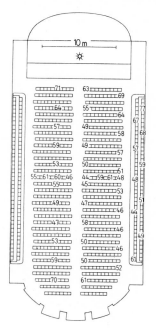

Abb. 33. Meßwerte von D am Meßort im Kurhaussaal, Wiesbaden.

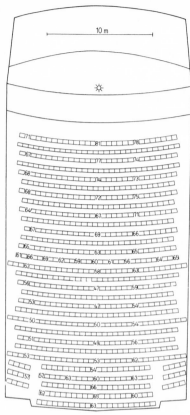

Abb. 34. Meßwerte von D am Meßort im Konzertsaal der Musikhochschule, Berlin.

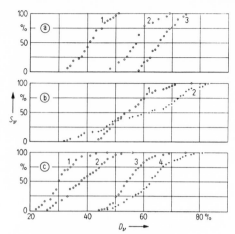

Abb. 35. Summenhäufigkeit der D-Werte;
  (a) 1: Jesus-Christus-Kirche, Berlin,
      2: Unterhaltungsstudio, Baden-Baden,
      3: Aula der schwedischen Handelshochschule,
          Helsinki;
  (b) 1: Kurhaussaal, Wiesbaden, Parkett,
      2: Konzertsaal, Turku;
  (c) 1: Studio 1, Frankfurt,
      2: Herkulessaal, München, Parkett,
      3: Royal Festival Hall, London,
      4: Konzertsaal der Musikhochschule Berlin,
          Parkett.

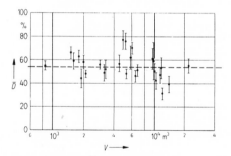

Abb. 36. Mittelwerte $\bar{D}$ und mittlere Streuung aus allen gemessenen Räumen.

gen Meßort eingetragen. In der Nähe der Schallquelle auf vorderen Plätzen ist naturgemäß der direkte Schall besonders groß. Daher erhält man dort in der Regel die höchsten Werte. Doch nehmen diese keineswegs kontinuierlich in Richtung auf rückwärtige Plätze ab; zwar wird der direkte Schall kleiner, doch bewirken die zahlreicheren Rückwürfe innerhalb der ersten 50 ms in den rückwärtigen Raumteilen eine gewisse Kompensation in der Weise, daß die D-Werte hinten wieder in der Größenordnung der vorderen Plätze liegen können (z. B. rückwärtiger Rang im Herkulessaal und unter dem Rang in den letzten Parkettreihen in der Musikhochschule Berlin). Die kleinsten Werte erhält man oft in der Raummitte, wie am Beispiel des Kurhaussaales in Wiesbaden (Abb. 33) besonders deutlich wird. Diese Beobachtung steht unter Umständen mit den häufigen ungünstigen subjektiven Beurteilungen der Hörsamkeit an den Mittelplätzen in vielen Räumen in Zusammenhang.

Der Bereich der D-Werte innerhalb eines Raumes läßt sich am besten an einer Summenhäufigkeits-

kurve übersehen. Bezeichnet man mit n_i die Anzahl der gemessenen Werte D_i, mit N die Gesamtzahl der Meßwerte, so hat man mit

$$S_\nu = \frac{100}{N} \sum_{i=1}^{\nu} n_i$$

den prozentualen Anteil aller Meßwerte bis zum Wert D_ν. Diese Summenhäufigkeit S_ν ist in Abb. 35 für einige Räume in Abhängigkeit von D_ν dargestellt. Man erkennt, daß sich die D-Werte, auch in Räumen mit vergleichbaren Volumen, durchaus über unterschiedliche Bereiche erstrecken. So ist z. B. das höchste D im Studio 1 in Frankfurt gleich dem niedrigsten im Konzertsaal der Musikhochschule in Berlin. Der Anstieg der S_v-Werte ist ein anschauliches Maß für die mittlere Streuung ΔD der Meßwerte des „50-ms-Energie-Anteils" in einem Raum:

$$\Delta D = \sqrt{\sum (D_i - \overline{D})^2 / N}.$$

$$\overline{D} = \sum_{i=1}^{N} D_i / N$$

ist der Mittelwert der im jeweiligen Raum gemessenen D-Werte. In Abb. 36 sind über dem Volumen der verschiedenen Räume, in denen D gemessen wurde, die Mittelwerte $\overline{D}$ zusammengestellt. Größter und kleinster Wert sind $\overline{D} = 77\%$ bzw. $\overline{D} = 31\%$. Die senkrechten Striche geben die mittlere Streuung D an. Sie liegt zwischen ± 3 und $\pm 16\%$. Eine charakteristische Volumenabhängigkeit besteht nicht. Da in allen diesen Räumen mit Ausnahme des Schillertheaters musikalische Aufführungen stattfinden, erscheint es sinnvoll, den Mittelwert über alle Räume zu bilden. Man erhält $\overline{D} = 54\%$ (waagerechte Linie in Abb. 36).

Erwähnt sei, daß D in 13 Räumen der Tabelle I auch dort gemessen wurde, wo von den Tonmeistern das Mikrophon für Rundfunkaufnahmen angebracht wird. An diesem Ort wurden gegenüber den anderen Plätzen stets die größten Werte erhalten. Sie liegen zwischen 38 und 76%. Der Mittelwert für diese Mikrophonorte beträgt 62%.

Es soll noch darauf hingewiesen werden, daß die mittlere Streuung ΔD für die akustische Beurteilung des Gesamtraumes von gewisser Bedeutung sein dürfte. In Rundfunkstudios mit kleinen ΔD-Werten ist z. B. nach einer Mitteilung von W. KUHL die Stellung des Aufnahmemikrophons, die von den Tonmeistern gewählt wird, weniger kritisch.

Allen Stellen, die uns bei der organisatorischen und technischen Durchführung der Messungen unterstützt haben, möchten wir unseren Dank aussprechen. Für die Möglichkeit der Untersuchungen in den ausländischen Räumen sind wir den Herren ARNI, CANAC und PARKIN zu besonderem Dank verpflichtet.

Der Deutschen Forschungsgemeinschaft verdanken wir die Bereitstellung der finanziellen Mittel für das Meßprogramm. Weitere Beihilfen gewährten in dankenswerter Weise die deutschen Rundfunkanstalten, der finnische Rundfunk, die Stadtverwaltung Turku und der London County Council.

(Eingegangen am 1. März 1956.)

Schrifttum

[1] HAAS, H., Über den Einfluß eines Einfachechos auf die Hörsamkeit von Sprache. Acustica **1** [1951], 49.

[2] SOMERVILLE, T. und GILFORD, C. L. S., Composite cathode ray oscillograph displays of acoustic phenomena and their interpretation. B.B.C. Quart. **7** [1952], 41.

[3] BOLT, R. H. und ROOP, R. W., Frequency response fluctuations in rooms. J. acoust. Soc. Amer. **22** [1950], 280.

[4] THIELE, R., Richtungsverteilung und Zeitfolge der Schallrückwürfe in Räumen. Acustica **3** [1953], 291.

[5] HARZ, H. und KÖSTERS, H., Ein neuer Gesichtspunkt für die Entwicklung von Lautsprechern? Techn. Hausmitt. NWDR **3** [1951], 205.

[6] PARKIN, P. H., SCHOLES, W. E. und DERBYSHIRE, A. G., The reverberation times of ten British concert halls. Acustica **2** [1952], 98.

[7] KUHL, W., Über Versuche zur Ermittlung der günstigsten Nachhallzeit großer Musikstudios. Acustica **4** [1954], 618.

[8] KNUDSEN, V. O., The absorption of sound in gases. J. acoust. Soc. Amer. **6** [1935], 199.
EVANS, E. J. und BAZLEY, E. N., The absorption of sound in air at audio frequencies. Acustica **6** [1956], 238.

[9] KUTTRUFF, H. und THIELE, R., Über die Frequenzabhängigkeit des Schalldrucks in Räumen. Acustica **4** [1954], 614.

[10] SCHRÖDER, M., Die statistischen Parameter der Frequenzkurven von großen Räumen. Acustica **4** [1954], 594.

[11] KUTTRUFF, H., Unveröffentlichte Diplomarbeit im III. Physikalischen Institut der Universität Göttingen (1954).

[12] CANAC, F., L'acoustique des théâtres antiques. La revue scientifique [1951], 151.

[13] MEYER, E. und BOHN, L., Schallreflexion an Flächen mit periodischer Struktur. Acustica **2** [1952], Beiheft 4, AB 195.

[14] MEYER, E. und SCHODDER, G. R., Über den Einfluß von Schallrückwürfen auf Richtungslokalisation und Lautstärke bei Sprache. Nachrichten der Akademie der Wissenschaften in Göttingen IIa. [1952], Nr. 6, S. 31.

[15] SCHODDER, G. R., Über die Verteilung der energiereicheren Schallrückwürfe in Sälen. Acustica **6** [1956], Beiheft 2, 445—465.

[16] VERMEULEN, R., Stereo-Nachhall. Philips' techn. Rdsch. **17** [1956], 229.

17

ACOUSTICAL INVESTIGATIONS IN NUMEROUS CONCERT HALLS AND BROADCAST STUDIOS USING NEW MEASUREMENT TECHNIQUES

E. Meyer and R. Thiele

This English summary was prepared expressly for this Benchmark volume by T. D. Northwood, National Research Council of Canada, from "Raumakustische Untersuchungen in Zahlreichen Konzertsälen und Rundfunkstudios unter Anwendung neuerer Messverfahren," in Acustica, **6**:425-444 (1956).

1. INTRODUCTION

Although reverberation time is a useful measure of acoustical performance of an auditorium, it does not account wholly for the range of subjective appraisals of such halls. To improve on this situation, the ideal procedure would be on the one hand to collect and organize subjective opinions for many halls, and on the other hand to relate them to possible new objective criteria. This paper deals mainly with the latter topic.

In 31 different rooms ranging in volume from 550 to 22,000 m^3, five different kinds of acoustical measurements were made. The objective was to examine several possible ways of describing the acoustical qualities of such rooms. The five methods are described in Sections 2 to 6.

2. REVERBERATION TIME AS A FUNCTION OF FREQUENCY

Table I lists the 31 rooms studied, their volumes, and average reverberation times for 0.5 and 1 kHz. Reverberation times were derived from slopes of decay curves taken on a graphic level recorder. In instances where double decays were observed, as in Figure 1, the best straight line for the range –5 and –35 dB was used. The rooms were unoccupied during the tests and varied in their furnishings as indicated in Table I. (See notes under Captions.)

The variation of reverberation time with frequency is shown in Figure 2. The curves vary considerably in shape especially at low frequencies. Figure 3 shows how reverberation time varies with room volume for three representative frequencies.

Room absorption was calculated from the modified Sabine's formula:

$$T = \frac{0.161V}{A + 4mV}$$

where the last term represents atmospheric absorption, which becomes increasingly important at frequencies above 1 kHz. For 8 kHz and for a temperature and relative

200

humidity of 20 degrees and 60 percent, $m = 1.86 \times 10^{-2} m^{-1}$. Absorption results for 1 and 8 kHz are shown in Figure 4, together with the calculated value of $4mV$ (broken line).

3. RECORDING OF REVERBERATION TIMES BY THE "IMPULSE GLIDE" METHOD

A refinement of the reverberation-time method, developed by T. Somerville [2], utilized a sequence of 200-millisecond tone bursts differing by small increments of frequency. The resulting decay curves were recorded in an overlapping sequence as illustrated in Figure 5. The patterns thus produced draw immediate attention to particular frequency regions where irregularities in the decays betray excitation of eigentones in the rooms (indicated by the arrows). In one instance (Figure 5f) a strong irregularity around 700 Hz was traced to excitation of a bell in the studio. In general, however, the glide patterns, though interesting, show effects that are too complex for easy interpretation, and this phase of the study was terminated.

4. FREQUENCY DEPENDENCE IN STEADY-STATE SOUND PRESSURE

If a room is excited by a steady sinusoidal sound whose frequency is slowly increased, the level observed at a given point in the room is found to fluctuate, yielding a series of maxima and minima. This phenomenon was expressed by Bolt and Roop [3] in terms of the "frequency irregularity,"

$$F = \frac{\Sigma p_{max} - \Sigma p_{min}}{\Delta \nu} \frac{dB}{Hz}$$

where p_{max} and p_{min} are successive maximum and minimum levels in the frequency interval $\Delta \nu$. Measurements of F were made in 19 rooms, including 14 of those listed in Table I. The results showed [9] that the average number, $\bar{N}$, of maxima in a given frequency range is proportional to reverberation time, and the average range between maxima and minima is about 10 dB, independent of the room under study. Thus $F = \bar{N} \cdot \bar{h}$ is proportional to reverberation time. This result is demonstrated in Figure 6 where, for a range of frequencies and room locations, the data fit the relation $F/T = 1.45$.

A similar conclusion was reached theoretically by Schroeder [10], for rooms and frequencies such that the eigenfrequencies lie sufficiently close together. The limiting frequency above which this result applied is defined by

$$\nu_g = 4000 V/T$$

Figure 7 shows the observed variation in $\bar{N}/\bar{T}$ and F/T with frequency in the vicinity of the cut-off frequency, for three different room conditions. It is of interest to note that the presence of diffusers in the room (Condition 3) makes little difference in the result.

It thus can be concluded that the frequency irregularity and similar steady-state phenomena are equivalent to reverberation time and yield no additional information.

5. DIRECTIONAL DISTRIBUTION OF REFLECTED SOUNDS

a) Apparatus

To provide a nondirectional source, the sound was radiated from the end of a 1.5-cm diameter tube, the signal being a warble-tone of 2000 ± 200 Hz, modulated 32 times per second. The sources was placed typically in the middle of the performing area in a hall or studio. The directional receiver consisted of a small microphone mounted at the focus of a 1.2-m diameter parabolic reflector, for which the response was down to the half-power point at 10 degrees from the axis. This was placed at various listening positions, and for each 10-degree interval in elevation a graphic record was taken for a 360-degree sweep in azimuth.

The results were displayed on a three-dimensional "hedgehog" consisting of a hemispherical base in which holes were drilled at 10-degree intervals. Inserted in the holes were rods whose lengths corresponded to the levels relative to the direct sound of the received sounds for the respective angles.

As a quantitative measure, the "directional diffusivity" was derived as follows [4]: The mean, M, of sound levels for all angular segments was calculated, and also the average deviation ΔM from the mean. The ratio of these, $m = \Delta M/M$, was then used to form the directional diffusivity,

$$d = 1 - (m/m_0)$$

where m_0 is the value of m for a room with no reflections. For such a room, $d = 0$. At the other extreme of an ideally diffuse sound field, $m = 0$ and $d = 1$.

b) Measurements

Sample hedgehog patterns, and the corresponding values of d, are given in Figures 9 through 18 for a number of rooms. Generally the diffusivity rises with distance from the source. Remarkably high diffusivities are observed for the studio depicted in Figure 17, when the room boundaries were broken up with hemicylindrical diffusers. A plot of all results, shown in Figure 19, shows a slight trend toward lower diffusion in larger halls.

Similar results were obtained for a model room of dimensions 2 X 1.2 X 0.8 m actuated by a directional source operating between 22.5 and 28.5 kHz. Figure 20 shows the effect of diffusing boundaries in the room, and Figure 21 the further effect of adding an absorbing layer on the floor to simulate audience absorption.

The significance of directional diffusivity as a measure of room quality cannot be determined without more subjective information. One surmises that a high value of diffusivity is desirable, but this does not take full account of the disturbing effect of particular strong reflections. It is probable that the temporal distribution of reflections is also important. It is known, for example [14], that the listener cannot resolve arrivals within 30 milliseconds of the direct sound regardless of direction of arrival. This would imply that the listener's impression of diffusivity would depend only on the later arrivals. A modified technique using a pulsed source would be needed to measure such a quantity.

6. MAGNITUDE AND TIME SEQUENCE OF REFLECTIONS

a) Measurement Procedure

The measurement procedure [4] utilized an impulsive signal from a spark source [15]. The sound arriving at each listening position was received with an omnidirectional microphone mounted at normal ear height. The direct sound and reflections were displayed on an oscilloscope screen and photographed. A microphone near the source was connected in parallel with the main microphone to provide a reference level and also a zero point for the time scale.

In addition to the oscilloscope display, the microphone signal was passed through a pair of integrating circuits that measure the total received energy and also the energy received in the first 50 milliseconds from the onset of the direct arrival. The ratio of these, the 50-millisecond fraction, is called the "Definition"

$$D = \frac{\int_0^{50} p^2(t)\, dt}{\int_0^{\infty} p^2(t)\, dt}$$

where $p(t)$ is the sound pressure level as a function of time.

b) Measurement Results

Figures 22 through 29 show some sample reflection-sequence records. Generally, D varies with the microphone position in the room, the direct sound dominating near the front, the reflections becoming increasingly important with distance from the source. (The open-air theater, Figure 29, is an obvious exception.) As a rule, many reflections merge into a more-or-less continuous "background," but frequently there are a few prominent discrete reflections. A particularly interesting collection of these is shown in Figure 28. Various statistical ways of evaluating these results were investigated by Schodder [15], but again there is lack of subjective evidence.

The definition, D, has been calculated for the rooms and measurement positions shown in Figures 30 through 34. As a rule the largest values of D are found near the front, where the direct sound is dominant; the next largest are for positions near the room boundaries, where the diminution in direct sound is partly compensated for by reflections arriving within the 50 milliseconds; lowest values are usually for positions in the middle of the room, a region that tends to be judged least satisfactory by audiences. It was also noted that microphone positions routinely used by sound engineers for broadcast purposes tended to correspond to the higher D values.

The variation in D for a given room may be seen from the cumulative distribution S_p, defined as the percentage of D values up to the value D_p. Representative plots of S_p are shown in Figure 35. It will be observed that they vary widely in form for different halls even with similar volumes.

A quantitative index for a particular hall was provided by the mean and standard deviation of the D values. These are plotted in Figure 36. Since all the halls considered here are for performance of music it is appropriate to note the grand mean, corresponding to the horizontal line $\bar{D} = 54\%$. No significant dependence on room volume was found.

TABLE AND FIGURE CAPTIONS

Table I Descriptive notes
1: room with variable reverberation time. 5, 11, 15–24, 28–31: upholstered seats. 13: with backdrops. 19: stage very reverberant at time of measurement. 25: wood seats with artificial leather upholstery. 26: wood benches. 27: cloth-covered seats. 29: upholstered seats with perforated plastic covering.

Figure 1 Examples of various forms of decay curves.

Figure 2 Reverberation time as a function of frequency. Figure 2i for Studio *i*, Bremen, shows reverberation characteristics for two positions of adjustable wall absorbers.

Figure 3 (a) Average of reverberation times at 0.5 and 1 kHz as a function of a room volume. (b) Circles are average reverberation times at 2 and 3 kHz; crosses are reverberation time for 8 kHz.

Figure 4 Absorption area for 1 kHz (o) and 8 kHz (+); ——— 4 mV at 20°C and 60% relative humidity.

Figure 5 Reverberation curves under gliding frequency.

Figure 6 Frequency variation of $F/\bar{T}$ for various frequency ranges.

Figure 7 (a) Ratio of frequency irregularity to reverberation time as a function of v/v_g (cubical room of 24 m³). Curve 1: empty room; Curve 2: damped room; Curve 3: room with diffusing wall structures; Curve 4: variation according to the eigenfrequency theory. (b) Frequency curve variations relative to the reverberation time as a function of v/v_g in a cubical room of 24 m³; Curve 1: empty room. Curve 2: damped room. Curve 3: room with structures.

Figure 8 Directional microphone characteristic ($f = 2000 \pm 200$ Hz).

Figure 9 Music studio in Baden-Baden. (a) Front edge of stage, $d = 46\%$. (b) Middle aisle, beside row 1, $d = 50\%$. (c) Middle aisle, beside row 8, $d = 57\%$. (d) Middle aisle, beside row 15, $d = 63\%$.

Figure 10 Kurhaussaal in Wiesbaden, orchestra. (a) Fourth row, right block of seats, middle seat, $d = 41\%$. (b) Middle aisle, beside row 12, $d = 56\%$. (c) Middle aisle, beside row 32, $d = 52\%$.

Figure 11 Broadcast hall of the Villa Berg in Stuttgart. (a) Second row, middle seat, $d = 43\%$. (b) Seventh row, middle seat, $d = 49\%$. (c) Thirteenth row, middle seat, $d = 63\%$. (Reflection from the ceiling is greater than the direct sound.)

Figure 12 Studio 1 in Frankfurt. (a) Middle aisle, beside row 5, $d = 54\%$. (b_1) and (b_2) Middle aisle, beside row 15, $d = 55\%$. (c) Middle aisle, beside row 29, $d = 59\%$.

Figure 13 Royal Festival Hall in London. (a) Stalls, last row, middle seat, $d = 45\%$. (b) Terrace stalls, last row, middle seat, $d = 54\%$. (c) Grand Tier, sixth row, middle seat, $d = 57\%$.

Figure 14 Herkulessaal in Munich, orchestra. (a) Seventh row, middle seat, $d = 38\%$, (b) Row 28, middle seat, $d = 63\%$.

Part III

SOUND INSULATION—SUBJECTIVE ASPECTS

Editor's Comments
on Papers 18 Through 24

One of the major objectives in building acoustics is to provide sufficient sound insulation between occupancies in a building. How much is sufficient can be determined only from the reactions of the occupants of the buildings, yet the decisions for a given building must be made when the building is designed rather than after it is occupied. Hence the problem is to relate the physical elements of the structure, and the consequent levels of background and intrusive noise, to the probable degree of disturbance of the inhabitants.

In a business environment, the problem is usually expressed in terms of speech intelligibility; and the necessary information is available for this kind of calculation. The separation of dwellings is, however, more complicated because of the variety of people's activities and because of the corresponding variation in disturbance from intrusive noise.

Part III deals with attempts to examine the subjective answers to the question of how much sound insulation is required. One line of attack is the social survey, in which the occupants of various types of dwellings are questioned about their satisfaction with their environment. One of the more successful surveys, done in Great Britain in 1952–1953 by P. G. Gray, Ann Cartwright, and P. H. Parkin, is reported in Paper 18. (The psychological problem of eliciting the desired information without overinfluencing the subject is discussed in the second part of the paper, not included here.) Other examples of the survey approach are included in Papers 21, 23, and 24, and also in the studies by C. Bitter and P. Van Weeren,[1] and by O. Brandt and I. Dalén.[2]

Another approach is to examine subjectively a laboratory simulation of a particular facet of the insulation problem. This is a more artificial situation, but it permits a study of one acoustical parameter at a time, rather than the mixture as encountered in actual constructions. Two examples, examining reactions to a variety of simulated sound insulation characteristics, are the studies by H. J. Rademacher[3] and by T. D. Northwood and D. M. Clark.[4]

In addition to actual subjective tests one can also calculate from typical noise spectra and sound-insulation characteristics the probable intrusiveness of noise in a given occupancy, and utilize this model to investigate the probable relation between occupant satisfaction and various objective ratings. Jan van den Eijk (Paper 19), for example, collected statistical data on typical radio use, and used this in deriving a criterion for the sound insulation between dwellings. Similarly, Northwood (Paper 20) considered typical household noises such as speech, radio sounds, and mechanical appliances, and related these to existing sound insulation rating schemes. The particular topic of speech, privacy, especially with respect to offices, was examined by Cavanaugh et al. (Paper 21), who calculated speech articulation for transmitted speech versus ambient noise. Subsequently the same material was reviewed by Young (Paper 22), who showed that A-weighted level was as good a measure of intrusive noise spectra as any of the more complicated ones.

In addition to airborne sounds, there is the equally serious problem of footstep sounds and other impact noises that begin as vibrations in the building structure itself. This has been a particularly difficult topic because there is no easy way of metering the input from the actuating mechanisms. Although there is an international standard test procedure, based on the noise transmitted by a standard tapping machine,[5] its validity as an objective test has been questioned by Mariner and Hehmann[6] and by Watters,[7] on the grounds that the machine does not simulate footstep impacts.

In particular, when the tapping machine is used on a hard surface it produces a higher proportion of high-frequency sound than is characteristic of real-life sounds such as footsteps. The standard rating method accentuates this defect by further emphasizing high frequencies. Subjective studies by W. Fasold for concrete slab floors[8] and by Olynyk and Northwood for a variety of floor constructions,[9,10] indicate that the limitation of the tapping machine method can be reduced by modifying the rating scheme to reduce the emphasis on high frequencies. This conclusion was reinforced by social surveys conducted by A. Coblentz and Josse (Paper 23). The search for a satisfactory objective method is, however, far from completed, as was pointed out by van den Eijk.[11]

Meanwhile, although there are several unsolved questions concerning sound insulation requirements, the urgent need for control measures has led national building authorities to establish regulations aimed at securing at least some measure of protection especially between dwelling units. In Paper 24, prepared in 1962, Ove Brandt relates these regulations to what was then available in the form of subjective information and objective test methods.

REFERENCE NOTES

1. C. Bitter and P. Van Weeren, *Sound Nuisance and Sound Insulation in Blocks of Dwellings*, Report No. 24 (Research Institute for Public Health Engineering, T.N.O., 1955.)
2. O. Brandt and I. Dalén, "Ar ljudisoleringen i vara bostadshus tillfredsstillande?" (Is the Sound Insulation in Our Dwellings Sufficient?) *Byggmästaren* 31:145–148 (1952).
3. H. J. Rademacher, "Subjektive Bewertung der Schalldammung, untersucht on elektrisch nachgebildeten Schalldammkurven," *Acustica* 5:19–27 (1955).
4. T. D. Northwood and D. M. Clark, "Frequency Considerations in the Subjective Assessment of Sound Insulation," *Proc. 6th Int. Cong. on Acoustics, Tokyo, August 1968*, P.E.–109–E–112.
5. International Organization for Standardization (ISO) Recommendation R140, "Field and Laboratory Measurements of Airborne and Impact Sound Transmission" (1960).
6. T. Mariner and H. W. W. Hehmann, "Impact Noise Rating of Various Floors," *J. Acoust. Soc. Am.* 41:206–214 (1967).
7. B. G. Watters, "Impact-Noise Characteristics of Female Hard-Heeled Foot Traffic," *J. Acoust. Soc. Am.* 37:619–630 (1965).
8. W. Fasold, "Untersuchungen über den Verlauf der Sollkurve für den Trittschallschutz im Wohnungsbau," *Acustica* 15:271–284 (1965).
9. D. Olynyk and T. D. Northwood, "Subjective Judgments of Footstep-Noise Transmission Through Floors, *J. Acoust. Soc. Am.* 38:1035–1039 (1965).
10. D. Olynyk and T. D. Northwood, "Assessment of Wood-Joist and Concrete Floors," *J. Acoust. Soc. Am.* 42:730–733 (1968).
11. J. van den Eijk, Some Problems in the Measurement and Rating of Impact Sound Insulation, *Appl. Acoust.* 2:269–277 (1969).

18

Reprinted with the permission of the Controller of Her Britannic Majesty's Stationery Office from pp. 1–26, and 34 of *National Building Studies Research Paper No. 27*, 1958, 61 pp.

NOISE IN THREE GROUPS OF FLATS WITH DIFFERENT FLOOR INSULATIONS

P. G. Gray, Ann Cartwright, and P. H. Parkin

INTRODUCTION

NOISE in dwellings falls into a certain general range of type and of loudness, but within that range it may vary considerably from time to time and from dwelling to dwelling. Consequently, it is not possible to approach the problem of standards of sound insulation for dwellings on the basis of measurements of the noise levels encountered and their reduction to some prescribed level that is judged to be acceptable to most people. The only method of approach is in fact to compare known standards of insulation with the disturbance experienced by people having those standards. The only satisfactory way of measuring the disturbance of people is by social survey on a large sample.

The investigation by social survey described here was conducted in 1952-53. So far as the design of dwellings is concerned, the results obtained from the survey have already been acted upon and broad standards of performance of sound insulation have been formulated.[4, 5] The details of the investigation are being published now because it is felt that much of the information may be of interest both to the designers of dwellings and to other investigators into similar social problems. General standards for wide application must of necessity be based on average conditions, and some designers may wish to know to what extent individual conditions or individual reactions may depart from the average, and also what are the specific noise problems encountered and how important are they relatively. For example, although it may not affect standards it is helpful to know that (as will be seen later) the middle floors of a block of flats are more disturbed by noise than either the top or bottom floors; it therefore follows that high blocks are at a disadvantage for sound insulation compared with lower blocks. Similarly, it will be seen that disturbance increases with the number of children in neighbouring flats, so that in blocks of flats without children (such as old people's flats and bachelor flats) the noise problem may be less acute. If all the noise problems are made known it becomes possible to prescribe and to plan in the best interests of sound insulation, in addition to applying average standards.

The inquiry reported here is the third which has been made into the problems of sound insulation in dwellings. The first inquiry[1] dealt with both flats and houses and showed, among other things, that there were many more complaints about noise in flats than there were about noise in houses. No attempt was made to relate the number of complaints in flats to the variation in actual sound insulation. The second inquiry (unpublished) dealt solely with semi-detached houses, and showed that the slightly higher sound insulation of an 11-in. cavity party-wall, compared with a 9-in. solid wall, produced no appreciable change in the proportion of complaints.

In this third inquiry, attention was turned to flats. It was known from numerous measurements (to be reported in detail in a further Research Paper) in flats all over the country that the airborne sound insulation (by which is meant the insulation against sounds originating in the air, such as voices, as distinct from sounds originating in the structure, such as footsteps) of party-walls in flats was usually about 50 dB, when averaged over the frequency range 100 to 3,200 cycles per second, and seldom fell as low as 45 dB. On the other hand, the insulation of party-floors varied from the 30–35 dB of the plain joist floor to the 50 dB of the floating concrete floor. It was also known that at the bottom end of this range, i.e. the plain joist floor, this insulation was so low as often to lead to vigorous complaints from the tenants to the authorities who sometimes had to carry out expensive modifications. Therefore in studying the effectiveness of different values of insulation there was little point in investigating this bottom end of the range: it was clearly unsatisfactory. There remained the problem of relating the rest of the range—from 40 to 50 dB—to the number of complaints from the tenants. It was decided to investigate three groups of flats covering this range of insulation.

Group I had party-floors with an airborne sound insulation of about 50 dB, Group II about 45 dB and Group III about 40 dB. The impact insulation of the three groups chosen varied in a corresponding manner, Group II being about 5 or 6 dB worse than Group I, and Group III about 6–8 dB worse than Group II at the lower and middle frequencies, though at the higher frequencies the impact insulation of Groups II and III was about equal. All groups had party-walls of about 50 dB insulation between living rooms and in most instances between bedrooms also, but in Group III a few of the party-walls between bedrooms had 45 dB insulation.

This enquiry, it should be observed, was confined to local-authority flats, and the results therefore relate only to the tenants of this class of flats. Other sections of the community may differ in their reactions to noise.

In selecting the flats for the investigation it was not possible (though it would have been desirable) to find groups that were identical apart from the sound insulation of the floors. However, there was no serious dissimilarity and the two most important requirements were met by all three groups; namely, they were on quiet sites, and they were planned so that living rooms were above living rooms, bedrooms above bedrooms, etc. The main differences in plan and construction between the three groups are summarized in Table I; other differences can be gleaned from the detailed description of the flats given in Appendix I. Perhaps the most unfortunate difference between the groups was in the number of storeys in blocks; Groups I and II had the same number but Group III had less, thus giving a varying proportion of middle flats in the three groups. The drawback has been overcome in all the tables of results that follow, which compare the relative disturbances of top, middle and bottom flats in the three groups by adjusting the results obtained from middle flats in Group III so as to give them equal weight within their group to the middle flats in the other groups, i.e. three middle flats to each top or bottom flat. The total number of flats varied in each group, but this is of little significance since all results are given as percentages or as averages.

Whole blocks of flats, rather than isolated flats, were chosen for the survey for three main reasons. First, the number of variations in plan and construction was thereby reduced; secondly, it was only possible to measure the sound insulation of a limited number of floors and walls at a limited number of sites, and these measured floors and walls had to be representative as regards sound insulation of all the floors and walls in the same group; and thirdly, it was necessary in considering the complaints from a particular flat to know the possible sources of noise in the surrounding flats, and it was clearly economical to be able to treat each flat visited as both a source of complaint and a source of noise.

TABLE I

MAIN CONSTRUCTIONAL DIFFERENCES BETWEEN THE THREE GROUPS OF FLATS

	Group I (highest insulation)	Group II	Group III (lowest insulation)
Floor construction	Floating concrete screed	Plain concrete	Pugged wood-joist
Number of flats in a tier	5*	5	3 or 4
Access	Stairs and lift	Stairs and outside balcony	Stairs
Number of flats with			
one bedroom	—	100	—
two bedrooms	60	226	219
three bedrooms	320	190	308
four bedrooms	—	44	72
Total numbers of flats	380	560	599
Total numbers in survey	377	545	569

*There were six storeys but the ground floor was devoted to communal services.

PART I—SURVEY RESULTS AND CONCLUSIONS

In this Part the replies to the survey questions relating to noise and allied matters are classified and discussed. The general criticisms levelled at the flats are given first. Then the housewives' answers to the more general questions about noise are given, followed by the results for particular noises and the disturbance they create. Factors affecting the incidence of noise or the liability to complain about noise are then considered, and family circumstances having a bearing on noise are also reviewed. Finally, the general conclusions reached are stated, and the standards of insulation that have been derived from a consideration of the main results are set out.

GENERAL CRITICISMS OF THE FLATS

At the beginning of the interview, after the interviewer had introduced the inquiry as a general one about flats without any reference to noise, the housewife was asked whether there were any things about the flat that she found unsatisfactory. The housewife was left to interpret this question in her own way and the interviewer was instructed to be careful not to suggest any possible source of complaint whatever. The interviewer merely asked whether there was "anything else" after each complaint was made until the informant had nothing more to say. In contrast to this the second question asked specifically about five points; the size of rooms, cupboards, the planning of the kitchen, the hot-water system, and the other heating arrangements. One further question followed before the interviewer introduced the subject of noise. This third question asked whether the housewife liked the neighbourhood. The answers to these three questions have been broadly classified and are given separately for the three groups of flats in Table II.

TABLE II

GENERAL CRITICISMS OF THE FLATS AND THE NEIGHBOURHOOD

	Group I (highest insulation)	Group II	Group III (lowest insulation)
Q. 1. *"Are there any things about this flat that you find unsatisfactory?"* *"Anything else?"* asked after each reply.	Proportion making complaint %	Proportion making complaint %	Proportion making complaint %
Any complaint about *noise* ...	22	36	21
Draughts	26	24	5
Dampness...	41	20	5
Size, layout or number of rooms	4	18	9
Other planning faults*	38	40	31
Space heating	37	18	7
Back-to-back grate unsatisfactory	—	—	10
Water heating	25	5	} 7
Any other complaints	26	16	
No complaints	7	14	42
Average number of these complaints per household	2·2	1·8	0·9

	Proportion thinking it unsatisfactory %	Proportion thinking it unsatisfactory %	Proportion thinking it unsatisfactory %
Q. 2. *"Do you think"*:			
The rooms are a reasonable size	17	42	7
There are enough cupboards ...	1	6	43
The kitchen is well planned ...	9	27	32
The hot-water system is satisfactory	62	28	7
The other heating arrangements are all right	74	50	24
Q. 3. *"Do you like this neighbourhood?"*			
Like the neighbourhood ...	70	74	74
If not, what don't you like about it?			
Reasons critical of neighbours ...	10	9	16
Other reasons	20	17	10
Number of interviews	377	545	569

* Includes a wide variety of complaints such as badly fitting windows and doors, lack of storage space, bad plumbing and windows that are difficult to clean.

In answer to the first question 22 per cent of the housewives in the flats with the highest sound insulation—Group I—made a complaint about noise in some form. For the intermediate Group II this proportion rose to 36 per cent but for the Group III housewives having the lowest sound insulation in their flats it is perhaps surprising to find that only 21 per cent made a complaint about noise. It seems however that the housewives in Group III flats were less inclined to complain about anything. As many as 42 per cent made no complaint at all compared with 14 per cent for Group II and 7 per cent for Group I. The average number of complaints per housewife was also lowest for this group. Furthermore almost every type of complaint, excluding those about the back-to-back grate which was peculiar to this group of flats, occurred less frequently for this group. It might be thought that apart from noise this group had less cause for complaint, but, as can be seen from Appendix IV this group of flats was in fact very crowded. Thus these results suggest that the housewives of Group III were less ready than the other groups to complain in similar housing conditions.

If only the relative proportions who made complaints about different things are considered, it will be seen that, whereas for Group I the proportion complaining about noise is exceeded by the proportions making six other types of complaints, with Groups II and III this proportion is only exceeded by the proportions making the very mixed collection of complaints concerning planning faults.

The main purpose of the second question was to provide a measure of the tendency to complain within each group of housewives, and there is no direct connection with noise. However, the results are included in Table II since they may be of some interest. It may be noted that the Group III flats had fewer cupboards than did the others.

There was little difference between the groups of housewives in the proportion who liked the neighbourhood, but of those who did not, a slightly higher proportion of those in Group III gave reasons critical of their neighbours.

INTERNAL NOISE IN GENERAL

In Table III are given the answers to some of the questions about noise. The percentage who thought too much noise came through the ceiling increases as the sound insulation decreases, being 61 per cent for Group I, 76 per cent for Group II and 78 per cent for the Group III flats. A similar but more pronounced trend exists for those who thought too much noise came through the floor, the proportions being 38 per cent in Group I, 57 per cent in Group II and 66 per cent in Group III.

As might be expected, the answers to similar questions dealing with walls did not vary in this way. Group II had the highest proportion who thought that too much noise came through the walls and also that the walls should be thicker. In a later section it will be shown that this is partly attributable to the different proportion within each group of flats with and without party walls.

The next question in Table III asked the housewife which she noticed most, noises from the flat above, noises from the flat below or noises from next door. (Obviously housewives in top flats had to choose only between the flat below and the flat next door, whilst those in bottom flats had to choose only between the flat above and the flat next door. An analysis of this question for housewives in top, middle and bottom flats will be given in a later section.) For each of the three groups noises from the flat above, which included impact as well as airborne noises, were noticed most by a majority of the housewives. It is clear from this that impact noise is very important. However, with the lower insulation of Group III the airborne noises would appear to be gaining in relative importance.

The housewife was then asked to sum up her opinion by saying whether she found noise from inside the building very disturbing, disturbing, a little disturbing or not at all disturbing. The distribution of answers showed that housewives in the Group II flats were considerably more disturbed by noises from inside the building than were those in the Group I flats. About twice as

TABLE III

OPINIONS ABOUT NOISE IN GENERAL

		Group I %	Group II %	Group III %
Do you think too much noise comes through the ceiling?*	Yes	61	76	78
	No	39	24	22
Do you think too much noise comes through the floor?†	Yes	38	57	66
	No	62	43	34
Do you think the floor and ceiling should be thicker than these	Yes	71	91	93
	No	29	9	7
Do you think too much noise comes through the walls?	Yes	32	41	26
	No	68	59	74
Do you think the walls between flats should be thicker than these?	Yes	50	72	67
	No	50	28	33
Of the noises coming from other flats which do you notice most:				
Noises from the flat above		56	64	57
Noises from the flat below		24	27	32
Noises from the flat next door		15	7	3
(No opinion given)		5	2	8
Could you sum up your opinion by saying whether you find noise from inside the building:				
Very disturbing		12 ⎱ 26	23 ⎱ 42	11 ⎱ 23
Disturbing		14 ⎰	19 ⎰	12 ⎰
A little disturbing		43 ⎱ 74	37 ⎱ 58	29 ⎱ 77
Not at all disturbing		31 ⎰	21 ⎰	48 ⎰
Do you feel you have to be careful about being quiet in flat so as not to disturb people in other flats?	Yes	71	72	57
	No	29	28	43
If yes				
Does that worry you very much?	Yes	42 ⎱ 71	46 ⎱ 72	28 ⎱ 57
	No.	29 ⎰	26 ⎰	29 ⎰
Total number of interviews		377	545	569

* The number of interviews excluding those in top flats were:
Group I 301, Group II 432, Group III 393.
† The number of interviews excluding those in bottom flats were:
Group I 302, Group II 434, Group III 384.

many said they were very disturbed. However, this question produced a quite different and unexpected distribution of answers for the least well insulated flats, Group III. The proportion saying they were very disturbed was 11 per cent—lower than that for Group I—while as many as 48 per cent said they found the noise not at all disturbing.

The last question in Table III reveals something of the housewives' attitude towards noise. They were asked if they felt they had to be careful about being quiet so as not to disturb people in other flats. If they said they felt they had to be careful about being quiet they were asked whether this worried them very much. It will be seen that there was close agreement between the answers for Group I and Group II but that appreciably fewer housewives in the Group III flats felt they had to be careful or were worried about making noise. Thus the housewives of Group III, while thinking that the insulation of their flats might be improved, were not particularly disturbed by noise or very worried about making it. This reluctance to be disturbed was also brought out in the answers of the Group III housewives to questions about specific noises.

PARTICULAR NOISES FROM OTHER FLATS

Table IV gives first the question dealing with the extent to which speech in adjacent flats could be understood. The proportion of housewives who said they could usually hear people talking in the flat above rose from 53 per cent for Group I to 87 per cent for Group III, while the proportion claiming to understand any of the words rose from 23 per cent to 49 per cent. Rather similar trends can be seen for speech from the flat below. The figures for speech from next door vary in a different way because of the different plans.

TABLE IV

PARTICULAR NOISES FROM THE ADJACENT FLATS

	From flat above			From flat below			From flat next door		
	Group I	Group II	Group III	Group I	Group II	Group III	Group I	Group II	Group III
	%	%	%	%	%	%	%	%	%
Q. 17. "Can you usually hear people when they are talking in the flat . . .?"									
Yes	53	72	87	48	59	86	21	30	11
No	47	28	13	52	41	14	79	70	89
When you hear people talking in the flat . . . can you understand:									
Some of the words ...	16	24	29	14	19	26	6	8	3
Most of the words ...	5	14	10	2	12	13	3	3	1
All the words	2	5	10	2	5	5	—	2	
None of the words ...	30	29	38	30	23	42	12	17	7
(Not heard)	47	28	13	52	41	14	79	70	89
Q. 11. "Do you hear . . . from the flat . . .?" Proportion hearing:									
Wireless or television ...	70	80	90	64	59	91	37	39	34
Baby crying	37	33	38	38	30	35	25	17	4
Child's voice	57	63	61	50	51	63	25	30	16
Grown-up's voice ...	63	77	91	58	64	86	28	37	17
Vacuum cleaning ...	48	34	28	28	13	12	19	14	4
Doors shutting	64	71	85	57	59	71	33	40	42
Footsteps	83	88	94	24	32	63	9	18	10
Banging or hammering	77	77	87	56	54	78	30	45	29
Fire being poked ...	66	82	87	27	48	70	28	54	6
Piano (or organ) ...	16	20	32	19	15	29	11	11	14
"Does . . . from the flat . . . disturb you?" Proportion disturbed by:									
Wireless or television ...	21	29	20	15	21	18	11	9	4
Baby crying	9	13	2	10	9	3	4	6	1
Child's voice	7	13	6	6	7	4	2	4	1
Grown-up's voice ...	6	19	9	8	12	7	3	6	1
Vacuum cleaning ...	10	12	6	3	2	2	3	1	1
Doors shutting ...	19	23	18	14	15	19	8	9	8
Footsteps	23	35	22	5	8	6	2	4	—
Banging or hammering	50	51	25	26	27	15	16	19	4
Fire being poked ...	12	29	4	5	10	3	3	16	—
Piano (or organ) ...	5	7	3	5	4	3	4	2	1
Number of interviews...	301	432	393	302	434	384	377	545	569

Speech was also to some extent covered by two of the ten noises listed in another question. These noises were chosen so as to be as precise as possible and at the same time to cover all noises mentioned at all frequently in the pilot inquiry. "Banging and hammering" is perhaps the least well defined of the noises, but it proved to be a very disturbing noise in the pilot inquiry, as indeed it did in the main inquiry also. Two noises, the W.C. and "other plumbing noises", were not included in this list since it was found in the pilot inquiry that the informants were not prepared to say where the noise came from. Even with the noises included, it is clear that in an appreciable number of cases the informants were mistaken as to their source. Thus more housewives claimed to hear vacuum cleaners from the flat above than were actually possessed by those flats. The ratio (vacuum cleaners heard from above to vacuum cleaners possessed) was 1.0 for Group I, 1.3 for Group II and 2.0 for Group III. Similar ratios for the piano or organ were 0.8, 1.2 and 1.3. It seems therefore that this ratio increased as the insulation decreased. An indication of where the noises, whose sources were mistaken, really came from was obtained by plotting on a block diagram both the known sources of noises and the claims to hear that type of noise. This suggested, for example, that the flat above was sometimes credited with noises coming from the flats to the side of the flat above and also on occasions with noises from the flat two storeys above. One might therefore expect this ratio to be affected both by floor insulation and by the number of flats in a block.

The proportion of households possessing vacuum cleaners varied considerably for the three groups of flats, and also there were appreciable differences in the number of children and babies in the Group II flats compared with the other two groups. However, apart from the vacuum cleaner, the baby crying, and a child's voice, all the other noises included in the list might be expected to occur with equal frequency in the three groups of flats. It is seen that the proportion of housewives saying they heard these noises, both from the flat above and from the flat below, increases as the sound insulation decreases. However, the proportion of housewives who heard a purely airborne noise such as adults' voices from either the flat above or the flat below had a greater variation between the three groups of flats than the proportion who heard a purely impact noise like footsteps from the flat above. Moreover the proportion who said they heard footsteps from the flat below, when in fact it is more likely to have been an airborne sound, shows the greater range associated with airborne noises.

A rather different picture is found when the proportion of housewives who are disturbed by the various noises from the flat above and the flat below is considered. In the main the proportion who were disturbed is higher in Group II than in Group I. However, the proportion in the Group III flats who said they were disturbed by these noises is lower than in either of the other two groups in all but one or two cases, in spite of the fact that the sound insulation was lowest for this group and more of the housewives said they actually heard the various noises.

No distinct trend is apparent between the three groups for the proportions who heard the various noises from the flat next door, which is to be expected as there was little difference in the sound insulation of the walls, and any variations are most probably due to the different layout of the flats in the three groups. It will be noticed however that the proportion who were also disturbed by each of the noises from the flat next door is again lowest for the housewives in the Group III flats.

Although for all three sources (above, below and next door) the Group III housewives appeared less willing to say they were disturbed by the noises they heard, the relative importance of different noises appears to be about the same for all three groups. Measured by the proportion who said they were disturbed by the noise from the flat above, the first four noises in importance are:

(1) Banging and hammering.
(2) Footsteps.
(3) Wireless or television.
(4) Doors shutting.

Two out of these four are impact noises; another (banging and hammering) is probably part airborne and part impact, and the remaining noise (the wireless or television) is airborne.

For noises from the flat below, the order of importance changes somewhat and Group III also becomes slightly different from the others. In all cases footsteps is no longer in this first four. Banging and hammering, the wireless and television, and doors shutting remain but some other airborne noise replaces footsteps.

Plumbing noise, which is partly airborne and partly impact, was dealt with by separate questions and not included in the list of ten noises. As mentioned before, the pilot inquiry had shown that housewives said they were unable to say where the sound of the W.C. and other plumbing noises came from. Whether these sounds are heard or not can be expected to vary not only with the sound insulation but also with the layout and size of the blocks of flats. It is seen from Table V that the proportion of housewives who said that they heard these noises is highest for Group II. The proportion who said they were disturbed follows the familiar pattern, with Group III having the smallest proportion.

The proportion disturbed by these plumbing noises has been compared with the proportion disturbed, from other flats in any direction, by the ten noises listed in Table IV. (A rough comparison can be made by taking the proportion disturbed from the flat above.) The four most disturbing noises listed above retain their leading positions, except for Group II where "other plumbing noises" takes third place.

TABLE V

THE W.C. AND OTHER PLUMBING NOISES

		Group I	Group II	Group III
		%	%	%
"Do you hear the W.C. flushing in other flats?"	Yes	80	87	84
	No	20	13	16
If yes, "Does it disturb you?"	Yes disturbs	13 } 80	28 } 87	8 } 84
	No	67	59	76
"Do you hear other plumbing noises?" ...	Yes	60	69	41
	No	40	31	59
If yes, "Do they disturb you?"	Yes disturbs	20 } 60	33 } 69	5 } 41
	No	40	36	36
Number of interviews		377	545	569

NOISES FROM THE STAIRS AND LIFT

So far only the noises arising in other flats have been considered. To complete the picture of noises arising from within the buildings, noises from the stairs, balcony and lift are now dealt with. Some idea of the importance of the stairs, balcony and lift as sources of noise can be obtained from Table VI. The house-wives were asked to say which they noticed most, noises from other flats, noises from the stairs or balcony, or noises from outside the building. As the floor insulation improves it is seen that the noises from the stairs assume a relatively more important position; with the insulation values of the Group I flats these noises would appear to be as important as those from other flats.

TABLE VI

NOISES FROM THE STAIRS AND LIFT

		Group I	Group II	Group III
		%	%	%
Which of these three sorts of noise would you say that you notice most:				
Noises from other flats		35	51	50
Noises from the stairs (or balcony)		35	31	14
Noises from outside the building		29	17	32
(No opinion given)		1	1	4
Do you hear footsteps on the stairs (or balcony)				
	Yes	79	84	90
	No	21	16	10
If yes, Do they disturb you?	Yes, disturb	30 } 79	38 } 84	9 } 90
	No	49	46	81
Do you hear voices from the stairs (or balcony)?				
	Yes	86	71	83
	No	14	29	17
If yes, Do they disturb you?	Yes, disturb	37 } 86	31 } 71	10 } 83
	No	49	40	73
Do you hear the lift?	Yes	96		
	No	4		
If yes, Does it disturb you?	Yes, disturb	39 } 96		
	No	57		
Number of interviews		377	545	569

Two questions similar to those for the W.C. and "other plumbing noises" were asked about footsteps on the stairs and voices on the stairs. They are included in Table VI. The answers to the question about footsteps on the stairs or balcony show the familiar pattern, the proportion hearing the noise increasing from Group I to Group III whilst the proportion disturbed is highest for Group II and lowest for Group III. For voices from the stairs or balcony the pattern is rather different, with the highest proportion both hearing this noise and being disturbed by it in Group I. This may well be due to the lifts in the Group I flats, although these noises will also be affected by the different layouts of the three groups of flats.

In the Group I flats a further question was asked about the lift. As many as 96 per cent of the housewives said they could hear the lift and 39 per cent said they were disturbed by it.

Measured by the proportion saying they were disturbed it is found that for Group I the three noises, the lift, voices from the stairs, and footsteps from the stairs, displace all but banging and hammering in the leading four positions in the list of noises. In Group II only one of these noises, footsteps from the stairs or balcony, manages to reach a position in the first four noises. There are of course no lifts. With the Group III flats the noises from the stairs offer no challenge to the first four noises. These results are consistent with those found for the first question in Table VI.

DISTURBANCE TO SLEEP

From Table VII it is seen that 15 per cent of the households in both Groups I and II, and rather more of Group III, had an adult member who got his main sleep during the day. The proportion who were said to have their sleep disturbed did not differ greatly between groups. In each case the proportion disturbed by noises from other flats was equalled or exceeded by one or both of the other two sources considered, namely, noise from outside and noises from the stairs or balcony.

TABLE VII
DISTURBANCE TO SLEEP

	Group I %	Group II %	Group III %
SLEEP DURING THE DAY			
Are there any adults in your household who get their main sleep during the day? ... ... Yes	15	15	23
No	85	85	77
If yes, those prevented from sleeping by:			
Noises from outside	7	6	8
Noises from stairs or balcony	7	7	3
Noises from other flats	6	7	6
SLEEP AT NIGHT			
Housewife's normal bedtime:			
Before 9 p.m.	—	1	—
9 — 9.59	4	6	4
10 —10.29	13	19	12
10.30—10.59	30	27	12
11 —11.29	37	32	35
11.30—11.59	8	7	9
Midnight or later	8	8	28
Anyone's sleep at night disturbed by:			
Noises from flat above	14	17	12
Noises from flat below	7	7	5
Noises from flat next door	5	5	2
The lift	30	—	—
Noises from stairs or balcony	26	25	6
Noises from outside	28	15	14
Number of interviews	377	545	569

A big difference is seen between Group III and the other two groups in the normal bedtime of the housewife. As many as 28 per cent of Group III house-wives said their normal bedtime was midnight or later. An even higher proportion, 36 per cent, had some member whose normal bedtime was so late.

As with other questions dealing with disturbance, a smaller proportion of the Group III households said their sleep was disturbed by noise from each source. Of the surrounding flats the one above was the most disturbing in each group, but for Groups I and II even more households were disturbed by noises from the stairs or balcony. As many as 30 per cent of the Group I households were said to have their sleep disturbed by the lift and 28 per cent by noises from outside.

THE MOST DISTURBING NOISE

So far noises have been judged by the number of households they disturb. The answers to the question asking which of all the noises from the inside building was thought to be the most disturbing will now be examined. These are given in Table VIII. The four most frequent answers were banging or doors shutting, radio or television, children playing, and footsteps, in that order. The radio or television is an airborne noise and footsteps an impact noise; children playing probably involves both airborne and impact noise and it is not known which of these elements was the chief cause of complaint; banging or doors shutting, which was most frequently named as the most disturbing noise, may also have airborne and impact elements but no doubt the impact element is the more offending. Thus both airborne and impact noises can be a major cause of nuisance, and the evidence as to which is the more troublesome is not conclusive, but there appears to be a strong tendency for impact or structure-borne noises to be considered the more disturbing. This view is strengthened by the fact that the flat above is by far the most frequently mentioned as the source of the most disturbing noise.

TABLE VIII

THE MOST DISTURBING NOISE WITHIN THE BUILDING AND ITS SOURCE

	Group I	Group II	Group III
Most disturbing noise	%	%	%
Banging, doors shutting...	31	21	15
Radio, television	9	15	13
Footsteps	7	12	9
Children	12	14	10
Plumbing noises	3	1	2
Poking fire	1	4	—
Adults' voices	5	4	5
Lift	14	—	—
Others	4	3	6
Not disturbed	14	16	40
Source of most disturbing noise	%	%	%
Flat above	27	33	30
Flat below	9	12	12
Flat next door	5	4	2
Stairs	12	8	3
Balcony	1	3	—
Lift	14	—	—
Others*	10	12	11
Don't know	8	12	2
Not disturbed	14	16	40
Number of interviews	377	545	569

* Mainly more than one source, of which the flat above is one.

It may be noted that a high proportion (40 per cent) of Group III said that no noise disturbed them, and also that an appreciable proportion of those in the Group I flats found that the most disturbing noise came from the lift. In general it will be seen that the results present a similar picture to that obtained from the questions already discussed dealing with individual noises.

FACTORS AFFECTING THE NOISE NUISANCE WITHIN GROUPS

All the results which enable comparisons to be made between different groups have now been set out and discussed, but results were also obtained from the survey which allow certain other factors to be considered for their effect on noise nuisance within a particular group. These considerations are (a) the effect of storey position, i.e. top, bottom or middle floors (all groups), (b) the effect of improving the insulation of party walls alone from Grade I to complete separation, (Group I), (c) the effect of distance from the stairs on the disturbance due to stairs or access balcony (Group II), (d) the effect of different floor coverings in the sitting room of the flat above, and (e) the effect of the number of children in surrounding flats. These factors will now be dealt with in turn.

Top, Middle and Bottom Floors

In the previous sections where comparisons have been made between the three groups of flats the results for the Group III flats have been re-weighted to give the same proportion of middle flats as the other two groups, that is, three middle to one top and one bottom. The necessity for this is at once apparent from a glance at the tables in this present section. Almost all the answers show a variation with the position of a flat in its tier: top, bottom or one of the middle flats. The results for the middle storeys of Group I and Group II, which had three flats between the top and bottom ones, do not, however, show any evidence of a trend from top to bottom and accordingly all these intermediate flats have been grouped as "middle".

TABLE IX

DIFFERENCES BETWEEN TOP, MIDDLE AND BOTTOM FLATS

	Group I			Group II			Group III		
	Top	Mid-dle	Bot-tom	Top	Mid-dle	Bot-tom	Top	Mid-dle	Bot-tom
	%	%	%	%	%	%	%	%	%
Proportion complaining of noise in answer to first question	13	19	32	20	41	25	11	25	16
Too much noise through ceiling ... ... Yes	—	60	64	—	79	70	—	78	76
No	—	40	36	—	21	30	—	22	24
Too much noise through floor ... ... Yes	39	38	—	62	55	—	66	66	—
No	61	62	—	38	45	—	34	34	—
Floor and ceiling should be thicker... ... Yes	57	73	75	85	91	93	86	94	93
No	43	27	25	15	9	7	14	6	7
Too much noise through walls ... ... Yes	34	31	33	42	41	40	31	24	23
No	66	69	67	58	59	60	69	76	77
Noise noticed most from:									
Flat above	—	68	75	—	76	91	—	63	97
Flat below	62	20	—	81	17	—	88	24	—
Flat next door ...	32	9	17	14	5	6	9	2	1
(No opinion given) ...	6	3	8	5	2	3	3	11	2

	Group I			Group II			Group III		
	Top	Mid- dle	Bot- tom	Top	Mid- dle	Bot- tom	Top	Mid- dle	Bot- tom
	%	%	%	%	%	%	%	%	%
Careful about being quiet:									
No	25	23	51	28	20	53	40	39	58
Careful, but not worried	32	31	21	24	24	32	29	30	23
Careful and worried...	43	46	28	48	56	15	31	31	19
Noise from inside building:									
Very disturbing ...	8	13	13	12	28	18	8	12	9
Disturbing	5	16	15	18	20	18	6	13	12
A little disturbing ...	41	42	51	41	36	39	26	29	31
Not at all disturbing ...	46	29	21	29	16	25	60	46	48
Number of interviews	76	226	75	112	323	110	176	208	185

Certain special factors affect the comparisons between floors of the Group III flats. It will be remembered that in some blocks the top flats were larger than those below, some of these top flats having two storeys. Furthermore the ground-floor flats as opposed to those on other floors were completely separated from their neighbours by the entrance hall which ran right through the block. Because of the sloping site, floors of the same storey were stepped up or down at some of the party walls. Thus a bottom-floor flat in one tier may have had part of its party wall common with a first-floor flat in the next tier.

The other two groups of flats did not have such differences between flats in the same tier, while floors of the same storey were not in general stepped up or down. Furthermore, no difference was found between storeys in the floor coverings and little difference in the number of children, these being two important factors affecting the complaints.

An analysis of the main questions dealing with noise in general by top, middle and bottom flats will be found for the three groups of flats in Table IX. It is seen that the proportion mentioning noise in the first question of the interview is lowest for the top flats in all groups, is highest for the middle flats in both Group II and Group III but highest for the bottom flats in Group I.

There is little difference between middle and bottom flats in those who thought too much noise came through the ceiling. Similarly there is little difference between top and middle flats in those who thought too much noise came through the floor. Again there is little difference for the question dealing with walls, except for the top flats of Group III, which tended to be larger than those below because they sometimes included extra bedrooms in an additional storey in the roof space. Thus the results suggest that the housewives' answers about noise from one direction were not appreciably influenced by the presence or absence of noise from other directions.

There is little difference between middle and bottom flats in those who thought the floor and ceiling should be made thicker. For all three groups of flats the proportion in the top flats who thought so is lower, being appreciably so in the case of the Group I flats.

When analysed separately for top, middle and bottom floors the question asking the housewife which she noticed most, noises from the flat above, the flat below or next door, shows more clearly than ever the nuisance of the flat above. The proportion choosing next door is of course affected by the layout of the flats as will be shown later.

From the last but one question in Table IX, which shows the proportion of housewives who felt they had to be careful about being quiet so as not to disturb people in other flats, it is found that there is little or no difference in this respect between those living in top and middle flats, but in bottom flats considerably fewer people felt they had to be careful about noise or were worried about making it. This emphasizes the importance of impact noise as it indicates that the housewife took it for granted that impact noises were the greater source of irritation to those around her.

When housewives were asked to sum up their opinion about noises inside the building it was found that housewives with no flat above them were appreciably less disturbed than the others. Also, those in middle flats found noise more disturbing than those in ground-floor flats, in Group II and Group III. In the Group I flats however there is no difference in this respect between those in the middle and bottom flats, but this may well be because those in the bottom flats here were considerably disturbed by noises from the stairs, as is seen from Table X.

TABLE X

SOME FURTHER DIFFERENCES BETWEEN TOP, MIDDLE AND BOTTOM FLATS

	Group I			Group II			Group III		
	Top	Mid-dle	Bot-tom	Top	Mid-dle	Bot-tom	Top	Mid-dle	Bot-tom
	%	%	%	%	%	%	%	%	%
Noise noticed most:									
Noises from other flats	36	40	20	50	57	33	40	56	40
Noises from stairs or balcony	25	30	60	31	31	29	18	10	19
Noises from outside ...	37	29	20	16	11	36	36	30	38
(No opinion given) ...	2	1	—	3	1	2	6	4	3
Proportion disturbed by:									
W.C. flushing... ...	9	12	17	21	32	24	9	7	9
Other plumbing noises	12	22	21	43	31	26	10	3	7
Voices from stairs or balcony	25	36	53	19	36	26	6	10	13
Footsteps from stairs or balcony	11	30	50	29	41	37	7	8	12
Lift	37	38	44	—	—	—	—	—	—
Number of interviews	76	226	75	113	321	111	176	208	185

Five particular noises which were found to vary with storey are also given in Table X. For all of these five the proportion of Group I saying they were disturbed by them increases from the top to the bottom storeys. The noises were reduced to four in Group II and Group III, which had no lifts. The pattern of the results is changed in Group II and it is the middle flats that are the most disturbed by all the noises except other plumbing noises, which were found most disturbing to the top flats probably beçause of the water storage tanks. With this exception the top flats are again the least disturbed. In the Group III flats the pattern of disturbance is rather more varied, but there is a tendency for disturbance to increase from top to bottom storeys, again with the exception of other plumbing noises which disturb the top flats most, no doubt for the same reason as is suggested for Group II, namely, the storage tanks.

At one stage in the interview the housewife was asked to say which floor in the block of flats she would choose for her family if she were able to choose. The results, including the reasons for the choice, are given in Table XI. A small number did not choose a particular floor but gave answers such as "any but the bottom". For this reason the percentages at the top of Table XI do not add across to one hundred. Similarly, the proportions of those choosing the top, middle or bottom flats for the different reasons given do not add down to one hundred per cent, because some housewives gave more than one reason for their choice.

TABLE XI

HOUSEWIVES' CHOICE OF FLOOR FOR HER FAMILY

	Group I Those choosing			Group II Those choosing			Group III Those choosing		
	Top	Mid-dle	Bot-tom	Top	Mid-dle	Bot-tom	Top	Mid-dle	Bot-tom
Percentage of housewives choosing this floor:	42	30	25	32	38	27	25	54	20
Reasons for choice	%	%	%	%	%	%	%	%	%
No noises overhead ...	69	—	—	76	—	—	68	—	—
No-one below to disturb	—	—	30	—	—	20	—	—	42
Away from outside noise	27	21	—	28	42	—	9	9	—
Can keep eye on children	—	3	36	—	3	21	—	1	24
Dislike of stairs ...	—	11	26	—	43	61	—	19	30
Warmer	1	11	—	—	5	—	1	42	3
Cleaner, healthier ...	22	16	—	24	20	—	28	37	1
Privacy	22	5	—	18	5	—	4	1	—
Not too high or low ...	—	38	—	—	11	—	—	14	—
Other reasons ...	8	12	17	2	6	9	14	13	24

The results are of some interest, because in the previous analyses of the preferences for particular storeys, which were based on noise disturbance alone, the top floor appeared to be favoured by all groups whilst in general the middle floors appeared to find least favour. The present results reveal the free choice of the housewife when other factors are balanced against noise. In Group I the preference still remains for the top floor, but it is transferred to the middle floors in Groups II and III.

The chief reasons given for the preferences now stated are interesting. The main reason for the choice of the top floor in Group I is freedom from overhead noise (which appears to imply a particular dislike of impact noise), but remoteness from outside noise is another frequent reason. Remoteness from outside noise is also one of the two main reasons for the majority choice of a middle floor by Group II, the other reason being dislike of stairs. Group III's decisive preference for the middle floor is largely for quite different reasons, namely, that this floor is warmer or that it is cleaner and healthier, though dislike of stairs is also frequently mentioned. If we consider the different reasons given for the choice made, the important factors appear to be noises overhead (presumably meaning impact noise since airborne noise is common to flats in other directions also), stairs, outside noise, and warmth. Taking all groups into consideration, objection to overhead noise appears to predominate, but dislike of stairs can do so in particular flats (as in Group II) when there are sufficient storeys and no lifts. If there had been lifts in Group II, the influence of the stairs on this group's choice of storey might well have been modified. The preference by Group III for the warmth or cosiness of the middle flats is felt to be due to special circumstances which may not be widely applicable. Outside noise, though important, seems to be mainly a reason for avoiding the bottom flat rather than for choosing either the top flat or a middle one, but the bottom flat has some popularity because it is free from stairs, because keeping an eye on children is easier and because there is no-one below to be disturbed. This last reason again brings in the question of impact noise.

In general, it appears that the bottom flat is least often preferred, whilst the choice of top or middle flats depends on the stairs; if there are lifts then the top flat would probably be preferred; if there are no lifts the middle flats appear to take preference, although this may not be so when there are less than five storeys.

The Party Wall in the Group I Flats

The arrangement of the flats within the blocks of Group I made it possible to illustrate the effects of the party walls. There were in all blocks four flats to a

TABLE XII
THE EFFECT OF THE PARTY WALL IN THE GROUP I FLATS

		End tiers	Middle tiers (party wall)	
		%	%	
Too much noise through ceiling	Yes	61	62	
	No	39	38	
Too much noise through floor	Yes	39	38	
	No	61	62	
Too much noise through walls	Yes	12	50	
	No	88	50	
Proportion hearing noise from next door:				
Wireless or television		8	63	
Banging or hammering		13	45	
Proportion disturbed by noise from next door:				
Wireless or television		2	19	
Banging or hammering		7	24	
			Top floors	Middle floors	Bottom floors
Noise noticed most from:					
Flat above		62	—	66	58
Flat below		33	37	14	—
Flat next door		1	56	17	32
(No opinion given)		4	7	3	10
Noise from inside building:					
Very disturbing		12	12		
Disturbing		11	16		
A little disturbing		46	42		
Not at all disturbing		31	30		
Number of interviews		178	199		

floor, the two end tiers being separated from the two inner tiers by the stairs and lift, while the two inner tiers shared a party wall. This party wall had the same airborne insulation (Grade I) as the party floors. Table XII shows some differences between the answers of housewives in the end tiers and in the middle tiers of the Group I flats.

As might be expected, the end flats heard less noise from other flats through the walls than did the flats in the middle tiers. The answers to the question whether the housewife noticed most noises from the flat above, the flat below or the flat next door are of some interest for the middle tiers of flats. In these flats (which shared a party wall with a flat next door) noise from next door was noticed more than noise from the flat below.

The Stairs and Balconies of the Group II Flats

With the larger blocks of Group II it is possible to examine the variation in disturbance with the distance from the stairs, as shown in Table XIII. The proportion disturbed by footsteps from the stairs or balcony is 61 per cent for those in tiers adjoining staircases, 22 per cent for those at the ends of blocks and 25 per cent for those in the remaining tiers. Similar figures for the proportion disturbed by voices from the stairs or balcony are 46 per cent for those next to staircases, 15 per cent for those at the ends of blocks and 23 per cent for the remainder. Similar trends are apparent for the percentages who said the most disturbing noise came from the stairs or balcony, and for the proportion whose sleep was disturbed by stair or balcony noise. It is clear from this that staircases and access balconies are important sources of noise. For comparison, the Table also gives the results for the Group I flats, where all the flats adjoin staircases. Those adjoining the staircases in Group II complained more than the housewives of Group I, while those at the end of blocks in Group II complained less than the housewives of Group I. It seems that those at the ends of blocks in Group II

TABLE XIII

STAIRS AND BALCONIES

	Tiers of Group II			Group I
	Ends of blocks	Not ends nor adjoining stairs	Adjoining stairs	All tiers adjoin stairs
	%	%	%	%
Proportion hearing:				
Footsteps from stairs or balcony ...	77	74	96	79
Voices from stairs or balcony	63	63	83	86
Proportion disturbed by:				
Footsteps from stairs or balcony ...	22	25	61	30
Voices from stairs or balcony	15	23	46	37
Most disturbing noise is from stairs or balcony	4	7	29	13
Sleep disturbed by noise from stairs or balcony	8	14	43	26
Number of interviews	98	222	225	377

derived some benefit in being away from the stairs, whilst the greater number of people using the stairs in the larger blocks of Group II caused more disturbance to those adjoining the stairs in this group than was caused in Group I.

It may be noted that 52 per cent of those adjoining the stairs in Group II chose the stairs or balcony rather than the other flats, or outside, as the source of noise they noticed most, compared with only 15 per cent for those not adjoining the stairs. The proportion finding noise from inside the building "very disturbing" is 26 per cent for those adjoining the stairs and 20 per cent for the remainder.

The Effect of Floor Coverings in the Sitting Room

Information about floor coverings was confined to the sitting room. These coverings have been classified into three groups, linoleum only (which includes a few cases with no covering at all), small rugs, and finally carpets or large rugs. The division between small rugs and carpets or large rugs was decided by the extent to which people walking about the room could be expected to walk on the carpet or rugs. The interviewer recorded whether she thought less than four out of every five steps would fall on the carpet or rugs; if so they have been classified as "small rugs."

TABLE XIV

THE EFFECT OF DIFFERENT FLOOR COVERINGS IN SITTING ROOM OF THE FLAT ABOVE

	Group I			Group II			Group III		
	Floor covering in Flat above			Floor covering in Flat above			Floor covering in Flat above		
	Carpets or large rugs	Small rugs	Lino-leum only	Carpets or large rugs	Small rugs	Lino-leum only	Carpets or large rugs	Small rugs	Lino-leum only
	%	%	%	%	%	%	%	%	%
Too much noise through ceiling Yes	60	63	75	73	77	89	77	75	92
No	40	37	25	27	23	11	23	25	8
Noise from inside building:									
Very disturbing ...	10	16	20	25	26	26	12	10	23
Disturbing ...	16	15	20	21	18	17	11	14	18
A little disturbing	44	46	35	34	37	42	27	31	30
Not at all disturbing	30	23	25	20	19	15	50	45	29
Proportion hearing noise from flat above:									
Footsteps	82	83	95	82	93	94	93	94	95
Banging and hammering	79	76	80	73	80	79	89	84	94
Proportion disturbed by noise from flat above:									
Footsteps	20	25	40	29	37	51	20	18	43
Banging and hammering	46	57	55	46	55	53	22	20	45
Number of interviews	166	113	20	187	175	47	241	325	93

In considering the results given in Table XIV it must be remembered that this classification only takes into account the floor covering in one room of the flats, the sitting room. Other rooms and passages may have different coverings.

The measurements of insulation made on linoleum-covered floors showed that the impact insulation of the Group I floors was 5 to 8 dB better than that of the Group II floors, which in turn was 6 to 8 dB better than the Group III floors at the lower frequencies, reducing to zero difference at the higher frequencies. The addition of carpets or rugs on a floor would be expected to add to the impact insulation (but not to the airborne insulation) and the improvement in any group would be of the order of up to 15 dB or so for a thick carpet on underfelt.

TABLE XV

CHILDREN IN OTHER FLATS

	Group I Number of children aged 3–14 in the flat *above*			Group II Number of children aged 3–14 in the flat *above*			Group III Number of children aged 3–14 in the flat *above*		
	None	One	Two or more	None	One	Two or more	None	One	Two or more
	%	%	%	%	%	%	%	%	%
Floor covering in the sitting room in flat above:									
Carpets or large rugs...	66	60	51	51	46	40	49	32	28
Small rugs	28	35	41	37	44	45	43	53	52
Linoleum	6	5	8	8	10	15	8	18	20
Too much noise comes through the ceiling:									
Yes	36	53	66	63	83	84	75	84	73
No	42	47	34	37	17	16	25	16	27
Proportion disturbed by:									
Footsteps from flat above	14	27	23	23	33	46	28	36	36
Banging and hammering from flat above	42	51	53	34	56	61	35	31	34
Noises from inside the building:									
Very disturbing ...	8	11	15	17	29	30	10	13	14
Disturbing	8	16	17	18	19	20	14	13	14
A little disturbing ...	51	41	44	39	35	35	28	31	28
Not at all disturbing ...	33	32	24	26	17	15	48	42	44
Number of interviews	36	81	182	131	124	159	243	189	284

	Number of children aged 3–14 in the flat *below*			Number of children aged 3–14 in the flat *below*			Number of children aged 3–14 in the flat *below*		
	None	One	Two or more	None	One	Two or more	None	One	Two or more
	%	%	%	%	%	%	%	%	%
Too much noise comes through the floor:									
Yes	38	41	37	47	59	65	70	61	66
No	62	59	63	53	41	35	30	39	34
Noises from inside the building:									
Very disturbing ...	6	14	12	26	25	21	12	13	10
Disturbing	13	11	14	13	20	23	16	9	13
A little disturbing ...	47	45	40	37	38	36	20	28	34
Not at all disturbing ...	34	30	34	24	17	20	52	50	43
Number of interviews	32	80	187	137	127	150	188	192	304

The first question in Table XIV shows that for comparable classes Group II and Group III showed little difference in the proportions who thought too much noise came through the ceiling while Group I were appreciably lower. In each group there is a reduction of about 15 per cent in this proportion from linoleum-covered floors to those with carpets or large rugs.

The second question in Table XIV, which sums up the housewife's opinion of noise from inside the building, shows trends which might be expected for Group I and Group III, the floor covering of the flat above having an appreciable influence on the level of disturbance. There is less evidence of a trend for Group II, the reason not being known. Of the particular noises, disturbance from footsteps in the flat above is also appreciably affected by the floor covering of the sitting room. Banging and hammering is less affected.

Children in Other Flats

In the top part of Table XV are shown some differences due to the number of children in the flat above. The bottom part of the Table shows the effect of children in the flat below. The age range 3-14 years has been used to define a child, and the range 0-2 years to define a baby.

The first item shows that the likelihood of a household having a carpet tended to fall as the number of children increased. Thus the effects shown in the previous section were partly due to differing numbers of children associated with the different floor coverings. Similarly the other results shown in the top part of Table XV are partly due to the difference in floor coverings. It can, however, be shown by further analysis that both the floor covering and the number of children in the flat above affected the complaint level independently of each other.

There are clear indications that the disturbance due to noise was affected by the number of children in the flat above, for example, from the question asking whether too much noise came through the ceiling, from the questions about footsteps and banging and hammering, and from the summing-up question about disturbance. The bottom half of the Table suggests that children in the flat below also affected the complaint level.

CONCLUSIONS AND STANDARDS

All the results of any significance have now been examined in some detail, but the information obtained is not all of equal importance. This section reviews the results in a broader manner, attempting to pick out those that are of value in revealing the broad practical issues of the problem of noise in flats. The main issues are summarized in the following paragraphs.

(i) *Value attached to sound insulation.* A useful guide to the relative importance of sound insulation is given by the complaints made about the flats before the subject of noise was introduced (Table II). If the relative proportions who made complaints about different things are considered it is found that whereas for Group I the proportion complaining about noise was exceeded by the proportions making six other types of complaint, with Groups II and III this proportion was only exceeded by the proportion making the very mixed collection of complaints concerning planning faults.

A further measure of the value attached to sound insulation is afforded by the answers to a question put to those who thought either the floors or the walls should be made thicker. When it was suggested that this would increase the cost of the building and informants were asked if they thought "it would be worth-while if it meant that the rent would then be a bit more", the proportion who thought the floors or walls should be thicker and who were also prepared to consider paying for it was 38 per cent for Group I, 79 per cent for Group II and 76 per cent for Group III. It should be pointed out that there was already an appreciable difference in rents between the three groups of flats, which may have affected their willingness to pay more. For a three-bedroom flat the weekly rents excluding rates were as follows:

> Group   I   16*s*.
> Group   II   12*s*. 3*d*.—14*s*. 9*d*.
> Group   III   7*s*. 9*d*.

(ii) *Relative importance of noise from different sources.* The housewives were asked to say which they noticed most, noises from other flats, noises from the stairs, or noises from outside the building (Table VI). With the Group II and III flats the majority said they noticed most the noise from other flats, but in the Group I flats as many said they noticed most the noises from the stairs as from the other flats, and nearly as many gave outside noise as the most noticeable. Also in Group I the lifts appeared to be a source of appreciable disturbance. It would seem that, when the floor insulation value of the Group I flats has been reached, sources of noise other than the surrounding flats become almost as important.

With staircases serving more than two flats from each landing by means of balconies as in the Group II flats there was an appreciable difference in the complaints coming from those in tiers adjoining the staircases and those in other tiers. (Table XIII). Those in tiers adjoining the staircase noticed noise from the staircase much more than noise from other flats.

In general, for those on the middle floors, noise was noticed more from the flat above than from the flat below and least from next door (Table IX). An exception to this was found in the case of those Group I flats which shared a party wall. Here noise from next door was noticed more than noise from below (Table XII), but noise from above was still the most noticed. It should be borne in mind that this case of the Group I flats which shared a party wall is the only one where noise from next door and noise from below can be properly compared, because only in this Group was the airborne sound insulation of both floors and walls the same, i.e. about 50 dB.

As might be expected because of impact noise, those in top flats complained less than those in bottom flats, who in turn complained less than those in the middle flats (Table IX). Thus increasing the number of floors of separate flats in a block tends to increase the proportion of tenants who are disturbed by noise.

(iii) *Airborne and impact noises.* Of noises arising within the buildings there seems little doubt that, for all three groups of flats, impact noises were at least as important as airborne noises (Tables IV, VIII, etc.). In each case a higher proportion of housewives said they were disturbed by "banging and hammering" than by any other noise. In each case impact noises were most frequently chosen as the most disturbing noise. This emphasis on impact noise is in agreement with the results of the earlier inquiry[1] and with the inquiry concerning semi-detached houses.

(iv) *Children in the surrounding flats.* Disturbance due to noise increased with the number of children in all the surrounding flats but was mainly affected by the number of children in the flat above (Table XV). Moreover, the likelihood of there being a carpet in the sitting room fell as the number of children increased which amplified the effect.

(v) *Variation in what is tolerable.* The proportion of households disturbed by noise was higher for Group II (44 dB insulation) than for Group I (49 dB insulation) as might be expected. Surprisingly, this proportion was no higher for Group III with the lowest insulation (39 dB) than it was for Group I with the highest insulation (49 dB). After examining all the measured differences between the groups it must be concluded that this tolerance of a higher noise level by the Group III tenants was born of earlier experience of still worse conditions, and the present overcrowding. There would appear to be as much as 10 dB difference between what disturbed the housewives of Group III and the housewives of the other two groups.

Standards for flats

This inquiry was intended to provide a guide, in association with other known factors, for the setting up of recommended standards of sound insulation for local authority flats. It could not on its own hope to be a complete guide because, for one thing, the sound insulations found in practice show many variations—in absolute levels, in dependence on frequency, and in the relation between airborne and impact sound insulation—and, for another thing, tenants show even more variations in their behaviour. The three groups of flats were chosen, of course, because their floors covered the interesting range of insulations, and it was hoped that the tenants would be reasonable samples of the whole population of local authority tenants. In the event, it has become obvious that the Group III tenants cannot fairly be regarded as typical: they are still influenced by their previous, untypical environment. Further, it is known that vigorous complaints are sometimes made when floors have an insulation the same as (or even slightly higher than) the Group III insulation; obviously the Group III tenants were nowhere near this stage of complaining. It appears, then, that the results for Group III should not be included when considering standards which are intended as a guide for the country as a whole.

Referring to Table II it is seen that, for the Group II tenants, noise was the biggest single factor that they thought unsatisfactory about their flats, while for the Group I tenants, noise came low on the list of complaints. Further, Table III shows that 26 per cent of the Group I tenants said that noise was disturbing or very disturbing, whilst the corresponding percentage for the Group II tenants was 42. Two other facts must be considered; first, the Group I floors had the highest insulation that it is possible to get at the present time in practicable constructions; secondly, the Group II floors were typical of most flat floors built in the 1920s and 1930s, and it is generally reckoned (e.g. in books discussing the design of flats) that noise from the neighbours, and lack of privacy of access and no garden, are the most commonly disliked facets of life in flats, but (to the best of the Building Research Station's knowledge) no local authority has ever found it necessary to remodel this type of floor.

For practical purposes, it appears that a fair summing-up of the position would be that, with floors of an insulation equal to the Group I floors, the neighbours' noise is only as disturbing as several other things; with floors of an insulation equal to the Group II floors, the neighbours' noise is to many of the tenants the worst thing about living in the flats, but at least half of the tenants are not seriously disturbed.

The grading system published[4, 5] by the Building Research Station is based on this assessment. Grade I floors have to have an insulation at least as good as that of the Group I floors, and Grade II floors have to have an insulation at least as good as that of the Group II floors. Figs. 1 and 2 show the measured insulation (airborne and impact respectively) of the Group I and II floors (corrected to a reverberation time of 0·5 seconds at all frequencies in the receiving room, in conformity with the standard measuring technique) and the grading curves based on these measurements. It is seen that, for airborne insulations, the grading curves follow the measured values closely, except at the highest frequencies where the grading curves are levelled off. This is because some constructions show an insulation graph of this shape, and it is thought that they should not be made to "fail" the grades only at these high frequencies, where the subjective effect is probably negligible. The two grades are separated by 5 dB at all frequencies.

Only in the Group I flats was it possible to obtain a controlled comparison of noise through the party floors and through the party walls, because the airborne sound insulation of floors and walls was the same. In Group I, those tenants who had living rooms on the party wall noticed noise more from the flat next door than from the flat below, though less than from the flat above, which is a source of impact as well as airborne noise (Table XII). It would seem from this that the wall insulation should be at least equal to the floor insulation. Thus,

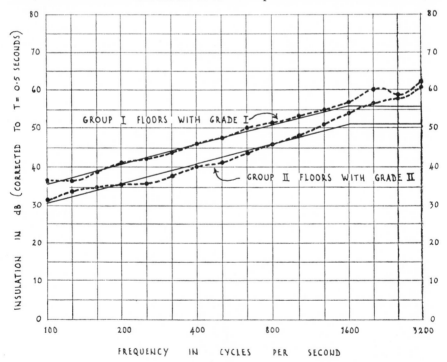

FIG. 1.   GRADING CURVES FOR AIRBORNE SOUND INSULATION; BROKEN LINES SHOW MEASURED VALUES FOR GROUP I AND GROUP II FLOORS

when the floors are Grade I the walls should be Grade I also; when the floors are Grade II the floors should be at least Grade II, and it might well be of some benefit to make them Grade I—as was the case in the Group II flats. In Group II, so far as can be ascertained from the results, the extra insulation of the party walls rendered noise from the flat next door less disturbing than noise from the flat below.

The impact sound insulation grades for floors follow the measured values as closely as possible, and the two grades are separated by 6 dB at all frequencies.

The detailed specifications of the grades are:

Airborne sound insulation

Grade I is 36 dB at 100 c/s, increasing (slope 5 dB per octave) to 56 dB at 1600 c/s and remaining at 56 dB up to 3200 c/s. Grade II is the same shape as Grade I but is 5 dB lower at all frequencies. To satisfy the particular grade, the insulation of the construction must be such that any deviations in the unfavourable direction should not exceed 1 dB when averaged over the whole frequency range. (The assumption is that any deficiency of insulation at any part of the frequency range will not, subjectively, be compensated for by adequate, or superfluous, insulation over the rest of the range). When the mean unfavourable divergence from Grade II does exceed 1 dB, then this mean divergence can be taken as a classification of the construction, i.e. *x* dB worse than Grade II.

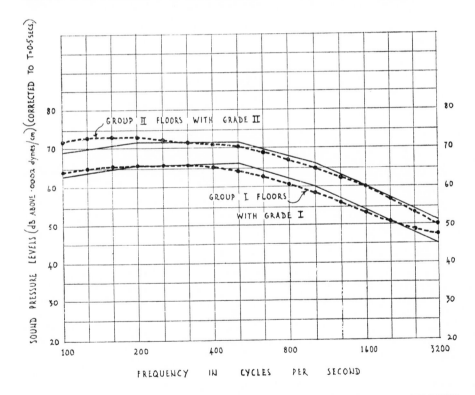

FIG. 2. GRADING CURVES FOR IMPACT SOUND INSULATION; BROKEN LINES SHOW MEASURED VALUES FOR GROUP I AND GROUP II FLOORS

Impact sound insulation

The sound-pressure levels in octave bands which must not be exceeded when the standard impact machine is operated on the bare floor above are: Grade I, 63 dB at 100 c/s increasing (slope 3 dB per octave) to 66 dB at 200 c/s, remaining at 66 dB up to 500 c/s, falling (slope 3 dB per octave) from 66 dB at 500 c/s to 63 dB at 1000 c/s, and then falling (slope 6 dB per octave) from 63 dB at 1000 c/s to 53 dB at 3200 c/s. Grade II is the same shape as Grade I but 6 dB higher at all frequencies. When measured with tenants' linoleum on the floor (as in this inquiry), then the grades are the same shape up to 500 c/s but at higher frequencies are: Grade I, 66 dB at 500 c/s falling (slope 6 dB per octave) to 60 dB at 1000 c/s, and then falling (slope 9 dB per octave) from 60 dB at 1000 c/s to 45 dB at 3200 c/s; Grade II as for Grade I but 6 dB higher at all frequencies. The same conditions for divergencies in the unfavourable direction apply as for airborne sound insulation.

As mentioned earlier, there is some variation in the measured insulations of supposedly identical floors, the standard deviations at individual frequencies being of the order of 2–3 dB although when taken over the whole frequency range the deviations are rather less. If the results for each floor are compared with the grades it is seen (Table XVI) that about 85 per cent of the floors satisfied the appropriate requirements. If, as it is proposed to do in future, a minimum number of four floors of each construction are measured before grading, then the variations between individual floors will have a negligible effect.

TABLE XVI

Floor	Grading					
	Airborne insulation			Impact insulation		
	Grade	Pass	Fail	Grade	Pass	Fail
Group I flats	I	26	1	I	13	3
Group II flats	II	36	5	II	24	7

The party floors which often cause vigorous complaints are plain joist floors, and these have an insulation of 8 dB worse than Grade II for both airborne and impact sounds. What is likely to happen between this level of insulation and Grade II can not be predicted with accuracy, but obviously the nearer the insulation drops to 8 dB worse than Grade II the more likely it is that there will be vigorous complaints.

This paper is being written two years after the first publication[4] of the grading system, and so far this system appears to be working satisfactorily. The gradings are intended as a general guide and no doubt there may be cases where, for example, Grade II insulation will not be enough, or where, as for the Group III tenants of this inquiry, a lower insulation than Grade II will be satisfactory. Each case must be decided on its own merits, bearing in mind such factors as the previous environment of the tenants. Lastly, two facts should be remembered: first, even with Grade I insulation a minority of the tenants will be seriously disturbed by noise; secondly, it is expensive to make any improvement, after construction, to sound insulation found to be too low.

[*Editor's Note:* Material has been omitted at this point.]

REFERENCES

1. CHAPMAN, A. "A Survey of Noise in British Homes". *National Building Studies, Technical Paper* No. 2, 1948.

4. PARKIN, P. H. and STACY, E. F. Recent Research on Sound Insulation in Houses and Flats, *Journal of the Royal Institute of British Architects,* **61**, 9, pp. 372–376, June, 1954.

5. Sound Insulation of Dwellings—I and II, *Building Research Station Digests* Nos. 88 and 89, May and June, 1956.

[*Editor's Note:* Appendixes I through V have been omitted due to limitations of space.]

19

Reprinted from *Int. Congr. Acoust., 3rd, Stuttgart, 1959, Proc.,* pp. 1041–1044

My Neigbour's Radio

J. VAN DEN EIJK

Department of Sound and Light, Research Institute for Public Health Engineering of the Organization for Applied Scientific Research (T.N.O.), Delft (The Netherlands)

AUTOMATIC ANALYZER FOR NONCONSTANT SOUND

In many cases the measurement of sound spectra is complicated by the fact that the sound pressure level varies considerably in time. Fig. 1 gives a diagram of the operation of an automatic analyzer for such fluctuating sounds.

The amplified microphone signal is fed to the joint input of eight octave band-pass filters. The output signal of each of these filters is fed via an amplifier, potentiometer and thyratron circuit to an column of six counting mechanisms, such as are used in automatic telephone systems.

The moment the sound level in a certain octave band exceeds a certain pre-set value, *e.g.* 65 db, one of the counters in the column belonging to that octave begins to function. As long as the level remains above the pre-set value the counter moves forward one number every second. If the sound level continues to rise and thus exceeds a second pre-set value of, say, 70 db, the next counter begins to function. This goes on until, at very high levels, all the counters in the column concerned are worling. When the level drops the number of counters operating decreases. This takes place simultaneously in all the eight octave bands.

The duration of a measurement may be *e.g.* a quarter of an hour or a full working day, depending on the nature of the research untertaken. After conclusion of the measurement one can, from the difference between the initial and final readings of the counters on one hand and the total duration on the other hand, very easily compute

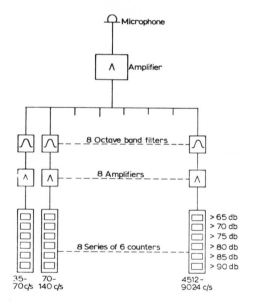

Fig. 1. Automatic analyzer for fluctuating sound.

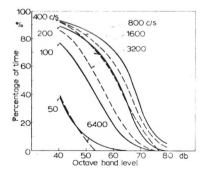

Fig. 2. Statistics of peak levels in radio-programmes. For each of eight octave bands the percentage of the time that certain octave band levels are surpassed is indicated. Based on about 55 hours of mixed radio-programmes.

239

the percentage of the total time during which the sound level in each of the eight octave bands was higher than, say, 65 db, 60 db, 75 db, etc.

The difference between the 6 pre-set levels can be varied in steps of 5 db wihtin a range of 50 db. The counting frequency can be stepped up to a maximum of 10 counting periods per second. Within each counting period the peak voltage maximum sound pressure is determined during an adjustable fraction of the counting period. In the measurements described here the peak voltage was determined every second during a period of 0.1 second.

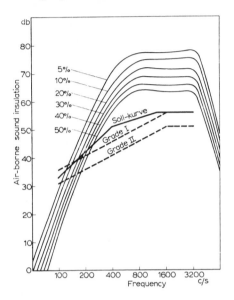

Fig. 3. Required airborne sound insulation based on a disturbing neighbour's radio level surpassing 0 phone during, in the mean, 5, 10, 20, 30, 40 or 50 per cent of the time. For comparison the German (Soll-Kurve) and the British (Grade I and II) requirements for dwellings are added.

THE NEIGHBOUR'S RADIO AND AIRBORNE SOUND INSULATION REQUIREMENTS

It would not seem unreasonable to suppose that, if the airborne sound insulation between two flats is great enough to reduce the annoyance caused by a neighbour radio to a bearable level, the annoyance caused by other airborne sounds from the neighbour's flat will in the vast majority of cases also be reduced to such an extent as to give no further cause for complaint.

It is said that many neighbours keep their radio working on a nonstop basis. We have followed this example. Without paying any attention to the nature of the programme we analyzed radio sounds for hours and hours on end. Fig. 2 shows the total result of 17 mornings and afternoons.

Based on these results Fig. 3 indicates how high the airborne sound insulation must be in order to ensure that the average level in my room produced by the neighbour's radio will not be above 0 phone (according to FLETCHER AND MUNSON) during more than 5, 10, 20, 30, 40 or 50 per cent of the time. For the sake of comparison the figure likewise shows the two sound insulation curves, used as standards by the British Building Research Station, as well as the standard value taken from the German specification *Vornorm DIN 4109* of 1959. It is evident that our curves for 0 phone are unattainably high.

Fig. 4 is based on a level of 20 phones (according to FLETCHER AND MUNSON) in the disturbed room. The curves are similar to those of Fig. 3 but they lie markedly lower. All the curves show a steep drop below 400 and above 4000 c/s. Between 400 and 4000 c/s the curves are nearly independent of frequency. Actual sound insulation curves do not show such a steep drop below 400 c/s and as a rule the insulation increases above 400 c/s.

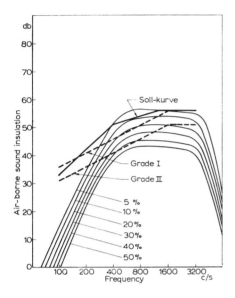

Fig. 4. Required airborne sound insulation based on a disturbing neighbour's radio level surpassing 20 phones during, in the mean, 5, 10, 20, 30, 40 or 50 per cent of the time. For comparison the German (Soll-Kurve) and the British (Grade I and II) requirements for dwellings are added.

In view of these facts one may ask: Is there any use in extending the standard requirements for airborne sound insulation between dwellings and the measurements to frequencies below 400 c/s and above 800 c/s?

RELIABILITY OF THE RESULTS

A great many factors of uncertainty, as *e.g.* type of program and position of volume-control and tone control, may still influence the exact location of the curves and, to some extent, also their form. But one may wonder whether there is any use in trying to achieve greater presision in this field. After all, the absolute level at which the standard airborne sound insulation values are to be set will largely be determined by economic considerations. But none of the uncertainties will alter the form of the curves so much that this would detract from the conclusion that it is the frequency range of 400 to 800 cycles per second which is of main importance for the abatement of the nuisance from my neighbour's radio!

20

Reprinted from pp. 493–500 of *J. Acoust. Soc. Am.* 34(4):493–501 (1962)

Sound-Insulation Ratings and the New ASTM Sound-Transmission Class

T. D. NORTHWOOD

Division of Building Research, National Research Council, Ottawa, Ontario, Canada

(Received December 18, 1961)

A survey is made of past and present systems for rating the sound-insulation value of building walls and floors. It is observed that in most countries ratings based on "average" transmission losses have been replaced by systems that compare transmission-loss characteristics with a standard contour. An example of this approach is the new sound-transmission class contained in ASTM E90-61T. This is discussed from various theoretical and practical viewpoints, and it is concluded that the sound-transmission class is a useful rating system for common architectural problems.

THE sound-insulation requirement for a partition depends on the occupancies it separates; more explicitly, it depends on the magnitude and frequency distribution of the noise produced on one side of the partition and the amount of intruding noise that will be tolerated on the other. In rare instances, the character of the noise may be predictable and constant, and the tolerance level of transmitted noise may be accurately specified. Then it may be possible to specify in a straightforward way the detailed transmission-loss requirements as a function of frequency. But more typical is the problem of designing an apartment or office building, where the required sound insulation varies widely with the individual occupants and their activities of the moment. Here the architect and his client need a simple figure of merit that will help them provide enough sound insulation to satisfy most of the building occupants most of the time.

It is the purpose of this paper to examine the problem of providing such a figure of merit, and in particular to describe the recently developed *Sound Transmission Class*, which appears in the revised standard ASTM E90-61T.[1]

The sound transmission class was designed primarily for assessing the sound-insulation value of walls and floors for use between dwelling units in apartment buildings and similar structures. It also has application in hotels, hospitals, schools, and other situations where the requirements are similar in character, though not necessarily in degree, to those encountered between dwellings. For partitions between offices it is usual to require only that transmitted speech be unintelligible. This suggests an approach slightly different from the dwelling problem, but the sound transmission class is again found to be a useful rating system.

The new classification system is an intermediate stage between the physical measurement of sound-transmission loss and the specification of minimum requirements for separating various occupancies. It might be noted that similar criteria are used in Britain and several European countries for specifying minimum requirements for dwelling separations. In Britain they constitute a *recommended* standard but in most other cases they are mandatory provisions of building codes. Actual minimum requirements are beyond the scope of the present paper and of ASTM E90-61T. Nevertheless, it is hoped that this discussion of the theoretical and practical bases of sound-insulation requirements will lead first to the general use of the sound transmission class as a rating system and ultimately to its use as the basis of minimum standards.

EMPIRICAL APPROACHES

For many years, both here and in Europe, the commonly used rating was the arithmetic average of the transmission losses in decibels measured at a specified series of test frequencies. On this continent the standard method of test, described in ASTM E90-55 and its counterpart American Standard Z24.19-1957, used the arithmetic average of the transmission losses measured at the nine frequencies 125, 175, 250, 350, 500, 700, 1000, 2000, and 4000 cps. Two objections have been raised to the use of such an average as a figure of merit: (1) it gives equal weight to all test frequencies regardless of their importance in sound-insulation problems (although by including the half-octaves below 1000 cps the U. S. nine-frequency average gives extra weight to the low-frequency range); (2) it gives equal weight to both high and low transmission losses, as if superlatively high values at some frequencies could compensate for deficiencies at other frequencies.

Attempts to meet the first objection have led to the introduction of several other "averages," obtained by altering the selection of frequencies included in the average. There is sometimes good reason for concentrating on a special frequency range but unfortunately it is rarely made clear, especially in trade literature, when or why a nonstandard average is being employed. As one step in producing an orderly presentation of information ASTM E90-61T now requires that only the "average" reported be the nine-frequency average, and that it be so labeled.

The second objection is based on the premise that a partition is no better than its lowest transmission loss. It led on this continent to the development of the "energy average,"[2] obtained by averaging the trans-

---

[1] ASTM E90-61T, "Tentative recommended practice for laboratory measurement of airborne sound transmission loss of building walls and floors" (American Society for Testing and Materials, 1916 Race Street, Philadelphia, 1961).

[2] R. V. Waterhouse, J. Acoust. Soc. Am. **29**, 544 (1957).

mission coefficients corresponding to the transmission losses and then taking the decibel equivalent of the average transmission coefficient. Such an average is dominated by the lowest values of transmission loss. Unfortunately, the lowest values are almost always at the lowest test frequencies, which are usually the least important in the evaluation of partition performance. Moreover, even in the laboratory it is difficult to make precise measurements of sound-transmission loss at the lowest test frequencies; thus, the energy average is usually based on the least reliable measurements.

Clearly, no simple average, or even a complicated one such as the energy average, can properly rate the performance of a partition unless it takes into account the variation of transmission loss with frequency. The usefulness of simple averages in the past may be attributed to the fact that the massive structures that predominated had rather simple transmission-loss curves, all rising in a similar way with frequency. But the present use of lightweight structures has resulted in both lower and more irregular transmission-loss curves, and necessitates a criterion that is more closely related to actual performance requirements.

Subjective Reactions and Standard Contours

In at least three countries, Holland, Sweden, and Britain, the dwelling separation requirement was investigated directly by canvassing the tenants of apartment buildings and row dwellings.[3–5] Each of the surveys was conducted without knowledge of the others, and it is of interest to examine the points of agreement and disagreement among their conclusions.

The British and Dutch surveys showed that noise from the floor above constituted the greatest disturbance, apparently because of the special importance of impact sounds such as footsteps and children playing. The Swedish survey also showed the importance of transmission through floors, but did not indicate that impact noise was as serious a problem as airborne sound. Nevertheless, current British and European standards, including the Swedish one, deal with both impact and airborne sound. On this continent there is as yet no standard test method for impact (although one is now being considered by ASTM Committee E-6). Hence this paper deals only with the airborne-sound problem, despite the evident importance of impact transmission.

The Swedish sample included similar structures located on noisy thoroughfares and on quiet residential streets. As might be expected these showed an inverse relationship between disturbance from traffic and disturbance from adjacent dwellings; the more tenants

were disturbed by traffic noise, the less they were disturbed by noise from their neighbors.

Correlation of tenant disturbance and airborne-sound insulation was in general very complex. For example, it is difficult to distinguish between reactions to airborne and impact sounds; tenants dissatisfied with one aspect of their dwellings tend to express dissatisfaction with other aspects; a tenant's past experience with very inferior housing may cause him to comment favorably on a slightly less inadequate environment; and of course there is the complicating effect of traffic noise, discussed in the foregoing. Nevertheless, both the British and the Swedish surveys showed a definite reduction in complaints for increased airborne-sound insulation. The Netherlands study, which involved seven different types of structure, yielded little correlation between tenant reactions and sound insulation.

The British survey included two main types of floor structure for which the transmission-loss characteristics were similar, with averages of 49 and 44 db.[6] In the 49-db structures, 22% of the tenants complained about noise problems, and noise was about equivalent to other sources of complaint. In the 44-db structures, 36% of the tenants complained about noise, and noise was the major complaint. These two transmission-loss characteristics, somewhat idealized, form the bases of the grade I and grade II curves shown in Fig. 1. They constitute recommended minimum values for party walls and floors in certain classes of apartments.[7]

The sound-insulation data from the three surveys are shown in Fig. 2, in which the abscissa is the ASTM sound-transmission class, which is described later; for the moment, it can be taken as roughly equivalent to average transmission loss. The results serve mainly to indicate the order of sound insulation required for acceptable dwelling separation. The one anomalous point in the British results was for a new block of apartments whose tenants were mostly refugees from a condemned-housing area; they are therefore not a representative sample. Some information was gleaned from the Dutch results by discarding two structural types for which impact insulation was exceptionally low, and by separating the rest into two groups with high and low insulation. Despite these cautionary remarks, the

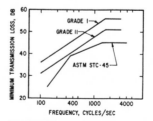

Fig. 1. British recommended minimum requirements for airborne-sound insulation between apartment dwellings. (A sound transmission class contour is shown for comparison.)

[3] C. Bitter and P. Van Weeren. "Sound nuisance and sound insulation in blocks of dwellings I," Rept. No. 24, Research Institute for Public Health Engineering T.N.O. (1955).

[4] O. Brandt and I. Dalén. Byggmästaren **31**, 145 (1952).

[5] P. G. Gray, A. Cartwright, and P. H. Parkin, "Noise in three groups of flats with different floor constructions," National Building Studies Research Paper No. 27, H.M.S.O. (1958).

[6] The standard European average is based on measurements at third-octave intervals from 100 to 3200 cps.

[7] *British Standard Code of Practice*, Chap. III: "Sound insulation and noise reduction," British Standards Institution (1960).

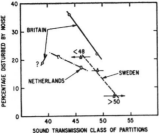

FIG. 2. Results of three surveys to determine the acceptability of existing sound insulation between apartment dwellings.

results suggest that a sound transmission class 45 is borderline for reasonably satisfactory sound insulation, and that a sound transmission class 55 represents about the maximum value that is worth attempting for ordinary housing.

A similar set of rating curves (Fig. 3) was incorporated in the German standard DIN 52211.[8] The same shape was subsequently adopted also by the Scandinavian countries as the basis of sound-insulation requirements for row-dwelling and apartment buildings.[9,10] The DIN curves, also based on the characteristics of actual walls, differ from the British curves in requiring slightly greater transmission loss in the middle frequencies (300 to 1000 cps) and slightly less at lower frequencies. It is to be noted that the standard requires a slightly higher value (2 db) for a laboratory test than for a field test. The difference is an allowance mainly for flanking transmission.

The use of a standard curve rather than a simple arithmetic average is based on the hypothesis that the transmission-loss requirement varies in a certain way with frequency and must therefore be specified for each frequency band. The direct evidence concerning the contribution of each band is rather sketchy although there are theoretical reasons for supporting this view. In a survey of occupants of row dwellings, the British were able to compare a 9-in. solid-brick wall with an 11-in. cavity wall; they found little superiority for the latter despite the fact that its average transmission loss was 5 db better. It was concluded that the lower and middle frequencies, for which there was little difference between the walls, were more critical than the high frequencies for which the superior transmission loss of the 11-in. wall is most evident. The transmission loss is presumably high enough at high frequencies in either case. Both walls just meet the British grade I requirement.

More extensive information is provided by the experiments of Rademacher,[11] who simulated wall-transmission characteristics electrically and had 20 subjects determine the subjective reduction in loudness provided by various transmission characteristics. For noise sources he used samples of music and speech (garbled to eliminate meaning) and several spectra of filtered white noise. His principal noise samples, including speech, music, and one sample of white noise, each had a broad maximum in the frequency range 200 to 800 cps.

Starting with the German DIN curve, he investigated the effect of decreasing the attenuation by 10 db in one octave band, as compared to a uniform attenuation over five octaves. He found that the 10-db decrease in one octave was equivalent to 1 to 4 db over five octaves, the greatest effect being, as might be expected, in the range of highest source power. In another series of tests, he compared four attenuation curves of different shapes, each corresponding to an average transmission loss of 48 db. One characteristic corresponded to the DIN curve, the second increased at 3 db per octave, the third increased at 6 db per octave, and the fourth had a plateau characteristic extending to 800 cycles and then a sharp increase in attenuation. The first three were rated subjectively at about 48 db (when tested with speech or music), but the fourth was rated at only 37 db. This result might be anticipated since only the fourth wall differed substantially in the range of maximum source power. Rademacher dismissed this fourth characteristic as physically unreal and concluded that a simple arithmetic average is adequate for rating most structures. Actually it is similar to many lightweight partitions, and it is to protect against transmission curves of this type that the standard curves were introduced.

Although the British and European standards emphasize the importance of the shape of the transmission-loss curve, they compromise somewhat in their interpretation of actual transmission characteristics. Generally, an average deficiency of 1 db (in some cases 2 db) is allowed relative to the prescribed curve. In both cases, the deficiency might conceivably take the form of a large deviation in a limited frequency band. It is commented[8] that this averaging arrangement is adopted in the absence of sufficient information about the contribution of individual narrow bands. In any case,

[8] "Bauakustiches Prufungen Schalldammzahl and Normtrittschallpegel," Deutschen Normenausschusses DIN 52211 (1953).
[9] F. Ingerslev and J. Kristensen, "Lydisolation I Boligbyggeri" (Sound insulation of dwellings) with English summary, Statensbyggeforskningsinstitut, Rapport 39, København (1960).
[10] "Anvisningar till byggnadsstadgan," BABS Stockholm (1960).
[11] H. J. Rademacher, Acustica 5, 19 (1955).

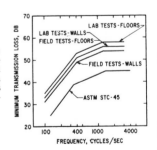

FIG. 3. German minimum requirements for airborne-sound insulation between apartment dwellings. (A sound-transmission class contour is shown for comparison.)

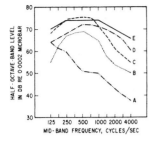

FIG. 4. Half-octave-band levels of typical household noises. Curve A—room air-conditioner, Curve B—vacuum cleaner, Curve C—normal speech (levels exceeded by 1% of speech peaks), Curve D—radio, television (peak levels, speech and music), Curve E—assumed "standard-household noise."

since excesses above the curve are disregarded, the allowable deficiency is actually rather limited.

Summarizing field experience and experimental evidence to date, the consensus is that a simple arithmetic average is not adequate, especially for rating lightweight structures which frequently have serious deficiencies in the most critical frequency range. Several countries are now using standard transmission-loss contours to represent minimum requirements for dwelling separations. Ideally, the transmission loss of a given partition should at no frequency fall below the standard contour; in practice, a limited amount of trading between frequency bands is permitted, although the consequences of this procedure are not yet fully understood.

THEORETICAL CONSIDERATIONS

The British and European standards were developed largely by informed interpretation of the comments of occupants of actual buildings. It is of interest to examine also various theoretical approaches, utilizing information on typical noises and various subjective criteria relating to loudness, noise, or annoyance. These provide some insight into the importance of various frequency bands and assist in selecting an optimum shape of transmission-loss characteristic.

Figure 4 shows half-octave band spectra for a few typical domestic noises. The trend in domestic appliances is toward control of the high-frequency components of noise, so that low- and medium-frequency components predominate in the residual noise. Speech, radio, and television noises are broadly peaked in the middle-frequency range. Speech intelligibility, as distinct from power, involves a slightly higher frequency range extending well beyond 4000 cycles, but this is irrelevant for dwelling separation since the transmission loss should be substantially greater than the amount required merely to reduce intelligibility. Musical instruments and high-fidelity record players will extend the range, especially toward the lower frequencies. Noting from the surveys the special importance of radio, television, and speech noises, it appears that one might consider a "standard household noise" spectrum flat from 250 to 1000 cps and diminishing by 4 to 6 db per octave below and above this frequency (Fig. 4). It should be noted that the curves of Fig. 4 are derived from several sources and are not strictly comparable;

they are intended to illustrate spectrum shapes, rather than absolute levels. Several important noises, particularly airborne noise resulting from footsteps and doors slamming, have been omitted for lack of information.

It is assumed that the optimum shape of a rating curve will not vary much with level in the range of application of the rating system, and that a family of parallel curves can be adopted. This is analogous to assuming that equal loudness contours are approximately parallel in the range 20 to 60 phons.

On the listening side, the most obvious procedure is to consider the loudness of the transmitted noise. The work of Stevens,[12] Quietzsch,[13] and Beranek[14] provides a good basis for such a criterion, although their results are more directly applicable to office and industrial problems than to dwelling separation. Stevens devised a method of calculating the loudness of complex sounds that, particularly in its latest form, agrees well with the subjective ratings obtained by Quietzsch for a wide variety of complex noises. Beranek used these data, supplemented by office-noise surveys, to develop the well-known noise criterion (NC) curves, which take into account both loudness and speech-interference levels approximately as they affect the acceptability of office noise. These criteria are used in later sections.

Assuming that the subjective reaction to noise is related to loudness, it is appropriate to determine the transmission-loss characteristic that reduces each band of "standard household noise" (Fig. 4) to a particular equal-loudness contour. For purposes of this paper the 0.5 sone contour, corresponding to a loudness level of 46 phons, is used. (Judging by tenant surveys, this is probably of the right order.) The transmission-loss characteristic that accomplishes this is shown as curve (a) in Fig. 5 (solid circles). This is the most efficient partition for achieving this transmitted loudness since all bands of transmitted sound are reduced to the point where they are equally important in determining over-all loudness.

Kryter[15] recently devised a slightly different series of "equal noisiness" curves, leading to "perceived noise levels" instead of loudness levels. The calculation is essentially the same as Stevens' except that Kryter provides greater weighting for high-pitched components of noise. Kryter claims for his curves only that they correlate better than loudness with subjective judgments of acceptability for certain high-pitched noises such as are produced by jet aircraft. It appears that the high-frequency components may be especially important when a noise is intrinsically obnoxious or alarming. Reviewing the most troublesome sources of dwelling noises, it seems likely that some of them fall into the intrinsically obnoxious category, and it might therefore be more plausible to use Kryter's criterion.

[12] S. S. Stevens, J. Acoust. Soc. Am. **33**, 1577 (1961).
[13] G. Quietzsch. Acustica **5**, 49 (1955).
[14] L. L. Beranek, J. Acoust. Soc. Am. **28**, 833 (1956).
[15] K. D. Kryter, J. Acoust. Soc. Am. **31**, 1415 (1959).

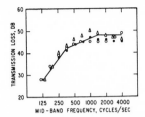

FIG. 5. Theoretically derived transmission-loss curves. Transmission loss required to reduce standard-household noise to: (a) 0.5 sone equal-loudness contour (solid circles), (b) 0.5 Noy equal-noisiness contour (open circles), (c) NC-25 contour (triangles). The solid line is the contour of sound transmission class 48.

Proceeding as before, one arrives at the required transmission loss defined by curve (b) of Fig. 5 (open circles).

The transmitted loudness or perceived noise level, *per se*, is not necessarily a good index of satisfactory sound insulation. More important is the degree to which the transmitted sound is masked by ambient noise existing on the listening side of a partition. For complete masking, a good approximation for complex noises is to require that the transmitted noise be no greater than this ambient level in any half-octave band.[16] Curves (a) and (b) could, on this basis, be regarded as possible spectra of ambient noise. Since Beranek's noise criteria are frequently used for similar purposes, a third transmission-loss characteristic, which reduces standard household noise to NC-25 contour, has also been shown as curve (c) in Fig. 5 (triangles).

Comparing these curves with the DIN curve, it is seen that the latter is reasonably similar in shape. Following these considerations, and with an eye to international standardization, it was decided to adopt the shape of the DIN curves as the basis for the ASTM sound transmission class contours (Fig. 3). Plotted on conventional semilog paper the STC contours consist of a horizontal segment from 1400 to 4000 cps, at a level corresponding to the sound transmission class; a middle segment that decreases 6 db from 1400 to 350 cps; and a low-frequency segment that decreases 14 db from 350 to 125 cps.

Equating the over-all loudnesses of transmitted noise and ambient noise would not result in complete masking if the loudness level of the transmitted sound were particularly high in a narrow band. But the distinction between the equal-loudness and masking criteria is not as great as might be anticipated, since the over-all loudness level of a complex sound depends markedly on the level in the loudest band. For example in Stevens' calculation for half-octave bands the loudest band is

weighted five times as much as the other bands. On either basis, it appears reasonably precise to require that a given transmission-loss requirement be met in each band, rather than just on the average.

For dwelling separation, however, where noise sources are highly variable, another approach is to consider the probability that noise will occur in a given frequency band. Both noisy and quiet conditions may fluctuate in the manner indicated in Fig. 4. A review of the spectra of the more troublesome noises suggests that high noise levels are more probable in the mid-frequency range than at high or low frequencies; hence this region should be given special attention. For this reason, in applying the proposed new rating, the following procedure is used: there shall be no deficiencies below the middle segment of the STC curve, but deficiencies averaging 1 db are allowed below the outer segments of the curve.

OFFICE PARTITIONS AND SPEECH PRIVACY

In the foregoing section, the sound transmission class was discussed primarily from the viewpoint of dwelling separation. It is of interest also to consider its applicability to the problem of office separation. This is the second large-scale problem confronting the designer, and differs in several ways from the dwelling separation problem.

The primary requirement for sound insulation between offices is to prevent the transmission of intelligible speech. This is so both for the speaker, who may wish to speak privately, and for the listener, since the distracting quality of speech noise is intensified when it begins to be intelligible. The most comprehensive recent study of factors affecting speech intelligibility was that of French and Steinberg,[17] who developed a straightforward method of calculating speech intelligibility from the properties of each link in a communication system. They first showed that speech intelligibility can be expressed rather precisely in terms of the available dynamic range in each of 20 equally important "critical" frequency bands. A maximum range of about 30 db (above either threshold of audibility or masking level set by noise) is required in each critical band to get the full contribution of the band. To a good approximation the contributions of the individual bands are independent of each other, although there are secondary effects due to masking by adjacent bands. The contribution of each critical band is expressed in terms of an "articulation index," and the average articulation index for the twenty critical bands provides a quantity that is related in a known way to the other common measures of speech intelligibility.

Beranek, in his earlier approach to the problem of specifying minimum noise requirements within spaces such as offices and conference rooms, concluded that ease of speech communication provided a good criterion.

---

[16] H. Fletcher, M. R. French and J. C. Steinberg, E. Zwicker, and others have demonsrated that calculations of loudness, masking, articulation index, and similar quantities involving perception of complex sounds are most precise when "critical bands" are used. Nevertheless, it is common practice to use data in octave, half-octave, or one-third octave bands. The approximation is valid if the spectra are reasonably smooth and continuous.

[17] M. R. French and J. C. Steinberg, J. Acoust. Soc. Am. **19**, 90 (1947).

Hence, he used the work of French and Steinberg to develop the criteria known as speech-interference levels,[18] obtained by considering speech levels relative to noise in the three octaves most important to intelligibility (300 to 2400 cps). Later, however, he found that acceptability of office noise is more closely related to loudness. This led to the NC curves mentioned earlier.[14]

The speech-privacy problem is the inverse of that considered by French and Steinberg and by Beranek, being concerned with the marginal condition in which speech is not quite intelligible. In this region, the simple linear approximation that works so well when intelligibility is high is no longer strictly applicable. Nevertheless, a linear approximation has been used successfully in office-separation problems by Hardy[19] and more completely by Cavanaugh et al.[20] The latter is probably the most useful method available when the problem is well defined: for example, when the room configuration is known and the ambient noise level known or specified. For low-intelligibility conditions, the effect of their linear approximation is to obtain articulation index values that are somewhat larger than those obtained by the French and Steinberg method. Since they deduce from office surveys a maximum acceptable value of this index, calculated by the same method, the discrepancy is perhaps safely canceled out again; but it should be noted that their results are not comparable with those presented below.

For the usual office-building design problem, a rating that ensures a reasonable probability of general satisfaction is again necessary. The uses and limitations of the sound-transmission class for this purpose are considered below, with the help of the French and Steinberg study and their nonlinear treatment for conditions of low intelligibility. Following Beranek's example,[18] the critical bands of French and Steinberg are replaced by bands of equal-frequency ratio, in this case by half-octave bands. The conversion is accomplished by weighting each half-octave by a factor proportional to the number of critical bands it contains.[16] The weighting factors are derived from composite studies of both male and female voices. The idealized speech spectrum of French and Steinberg (Fig. 4) is assumed to exist in a source room separated from a listening room by a partition having a transmission loss defined by an STC contour. The transmitted-speech level in the listening room will depend on the room absorptions and the partition area; room absorptions of 100 sabins each and a partition area of 100 sq ft will be assumed. It will further be assumed that the ambient level in the listening room is defined by an NC curve. Then the signal-to-noise ratio in each band is dependent on the sum of the STC and NC values (assuming that the NC curves of interest are all parallel to NC-30). The articulation index in each band is determined by the speech level (exceeded by 1% of peaks) relative to noise level; for level differences less than 12 db, the band-articulation index is shown in Table I, and for level differences greater than 12 db, the value is $W = (E-6)/30$, where E is the level difference. This procedure applies when the ambient level is moderately above the threshold of audibility, but not so high that nonlinear effects become significant.

The relationship between articulation index and the sum of STC and NC numbers is shown in Fig. 6. It is seen that for an articulation index of 0.03 and for $S/(A_1 A_2) = 0.01$, STC plus NC should equal 68. For example, assuming an ambient level corresponding to NC-35, the transmission loss should then be equal to or greater than STC-33.

Now, consider the variables affecting this result. Differences between individual speakers may affect the speech level by up to ±10 db; differences in voice usage between a small office and a large conference room might introduce a similar change of ±10 db, although the two effects are probably not cumulative (i.e., to some extent a loud talker is one who habitually declaims). Differences in room absorption and partition size from the assumed values will affect the transmitted level by changing the factor $10 \log[S/(A_1 A_2)]$, but, in any case, this factor, derived from reverberant-room theory, is a crude approximation in modern offices with absorbent ceilings.

TABLE I. Band-articulation index for small differences between the level exceeded by 1% of speech peaks and the ambient level. (Adapted from reference 17.)

Level diff	30W[a]	Level diff	30W
1	0.2	7	2.4
2	0.4	8	3.0
3	0.7	9	3.6
4	1.1	10	4.4
5	1.5	11	5.1
6	1.9	12	6.0

[a] 30W rather than W is tabulated for convenience in calculations.

[18] L. L. Beranek, Proc. Inst. Radio Engrs. **35**, 881 (1947).
[19] H. C. Hardy and J. E. Ancell, Noise Control **4**, 9 (1958).
[20] W. J. Cavanaugh, W. R. Farrell, P. W. Hirtle, B. G. Watters, "Speech privacy in buildings," J. Acoust. Soc. Am. **34**, 475 (1962).

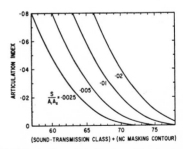

FIG. 6. Combined effect of idealized partition (conforming to a given sound transmission class contour) and a given ambient-noise level (conforming to modified NC contour) on articulation index of transmitted speech. (S—Area of partition; A_1 and A_2 are absorptions of source and receiving rooms.)

TABLE II. Analysis of four walls.

| Wall | Sound transmission class | Equivalent sound transmission class | | | "Averages" | | |
		Test 1 Equal loudness	Test 2 Equal masking by ambient noise	Test 3 Equal artic. index	9-Freq. arithmetic	11-Freq. arithmetic	Energy
A	47	51	49	50	46	48	40
B	36	37	35	39	32	34	26
C	30	36	31	32	33	36	27
D	20	27	23	25	21	22	21

Deviations of the ambient-noise spectrum from the assumed NC-curve will also affect the result. In the calculation leading to Fig. 6 it was found that the articulation index depends on the bands from 350 to 1000 cycles, with a lesser contribution from the 250- and 1400-cycle bands; hence the NC-curve should be matched to the noise in this range unless the noise is concentrated in other bands.

Finally, there is the error introduced by the method of matching a transmission-loss curve to the STC contours. Typically, an actual curve will fall on the STC curve at one or two frequencies in the middle range and be above at the other 3 or 4 frequencies. Consequently, a partition with a sharp mid-frequency dip will provide a lower articulation index than its STC rating would indicate. In a few cases calculated for actual partitions the standard matching procedure was too conservative by 0 to 6 db. This seems small enough for a general rating of this type in view of the other variables in the calculation. Moreover, it is suspected that although speech intelligibility is a primary consideration in office separation, loudness is probably a secondary but significant one. This is illustrated, for example, by experiments of Cavanaugh et al.,[20] who used narrow-band transmission in a study similar to those of Rademacher, but with speech privacy as a criterion. It was found that a narrow-transmission band in the region of maximum speech power had an annoying effect out of proportion to the intelligibility it carried.

The average error due to matching curves can be minimized by relaxing the requirement slightly; it is therefore suggested that STC plus NC values should total 66 for room conditions corresponding to $S/(A_1A_2)$

$=0.01$. More-precise values can be determined for other values of $S/(A_1A_2)$ by referring to the appropriate curves of Fig. 6.

RATINGS OF ACTUAL WALLS

To illustrate the use of the ratings, a detailed analysis is made of the performance of the partitions whose transmission characteristics are shown in Fig. 7. Wall A is a solid concrete wall 3 in. thick. The coincidence frequency for this wall is below the test range and it therefore has a smooth, steeply rising characteristic. Wall B is a 2- by 4-in. stud and plaster wall. Such walls have a characteristic high-frequency dip and a low-frequency dip that varies in detail from sample to sample; this particular curve was taken from reference 20, Fig. 16. Walls C and D are two office partitions that have pronounced deficiencies in the middle range.

The sound-transmission class, determined as prescribed in ASTM E90-61T, is given in the second column of Table II. The reliability of the class rating was tested in three ways described in the following. Since the standard test frequencies form a half-octave series, the analysis was made on a half-octave basis throughout. A sample set of calculations is given in Appendix A.

Test 1

Loudness levels were calculated using the method (Mark VI) given by Stevens[12] for the fraction of standard-household noise (Fig. 3) transmitted by each wall. Similar calculations were made for a range of sound-transmission class contours in order to obtain the curve of Fig. 8. From this it was possible to determine

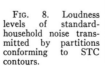

FIG. 7. Transmission losses of four typical walls. Curve A—3 in. concrete wall, Curve B—2×4 in. stud wall (Reference 20, Fig. 16), Curves C and D— Office partitions.

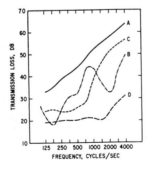

FIG. 8. Loudness levels of standard-household noise transmitted by partitions conforming to STC contours.

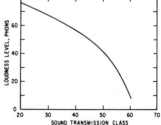

the STC contour that would transmit the same loudness as each wall. This "equivalent sound-transmission class" is given in the third column of Table II. This is a test of the system on the assumption that the absolute loudness of transmitted sound is an index of wall performance. (A similar calculation, using perceived-noise levels instead of loudness levels, produced almost identical results.)

Test 2

A better criterion is to require that transmitted noise be low enough to be masked by ambient noise. This condition would be approximately attained if peak levels of transmitted noise were not greater than the ambient level in any half-octave band.[16] Assuming an NC contour of ambient noise (modified for half-octaves) the NC curve that just-masked transmitted noise was determined for each wall. Then the STC contour just masked by the same NC curve was also determined. This is given in the fourth column of Table II.

Test 3

Articulation indices were calculated for each wall, on the assumption that ambient noise corresponding to an appropriate NC curve was present. (The NC curve was chosen so that the STC rating of the wall plus the NC value totalled 66, corresponding to the speech-privacy requirement discussed earlier, but the exact criterion used is not important in the calculation.) The actual articulation index obtained was used to deduce from Fig. 6 the equivalent sound-transmission class for speech, i.e., an actual STC contour that would combine with the assumed NC curve to give the observed articulation index.

The three equivalent sound-transmission classes, based on the three criteria described above, are shown in Table II. Since the masking criterion (test 2) depends on the highest band of noise, relative to the NC curves, it agrees closely with the actual sound-transmission class, which is determined in a similar way. This is the safest criterion, since it is a measure of the probability that the transmitted noise will be masked by ambient noise and thus unnoticed. As might be expected the other two, since they tend to average out peaks, fall slightly above the sound-transmission classes found by the standard procedure. But, apart from a slight shift in scale, the sound-transmission class accurately rates partitions in comparison with any of the three tests considered here.

For comparison, three "averages," the 9-frequency average, an 11-frequency average (including data for 1400 and 2800 cps), and the energy average are shown in the last three columns. These serve to illustrate the inconsistencies that can arise with simple averages. All three show considerable scatter relative to the sound-transmission class or to any of the three tests; none of the three gives the proper ranking for walls B and C;

the energy average, unduly influenced by low-frequency transmission losses, grossly under-rates walls A and B.

SUMMARY

Evidence has been presented showing that a simple average is an unreliable index of the sound-insulation value of a partition. It is noted that in many other countries the simple average has been replaced by a standard contour that defines transmission loss as a function of frequency. The significance of such contours has been examined theoretically from the viewpoint of dwelling and office separation, and it is shown that the sound-transmission class now incorporated in ASTM E90-61T provides a simple and accurate rating system.

ACKNOWLEDGMENTS

The author is grateful for considerable help from the other members of the Writing Group responsible for ASTM E90-61T, and especially for comments and criticisms received from Ralph Huntley, William Jack, Thomas Mariner, T. J. Schultz, R. W. Young, and Bill Watters. This paper is a contribution from the Division of Building Research of the National Research Council, Canada, and is published with the approval of the Director of the Division.

[*Editor's Note:* Appendix A, sample calculations of wall performance, has been omitted due to limitations of space.]

21

Reprinted from *J. Acoust. Soc. Am.* 34(4):475–492 (1962)

Speech Privacy in Buildings

W. J. Cavanaugh, W. R. Farrell, P. W. Hirtle, and B. G. Watters

Bolt Beranek and Newman Inc., Cambridge, Massachusetts

(Received January 10, 1962)

Obtaining adequate speech privacy in modern buildings is one of the important goals of the architect and consultant. This paper deals with the development of a rating method which takes into account the several factors influencing speech privacy. Our work in this area began with a brief laboratory study. The results indicated that speech privacy is related to speech *intelligibility* rather than to *level*. The initial experiments were supplemented with an analysis of about 40 case histories representing about 400 pairs of spaces in different kinds of buildings. There appears to be good correlation between the articulation index of intruding speech sound and the reactions of building occupants.

I. INTRODUCTION

BECAUSE structural and architectural needs can usually be met more cheaply with thin, lightweight, often partial-height partitions, the need for acoustical isolation is perhaps the prime reason for specifying heavy, solid partitions in today's buildings. However, architects often are required, for economic reasons, to specify relatively lightweight partitions, particularly in multistory structures, even at the sacrifice of adequate acoustical isolation. Also the demands for flexibility and movability of partitioning in many buildings often conflict with the acoustical isolation requirements. The prime acoustical problem in many spaces in these buildings is that of isolating speech sounds. In considering this need, we have surveyed the various schemes currently available to the architect and acoustical consultant for designing isolation for speech

privacy. These are described briefly in Sec. II, and their shortcomings for the present purpose are discussed in Secs. III and IV. We then describe in Sec. V laboratory experiments which show a strong relationship between speech intelligibility and a feeling of speech privacy. Taking this fact as basic to a design procedure for providing speech privacy, we develop in Secs. VI, VII, and VIII one-number rating schemes for background-noise levels and noise-reduction values that are specifically geared to the intelligibility of intruding speech. These are combined with factors accounting for source room size, speech effort, and privacy requirement to give a one-number over-all rating of speech privacy. Finally, in Sec. IX we apply this rating scheme to 37 case histories and find that it correlates well with the observed reactions of the occupants.

II. AVAILABLE DESIGN SCHEMES

The most commonly used design technique currently available to the architect we will call the "categorization scheme."[1–5] It is typified in Table I.

The strength of this scheme lies in the body of experience from which it grew; its weakness is that *all* acoustical aspects of every space in each category are assumed to be identical.

Sometimes the "categorization scheme" is supplemented[1,3,5] or supplanted[6] by what we will call the "acceptable-level scheme." This scheme sets an acceptable or criterion level of sound in the listening space and specifies that the transmission loss be equal to the difference between the source-sound level and the acceptable level. Typical recommendations are shown in Table II.

The acceptable level scheme assumes that people's reactions to a sound are uniquely related to its level (or loudness). While it may be true that the annoyance and sound-pressure level are related for certain sounds such as air conditioning or traffic noise, we have come to believe, for reasons which are discussed later, that annoyance due to information-carrying sounds, such as intruding speech, is determined primarily by the *intelligibility* and, to first order, has nothing at all to do with the level.

Several writers[2,5,6] mention the idea of "masking," i.e., of providing enough transmission loss to reduce the intruding sound below the level of the background noise *which actually exists* (not the background-noise criterion) in the listening space, but none, as far as we know, has organized this concept consistently enough to be useful as a design tool for the average designer.

The lack of precision of the available design procedures has been recognized. For example, one notes the strong qualifications which are found in many of the

TABLE I. Example of commonly used "categorization scheme." (Do not use for design!)

Type of spaces to be isolated	Recommended average transmission loss db
Private offices	35 to 45
Low-cost hotels	40 or greater
First-class hotels	45 or greater
School classrooms	35 to 45
Apartments, hotels, etc.	45 to 55

[1] V. O. Knudsen and C. M. Harris, *Acoustical Designing in Architecture* (John Wiley & Sons, Inc., New York, 1950).
[2] P. H. Parkin and H. R. Humphreys, *Acoustics, Noise and Buildings* (Faber and Faber, Ltd., London, 1958).
[3] R. H. Bolt and R. B. Newman, "Architectural acoustics," Architectural Record, April, June, September, and November, 1950.
[4] *Sound Control in Design*, U. S. Gypsum Company (1959).
[5] H. J. Sabine, *Less Noise, Better Hearing*, The Celotex Corporation, 1941 and 1950.
[6] S. Edelman, R. V. Waterhouse, and H. J. Lunback, Jr., *Sound Insulation of Wall and Floor Constructions*, United States Department of Commerce, National Bureau of Standards, Building Materials and Structures Rept. 144 (1955).

TABLE II. Recommended acceptable average noise levels in unoccupied rooms. (Do not use for design!)

Category of space	"A" scale levels db
Hospitals	35–40
Apartments, hotels, homes	35–45
Conference rooms, small offices	40–45
Private offices	40–45
Restaurants	50–55

above references:

". . . This table simplifies a large body of interrelated information obtained from practical experience. Though very useful, this material should be used only with caution and analysis of each problem."[3]

". . . This rule is not rigid, and in case of doubt it is advisable to provide a wall 5 or 10 db better than the rule indicates. . . . Actual experience, however, is of the most value in estimating sound insulating requirements."[1]

Unfortunately, a possible overdesign of the magnitude suggested by the latter qualification often cannot be afforded throughout a structure as extensive as a modern multistory office building.

In summary, two basic design schemes appear in the literature. There is the "acceptable level" scheme which is inadequate because it equates all sounds of the same level, regardless of their character. More common and probably more often used by the average designer is the "categorization" scheme, which prescribes, for example, a 45-db partition between rooms of all first-class hotels. Unfortunately, there are all kinds of first-class hotels; many of them do not require as much as a 45-db partition; some of them require a much better one.

III. SOME CASE HISTORIES

Figure 1 demonstrates the limitations of the categorization scheme, which uses only the average trans-

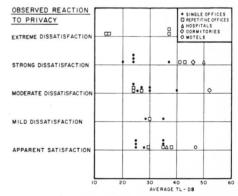

FIG. 1. Plot of the subjective reactions observed in 37 case histories of speech privacy versus the average TL rating of the isolating wall. For the most part, published average TL values were used; where the wall was flanked by other sound-transmission paths, measured values were used.

mission loss[7] of the isolating partition. On this figure are plotted data from 37 carefully documented case histories. Each of the open data points represents a building with a large number of more-or-less identical rooms. We estimate that the data actually represent well over 400 pairs of rooms.

The average transmission-loss (av TL) values shown are, wherever possible, those available to the architect (e.g., advertised values). In those cases where published data were not available or where field listening showed that the partition was seriously flanked, we have used the field data.

The subjective evaluations were made after inspecting the buildings and talking with owners and occupants. For the most part, these subjective evaluations represent a long-term evaluation. They were divided into five categories ranging from complete satisfaction to the most serious dissatisfaction and were made using the following definitions:

Extreme Dissatisfaction

The cases included here evidenced complete intolerance to the privacy situation on the part of the occupants or owners. For example, executives of a large insurance company found inter-office privacy so inadequate upon moving into their new quarters that it was necessary in some instances to vacate every other office. More adequate partitions were designed and are being installed at great additional expense.

In these "most serious" cases the owner is willing to pay almost any price for a solution to the problem. A tenant under these circumstances would probably break a lease and find new quarters if the owner refused to provide a remedy.

Strong Dissatisfaction

Cases included in this category are those with numerous and continuous strong complaints. The complaints are accompanied by such action as finding new quarters at the end of a lease, expenditure of moderate amounts of money to effect a solution, etc. A typical case history in this category is a college dormitory building where complaints by students have been persistent since its initial occupancy. The high cost of improving privacy conditions led authorities to defer action each year with the hope that the next crop of students would be more tolerant. After some five years of such hopeful waiting remedial action has now been undertaken.

Moderate Dissatisfaction

Cases here would fall midway between the extremes. Complaints may be less numerous than for the previous category, but are still persistent. An owner may be willing to make small expenditures to correct matters, but would feel that conditions justify no real inconvenience. A case history here includes engineering offices where "leaky" movable partitions gave rise to continuous complaints, but complaints all but disappeared after the partition performance was slightly improved.

Mild Dissatisfaction

Situations where complaints are sporadic and perhaps not serious enough to "call in a consultant" would fall in this category.

[7] The numerical average of the measured transmission loss at the following test frequencies: 125, 175, 250, 350, 500, 700, 1000, 2000, and 4000 cps.

Cases of this sort include those triggered by an excessively loud neighbor or a particularly fussy occupant. Little if any corrective action is usually taken in these cases, and the complaints most likely disappear with time.

Apparent Satisfaction

There is no apparent awareness of speech privacy as a problem.

If the categorization design scheme were accurate, we should expect the occupants of these spaces to be satisfied when the average transmission-loss values exceeded a recommended value. For values below the recommendation, there should be some degree of dissatisfaction. Since most of the data in Fig. 1 are for private offices, Table I suggests a minimum recommended TL of 35 db; a somewhat higher criterion applies to the few motel, dormitory, and hospital data. But it is clear from Fig. 1 that an average transmission loss of 35 db is not the demarcation between satisfaction and dissatisfaction. In 7 of the 11 office cases where the transmission loss was 35 db or above, the occupants experienced significant dissatisfaction. Perhaps of more importance, in 5 of the 9 office cases where the occupants were apparently satisfied with the privacy, the average transmission loss was *less* than 35 db.

In the sections which follow we hope to show why there is no clear trend to the data of Fig. 1 and to develop a more comprehensive rating scheme in terms of which the case history data are reanalyzed.

IV. OTHER FACTORS IMPORTANT TO SPEECH PRIVACY

Figure 1 shows that the categorization scheme does not permit an accurate design for speech privacy, i.e., one which just yields a reaction of apparent satisfaction. The reason for failure of this scheme is that there are acoustical factors other than the transmission loss that help determine the over-all occupant satisfaction. Here are some of these factors along with an estimated range, in decibels, of their possible influence on speech privacy.

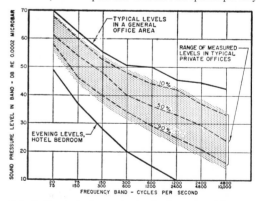

FIG. 2. Steady background-noise levels measured in spaces where speech privacy is important. The dashed lines represent measurements in 62 private offices in which there was no complaint about the noise (reference 8). Octave-band levels in the given percent of offices exceeded the dashed contour shown.

(1) Background Noise

The intelligibility of intruding speech is destroyed when the speech peaks are submerged in steady background noise (or when reduced below the threshold of hearing—a rare instance). *Thus, an increase in the background level has the same effect on intelligibility as an increase in the transmission loss.*

Figure 2 shows typical measurements of steady background noise in various spaces. The lowest curve gives average evening noise levels in a hotel in a quiet section of a small city. Similarly low levels are frequently found in un-air-conditioned hospitals, dormitories, office or "professional" buildings, etc. These levels are approximately equal to the average threshold of hearing for broad-band noise. The upper curve shows measured levels typical of a general office area. Between these two extremes are shown the results of a recent survey[8] of the background-noise levels in 62 private offices. (In none of these cases was there a complaint because of the noise.) The numbers on the three dashed curves give the percentage of the data points that exceeded the respective curves in each octave band.

Figure 2 shows that background levels in the speech-frequency bands vary from building to building by 30 db or more. Thus, we would expect a comparable variation in the "effectiveness" of a given isolating structure built in these different locations. An accurate design scheme must include the accurate prediction of background levels in the completed building. While there are many sources of background noise, only a few can be relied upon to be continuously present. Most important of these are traffic ("city") noise[9] and air-conditioning terminal-device noise.

(2) Published TL Data

Unfortunately, the published TL data for sound-isolating barriers are often significantly higher than the corresponding field performance. For example, it is not uncommon to find differences in the order of 10 db or more between the advertised transmission-loss values of a partition, and the data obtained from a valid field measurement (i.e., one in which the structure in question is not flanked by other sound-transmission paths). Obviously, this degree of uncertainty cannot be permitted in any scheme which must accurately design for speech privacy in buildings. Admittedly there are knotty problems being explored by acoustical researchers to determine and define the significance of differences in TL measurements both in the laboratory and in the field.[10–12] However, since these differences do

in fact exist, it seems appropriate to reflect on some of the causes for optimistically high published values.

(A) Incomplete Test Sample

An important example is the prefabricated partition which typically consists of not only the main panels but also the base plates, headers, connectors, trimstrips, fillers, edge gaskets, etc., all of which are adjusted in place to meet the usual variances in building construction. The differences in TL of the panel and that of the entire partition assembly can easily be 10 or more db.[13] With such panel assemblies, useful TL data for building design can be derived only from a test specimen large enough to incorporate all of the essential elements of the construction in their normal proportion and with a careful simulation of the usual building irregularities at the perimeter.

(B) Test-Panel Size and Boundary Damping

The transmission loss of many walls, especially rigid ones, is influenced by the size and edge damping.[14,15] Until an adequate theoretical basis exists for relating such effects on the transmission loss of a barrier, the prudent course for the designer interested in field performance is to use test data on walls which are tested under essentially the *same* size and edge conditions as will be used in the field.

(C) Noise-Reduction or Attenuation Data Reported as Transmission Loss

Much of the manufacturers' literature contains noise-reduction or sound-reduction data taken under rather special conditions (i.e., the source or receiver room is unusual in size or sound absorption). Generally, such data are qualified in the original test report, but these qualifications are often not fully understood by the test sponsor. Such tests may be quite useful as "pilot" data in guiding the development of a structure, but the indiscriminate use of such data has resulted in many unsatisfactory situations in finished buildings.

(D) Obsolete Data

In the past, there has been no means of recalling data which have become obsolete. As a result the body of data

mission through suspended ceiling systems," J. Acoust. Soc. Am. 33, 1523–1530 (1961).

[11] R. N. Lane and E. E. Mikeska, "Problems of field measurement of transmission loss as illustrated by data on lightweight partitions used in music buildings," J. Acoust. Soc. Am. 33, 1531–1535 (1961).

[12] T. Mariner, "Critique of the reverberant room method of measuring airborne sound transmission loss," J. Acoust. Soc. Am. 33, 1131–1139 (1961).

[13] R. N. Hamme, "Understanding sound transmission loss of lightweight partitions," Noise Control 6, 13–17 (1960).

[14] M. Heckl and K. Seifert, "Untersuchungen über den Einfluss der Eigenresonanzen der Messräume auf die Ergebnisse von Schalldämmessungen," Acustica 8, 212–220 (1958).

[15] B. G. Watters, "In-place flexural damping of walls," Session N, 62nd Meeting of the Acoustical Society of America, November, 1961.

[8] W. R. Farrell, "Evaluation of the effectiveness and acceptability of masking noise for providing speech privacy in buildings," Paper T7 presented at the 60th Meeting of the Acoustical Society of America, 1960.

[9] L. N. Miller, "A sampling of New York City traffic noise," Noise Control 6, 39–43 (1960).

[10] R. N. Hamme, "Laboratory measurements of sound trans-

with which the designer must work is a hodgepodge. Refinements in our understanding of transmission loss have resulted in improvement of testing techniques and in some cases the building of new test facilities. Differences between the test values of barriers tested in new and obsolete facilities have been 5 db or more, yet many manufacturers continue to publish the obsolete (and usually higher) values.

Some means is needed to keep test data current and abreast of the times. One positive step in this direction is that adopted by the Acoustical Door Institute in conjunction with the Riverbank Acoustical Laboratories which regards data four or more years old as automatically obsolete.

Solutions to these problems and others related to obtaining adequate TL data in terms of field performance will require the cooperation of manufacturers, consultants, acoustical researchers, testing laboratories, and standards committees alike. Meanwhile, the burden of providing the impetus for solution falls on the data users themselves who must be critical of the data used in actual building design. Clearly they must demand as a minimum the complete details of the test data of the structure under consideration including a full description of the test sample and installation details, the test procedure, date of test, etc. Only those data which are truly representative of "field-like" TL performance can be useful in an accurate design scheme.

(3) Size, Shape, and Sound Absorption of the Rooms to be Isolated

For two adjacent rooms not too large or too "dead," the sound-pressure level incident upon the isolating wall is proportional to the average level in the source room, SPL_1

$$SPL_1 = PWL - 10 \log(S\bar{a})_1 + 6 \text{ db.}$$

In similar fashion, the intruding-speech level, SPL_2, is given by

$$SPL_2 = SPL_1 - TL + 10 \log S_w - 10 \log(S\bar{a})_2,$$

where $S\bar{a} = (0.049V/\bar{T})$ sq ft, $V =$ the volume of the room in cu ft, $\bar{T} =$ the reverberation time in sec, $S_w =$ the area of the transmitting wall in sq ft, $TL =$ the random incidence sound-transmission loss of the wall in db, SPL_1 and $SPL_2 =$ the rms sound pressures averaged over the volume of the source room and receiving room, expressed in db re 2×10^{-4} d/cm², $PWL_1 =$ the sound-power level of the speech source in db re 10^{-13} w.

The sound-absorbing area $S\bar{a}$ is determined not only by the absorption coefficient and area of acoustical materials in the room, but also by the location of the materials,[16,17] the absorption of carpets, drapes, furnish-

[16] L. L. Beranek, "Acoustic impedance of commercial materials and the performance of rectangular rooms with one treated surface," J. Acoust. Soc. Am. **12**, 14–23 (1940).
[17] D. Fitzroy, "Reverberation formula which seems to be more accurate with nonuniform distribution of absorption," J. Acoust. Soc. Am., **31**, 893–897 (1959).

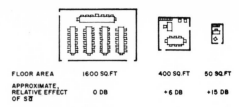

FLOOR AREA	1600 SQ.FT	400 SQ.FT	50 SQ.FT
APPROXIMATE, RELATIVE EFFECT OF $S\bar{a}$	0 DB	+6 DB	+15 DB

FIG. 3. The buildup of speech levels in rooms of various sizes.

ings, and people, and by the coupling of the various room modes provided by large objects such as desks, chairs, and file cabinets. To our knowledge, no practical theory presently available permits the calculation of $S\bar{a}$ taking into account all of the above factors. *Fortunately, if unusually "live" and unusually "dead" rooms are neglected, field data show $S\bar{a}$ to be approximately equal to the floor area of the room, in the speech frequency range, for rooms of moderate size.*

Using this result, it is easy to see the effect of neglecting the room size when calculating the intruding-speech level. Figure 3 shows a range of talker-room sizes from a small 6 ft×8 ft office or interview room to a 40 ft×40 ft classroom or board room. Assuming a constant speech power, the corresponding range in reverberant speech levels is 15 db. An accurate analysis scheme must take this effect into account.

Figure 4 shows a range of *ratios* of wall area to listener-room floor area from $\frac{1}{2}$ to 2; the corresponding variation in transmitted levels is 6 db. An accurate design scheme must take this additional effect into account as well.

(4) Expected Speech Activity

As talkers move from one environment to another, we should expect changes in their *average* speech level. For example, the average talker might be expected to use "normal" or "conversational" speech effort when talking to someone a few feet away (as in a small office or in a hotel room). The same average talker would

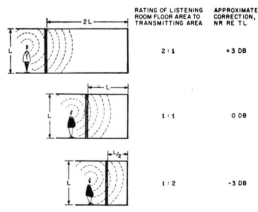

	RATING OF LISTENING ROOM FLOOR AREA TO TRANSMITTING AREA	APPROXIMATE CORRECTION, NR RE TL
	2:1	+3 DB
	1:1	0 DB
	1:2	-3 DB

FIG. 4. The relationship of NR and TL in rooms of various shapes.

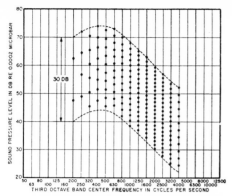

FIG. 5. Graphical representation of normal speech levels in a 100 sabine room. Number of dots in each third-octave band signifies relative contribution to articulation index. The data are obtained from Fig. 2 of reference 22 using the directivity index data given by Beranek (reference 21) after H. K. Dunn and D. W. Farnsworth [J. Acoust. Soc. Am. **10**, 184 (1939)]. The peak factor of speech is taken to be +12 db at all frequencies.

probably increase his speech effort to a "raised voice" when speaking to a group of people in a conference room. In some unusual conditions (for example, in psychiatric interviewing offices), the speech effort might occasionally increase further to a "loud voice." The sound levels for a "raised voice" are about 6 db higher than those for conversational speech; a "loud voice" is about 12 db higher.[18]

(5) Kind of Privacy Required

The degree of speech privacy required by the occupant of a room depends on his activity. As an example, consider the case of an engineer or other technical person. During most of his work day, his desire for speech isolation is set by his wish for freedom from distraction. We have called this "normal" privacy. However, if he should be called into the office of his supervisor or employer to discuss salary or personal matters, the need for speech isolation is different. It no longer is the freedom from distraction, but now becomes the assurance of not being overheard. This kind of privacy we have called "confidential." Let us further imagine that a part of his work concerns a highly classified project. Conferences he may have in this connection may need to be truly secret.[19]

In laboratory tests described in Sec. V, where a private office environment was simulated, the more critical occupants desired a fairly low intelligibility of intruding speech when their work was confidential and permitted a higher intelligibility when their work required only freedom from distraction. Generally, this amounted to a 6-db difference in the ratio of speech signal to background noise.

V. EXPERIMENTS ON THE RELATIONSHIP BETWEEN SPEECH INTELLIGIBILITY AND SPEECH PRIVACY

It is a matter of simple observation that even though the background-noise level around us is well above our threshold of hearing, we almost invariably become accustomed to it, accept it, and, most of the time, are altogether unaware of it. On the other hand, we sometimes express strong dissatisfaction at intruding speech whose rms levels are no greater than those of the background noise.

One possible explanation for the above observations is that speech privacy is related to speech intelligibility. Speech intelligibility, we know, is determined not by the level of the speech but rather by the ratio of speech to noise. Our experimental work has looked for confirmation of this assumed relationship. We found that each subject had a precise personal criterion for the speech-to-noise ratio which for him just constituted privacy, that there is a wide variation (10 db or so) in the criteria of various subjects, but that the assumed relationship of intelligibility and privacy seemed to be consistent with the experimental results.

A. Review of Speech-Intelligibility Theory

A general theory of speech intelligibility was developed about 20 years ago by the Bell Telephone Laboratories[20] and has been expanded by Beranek[21] and

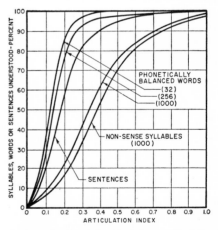

FIG. 6. Approximate relationship between articulation index and intelligibility for skilled talkers and listeners. The numbers in parentheses give the size of the test vocabulary.

---

[18] L. L. Beranek, *Acoustics* (McGraw-Hill Book Company, Inc., New York, 1954), p. 338.

[19] True secrecy is much more complicated than the "confidential" requirement just mentioned. For example, if an eavesdropper is free to place his ear to the wall and thus shut out the airborne background noise, the problem becomes one of the signal-to-noise ratio of the vibration levels in the wall structure. True secrecy is not considered here.

[20] N. R. French and J. C. Steinberg, "Factors governing the intelligibility of speech sounds," J. Acoust. Soc. Am. **19**, 90 (1947).

[21] L. L. Beranek, "Design of speech systems," Proc. IRE, **35**, 880–890 (1947).

others. Although more recent studies[22] have cast doubt on its exactness, it remains the most generally useful theory available today.

Some important aspects of this theory are:

(1) The intelligible part of speech energy lies roughly between 200 and 6000 cps.

(2) Most of the energy of speech is in the frequency range below 800 cps; most of the contribution to intelligibility above 800 cps.

(3) In each frequency band, speech has a dynamic range of about 30 db; the peak values lie about 12 db above the long-time rms levels.

(4) Any frequency band in the range 200 to 6000 cps may be considered to make a contribution to intelligibility that is proportional (a) to the fraction of its 30-db dynamic range which is greater than the masking noise (or threshold of hearing), (b) to the bandwidth, and (c) to the "importance function" for that band. The importance function is a maximum at about 2000 cps.

These facts are symbolized in Fig. 5, where the useful speech signal is shown as a dot field beginning at 200 cps and extending to 6000 cps. Each dot signifies a possible $\frac{1}{2}\%$ contribution to the articulation index. The field is 30 db "high" and the greatest density of dots is at 2000 cps. The dot field is drawn for an average talker using "conversational" speech effort. The upper envelope of the field gives the approximate peak sound-pressure level in a room with 100 sabines of sound absorption.

If on Fig. 5 we plot the background-noise level in $\frac{1}{3}$-octave bands then the ratio of the number of dots which lie above the plot to the total number is approxi-

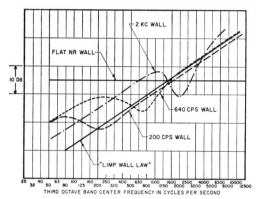

FIG. 8. Relative noise-reduction characteristics of test walls.

mately equal to the articulation index (AI). This objective quantity has been empirically related to the intelligibility of various kinds of speech (i.e., unrelated words, sentences, etc.) as shown in Fig. 6. Our study tried to determine if this same quantity could also be related to the feeling of speech privacy.

B. Some Experimental Results

Most of our experiments were conducted in simulated private-office environments having the elements shown in Fig. 7. Usually the test subject was provided with a desk, chair, and reading matter.

The intruding speech was generated with a loudspeaker and tape recorder in the adjacent room. The speech signal used was a shorthand-training recording;[23] the talker gave dictation at a rate of 100 wpm at a fairly constant level. The distribution of average levels existing in 150-msec intervals was measured.

The isolation was provided partly by the actual physical wall separating the two test rooms and partly by a spectrum-shaping $\frac{1}{3}$-octave filter and a 2 db-per-step attenuator controlled by the experimenter. In one set of experiments, for example, the filter was adjusted so that the total isolation was equal, successively, to each of the five curves in Fig. 8.

These curves are representative of commonly used partitions. The "flat" curve is often found when a partition is seriously flanked by air leaks. The 6-db per octave "limp" wall curve is approximated by single, thin sheets of steel, lead, or plastic. The 200-, 640-, and 2000-cps curves are characteristic of walls having these critical frequencies. Hollow masonry block, solid studless plaster, and plasterboard or plywood are examples, respectively.

Care was taken to ensure a steady, consistent background noise. In one experiment, for example, the noise was established by a 2-channel tape recording of traffic and diffuser noise. The diffuser-noise loudspeaker was

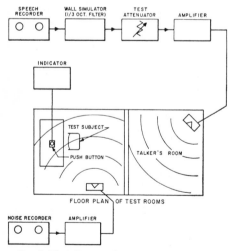

FIG. 7. Block diagram of speech-privacy test setup simulating private-office environment.

[22] J. C. R. Licklider, "Three auditory theories," in *Psychology: A Study of a Science*, edited by S. Koch (McGraw-Hill Book Company, Inc., New York, 1959).

[23] Sustained Dictation Record, No. SD-2, Gregg Publishing Division, Bus. Ed. Div., McGraw-Hill Book Company, Inc., New York, Dictation Speed, 100 wpm.

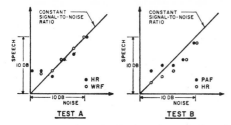

FIG. 9. Results of test showing dependence of tolerated speech levels on background-noise level.

hidden in the ceiling above a real air diffuser; the traffic-noise loudspeaker was placed near a window. The noise levels approximated the NC-35 contour in the speech-frequency range and were held constant throughout the experiment. The reaction of the test subject was generally transmitted by a push botton mounted on his desk. The sound-pressure levels of both the background noise and of the intruding speech were measured in $\frac{1}{3}$-octave bands using a nondirectional microphone at several points near where the test subject was to be seated.

Two sets of written instructions were normally given to the test subjects as shown below. The instructions for Test B were given to the subject after completion of Test A.

The purpose of these tests is to measure the amount of sound isolation people consider adequate for their offices.

Test A

In the office adjacent to the room you are in, we will reproduce through a loudspeaker various types of speech signals that will gradually increase in intensity until you can hear them. When the speech reaches a level that you consider bothersome, please push the button-switch on the table before you. During this test, imagine, if you will, that you are doing your normal work, including conferences, in your own office. In other words, push the switch when you believe the speech from the office adjacent to yours first reaches a level that, day in and day out, would interfere with the performance of your average, normal work.

As soon as you have pressed the switch, the intensity of the speech signal will again be reduced to an inaudible level and then slowly increased until you can hear it. Each time, please press the switch when the speech just reaches the level you feel would be bothersome to your work routine. This sequence will be repeated a number of times. You will be told by the experimenter when test A is completed.

Test B

During test B, respond to the speech coming from the office next door in somewhat the same way as in the previous test. However, this time, please judge the privacy that you require for the most sensitive or confidential work you do in your office. For example, the discussion of company-classified material, personnel matters, etc. For test B bear in mind that conversation in your office will be heard outside to the same degree you are able to hear speech coming from the adjacent office.

As soon as you have pressed the switch, the intensity of the speech signal will be reduced to an inaudible level and then slowly increased until you can again hear it. Please press the switch again

at what you consider the proper time. This sequence will again be repeated several times. You will be told by the experimenter when test B is completed.

A brief preliminary test was run to see if the absolute level of the background noise influenced the test results. The data of Fig. 9 show that on the average our test subjects responded with a constant signal-to-noise ratio as the background levels varied. The fairly narrow range of test background levels in this case was imposed by the real background (about NC-20) and the desire to keep the test levels below about NC-35.

As stated in the instructions, the 2-db per step gain control which controlled the level of the intruding speech was increased in a cyclic fashion; beginning well below the response level of the subject, increasing gradually (in 2-db steps), and remaining constant for 10 sec on at least the last step below the setting which invoked a response.

The short-time variability of almost all subjects was small. For example, during one day of testing, 13 subjects made 356 judgments of privacy. The rms variation from their average responses was 1.3 db. A significant part of this variation probably is the result of the fact that the speech levels could not be varied in steps smaller than 2 db.

The learning process of most subjects appeared to be very rapid. In general, the testing was continued until the subject appeared to give a consistent response; about four responses were normally found to be adequate. The 1.3-db variation above includes any learning effect.

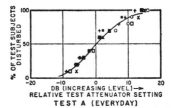

	ZERO DB CORRESPONDS TO ATTENUATOR SETTING OF	
	TEST A	TEST B
• FLAT	32 DB	26 DB
+ LIMP	29 DB	21 DB
o 200	28 DB	22 DB
x 640	27 DB	21 DB
□ 2 KC	30 DB	23 DB

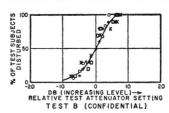

FIG. 10. Relative speech levels required to just-cause annoyance for the (A) (everyday) and (B) (confidential) tests.

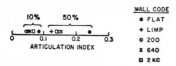

WALL CODE
● FLAT
+ LIMP
○ 200
✗ 640
▣ 2 KC

Fig. 11. Calculated articulation index of normal speech transmitted through the five walls of Fig. 17. The speech level is that required to induce a reaction by 10 and 50% of the test subjects.

We have little data on the long-time variability of response. In one experiment with two test subjects, a repeat test after an hour's delay gave no change in response. For a second subject, a day's delay produced an apparent 2-db shift. The response level of one of the authors was found to be within about 2 db of its initial value after an 18-month time lapse.

While any one subject had a definite idea of what constituted speech privacy, not all subjects agreed on the *same* definition. Figure 10 shows the variation in response levels of a group of about 10 test subjects drawn from a business-office environment. The five symbols in the figure are for the five different simulated transmission-loss curves shown in Fig. 8. Each subject was tested with speech intruding through each of the five simulated walls, first following the instructions for test A (everyday privacy) and then assuming the need for confidential privacy (test B). The data for each transmission-loss curve have been shifted arbitrarily as a set so that all data coincide at the 50% level of reaction.

When asked to define "normal" or everyday privacy, the least critical subject allowed 20-db higher speech levels than the most critical. When defining "confidential" privacy, the maximum variation among subjects was about 10 db. There was a shift of about 6 db between the definitions of "normal" and "confidential" privacy as given by the most critical subjects.

Figure 10 is characteristic of all the data we have obtained. However, just as the various subjects for this test differed in their concept of privacy, various groups will differ. The subjects for Fig. 10 were somewhat less critical than others we have tested, and the slopes of the curves are more gradual than for other groups.

The different transmission-loss curves of Fig. 8 result in different spectra for the intruding speech. Since our subjects were tested with these different spectra, we can check the computed articulation index for each spectrum against the subjects' response levels. Such a comparison is shown in Fig. 11 for the 10 and 50% levels of annoyance for the subjects of Fig. 10 tested for "confidential" privacy. It appears that both the most critical 10% of the subjects and the average subject tolerated a higher articulation index for the "flat" wall, the wall with relatively weak low-frequency speech components. However, it should be noted that although the articulation index for the "flat" wall is nearly double that for the other walls, a reduction of only 3 db in the "flat" speech levels would have reduced the articulation

index to that for the other walls. The over-all precision of the experiment is probably no greater than 3 db.

Although the results of Fig. 11 seem to be fairly well explainable in terms of the calculated intelligibility of the intruding speech, it is interesting to speculate on the actual basis for judgment by our subjects. To this end, we talked with some of the subjects after the experiment was completed. It is clear that real or apparent intelligibility is involved in the judgments of many if not most of them. A typical comment was that the subject reacted when he could just begin to understand the intruding speech. Some of the less critical subjects said that they reacted when the intruding speech distracted them from the work (reading) they were performing. Usually this occurred at a fairly high level of intelligibility.

Some of the most critical subjects appeared to react when the speech sounds became recognizable as such, even though the intelligibility of the speech was virtually zero. One explanation put forth was that the very tone of voice used by the talker (if he were someone well known to the listener) could convey intelligibility. More or less at the opposite extreme, some subjects accepted a moderately high degree of intelligibility of intruding speech even for the "B" or confidential test. One subject told us that even though the intruding speech levels were clearly intelligible, he knew that he could carry on a private conversation in his own office by the simple

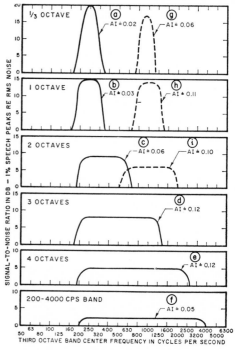

Fig. 12. Approximate 1% peak speech levels judged by one subject to be equally annoying. (Background noise approximately N33, see Sec. VII.)

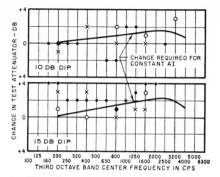

FIG. 13. Change in test attenuator required to compensate for ⅓-octave-wide TL dip. Intruding speech spectrum otherwise has equal signal-to-noise ratio in all bands.

expedient of lowering his voice, and apparently was willing to do so.

A critical question arises concerning the importance of sharp dips in the TL curve of a sound-isolating structure.[24] A good example is the plaster partition of Fig. 16. We conducted some exploratory experiments to see if the AI relates to structures with sharp dips in the same way it relates to the more usual structure. These experiments were conducted along the lines described earlier. The Gregg-shorthand recording was played through a loudspeaker in an adjoining room after first being passed through a ⅓-octave spectrum-shaping filter. The background noise for these tests was generated by a ceiling air diffuser.

We began with the intruding-speech sound filtered through the single 250-cps ⅓-octave as shown in Fig. 12 [spectrum (a)]. The over-all level was increased until the subject reacted. The bandwidth was then increased to a full octave, centered on 250 cps as shown by spectrum (b). The signal-to-noise ratio throughout the octave was approximately independent of frequency. The over-all level was again adjusted until the subject reacted. As shown by spectra (c) through (f), the bandwidth was further increased in steps until the full range from 200 to 4000 cps was covered. The signal-to-noise ratio, expressed in decibels, was determined for each bandwidth. A similar experiment began with the 1000 cps, ⅓-octave band [spectra (g) through (i)]. One of the authors was the subject tested.

While this was purely a preliminary experiment, some interesting points may be noted. First, for the single ⅓-octave bands and for the octave band centered on 250 cps, the subject found the intruding speech signal

to be almost completely unintelligible although it could be recognized as speech. There was increasing intelligibility for the wider-band tests. Thus, we do find annoyance caused by unintelligible, narrow-band speech, especially at the low frequencies. The calculated AI's for the various bandwidths confirm this. The AI for the 250 cps, ⅓-octave was only 2% when the subject responded. This compares with 5–10% for the wider bandwidths.

The more compelling point, however, is *the relatively small importance of a narrow band of intruding speech.* While the observed importance is greater than given by the articulation index, we found that the signal-to-noise ratio for the single 250-cps ⅓-octave band could be about *18 db greater* than for the full frequency range.

A further experiment was designed to more nearly simulate the conditions of a wall with a sharp TL dip. In this experiment, the spectrum-shaping filter was first adjusted for equal signal-to-noise ratio in all bands from 200 through 4000 cps. The test attenuator was then adjusted until the subject responded. Next, the level of one one-third octave band was raised 10 db above its previous setting. As can be recognized, this 10-db adjustment corresponds to a 10-db dip in the TL of a structure which otherwise provides an equal signal-to-noise ratio throughout the entire range of speech-intelligibility frequencies. Finally, the test attenuator was again adjusted until the subject responded. Three of the authors were subjects for this experiment. The data are given in the upper part of Fig. 13. As may be seen, a 10-db dip only one-third octave wide was judged to be relatively unimportant. Indeed, the importance was fairly well predicted by the contribution to the articulation index of the dip.

The experiment was again run but with a 15-db dip. The results are given in the lower portion of Fig. 13. As may be seen, there is a response which at the lower frequencies is not predicted by the change in the articulation index caused by the dip. During the lower-frequency tests, the subjects were aware of two separate sources of annoyance. One was the just-intelligible broad-band speech signal; the other was a narrow band of noise, which could, however, be recognized as being caused by the speech signal. It was also noted that the subjects most critical of wide-band intruding speech gave more importance to low-frequency TL dips.

It is clear that dips in the TL curve of a structure can, if they are deep enough, cause annoyance which is not explainable in terms of speech intelligibility. Stated another way, the function which relates *annoyance* to signal-to-noise ratio is larger, in the case of low-frequency narrow bands of speech, than the function which relates *intelligibility* to the signal-to-noise ratio. This effect deserves more-careful study. However, it is even more apparent that the effect of a narrow TL dip is small, and, for the great majority of TL curves encountered, the effect is accounted for by the articulation-index rating scheme proposed here.

---

[24] Another reason for considering this problem is to determine the relevance to the speech-privacy problem of single-number rating schemes which may rate a wall solely upon the TL measured at the depth of such a dip. The new Sound Transmission Class (STC) scheme as described in the ASTM standard E90-61T tends to give a very high weighting to sharp dips in TL curves. In contrast, the British "grade" scheme and the German DIN rating method tend to smooth or to average out dips in the TL curve. The AI scheme developed in this paper uses a weighted averaging technique.

Table III. Characteristic noise levels, db.

Octave band	150 300	300 600	600 1200	1200 2400	2400 4800 cps									
N 30	43	37	32	30	28									
L 30	43	37	32	24	16									
M 30	31	31	32	24	16									
H 30	31	31	32	30	28									
½-octave center frequency	250	354	500	707	1000	1400	2000	2800	4000 cps					
N 30	38	35	32	30	28	27	26	25	24					
L 30	38	35	32	30	27	22	18	14	10					
M 30	28	28	29	29	27	22	18	14	10					
H 30	28	28	29	29	28	27	26	25	24					
⅓-octave center frequency	200	250	320	400	500	640	800	1000	1250	1600	2000	2500	3200	4000
N 30	38	36	34	32	30	29	27	26	26	25	24	24	23	22
L 30	38	36	34	32	30	29	27	25	22	19	16	13	11	8
M 30	26	26	26	26	27	28	27	25	22	19	16	13	11	8
H 30	26	26	26	26	27	28	27	26	26	25	24	24	23	22

VI. ARTICULATION INDEX. A BASIS FOR A SPEECH-PRIVACY RATING SCHEME

The studies described in Sec. V show that:

(1) The typical test subject's feeling of privacy appears to be related to the *ratio* of the intruding speech to the ambient or background noise rather than to the level of the speech itself.

(2) More particularly, the subject reaction seems to be correlated with the articulation index which is a "weighted" signal-to-noise ratio.

(3) For the particular group of subjects whose reactions are shown in Figs. 10 and 11, the most critical 10% of the subjects began to feel a lack of privacy when the articulation index reached 0.05.

Although one could analyze speech privacy by exactly calculating the articulation index, we have made use of a slightly less accurate but much more convenient scheme. This convenience is even more compelling when incorporated into a scheme for designing rather than for analyzing buildings.[25]

This scheme condenses the spectrum-wide data for each of the important acoustical factors into single numbers. In particular, the frequency characteristics of the data are "weighted" to be proportional to speech intelligibility. The background noise and the noise reduction are the two quantities whose frequency characteristics are most likely to vary widely from one case to another. They are dealt with in Secs. VII and VIII. Although the frequency characteristics of other quantities (such as the sound absorption) may also change, these changes are believed to be relatively small and average values have been assumed.

The following sections, then, describe the proposed rating scheme, after which the case histories mentioned in the foregoing are re-evaluated in terms of this scheme.

VII. RATING FOR BACKGROUND NOISE

Most noise-rating schemes (e.g., the NC criteria,[26] and the A-, B-, or C-scale readings of a sound-level meter) are concerned with the loudness of or the annoyance caused by *noise*. Our concern with noise is different. We want to know the amount of masking of intruding speech sounds it provides. Since intelligible speech sounds occur in the entire range from about 200 to about 6000 cps, the noise rating must contain information about the noise levels, as a function of frequency, throughout this range.

We have set up four characteristic noise-spectrum *shapes* as shown in Fig. 14. In rating a particular noise curve one must:

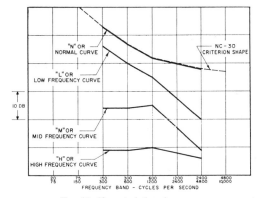

Fig. 14. Characteristic-noise curves.

[25] Such a design scheme is used in the "Speech privacy analyzer" published by the Owens-Corning Fiberglas Corporation.

[26] L. L. Beranek, "Revised criteria for noise in buildings," Noise Control **3**, 19–27 (1957).

TABLE IV. Weighting factors for rating noise curves.

	150 300	300 600	600 1200	1200 2400	2400 4800 cps
Weighting factor	0.4	0.6	0.8	1.0	0.5

(1) Identify the *shape* of the noise curve with one of the four characteristic curves.

(2) (Approximately) determine the NC rating of the "best-fit" characteristic curve.

Thus, the rating proposed here for background noise consists of two parts. One part is a letter (N, L, M, or H) which corresponds to one of the characteristic-spectrum shapes and which describes the tonal character of the noise. The second part is a number which is approximately the NC rating of the noise.

As may be seen in Fig. 14, the four characteristic curves are defined only in the five octave bands beginning at 150 cps. (Table III tabulates the four characteristic noise curves at the 30-db level, given for full-, half-, and third-octave bands.) The upper solid curve is the spectrum shape most commonly found in buildings[8] and is called a "normal" or N curve. The "low-frequency" or L curve matches the N curve (and thus the NC-30 spectrum shape) in the low frequencies but falls off below it at 6 db/octave at the high frequencies. The "high-frequency" or H curve matches the NC-30 spectrum shape in the upper frequencies but falls off below it at 6 db/octave at the low frequencies. Finally,

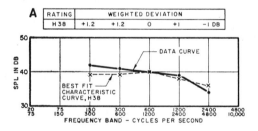

A	RATING	WEIGHTED DEVIATION			
	H 38	+1.2	+1.2	0	+1	−1 DB

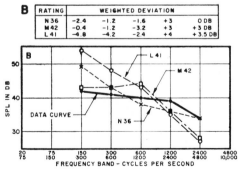

B	RATING	WEIGHTED DEVIATION			
	N 36	−2.4	−1.2	−1.6	+3	0 DB
	M 42	−0.4	−1.2	−3.2	+3	+3 DB
	L 41	−4.8	−4.2	−2.4	+4	+3.5 DB

FIG. 15. Examples of fitting data curves
to characteristic-noise curves.

there is a "mid-frequency" characteristic curve which touches the NC-30 spectrum shape only in the 600- to 1200-cps band and falls below it at a rate of 6 db/octave at both low and high frequencies. All curves of a given category regardless of their level are, by definition, parallel. This is in contrast to the NC-criterion curves, which change shape slightly with changing level.[27]

To provide a nonarbitrary method of dealing with noise spectra which do not precisely fit any one of the characteristic spectra, the weighting factors listed in Table IV were developed. These factors take into account not only the probability that a given frequency band will have a positive ratio of intruding speech to noise, but also the contribution to intelligibility of a given positive ratio for each frequency band.

To illustrate the use of the proposed noise-rating scheme, Fig. 15(A) shows a measured background-noise curve and its rating. The data are for an under-window air-induction unit operated at 40 cubic ft/min and with 3-in. H_2O pressure drop across the damper valve. The resulting SPL in a typical installation is shown along with the best-fit characteristic curve. The "best-fit" condition is the one which gives the smallest maximum weighted deviation in any octave band. In the example of Fig. 15(A) the H38 contour gives the best fit. The weighted deviation is 1.2 db. The maximum weighted deviation between the data curve and an H37 contour is 2 db; between the data and an H39 contour is 1.5 db. In the event that the maximum weighted deviation for one of the other characteristic curves had also been 1.2 db, we would refer to the *second* largest weighted deviation in an octave band as the basis for assigning the rating.

Figure 15(B) shows the N, L, and M contours which best fit the same measured data. The H38 curve gives a smaller maximum, weighted deviation than any of these. Thus, the induction-unit noise is assigned a rating of H38.

There may, of course, be some unusual spectra which will be considerably different from any of the four characteristic shapes. These are considered to be outside of the scope of the rating scheme.

VIII. RATING FOR NOISE REDUCTION AND TRANSMISSION LOSS

An NR rating number should be related to the structure's ability to prevent annoyance due to intruding sounds. In Sec. V we show that our test subjects' reaction was correlated with the articulation index of intruding speech sounds. A difficulty, however, is that

[27] Although the NC curves do change shape somewhat with level, the change is small in the frequency range above 150 cps for the range of levels commonly encountered in typical building situations (NC 20 to NC 50). The maximum variation in shape from the NC 30 curve is from −2 to +3 db and occurs mostly in the 150- to 300-cps band. Thus, within this tolerance, the proposed rating is indeed an extension of the NC scheme, being composed of the NC-rating number, plus a letter denoting the tonal character of the noise.

TABLE V. Sound-power level (db re 10^{-13} w) and importance function of normal speech.

$\frac{1}{3}$-octave-band center Frequency	200	250	320	400	500	640	800	1000	1250	1600	2000	2500	3200	4000	cps
PWL, 1% Non-directive speech peaks	84	86	87	88	88	87	85	82	80	77	74	71	68	66	db
Importance function	3.3	6.7	10	13.2	16.6	20	23.3	26.7	30	33.3	36.7	35	30	20	$\times 10^{-4}$
IF×6000[a]	2	4	6	8	10	12	14	16	18	20	22	21	18	12	

$\frac{1}{2}$-octave-band center Frequency	250	354	500	707	1000	1400	2000	2800	4000	cps
PWL, 1% Non-directive speech peaks	88	89	89	89	84	80	76	71	68	db
IF×6000[a]	8	11	15	20	25	29	32	29	20	

[a] The factor of 6000 has been introduced to avoid awkward fractions; it must be divided out later, as illustrated in the example of Table VI.

the NR vs frequency curves for different partitions do not all have the same shape. Thus, the rate of growth of the AI with the *over-all* level of the source speech signal is different for various structures.

This difficulty is illustrated in Fig. 16. Plotted here are measured NR data for a common plaster partition. We recall that the AI is a "weighted" signal-to-noise ratio and is defined here by

$$AI = \sum_{200 \text{ cps}}^{6000 \text{ cps}} (SPL_2 - SPL_{Bkgrd}) \times IF,$$

where SPL_2 are the 1% peak speech levels in the listening room, SPL_{Bkgrd} are the rms background-noise levels in the listening room, and IF is the "importance function" or weighting function given in Table I of reference 21 (see also Table V of the present paper). Consequently, AI is related to the *source-room* speech levels SPL_1 by

$$AI = \sum_{200 \text{ cps}}^{6000 \text{ cps}} (SPL_1 - NR - SPL_{Bkgrd}) \times IF$$

This calculation can be made graphically as shown in Fig. 16. The dot field superimposed over the NR curve represents the difference $SPL_1 - SPL_{Bkgrd}$ and is the data of Fig. 5 modified by the "N" background-noise spectrum shape. The dot field (normally a transparent overlay) has been adjusted vertically on the graph of Fig. 16 so that 5% of the total number of dots lie above the NR curve, corresponding to an AI of 0.05. As may be seen, all frequency bands from 200 through 630 cps contribute to AI. The dip in the NR curve at 2 kc is just shallow enough to be of no importance *at this value of AI*. However, it is clear that at some higher AI value, the dot field would lie higher upon the graph and the contribution from the 2-kc frequency range would be quite significant. Conversely, at a lower AI value, only the dip at 200 cps is important.

If we plot the growth in AI as the over-all signal level is increased, Fig. 17 results. Zero db has been taken to be the over-all level which just yields an AI of 0.05. We note that, because of the dip at 200 cps, AI becomes greater than zero when the relative speech level is -12 db but that it grows rather slowly as the level is increased. By contrast, the AI for an NR curve having a shape identical with the envelope of the dot field would increase very rapidly, also shown in Fig. 17. A study of the NR curves for 15 or so common wall constructions produced a range in AI growth rates shown by the shaded region.

It is clear that a single-number rating scheme based on the articulation index and having units proportional to decibels is accurate *only at one value of AI*. It is important, therefore, that the proper reference value be chosen. Thus, the 15 or so NR curves for Fig. 17 are equivalent despite their differing shapes, but *only* when the relative speech level is zero db (and the chosen reference AI is 0.05). The experimental work of Sec. V

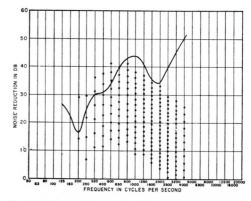

FIG. 16. Measured NR of wall, 2×4 wood studs, gypsum lath, and $\frac{1}{2}$-in. sand plaster. Superimposed dot field is for "N" background-noise spectrum shape and has been adjusted so that 5% of dots lie above NR curve.

TABLE VI. Use of NR rating scheme for a plaster partition. (Negative values shown in parentheses).

Frequency	200	250	320	400	500	640	800	1000	1250	1600	2000	2500	3200	4000	
[a] PWL, 1% Non-directive speech peaks	84	86	87	88	88	87	85	82	80	77	74	71	68	66	db
[b] Measured NR	16	25	31	30	32	41	42	44	41	35	34	41	45	50	
[c] $\Delta_1 = $ PWL $-$ NR	68	61	56	58	56	46	43	38	39	42	40	30	23	16	
[d] N46	54	52	50	48	46	45	43	42	42	41	40	40	39	38	
[e] $\Delta_2 = \Delta_1 - $ N46	14	9	6	10	10	1	0	(4)	(3)	1	0	(10)	(16)	(22)	
[f] $\Delta_2 \times $ IF $\times 6000$	28	36	36	80	100	12	...	...	...	20	...	...	...	...	

[g] AI $= \Sigma/6000 = 312/6000 = 0.052$
Speech-privacy rating with N background noise $= 100 - 46 = $ N54

[a] Enter PWL, 1% Nondirective speech peaks from Table V.
[b] Enter measured NR.
[c] Subtract NR from PWL.
[d] Enter levels for N46 background noise obtained from Table III by adding 16 db to the N30 ½-octave-band curve. This represents the last step in a trial-and-error process which tested various "N" background noises at different levels till this one yielded AI ≐0.05 (see below).
[e] Subtract background-noise levels from (c).
[f] Multiply e times (IF ×6000) from Table V.
[g] Sum f. In this example, $\Sigma/6000 = 0.052 \cong 0.05$; thus NR rating (for N background noise spectrum) $= 100 - 46$ or N54.
In similar fashion, the NR ratings in the presence of L, M, and H background-noise spectra are L50, M47, and H48.

shows that the most critical test subjects felt an AI of around 0.05 to be the "break" point between adequate and inadequate "confidential" privacy. On the other hand, with "everyday" privacy requirements, even the more critical subjects were satisfied with an AI somewhat greater than 0.10. Considering the variations shown in Fig. 17, we have chosen an AI of 0.05 as being a suitable average value for the more critical subjects.

The NR rating number we propose for a room-to-room situation then, is given by

$$\text{NR rating} = \left\{ \begin{array}{c} \text{A suitable one-number} \\ \text{rating of the source-room} \\ \text{speech levels, in db} \end{array} \right\} - \left\{ \begin{array}{c} \text{A suitable one-number} \\ \text{rating for a listener's room} \\ \text{background level, in db,} \\ \text{chosen to yield an articula-} \\ \text{tion index of 0.05 for the} \\ \text{chosen source-room speech} \\ \text{level} \end{array} \right\} + \text{a constant, in db.}$$

The absolute values of the speech and background levels in this equation are irrelevant, so long as the AI, defined previously, is equal to 0.05. We have deliberately chosen the constant above so that a given rating number is appreciably higher than the corresponding average TL. In this way we hope to avoid confusion with the average TL, STC, and other rating schemes.

In particular, if when adjusting the AI to be equal to 0.05 one imagines the source-room speech levels to be equal, band by band, to the PWL of normal speech;

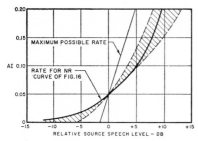

FIG. 17. Rate of growth of AI with source-room speech level for NR curve of Fig. 16. Shaded region gives range found for more than 15 common walls.

and if one rates the required background noise using the scheme presented in Sec. VII, the NR rating for the partition is given by

NR rating
$$= 100 - \text{noise rating of required background noise}$$

The use of the scheme is illustrated in Table VI by the following example, only the final step of which is shown. The background-noise levels (in this example, the "N" spectrum shape was selected) have been chosen, by trial and error, so that AI≅0.05. In the alternative graphical computation, the "N dot field" transparency is moved up or down over the NR curve until 5% of the dots lie above the curve. The sound power level (in db re 10^{-13} w) and the importance function for normal, male speech are shown in Table V. (Speech levels for men are on the average higher than for women and thus constitute the more stringent problem.) The data in Table V have been derived by smoothing somewhat the data of Table I in reference 21. To avoid working with fractional numbers, the importance function is also given after being multiplied by 6000. This is divided out later.

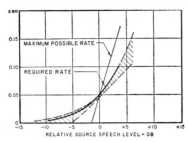

FIG. 18. Rate of growth of AI required if the summation formula described in Sec. VIII is to be valid.

It is clearly desirable from the designer's point of view to "prescribe" various sound-isolating structures for use in speech-privacy analysis. In order to do this, we must assume $S\bar{a}$ in the receiving room to be independent of frequency. Then the entire procedure just described in terms of NR could just as well be carried out in terms of TL of the individual elements (walls, ceiling, etc.) of the construction. A correction would then be applied later to account for different values of $S\bar{a}$ and for the areas of the element.

In the usual design problem, the net or resultant NR between two spaces is determined jointly by the several sound-isolating structures which separate them. The architect or consultant should have data on all of these paths—the wall, ceiling, doors and corridors, ductwork, etc. By the usual method of power summation, we can calculate the total NR, frequency band by frequency band, from the several sets of published data, and rate the whole situation using the above scheme. Since this procedure is tedious at best, we have considered the implications of combining the TL-rating numbers for the various paths (suitably corrected for their relative

areas) much as one would calculate, at any one frequency, the effective TL for a number of sound-transmission paths.

Namely, we have asked ourselves is the following summation formula valid?

Effective NR rating for a set of paths $\cong$

$$10\log_{10}\left[\sum\left(\text{antilog}_{10}\frac{\text{NR Rating}_1}{10}\right)^{-1} \right.$$
$$\left. +\left(\text{antilog}_{10}\frac{\text{NR Rating}_2}{10}\right)^{-1}+\cdots\right]^{-1}.$$

Clearly, if the several NR curves have exactly the same shape then this formula is precisely valid. Further, if the several curves "show dots," i.e., contribute to the AI, in substantially the same frequency bands, then the summation formula should be substantially valid.

If however, each of the several curves shows dots in different frequency bands, we must consider the rate of growth of AI vs speech level for each NR curve. In particular, if for each curve

$$\text{AI}\sim\text{antilog}_{10}(\text{SPL speech}/10)$$

then the summation formula is again exactly valid. In Fig. 18 we plot this growth rate along with the range in rates for 15 or so common walls. We note that not only is the relation $\text{AI}\sim\text{antilog}_{10}(\text{SPL speech}/10)$ a good fit to most of these data, but the error, in the case of NR curves with rapid growth rates, is a conservative one. That is, the true, effective NR rating in most cases would be at worst somewhat higher than the rating given by the approximation above.

FIG. 19. Work sheet for compiling case-history data.

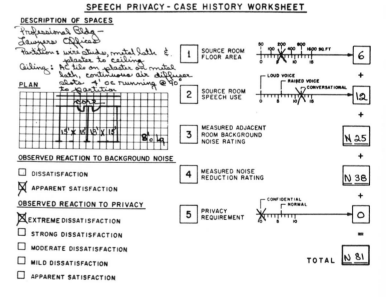

TABLE VII. Summary of case-history data.

Description of space and isolating construction[a]	Item 1 (source rm. size)	Item 2 (speech use)	Rating factors[b] Item 3 (bkgrd.)	Item 4 (NR)	Item 5 (priv. req.)	Total	Observed reaction to privacy
(1) Professional bldg. lawyers offices (wire stud, met. lath plaster, limited by continuous slots in clg. for air distribution from plenum)	6	12	N25	N38	0	N81	Extreme D
(2) Professional bldg. exec. offices (same constr. as 1)	6	12	N25	N42	0	N84	Extreme D
(3) Private offices to Admin. area, home office bldg. of large ins. co. (lightwt. mov. wood part. 20% glass)	6	6	N41	N30	0	N83	Extreme D
(4) Exec. offices (same constr. as 3)	6	6	N33	N32	0	N77	Extreme D
(5) Law-school dorm (4 in. painted CB part,—no air cond.)	5	12	N21	N45	6	N89	Strong D
(6) Exec. engineer's office (mov. met. part. to porous lightwt. susp. lum. clg.)	6	12	L38	L31	6	L93	Strong D
(7) Textile mfg. co. exec. office (4 in. plastered gyp. blk.)	6	12	N24	N44	0	N86	Strong D
(8) Chem. mfg. co. vice-pres. office to corridor (mov. met. part., 20% glass, door limited)	8	0	N37	N48	0	N93	Strong D
(9) Chem. mfg. co. personnel exec. office (mov. met. part., 20% glass)	5	12	N31	N44	0	N94	Strong D
(10) Personnel exec. office to sec. office (mov. met. part., 20% glass, door limited)	7	12	N31	N41	0	N91	Strong D
(11) Small hospital private rooms (4 ft plastered CB—no air cond.)	5	9	N15	N48	6	N83	Strong D
(12) Small town professional office bldg. (plaster, gyp. lath, wood stud part.)	6	6	N27	N58	0	N97	Strong D
(13) Engineering exec. office to sec. office (gyp. bd. on 2×4's limited by gasketed door and porous susp. clg.)	5	12	N31	N41	0	N89	Strong D
(14) Exec. engineer's office (mov. met. part. to plaster plenum closer)	6	12	L38	L38	6	L100	Mod D
(15) College dorm (6 in. plastered CB, —no air cond.)	5	12	N15	N55	6	N93	Mod D
(16) Personnel exec. office (mov. met. part., 20% glass	5	12	M42	M40	0	M99	Mod D
(17) Engineering offices (mov. met. part. to porous susp. clg.)	5	12	N28	N39	6	N90	Mod D
(18) Large high-quality speculative offices for top execs. (wood mov. part.)	9	6	N32	N45	0	N92	Mod D
(19) Private offices, large mfg. co. in rural setting (mov. met. part. to porous susp. clg.)	5	12	N31	N39	6	N93	Mod D
(20) Exec. office to corridor sec. area (gyp. bd. on studs, door limited)	7	12	L32	L41	0	L92	Mod D
(21) Research consulting exec. office (gyp. bd. on 2×4's, solid wood gasketed door in part.)	5	12	H35	H40	0	H92	Mod D
(22) Consulting engineers exec. office (gyp. bd. on 2×4's to susp. min. fiber title clg.)	5	12	N29	N41	6	N93	Mod D
(23) Consulting engineers office (same constr. as 22)	5	12	N29	N44	6	N96	Mod D
(24) Case 23 after modifications (part. carried to clg.+part. leaks sealed)	5	12	N31	N45	6	N99	Mild D
(25) Business exec. to conf. room (same as 33 except clg. has gyp. bd. above min. fiber tile)	5	12	H33	H40	6	H96	Mild D
(26) Engineering exec. offices (gyp. bd. on 2×4's, door limited)	7	12	L39	L43	0	L101	Mild D
(27) Quality motel (double stud gyp. lath+plast. constr. packed w/ fibrous matl.—no air cond.)	9	12	L21	L65	0	L107	App Sat
(28) Research engineer's office (1½ in. plast. bd. both sides 2×4's)	5	12	L34	L54	0	L105	App Sat

TABLE VII (*continued*)

Description of space and isolating construction[a]	Item 1 (source rm. size)	Item 2 (speech use)	Item 3 (bkgrd.)	Item 4 (NR)	Item 5 (Priv. req.)	Total	Observed reaction to privacy
				Rating factors[b]			
(29) Business office (mov. met. part., 20% glass)	5	12	N33	N44	6	N100	App Sat
(30) Business office (mov. met. part., 20% glass)	5	12	N30	N48	6	N101	App Sat
(31) Hospital-psychiatric interview offices (wire stud with resil. clips, gyp. lath plaster both sides)	3	6	H34	H58	0	H101	App Sat
(32) Consulting engineers office (2×4's, ½ in. plasterboard both sides)	5	12	L26	L52	6	L101	App Sat
(33) Private offices (mov. met. partitions)	6	12	H37	H44	6	H105	App Sat
(34) Private offices-large shipping firm (laminated gyp. bd. mov. part.)	5	12	L28	L49	6	L100	App Sat
(35) Exec. offices, consulting engineers (gyp. bd. on wood studs part., solid wood door-fully gasketed)	7	12	N30	N54	0	N103	App Sat
(36) Consulting eng. office (gyp. bd. on 2×4 studs to min. fiber tile clg.)	3	12	N29	N42	6	N92	App Sat
(37) Consulting engineers office (same constr. as 36)	3	12	L38	L41	6	L100	App Sat

[a] Except where otherwise noted, the type of space listed is isolated from a space of the same type. The parenthesis under each space includes a brief description of the partition construction. If the partition construction is limited by other flanking paths, these are also described. All spaces are air conditioned unless otherwise noted.
[b] See Fig. 19 and text for description of rating factors.

This, in effect, means that the data user can confidently work with the *single-number ratings* associated with a number of independent sound paths instead of having to use the entire set of "NR-vs-frequency" data for each path. *Since the ratings can be provided by testing laboratories along with the TL data for the structures, the user need never go through the trial-and-error procedure of establishing a privacy rating himself, even though he deals with complex composite structures.*

IX. A COMPARISON OF THE PROPOSED RATING SCHEME AND FIELD REACTIONS

Experience suggests that a "real-life" environment is much more complex than the controlled environment of a laboratory situation. For this reason, we would hesitate to offer an annoyance-analysis scheme without first having checked its performance against field experience. The results of our preliminary comparisons follow.

An example of the work sheet used to gather the relevant data is shown in Fig. 19. It is filled out for the privacy situation of item I in Table VII. Noting the five items on the right side of the form:

(1) The sound absorbing area of the talker's room has been taken into account by noting the room size and assuming that $S\bar{a}$ = floor area.

(2) The kind of speech activity was observed and categorized into conversational, raised, or loud voice.

(3) The actual background noise has been measured and rated, using the scheme described in Sec. VII. (The number is the approximate NC rating; the letter defines the tonal character and corresponds to one of the four characteristic curves.)

(4) The actual noise reduction was measured and rated using the scheme described in Sec. VIII.

(5) The need for privacy has been observed and categorized into "normal" or "confidential."

The numbers given for the five items may be totaled. Since an increase in the value for any of the five items represents an improvement in the expected privacy, we should be able to relate the total to the over-all degree of privacy.

Figure 20 plots the totals for the re-analyzed data for our 37 case histories against the distributions of the observed subjective reaction by the occupants of these buildings. The most significant fact about Fig. 20 is the fairly tight grouping of the data. Certainly the grouping

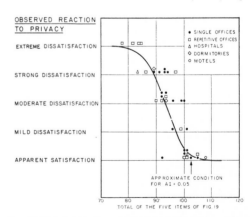

FIG. 20. Plot of subjective reactions observed in 37 case histories versus the total rating computed from the proposed rating scheme.

is much better than for the "average TL" analysis which led to Fig. 1.

It is also significant that situations with an articulation index of 0.05 are more or less on the borderline between apparent satisfaction and some degree of dissatisfaction. Thus, if economy is of utmost importance, the design criterion should be based on this value.

It should be observed that the slope of the curve in Fig. 20 is quite steep. This means that a small change in the noise reduction or background noise can make a significant variation in the occupant's reaction. When the rating schemes presented here are used for the design of a building, it is clearly desirable that the probable total uncertainty in estimating the five items of Fig. 20 be less than ±4 or 5 db. To achieve this over-all accuracy will demand an accuracy for each component of ±2 or so db.

It is interesting that while apparent satisfaction appears just to exist in situations where the articulation index is 0.05, *on the average*, an increase in the signal-to-noise ratio of only 10 or so db would be expected to result in fairly complete sentence intelligibility. Since occasional increases of this order are to be expected (because of the occasional raised voice or occasional loud talker), it appears that most occupants desire privacy *on the average* but may be willing to make allowances for the occasional loud talker and do not expect privacy when they themselves talk loudly.

Readers who wish to analyze our data further will find a detailed listing in Table VII.

X. CONCLUSIONS

While both the laboratory and field studies reported in this paper have led to some meaningful conclusions regarding speech privacy and, we believe, to a design scheme which can predict with reasonable accuracy the reactions of building occupants to intruding speech, there remain some qualifications of this analysis. More important, there is a continuing need for judgment in applying these results.

For example, the studies have been strictly limited to speech as the intruding signal. The laboratory studies were performed in the context of a typical office situation. The field study and case histories were for the most part office situations although the few hospital private rooms, dormitories, and motels included in the case histories have shown good agreement with the results for offices. More laboratory and field studies are needed however, to verify the applicability to other kinds of spaces. In particular, studies are needed of apartments where the intelligibility of intruding speech may not prove to be as significant as it seems to be with "private" offices and similar spaces.

In addition, our studies have been restricted to buildings in this country. We hope that studies in other countries would be in general agreement with our results, but it is not unreasonable to expect an average response curve such as that of Fig. 20 to shift laterally by a few units. Indeed, as the experience of other investigators in this country are added to the field cases we have studied, there may be reason to modify the average response curve we have indicated.

Our present studies can be summarized as follows:

(1) Laboratory and field studies generally show a strong relationship between the intelligibility of intruding speech sounds and peoples' judgments of speech privacy.

(2) The use of average TL values as the sole parameter in designing for speech privacy can lead to grossly inaccurate results.

(3) The proposed scheme for designing for speech privacy and/or evaluating speech-privacy problems has proved to be more accurate because:

(a) It rates the TL of sound-isolating structures according to their ability to reduce the intelligibility of intruding speech.

(b) It recognizes the several other factors (such as background noise and room size) which combine with the TL to determine the intelligibility.

ACKNOWLEDGMENTS

The authors wish to express their appreciation for the advice and assistance offered throughout the course of these studies from their colleagues at Bolt Beranek and Newman Inc., in particular to Dr. T. J. Schultz whose assistance during the preparation of this paper has been invaluable. In addition, a special word of thanks to Walter F. Young, Jr., James Porter, Myron Olson, Donald Huggins, Roger Johnson, and others from the Owens-Corning Fiberglas Corporation.

22

Reprinted from *J. Acoust. Soc. Am.* **38**(4):524–530 (1965)

Re-Vision of the Speech-Privacy Calculation

ROBERT W. YOUNG

U. S. Navy Electronics Laboratory, San Diego, California 92152

The articulation index used in telephone communication, the listening equation for sonar, and the acoustical-privacy calculation of architectural acoustics are all founded on a computation of an excess of signal level over noise level just sufficient to permit some stated detectability. From this viewpoint, data compiled for "Speech Privacy in Buildings" by Cavanaugh, Farrell, Hirtle, and Watters [J. Acoust. Soc. Am. **34**, 475–492 (1962)] have been reviewed for possible simplications in their procedure for estimating acoustical privacy. Interrelations are demonstrated among various current methods for rating noise and sound insulation, and a general equation for acoustical privacy is derived. As an example, for the case of "confidential" privacy, 200-ft² floor area, and raised voice, satisfactory acoustical privacy is to be expected if $D+N_A=85$ dB, where D is the sound isolation between the rooms (the reduction of sound from one room to the next) as rated by a procedure like that for the sound-transmission class, and N_A is the background-noise level in the receiving room measured with a sound-level meter on A-weighting.

INTRODUCTION

FOR practical engineering of acoustical privacy, ratings are needed, respectively, for the sound source, the sound-insulating structure, and the masking contributed by ambient noise. For each of these, several ratings are presently in use. Relationships among such ratings are here analyzed and arranged in a form for easy calculation such that each element can be checked readily by direct measurement.

In a general sense, the articulation index (AI), as it is known in telephone communication, the listening equation, as it is called in sonar design, and the speech privacy[1,2] calculation, as it has been dubbed in architectural acoustics, are all founded on a calculation of an excess of signal level over noise level just sufficient to permit some stated detectability of the "signal." Usually, the signal (such as a voice sound) originates at some distant point, and it experiences some kind of loss in transmission to the point at which it is heard.

I. LISTENING EQUATION

The listening equation is simply a statement[3] that, if L be the signal level in decibels at some given point

in the vicinity of a source and H the loss in sound-pressure level en route from that point, the signal at the listening location is $(L-H)$, the excess of signal level over ambient-noise level N at that location is $(L-H)-N$, and the signal will just be heard if

$$(L-H)-N=\Delta, \qquad (1)$$

where the recognition differential Δ is defined[4] as the number of decibels by which the signal level, as received at the point of detection, must exceed the noise level in order to be detected in some specified fraction of trials, such as 50%. The recognition differential is an omnibus psychoacoustical allowance that depends upon the nature of the signal and of the noise, as well as on the bandwidth of the transmission system. For machinery noise heard[5] over a wide-band audio system, Δ is something like -10 or -15 dB. The success of the supplication of the recognition differential depends largely on the appropriateness of the method of measurement for signal and noise in relation to the detection system.

The AI is similarly calculated. It is based on the idea that the total intelligibility of speech is the sum of

[1] B. G. Watters, in *Noise Reduction*, L. L. Beranek, Ed. (John Wiley & Sons, Inc., New York, 1960), pp. 533–537.

[2] W. J. Cavanaugh, W. R. Farrell, P. W. Hirtle, and B. G. Watters, "Speech Privacy in Buildings," J. Acoust. Soc. Am. **34**, 475–492 (1962).

[3] R. W. Young, "Method for Calculating the Number of Hydro-

phones Needed for Listening Posts," Univ. Calif. Div. War Res. (1942 unpublished). The term was, first, *listening ratio*, and was later called *listening differential* and then *recognition differential*.

[4] American Standard Acoustical Terminology S1.1-1960, Def. 12.27.

[5] R. S. Gales, J. Acoust. Soc. Am. **19**, 736(A) (1947). For details, see "Recognition of Underwater Sound," Natl. Defense Res. Com. Div. 6, Sum. Tech. Rept., **6**, p. 48 (1946), PB 145086.

weighted contributions from different frequency bands of the speech. More specifically, the AI is the ratio to 30 dB of a weighted peak-signal-level-over-noise-level excess given by

$$\sum w_i(L'_i - N_i)/30 = AI, \tag{2}$$

where L'_i is a certain peak-signal level at the listener, in the ith band, and N_i is the noise level in that band.

The weighting w_i reflects the relative importance of the speech components in different frequency bands; examples of w_i are to be found in Table I. The values listed on line 1 of Table I are the weights for octave bands given by Kryter,[6] but multiplied by 30; those on the second line are the importance functions as used[2] for the speech-privacy calculation, but summed by octaves and divided by 200. The weights in line 3 are those assumed heuristically for the present study, after having been found superior to the weights in line 2 as to constancy of rating for the equally speech-interfering noises of different spectra studied by Webster and Klumpp,[7] who commented on the generally downward shift in the frequency of central importance as threshold is approached.

To illustrate that the calculation of a speech interference level (SIL) is of the same nature as that of the AI, in lines 4 and 5 of Table I are given the weights that lead to the particular four-band and three-band SIL's found by Klumpp and Webster[8] as effective for predicting speech intelligibility where there is strong noise interference. It will be seen that all methods attribute about one-fourth the intelligibility to a band near 1000 cps, and differ principally in the abruptness of the drop to zero importance.

For the AI as it is usually calculated,[6] the peak voice sounds are customarily obtained by adding 12 dB to the maximum levels read on a sound-level meter; also, it is assumed that there is no contribution for bands in which the peak-signal-excess-over-noise level is less than zero. There is reason[9] to believe, however, that a peak only 6 dB above the root-mean-square voice level in a band contributes something; original work[10,11] on validation of the AI was based principally on bands of equal contributions to intelligibility for a wide-band speech signal well above the noise; here, however, octave bands wider than critical bands are involved, and the signal is near the noise; for small speech-over-noise excess, omission of limits was found[7] to result in an AI more nearly constant for equally speech-inter-

TABLE I. Various importance factors for AI and SIL calculations, for sound measured in octave bands, normalized so that their sum is unity.

Line		Center frequency (cps)					
		125	250	500	1000	2000	4000
1	Kryter[a]	0	0.07	0.15	0.22	0.32	0.23
2	CFHW[b]	0	0.06	0.15	0.24	0.32	0.15
3	RWY[c]	0.06	0.15	0.23	0.24	0.21	0.11
4	Four-band SIL[d]	0	0.25	0.25	0.25	0.25	0
5	Three-band SIL[e]	0	0	0.33	0.33	0.33	0

[a] Ref. 6.
[b] Ref. 2.
[c] Present study.
[d] Ref. 8.
[e] For the 3-band SIL within the limits 600–4800 cps, the center of emphasis is about ⅔ oct higher than here indicated.

fering noises of different spectra and, thus, better than the use of the traditional limit. Accordingly, the lower limit is not invoked in the present analysis. This omission makes it possible to separate the summations in Eq. 2.

For an estimate of the sound-pressure level in a room where a person is talking, it is convenient to resort to the relation of steady-state sound-pressure level to room absorption. Even though direct sound striking a wall may be of primary importance, there is likely to be a reduced sound-pressure level for increasing room size, which, for furnished rooms, is roughly proportional to the absorption present. Thus, the voice level in the source room, in the ith band, is estimated from

$$L_i = S_i + F_i, \tag{3}$$

where S_i is the *source level* (the root-mean-square sound-pressure level at 1 m from an equivalent point source), and

$$F_i = 10 \log 16\pi - 10 \log(A_i/10.8) \tag{4}$$

is a room function in which A_i ft² is the sound absorption in the source room in sabins (square foot), and 10.8 ft² = 1 m² is a reference area. In brief, the room function is the amount by which the level of the sound throughout most of the room exceeds the sound-pressure level that would exist at a distance of 3.3 ft (1 m) from the source in a free field.

Now, with H_i as the transmission loss for the ith band and $L'_i = L_i - H_i + 12$, Eq. 2 may be rewritten as

$$S + F - H - N = 30 \, AI - 12, \tag{5}$$

where

$$S = \sum w_i S_i, \tag{6}$$

$$F = \sum w_i F_i, \tag{7}$$

$$H = \sum w_i H_i = D, \tag{8}$$

$$N = \sum w_i N_i. \tag{9}$$

For adequate speech privacy, an AI not greater than 0.05 was taken[2] as an acceptable condition, so that

$$[(S + F) - H] - N = -10.5. \tag{10}$$

[6] K. D. Kryter, J. Acoust. Soc. Am. 34, 1689–1697 (1962). The weights have here been very slightly modified in accordance with private communication (1965).

[7] J. C. Webster and R. G. Klumpp, J. Acoust. Soc. Am. 35, 1339–1344 (1963).

[8] R. G. Klumpp and J. C. Webster, J. Acoust. Soc. Am. 35, 1328–1338 (1963).

[9] J. C. Steinberg, private communication (1964).

[10] N. R. French and J. C. Steinberg, J. Acoust. Soc. Am. 19, 94–119 (1947).

[11] L. L. Beranek, Proc. IRE 35, 880–890 (1947).

TABLE II. Elements of speech-privacy rating, in dB.

Case	Source CFHW	Source S_A	Room size CFHW	Room size F	Privacy CFHW	Privacy P	Isolation D_{CFHW}	Isolation D_{64}	Noise N_{CFHW}	Noise N_A	Total sum	Total X	Judgment
1	12	60	6	4	0	15	N38	24	N25	35	N81	20	ED[a]
2	12	60	6	4	0	15	N42	29	N25	35	N85	15	ED
3	6	66	6	4	0	15	N30	15	N41	49	N83	21	ED
4	6	66	6	4	0	15	N32	18	N33	42	N77	25	ED
5	12	60	5	5	6	9	N45	30	N21	30	N89	14	SD[b]
6	12	60	6	4	6	9	L31	19	L38	44	L93	10	SD
7	12	60	6	4	0	15	N44	34	N24	33	N86	12	SD
8	0	72	8	2	0	15	N48	33	N37	46	N93	10	SD
9	12	60	5	5	0	15	N44	29	N31	41	N94	10	SD
10	12	60	7	3	0	15	N41	26	N31	42	N91	10	SD
11	9	63	5	5	6	9	N48	34	N15	24	N83	19	SD
12	6	66	6	4	0	15	N58	41	N27	38	N97	6	SD
13	12	60	5	5	0	15	N41	27	N31	39	N89	14	SD
14	12	60	6	4	6	9	L38	25	L38	44	L100	4	MoD[c]
15													
16	12	60	5	5	0	15	M40	29	M42	46	M99	5	MoD
17	12	60	5	5	6	9	N39	25	N28	36	N90	13	MoD
18	6	66	9	1	0	15	N45	32	N32	42	N92	8	MoD
19	12	60	5	5	6	9	N39	25	N31	41	N93	8	MoD
20	12	60	7	3	0	15	L41	28	L32	41	L92	9	MoD
21	12	60	5	5	0	15	H40	29	H35	42	H92	9	MoD
22	12	60	5	5	6	9	N41	27	N29	38	N93	9	MoD
23	12	60	5	5	6	9	N44	30	N29	38	N96	6	MoD
24	12	60	5	5	6	9	N45	31	N31	41	N99	2	MiD[d]
25	12	60	5	5	6	9	H40	33	H33	39	H96	2	MiD
26	12	60	7	3	0	15	L43	30	L39	48	L101	0	MiD
27	12	60	9	1	0	15	L65	52	L21	27	L107	−3	Sat[e]
28													
29	12	60	5	5	6	9	N44	29	N33	44	N100	+1	Sat
30	12	60	5	5	6	9	N48	30	N30	40	N101	4	Sat
31	6	66	3	7	0	15	H58	45	H34	40	H101	3	Sat
32	12	60	5	5	6	9	L52	36	L26	35	L101	3	Sat
33	12	60	6	4	6	9	H44	34	H37	44	H105	−5	Sat
34	12	60	5	5	6	9	L49	36	L28	38	L100	0	Sat
35	12	60	7	3	0	15	N54	40	N30	40	N103	−2	Sat
36	12	60	3	7	6	9	N42	29	N29	38	N92	9	Sat
37	12	60	3	7	6	9	L41	29	L38	43	L100	4	Sat

[a] ED: extreme dissatisfaction.
[b] SD: strong dissatisfaction.
[c] MoD: moderate dissatisfaction.
[d] MiD: mild dissatisfaction.
[e] Sat: apparent satisfaction.

This equation is of the same form and significance as Eq. 1, for a recognition differential $\Delta = -10.5$.

II. CFHW DATA

Cavanaugh, Farrell, Hirtle, and Watters[2] (CFHW) performed a laboratory study on speech privacy and analyzed case histories for some 37 offices, hospitals, and motels, and developed procedures for rating and evaluating speech-privacy problems. In this second look at—and hence re-vision of—their speech-privacy calculation, the original data (plus unpublished details about noise reduction kindly made available for the present investigation) are used to test different combinations of single-number ratings of sound isolation and interfering noise in the theoretical framework just outlined.

An extract of essential information compiled by CFHW appears in Table II. The case numbers are the same as those in the original paper,[2] where explantion of the circumstances is given. Under "Source," CFHW assigned the numbers 12 for ordinary conversational speech and 6 for raised voice.

Their number under "Room size" is 10 times the logarithm of the ratio of the floor area to 50 ft², coming from the realistic assumption that the absorption in a furnished apartment of office is about equal to its floor area. CFHW assigned 6 to a location where "normal" privacy was desired, or zero for "confidential" privacy.

The reduction of sound from one room to the next was measured by CFHW with octave bands of noise, and rated by a technique based on the AI, here tabulated in Table II in the column headed D_{CFHW}. The background noise in the receiving room was rated by a quasiaveraging and graphical method intended to characterize the noise-spectrum shape and NC-level. The signs of the various ratings and constants were chosen so that the sum of the individual ratings would serve as a single measure of acoustical privacy, this *sum* being that in the general column headed "Total." The final column in Table II lists the judgments of acoustical privacy as recorded by CFHW in the five categories of extreme dissatisfaction (ED), strong dissatisfaction (SD), moderate dissatisfaction (MoD), mild dissatisfaction (MiD), and apparent satisfaction (Sat).

Figure 1(a) shows how the subjective judgments related to the sum of the physical measures. In spite of

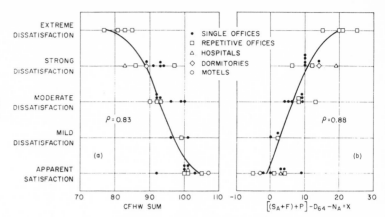

FIG. 1. Judgments of satisfaction with speech privacy vs the sum of elements used by CFHW and vs the sound excess X.

scatter, the correlation is evident; there is a rank-order correlation coefficient of 0.83.

III. PARALLEL PROCEDURE

In effect, the CFHW procedure is an application of Eq. 5, the weights w_i being those in line 2 of Table I. Various convenient approximations are available for the several terms in Eq. 5, and results of using these approximations are now explored.

Sound isolation between offices was rated by the procedure indicated in Fig. 2. As explained under the Figure, this curve is to be shifted until it fits experimental data until the average deficiency of the measured values, below those corresponding to the rating curve, is not more than 2 dB, and only positive deficiencies being counted. The basic idea has been in use for a number of years (as an example, see German Standard DIN 52211), but this specific routine is a modification of a proposal for international standardization in International Organization for Standardization document 43 (Secretariat 222)346 of August 1964, so the rating is here called an ISO index and identified by the subscript "64." It is convenient to apply the procedure both to the sound insulation of a wall (which is called *sound-transmission loss* in American terminology and *sound-reduction index* in British terminology) and to the sound isolation between rooms (American: *noise reduction;* British: *transmission loss;* and "level difference" D in an international standard). The results of applying this procedure to the CFHW data are given in the column headed D_{64} in Table II; since the original data were obtained with different octave bands than those designated in Table I, the appropriate values were obtained by linear interpolation.

Other experience[8,12] with room noise indicates that sound level (A) correlates well with the disturbance produced by the background noise, so this sound level was calculated with the standard[13] A-weighting applied to

the octave-band data for room noise, graciously furnished by CFHW, with the results tabulated under N_A in Table II.

In accordance with Eq. 5, and for consistency with the use of sound level (A) for the background noise, plausible values of source level similarly measured with the A-weighting were inserted under S_A in Table II, 60 dB being assigned for conversational voice [this being a likely sound level (A) at a distance of 1 m], 66 dB for raised voice, and 72 dB for loud voice. These numbers increase in the opposite direction to those in the column headed CFHW, but have the same increments. Similarly, the numbers under F come from Eq. 4 but with the increments of CFHW. Instead of a zero to signify "confidential" privacy, under P, the number 15 has been placed, and 9 for "normal" privacy, so that the numbers increase as the desired privacy increases. Thus, as to rank ordering, the numbers under S, F, and P have the same significance as the corresponding CFHW data from which they were derived; under D_{64} and N_A, there is correlation but not exact correspondence with the respective CFHW values.

Then, from the logic previously outlined for the privacy calculation, the excess of expected speech "signal" over noise level was obtained for each case from

$$[(S_A+F)+P-D_{64}]-N_A=X. \tag{11}$$

The results are listed in Table II, and in Fig. 1(b) the original judgments are plotted against X.

It is evident that dissatisfaction increased as X increased. The rank-order correlation coefficient is 0.88. If one takes $X=0$ as corresponding to an acceptable situation, it follows that the requirement for speech privacy is

$$[(S_A+F)-D_{64}]-N_A=-P. \tag{12}$$

Since (S_A+F) is the sound level (A) throughout the source room, and D_{64} is the typical transmission loss from one room to the next, it is evident on comparison

[12] R. W. Young, J. Acoust. Soc. Am. **36**, 389–295 (1964).
[13] American Standard Specification for General-Purpose Sound Level Meters, S1.4-1961.

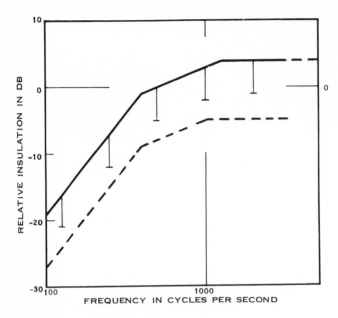

RELATIVE INSULATION IN DB

FREQUENCY IN CYCLES PER SECOND

Fig. 2. Airborne-sound-insulation rating curve. Place on graph of measured sound insulation of a partition or sound isolation between rooms, and shift vertically until the mean deficiency in the measured values from the rating curve does not exceed 2 dB and so that the maximum deficiency is not more than 5 dB for tests by octave bands, or 8 dB for $\frac{1}{3}$-oct bands; read the insulation rating on the measured graph to match the 0-dB line on this overlay. Thus, for the 5 octave bands within the rating interval, the sum of the deficiencies is to be 10 dB, or 32 dB for the 16 $\frac{1}{3}$-oct bands.

with Eq. 1 that the speech-privacy allowance P is simply the negative of the recognition differential.

Equation 12 thus represents a general condition for acoustical privacy, in which measured values may be inserted as available. Values of P may be assigned for different kinds of privacy against different kinds of sound sources; constant differences in the ratings for the other three quantities can be absorbed in P.

For a common privacy situation, one may plausibly assume a raised voice ($S_A = 66$ dB), $F = 4$ for a 200-ft^2 office, and $P = 15$ for "confidential" privacy. Then, the sound isolation and ambient noise are related by

$$D_{64} + N_A = 85. \qquad (13)$$

If other constants are appropriate, as when lesser privacy is acceptable, or "mild dissatisfaction" in the terms of Fig. 1, then

$$D_{64} + N_A = 80. \qquad (14)$$

IV. RELATIONS AMONG RATINGS

Another rating of sound insulation is the sound-transmission class (STC). It too was established after consideration[14] of the AI principle; it differs from the procedure in Fig. 2 principally in the method of averaging deficiencies and the maximum deficiency allowed. For the present octave-band data, the correlation with D_{64} is good: STC$= D_{64} - 0.6$, with a standard deviation of 0.9 dB.

The procedure of Fig. 2 differs principally from that used by CFHW in that, for Fig. 2, the deficiency is un-

[14] T. D. Northwood, J. Acoust. Soc. Am. **34**, 493–501 (1962). See also ASTM Designation: E90-61T.

weighted, whereas CFHW employed a weighted mean deficiency obtained by a graphical means or by the computation described in the original paper[2]; there is, nevertheless, a correlation represented by $D_{CFHW} = D_{64} + 13.5$ with a standard deviation of 2.2 dB. Likewise, the weighed mean defined by Eq. 8 is of the same nature, but it allows weighted credit for all sound isolation; for these data, with D_1 representing values obtained with the weights on line 1 of Table I, it turns out that $D_1 = D_{64} - 1.0$, with a standard deviation of 1.7 dB; also, $D_3 = D_{64} - 2.4$, with a standard deviation of 0.9 dB.

According to a proposed revision of the STC rating system, $\frac{1}{3}$-oct-band data from 125 to 4000 cps would be considered in relation to a curve like that in Fig. 2. With such a rating (STC'), but applied to the octave-band data here available, STC'$= D_{64} - 0.5$ with a standard deviation of 1.1 dB.

For identification of background-noise ratings, let N_1 be the weighted sum defined by Eq. 9, with the weights given in line 1 of Table I, and N_3 the similar sum with the weights in line 3. Then, $N_1 = N_A - 10.3$, with a standard deviation of 1.5 dB; $N_{CFHW} = N_A - 8.5$, with a standard deviation of 1.8 dB; $N_3 = N_A - 5.4$ dB, with a standard deviation of 0.8 dB.

Combination	ρ
CFHW sum	0.83
$D_{64} + N_A$	0.88
STC$+ N_A$	0.84
STC'$+ N_A$	0.86
$D_1 + N_A$	0.85
$D_1 + N_1$	0.84
$D_3 + N_A$	0.87
$D_3 + N_3$	0.88

TABLE III. Rank-order correlation coefficients for judged speech privacy versus different ratings of sound isolation and background noise.

In view of the correlation among the sound isolation and noise ratings, respectively, it is to be expected that their sums will be correlated when the appropriate numbers are inserted in Eq. 11. The correlation of the judgments of acoustical privacy against the sound excess X, for different combinations of ratings, is indicated by the rank-order correlation coefficients listed in Table III. It is evident that essentially the same correlation results from any one of the combinations, although there is a suggestion of superiority for ratings that emphasize sound in the region below 1000 cps, as do the weights in line 3 of Table I.

It should be remembered that the numbers for speech activity and need for privacy, listed in Table II, came from judgments of the original investigators. Since the same judgments were used for all the calculations for Table III, the correlation coefficients given there serve to show the relative merits of various combinations of ratings of sound insulation and of noise. There has not been, however, a complete check on Eq. 11. Also, the available data do not include many examples of good sound isolation between rooms.

V. SOUND-INSULATION RATINGS

A simple average insulation was once employed generally as a rating, but it is now looked upon with disfavor because it provides no warning of the existence of a deep dip in the insulation. The limit feature of Fig. 2 could be applied in the average insulation method by implicit, rather than explicit, use of the rating curve. From an analysis of mean differences, much like Eq. 5, it follows that this could be accomplished, for octave-band measurements to which the 5-dB limit is appropriate, by taking as the insulation rating the smallest one of the following: the average insulation plus 4 dB or the insulation at 125, 250, 500, 1000, and 2000 cps plus 21, 12, 5, 2, and 1 dB, respectively. Such a rule leads to nearly the same result as Fig. 2, but without the inconvenience of a trial calcula-

tion to establish the point below which deficiencies are to be counted.

The sound-isolation rating obtained by the CFHW method was related to the shape of the background-noise spectrum, rated according to whether the spectrum was of the "NC shape" and thus (N)ormal, or matched it at the (L)ow, (M)iddle, or (H)igh frequencies. The rating was calculated without any credit for sound isolation in excess of a certain amount; these features were in accordance with the traditional AI calculation. The fact that speech-privacy calculations can be made equally well without this limitation, at least as here tested against these particular case histories, encourages one to consider a weighted-mean sound insulation per Eq. 8 as the single-number rating of the sound insulation of a wall or the sound isolation between rooms. If necessary, one could apply the limit rule, as just described, for the average insulation calculation.

VI. SUMMARY

This re-vision of the work of Cavanaugh, Farrell, Hirtle, and Watters on speech privacy in buildings, in relation to the underlying theory, has indicated that, at least for the case histories included in this study, it is feasible to calculate separate ratings for sound insulation and noise without specific consideration of spectrum shape. Relationships among the various available ratings of sound insulation and noise have been developed so that they may be used in the basic equation for speech privacy. The possibility of rating sound insulation as a weighted mean has been investigated. Some evidence has been found favoring the use of importance factors for the acoustical-privacy calculation shifted downward in frequency about an octave from those often employed for the calculation of the articulation index. It has been shown that, with validity equal to that attained by the original method, speech privacy can be predicted by use of an often-available rating of sound insulation, such as sound-transmission class, and the background-noise level (A) measured directly with the sound-level meter.

Appendix A

The practical prediction of acoustical privacy, described above, consists in combining estimates, or measured values, of the following quantities:

S is the source level, the typical sound level (A), in decibels, at a distance of 3.3 ft (1 m) from the sound source; use 60 for conversational voice, 66 for raised voice, 72 for loud voice.

F is the room function, the number to be added to the source level (A) to obtain the typical sound level throughout a room; estimate from Fig. A-1(a).

P is the privacy allowance; use 15 for "confidential," 9 for "normal," or other value selected in consideration of the nature of the sound involved.

I is the rating of sound insulation of a partition, such as the sound-transmission class (STC) or ISO index.

K is the quantity to be added to the sound insulation I of the common partition (the wall or floor–ceiling separating two rooms) to obtain effective sound isolation between rooms; from ratio of area of common partition to that of floor of receiving room, estimate K from Fig. A-1(b).

$D = I + K$ is the sound isolation between rooms.

N is the background-noise level (A) in receiving room.

273

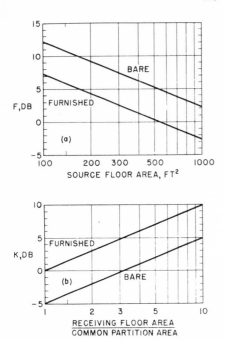

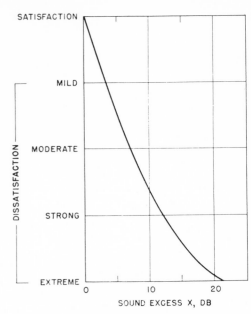

Fig. A-2. Satisfaction with acoustical privacy vs sound excess X.

Fig. A-1. Estimates of F and K from room dimensions, for stud-and-plaster construction and room heights like 10 ft.

Compute sound excess

$$X = (S+F)+P-D-N. \qquad (A1)$$

Read expected degree satisfaction with acoustical privacy from Fig. A-2.

Values for K and for the room function F, as read from Fig. A-1(a) and (b), are based on the approximations that the absorption is equal to the floor area of a furnished room (stud-and-plaster construction), and that the reverberation time of a bare room is about three times that of a furnished room. Both of these stem from field experience. The first is consistent with the observation that the reverberation time T, in many furnished rooms, tends[A-1] to be near 0.5 sec; if typical room height were 10 ft, it follows that, in a furnished room of floor surface S (ft²), the Sabine absorption A is approximately equal to the floor area, because

$$A = 0.049V/T = 0.049 \times 10 \times S/0.5 \approx S. \qquad (A2)$$

The "rule of thumb" may be remembered by noting that the room function is zero for a well-furnished room of floor area 500 ft² (50 m²), and 5 dB greater for a bare room, and that a doubling of the floor area reduces the room function by 3 dB.

A-1 O. Brandt, in *Proceedings of the Fourth International Congress on Acoustics, Copenhagen, 1962* (Organization Committee of the 4th ICA and Harlang & Toksvig, Copenhagen, 1963), Vol. 2, pp. 31–54, esp. p. 45.

ERRATUM

Page 524, column 2, line 16 should read: ". . . The success of the application . . ."

23

Reprinted from *Centre Sci. Tech. Bâtiment Cahiers* **92**(800):1–9 (June 1968)

A PROPOS DES BRUITS D'IMPACT

A. COBLENTZ, Association d'Anthropologie Appliquée
R. JOSSE, division acoustique du C.S.T.B.

Dès que les acousticiens ont dû se préoccuper des problèmes d'isolation acoustique dans les bâtiments d'habitation collective, ils ont été amenés à inventer des mesures permettant de juger la qualité de l'isolation.

C'est ainsi qu'en 1932 Reiher [1] (1) construisait une première machine devant permettre de tester la sonorité des planchers sous les impacts résultant de la marche des personnes. Elle se composait d'un marteau en bois de masse 280 g tombant de 3 cm de haut à la cadence de la marche normale. Masse et hauteur de chute avaient été choisies de manière que le bruit ressemble à celui provoqué par la marche d'un adulte dont les chaussures seraient munies de talons en cuir.

Cette machine, qui produisait un bruit trop faible et trop discontinu pour être facilement mesuré, fut rapidement abandonnée au profit de celle que nous connaissons bien maintenant (norme AFNOR S 31-002 : marteaux en laiton de 500 g chacun, terminés par un élément sphérique de rayon 50 cm, tombant de 4 cm de haut à raison de dix coups par seconde) et dont la première description figurait, dès 1938, dans la norme allemande DIN 4110 (2).

La nouvelle machine ne présente pas les inconvénients de la précédente, mais elle a celui de produire des chocs qui, apparemment, ont peu de rapport avec ceux résultant de la marche normale. Malgré ce défaut de conception, elle sert de base à la plupart des normes nationales concernant la mesure de la sonorité des planchers sous les impacts.

Une étude est en cours au C.S.T.B. pour rechercher une nouvelle machine qui, sans être aussi simplifiée que celle de Reiher, produirait des impacts ressemblant plus à ceux de la marche humaine, ceci naturellement au détriment de l'intensité.

Dans le cadre général d'une étude sur le bruit dans l'habitat collectif, financée par la D.G.R.S.T. (3) et exécutée par le C.S.T.B. et l'Association d'Anthropologie Appliquée, une recherche a été exécutée pour voir s'il existe un lien entre la gêne ressentie par les habitants, résultant de la marche des occupants de l'étage supérieur, et les résultats des tests physiques qui sont faits à l'aide de la machine normalisée, malgré l'imperfection de celle-ci.

Cette recherche a eu pour support deux enquêtes exécutées par l'Association d'Anthropologie Appliquée :

— une première enquête (1963), menée auprès de 266 familles dans six types d'immeubles différents, qui visait le confort acoustique en général dans les habitations,

— une enquête complémentaire (1965), portant sur 142 familles dans trois types d'immeubles différents, destinée à l'étude spécifique de la gêne causée par les bruits d'impact.

L'enquête complémentaire a été décidée parce qu'il était apparu que les résultats de la première enquête ne permettaient pas d'établir la corrélation existant entre la qualité acoustique de la protection contre les impacts

1. Les chiffres entre crochets renvoient à la bibliographie.
2. DIN = Deutsche Industrie Normen.
3. Délégation Générale à la Recherche Scientifique et Technique.

telle qu'elle est jugée à partir des résultats de mesures et le degré de satisfaction correspondant exprimé par les occupants.

1. LA MÉTHODE

1,1. Les mesures acoustiques.

La qualité de la protection contre les bruits d'impact sur les sols des logements d'un immeuble a été jugée d'après des mesures acoustiques objectives effectuées conformément à la méthode décrite en détail dans la Notice technique du C.S.T.B. du 1er décembre 1958 et ses mises à jour successives (4).

Cette méthode consiste à frapper le sol d'un logement à l'aide de la machine à chocs normalisés et à mesurer le niveau du bruit ainsi produit dans le local sous-jacent appartenant en général à un autre logement.

Pour un type de sol donné, la mesure est répétée dans plusieurs logements afin d'obtenir une valeur moyenne.

La moyenne des résultats de mesures est comparée aux valeurs indiquées dans la notice technique comme ne devant pas être dépassées, c'est-à-dire, par 1/3 d'octave, pour les chambres et séjours :

— 66 dB aux fréquences graves,

— 62 dB aux fréquences moyennes,

— 51 dB aux fréquences aiguës.

L'écart existant, pour une bande de fréquences, entre les résultats de mesures et les limites précisées dans la notice technique est dénommé indice de qualité (I.Q. en abrégé) de la protection contre les impacts, pour la bande de fréquence considérée. Le plus petit des trois indices (fréquences graves, moyennes et aiguës) est l'indice de qualité global retenu pour l'ensemble des fréquences. On le désigne par l'abréviation I.Q.I. (Indice de qualité pour les impacts). Si, par exemple, les niveaux mesurés sont de 64, 52, 37 dB respectivement aux fréquences graves, moyennes et aiguës, les IQ correspondants sont 2, 10, 14 dB et l'indice de qualité global est I.Q.I. = 2 dB. Un indice de qualité global nul ou positif indique que la protection est conforme à ce que recommande la notice technique. Un indice négatif indique une non conformité.

Les valeurs limites réglementaires ainsi que la manière d'estimer l'I.Q.I. sont inspirées de la méthode allemande (DIN 4109), utilisée depuis plusieurs années par le C.S.T.B.

L'Organisation Internationale de Normalisation a en projet (projet nº 880) une recommandation décrivant une méthode d'appréciation de la qualité de la protection acoustique contre les impacts. Les mesures physiques correspondant à cette méthode sont identiques à celles décrites dans la notice technique du C.S.T.B. Seule la manière d'apprécier la qualité d'après les résultats des

4. Titre VI et Annexe nº 4, *Cahier du C.S.T.B.*, nº 554 de la livraison nº 66 (février 1964).

[*Editor's Note:* An English summary of this article prepared by T. D. Northwood follows.]

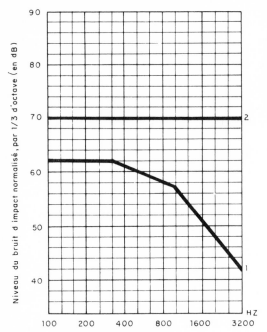

Fig. 1. — **Courbes limites concernant les bruits d'impact.**
1. Projet de norme ISO (n° 880).
2. Méthode canadienne.

mesures diffère : la protection est jugée satisfaisante si le spectre du bruit de chocs ne dépasse pas la courbe limite indiquée sur la figure 1 :

— de plus de 2 dB, en moyenne (pour les seize fréquences de mesure),
— de plus de 8 dB, à chacune des fréquences de mesure.

1,2. L'appréciation subjective des occupants.

Les questionnaires destinés aux deux enquêtes ont été rédigés de manière à apprécier *la gêne de jour* par les réponses à la question suivante :

Question J	Réponse
Entendez-vous dans la journée et dans la soirée, (venant de l'appartement du dessus) les personnes qui marchent dans leur salle de séjour et dans leur chambre ?	Non perçu simplement entendu, entendu et gênant.

La *gêne de nuit* a été appréciée par les réponses aux questions suivantes :

Questions N	Réponse
La nuit, est-ce que les gens qui marchent au-dessus de vous : — vous empêchent de vous endormir ? — vous réveillent ?	 oui ou non oui ou non

Le degré de gêne de jour peut donc, pour un immeuble, être représenté par le pourcentage de gens entendant le bruit et le trouvant gênant tandis que le degré de gêne de nuit peut être représenté par la moyenne des pourcentages correspondant aux réponses oui aux deux questions (s'endormir et réveil).

1,3. La comparaison des résultats de mesures aux appréciations subjectives.

Cette comparaison revient à rechercher s'il existe une corrélation entre les I.Q. mesurés et les degrés de gêne enregistrés.

2. L'ÉCHANTILLON

2,1. Les bâtiments.

Les enquêtes et les mesures acoustiques correspondantes ont porté sur neuf sortes d'immeubles de qualité différente situés à :

— Athis-Mons, Blanc-Mesnil, Choisy, Clichy-sous-Bois, Créteil, Thionville pour la première enquête,
— Creil, Strasbourg-Meinau, Strasbourg-Esplanade, pour l'enquête complémentaire.

Tous sont des immeubles construits après la dernière guerre et pour la plupart H.L.M.

2,11. La composition des planchers.

La composition des planchers des pièces principales (chambres et séjours) de ces immeubles, seules pièces auxquelles nous nous soyons intéressés, est la suivante :

Athis-Mons.

— Plastique collé (chambres), parquet mosaïque collé (séjours);
— Dalle pleine en béton armé de 14 cm d'épaisseur.

Blanc-Mesnil.

— Enduction plastique sur feutre;
— Dalle pleine en béton armé de 14 cm d'épaisseur.

Choisy.

— Parquet bois 8 mm collé;
— Dalle pleine en béton armé de 14 cm d'épaisseur.

Clichy-sous-Bois.

— Enduction plastique sur feutre;
— Dalle pleine en béton armé de 13 cm d'épaisseur.

Créteil.

— Enduction plastique sur feutre;
— Dalle pleine en béton armé de 13 cm d'épaisseur.

Thionville.

— Parquet bois agrafé sur panneaux isolants en fibres de bois collés (séjours) ou enduction plastique sur feutre (chambres);
— Dalle pleine en béton armé de 14 cm d'épaisseur.

Creil.

— Parquet chêne d'épaisseur 2,3 cm cloué sur solives en bois de 11 cm d'épaisseur espacés tous les 32 cm;
— Plafond en Isorel d'épaisseur 1 cm cloué sous les solives.

Strasbourg-Meinau.

— Linoléum collé;
— Plancher constitué par des poutrelles préfabriquées, en béton, supportant des hourdis sur lesquels une dalle (4 cm) de béton de ciment armé est coulée;
— Plafond plâtre.

Strasbourg-Esplanade.

— Parquet mosaïque en chêne de 8 mm d'épaisseur, collé;
— Dalle flottante en béton de 3 cm d'épaisseur coulée sur un matelas constitué de grains de liège collés sur du papier kraft bitumé (épaisseur 6 mm);
— Dalle pleine en béton armé de 14 cm d'épaisseur.

La composition de l'échantillon est donc la suivante :

Planchers.

Dalles pleines en béton :	7
Poutrelles et hourdis creux :	1
Solives bois avec parquet chêne et plafond Isorel :	1

Revêtements de sol.

Plastique collé :	1
Parquet bois collé :	2
Linoléum collé :	1
Enduction plastique sur feutre :	4
Parquet agrafé sur panneaux isolants collés :	1
Dalle flottante avec parquet collé :	1

On constate une nette prédominance des planchers dalles pleines et des revêtements enduction plastique feutre. C'est le reflet de la construction actuelle.

2,12. Résultats des mesures acoustiques.

Dans chacun des bâtiments objet de l'enquête, plusieurs planchers de pièces principales ont été testés à l'aide de la machine à chocs normalisés ainsi qu'il a été dit plus haut.

Sur la figure 2 (page 4), nous avons porté la moyenne des résultats par 1/3 d'octave pour chaque bâtiment.

Les valeurs moyennes correspondantes du niveau des bruits d'impact normalisés sont (arrondies au décibel) :

	Graves	Moyennes	Aiguës
Athis-Mons	66	70	67
Blanc-Mesnil	64	65	51
Choisy	66	70	67
Clichy-sous-Bois	65	65	48
Creil	82	72	52
Créteil	67	65	47
Strasbourg-Meinau	66	72	75
Strasbourg-Esplanade	64	52	37
Thionville	65	57	37
Rappel des valeurs limites	66	62	51

Pour Athis-Mons et Thionville, où les revêtements de sol des séjours et chambres sont différents mais d'efficacités comparables, c'est la moyenne des niveaux sonores correspondants qui a été indiquée.

Pour chacun des bâtiments et chaque type de planchers, trois mesures, au moins, ont été exécutées.

A Strasbourg-Meinau, où soixante-deux planchers furent testés, on a constaté que les résultats se répartissent d'une manière gaussienne avec un écart quadratique

moyen de 2 dB aux fréquences graves et aux fréquences moyennes et de 2,6 dB aux fréquences aiguës.

2,2. Les enquêtes.

L'ensemble des deux enquêtes a porté sur 408 logements répartis comme suit :

Première enquête.

Athis-Mons :	20
Blanc-Mesnil :	38
Choisy :	18
Clichy-sous-Bois :	38
Créteil :	76
Thionville :	76

Enquête complémentaire.

Creil :	28
Strasbourg-Meinau :	73
Strasbourg-Esplanade :	41

Alors que la première enquête ne tenait pas compte du nombre de personnes occupant le logement où sont produits les impacts, l'enquête complémentaire ne porte que sur des logements au-dessus desquels vivent au moins deux enfants. Une enquête préalable à l'enquête complémentaire a mis en évidence, en effet, que les bruits d'impact ne sont réellement gênants que si plusieurs enfants vivent dans le logement du dessus.

On trouvera une présentation détaillée de l'échantillon dans les rapports 01/65 et 10/65 de l'Association d'Anthropologie Appliquée ([5]).

3. LES RÉSULTATS

3,1. Résultats de l'enquête.

Les réponses à la question J (Entendez-vous, dans la journée et dans la soirée, venant de l'appartement du dessus, le bruit que font en marchant les personnes qui se déplacent dans leur salle de séjour et dans leur chambre ?), sont en pourcentage les suivantes :

Lieu	Degré de gêne		
	X	E	G
Athis-Mons	46	31	23
Blanc-Mesnil	48	42	10
Choisy	53	35	12
Clichy-sous-Bois	44	44	12
Creil	7	32	61
Créteil	82	15	3
Strasbourg-Meineau	30	40	30
Strasbourg-Esplanade	91	7	2
Thionville	47	41	12

X signifie non entendu.
E signifie entendu sans être gênant.
G signifie gênant.

5. Ces rapports peuvent être consultés à la Bibliothèque du C.S.T.B.

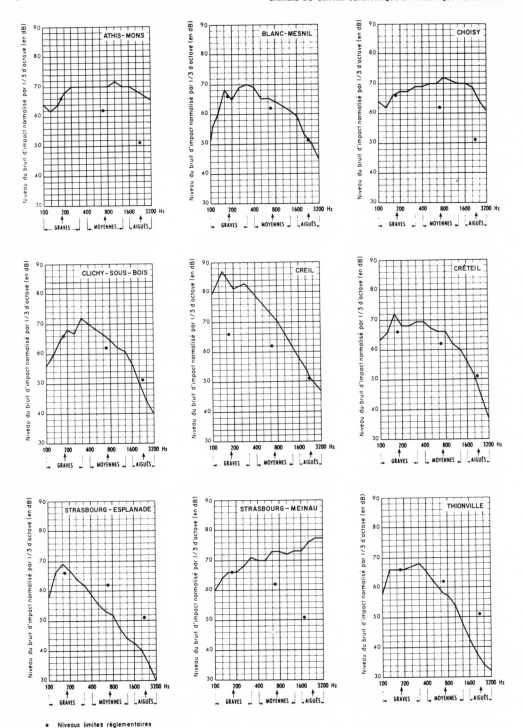

● Niveaux limites réglementaires

Fig. 2. — **Niveau du bruit d'impact normalisé pour chacun des types de bâtiments étudiés (valeur moyenne).**

Les réponses aux questions N (la nuit, est-ce que les gens qui marchent au-dessus de vous vous empêchent de vous endormir, vous réveillent ?) sont les suivantes :

	Réponses oui (pourcentage)	
	Empêchent de s'endormir	Réveillent
Athis-Mons	15	5
Blanc-Mesnil	8	8
Choisy	17	11
Clichy-sous-Bois	8	11
Creil	43	43
Créteil	0	0
Strasbourg-Meineau	18	21
Strasbourg-Esplanade	5	0
Thionville	12	9

3,2. Corrélation entre les résultats des mesures et les résultats des enquêtes.

3,21. Gêne de jour.

3,211. *Évaluation de la qualité de l'isolation suivant la méthode décrite dans la notice technique du C.S.T.B.*

Notre premier objectif étant de savoir s'il existe une corrélation entre la gêne exprimée par le pourcentage des habitants qui s'estiment gênés par les bruits d'impact et l'indice de qualité global I.Q.I. résultant des mesures, nous avons calculé à partir des résultats donnés au paragraphe 2,12 les indices de qualité relatifs à chaque bâtiment :

	Indices de qualité par bandes de fréquences			Indice de qualité global (I.Q.I.)
	Graves	Moyennes	Aiguës	
Athis-Mons	0	— 8	— 16	— 16
Blanc-Mesnil	2	— 3	0	— 3
Choisy	0	— 8	— 16	— 16
Clichy	1	— 3	3	— 3
Creil	— 16	— 9	— 1	— 16
Créteil	— 1	— 3	4	— 3
Strasbourg-Meineau	0	— 10	— 25	— 25
Strasbourg-Espanade	2	10	14	2
Thionville	1	5	14	1

Sur la figure 3, nous avons porté les résultats correspondants. On constate que, si l'on faisait abstraction des résultats relatifs à Creil, il existerait une certaine corrélation entre l'indice I.Q.I. et le pourcentage des gens gênés, ce dernier croissant modérément lorsque l'indice de qualité diminue.

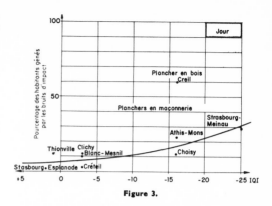

Figure 3.

Tous les planchers correspondants sont en maçonnerie et sont, sauf un (Strasbourg-Meinau), constitués par une dalle pleine en béton de 13 ou 14 cm d'épaisseur recouverte par différents revêtements de sol.

Malgré la légère croissance de la gêne lorsque l'indice de qualité diminue (planchers en maçonnerie exclusivement) ce dernier est mal approprié à l'expression de la qualité de l'isolation. On constate ce fait en observant que le taux de croissance de la courbe moyenne représentant la relation entre gêne et indice de qualité global est extrêmement faible en regard des écarts existant entre les pourcentages de gens gênés vivant dans des habitations équivalentes (au point de vue des bruits d'impact uniquement, c'est-à-dire ayant le même I.Q.I.). On observe, par exemple, qu'à Thionville et Choisy, la gêne est la même, alors que les indices de qualité diffèrent de 17 dB. La sensibilité de l'indice de qualité paraît donc exagérée lorsqu'il s'agit de comparer des planchers en maçonnerie entre eux. Cette exagération de la sensibilité semble provenir de ce que l'on donne trop d'importance aux fréquences aiguës dans le test tel qu'on le pratique à l'aide de la machine à chocs. En effet, les trois planchers en maçonnerie relatifs aux points situés dans la partie droite de la figure 3 (Athis-Mons, Choisy, Strasbourg-Meinau) sont cotés — 15, — 13, — 27, parce qu'ils sont très sonores aux fréquences aiguës eu égard à la limite maximale fixée (notice technique du C.S.T.B.) à ces fréquences.

Il apparaît donc que ce fort accroissement de sonorité aux fréquences aiguës, en comparaison de la sonorité des planchers relatifs aux autres immeubles, ne constitue qu'un élément de gêne complémentaire relativement limité.

Si l'on ne fait pas abstraction des résultats de Creil, lesquels correspondent à un plancher extrêmement léger entièrement en bois, on constate que l'indice de qualité global ne convient pas du tout pour évaluer la gêne : à un même indice de qualité (— 16 dB pour Choisy et Creil correspondent des pourcentages de personnes gênées différant de 50 %.

Cette différence importante doit résulter, en partie, de la trop grande importance que l'on attache à une sonorité excessive aux fréquences aiguës comparativement à ce que l'on fait pour les fréquences basses ou moyennes. Autrement dit, 10 ou 15 décibels de trop aux fréquences basses ne conduisent pas aux mêmes conséquences, pour les habitants, que 10 ou 15 décibels de trop aux fréquences aiguës.

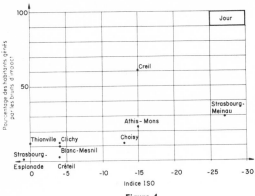

Figure 4.

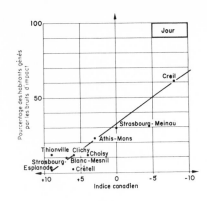

Figure 5.

Voyons si nous pouvons trouver une justification à l'hypothèse précédente en recherchant quelle différence de sonorité peut exister entre les planchers de Creil et de Choisy sous l'effet de la *marche réelle*.

Par des expériences en cours au C.S.T.B. ainsi que par d'autres effectuées dans différents laboratoires, on sait que le bruit provoqué par la marche normale d'un homme équipé de chaussures à talons en plastique ou en caoutchouc, sur un plancher comportant une dalle pleine tel que celui de Choisy, est d'un niveau relativement bas (20 à 40 dB A) en comparaison du niveau de bruit ambiant qui règne en général, de jour, dans les logements. Par suite, ce bruit est relativement peu audible et il est essentiellement composé de sons de fréquences graves. Par contre, la même marche sur un plancher tel que celui de Creil, ébranle fortement ce plancher par suite de la grande flexibilité et de la faible masse de celui-ci. L'ébranlement, qui peut aller jusqu'à faire bouger les lustres fixés au plafond de l'étage au-dessous, se traduit par un bruit sourd dont le niveau peut-être estimé (d'après les résultats obtenus à l'aide de la machine à chocs, aux fréquences graves), à environ 20 décibels de plus que dans le cas du plancher comportant une dalle pleine.

Si, au lieu d'un homme qui marche, c'est une femme équipée de chaussures à talons hauts, le bruit produit sous la dalle pleine de Choisy devient alors très audible par suite de la présence de composantes de fréquences aiguës d'autant plus intenses que la démarche est plus « sèche ».

Le bruit résultant de la marche de la même personne, sous le plancher de Creil, sera bien moins riche en composantes aiguës (d'après les résultats des mesures à l'aide de la machine à chocs) mais aura, comme dans le cas du marcheur masculin, des composantes aux fréquences graves de niveau élevé.

En résumé, à Creil, l'effet de la marche d'hommes ou de femmes sera fortement ressenti sous la forme de bruits sourds tandis qu'à Choisy seuls les bruits résultant d'impacts assez « secs » seront bien perçus. Ce dernier type de bruit est fortement lié au mode de vie des occupants et peut, dans le cas de personnes se déchaussant dès qu'elle sont chez elles (femmes surtout), devenir assez rare.

Ces considérations semblent suffisantes pour expliquer qu'à intensités identiques un excès de sonorité d'un plancher est nettement moins gênant aux fréquences aiguës qu'aux fréquences graves.

Bien que l'indice I.Q.I. cesse d'être approprié à l'évaluation du degré de gêne lorsqu'il devient fortement négatif, il ressort de la figure 1 qu'un plancher d'indice positif ou faiblement négatif (supérieure à — 5 dB) donne satisfaction à au moins 90 % des occupants ce qui peut être considéré comme assurant un confort suffisant.

3,212. *Autres méthodes d'évaluation de la qualité de la protection contre les bruits d'impact.*

a) Projet ISO nº 880 ([6]).

Comme il l'a été dit plus haut, l'Organisation Internationale de Normalisation envisage de recommander une méthode basée sur l'utilisation d'une courbe limite (figure 1).

Cette méthode est pratiquement identique à celle normalisée en Allemagne fédérale depuis plusieurs années (DIN 4109).

Nous avons, à partir des résultats de nos mesures par 1/3 d'octave, calculé les indices ISO correspondants. Les résultats sont portés sur la figure 4. La corrélation entre l'indice ISO et la gêne n'apparaît pas meilleure qu'entre

6. ISO = International Standard Organization.

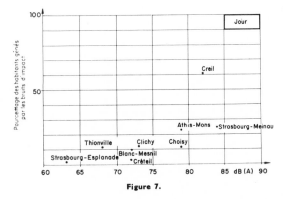

Figure 7.

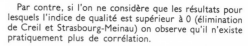

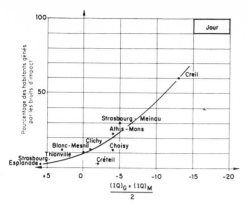

Figure 6.

celle-ci et l'indice I.Q.I. Au contraire, l'écart entre Strasbourg-Meinau et Creil se trouve accentué et le taux de croissance de la courbe relative aux planchers en maçonnerie est inférieur à celui signalé plus haut.

Cette méthode d'évaluation de la qualité de la protection contre les bruits d'impact est donc aussi mauvaise que la précédente et il paraîtrait judicieux de ne pas en recommander la normalisation.

b) Méthode canadienne.

Nous désignons par cette appellation la méthode préconisée par Olynyk et Northwood [2]. Le principe en est comparable à celui de la méthode ISO, à cette seule différence près que la courbe limite est la droite horizontale correspondant à un niveau de 70 dB par 1/3 d'octave. L'indice 0 correspond sensiblement à une dalle de béton nue.

Les résultats qui intéressent notre étude sont portés sur la figure 5.

On constate, si l'on prend en compte tous les résultats, une bien meilleure corrélation entre indice de qualité et gêne avec cette méthode qu'avec les précédentes. La droite définie par les points représentatifs des résultats de Creil et Strasbourg-Esplanade passe près de l'ensemble des résultats.

Par contre, si l'on ne considère que les résultats pour lesquels l'indice de qualité est supérieur à 0 (élimination de Creil et Strasbourg-Meinau) on observe qu'il n'existe pratiquement plus de corrélation.

Ceci revient à dire qu'on ne peut pas affirmer qu'un plancher d'indice + 10 est plus satisfaisant qu'un plancher d'indice + 5. Par contre, il semble qu'on puisse dire qu'à partir de l'indice + 5, moins de 15 % des habitants s'estiment gênés.

Malgré ce manque de sensibilité pour les indices positifs, cette méthode, qui élimine la pondération suivant les fréquences lors des tests à la machine à chocs, semble — tout au moins dans le cas de notre échantillon — convenir à l'évaluation de la qualité de l'isolation.

c) Méthode de la moyenne des indices de qualité aux fréquences graves et aux fréquences moyennes.

Après avoir constaté que l'on donne trop d'importance, dans la méthode utilisée actuellement en France et dans celle en préparation à l'ISO, aux fréquences aiguës, nous avons pensé à éliminer toute condition aux fréquences aiguës et à ne retenir comme indice de qualité que la moyenne des indices relatifs aux fréquences graves et aux fréquences moyennes. Les résultats obtenus avec cette méthode sont portés sur la figure 6. On constate que, par rapport à la méthode de l'I.Q.I. actuelle la corrélation est bien meilleure. On relève que le degré de corrélation est à peu près le même que dans le cas de la méthode canadienne. Un indice supérieur à — 2 dB correspond à moins de 15 % de personnes gênées.

d) Méthode fondée sur l'utilisation des décibels A, B et C.

A partir des résultats des mesures par 1/3 d'octave nous avons calculé les niveaux de bruit globaux en tenant compte des pondérations correspondant aux filtres A, B et C des sonomètres.

Ces niveaux sont sensiblement ceux que l'on mesurerait d'une manière globale, dans les locaux lorsque la machine fonctionne, à condition que la durée de réverbération de ces locaux soit égale à 0,5 s à toutes les fréquences.

Les résultats sont donnés sur les figures 7, 8 et 9.

On constate que le décibel (A) présente les mêmes inconvénients que l'indice I.Q.I. ou l'indice ISO. Par contre le décibel (B) et le décibel (C) semblent convenir aussi bien que l'indice canadien ou l'indice moyen relatif aux fréquences graves et moyennes.

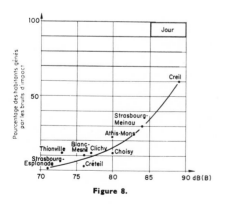

Figure 8.

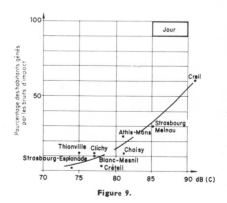

Figure 9.

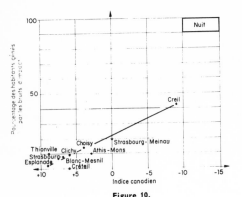

Figure 10.

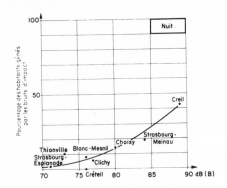

Figure 12.

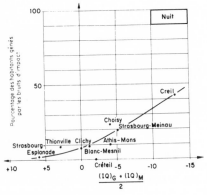

Figure 11.

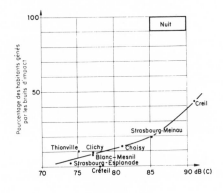

Figure 13.

3,22. Gêne de nuit.

Comme nous l'avons dit plus haut, c'est la moyenne des pourcentages de réponses « oui » aux questions N qui est retenue comme représentant la gêne de nuit.

Comme dans le cas de la gêne de jour, on constate qu'il n'y a pas une relation étroite entre le degré de gêne et l'indice de qualité de la protection contre les impacts.

Les conclusions auxquelles nous avons abouti précédemment en ce qui concerne la gêne de jour s'appliquent aussi à la gêne de nuit : les quatre méthodes retenues précédemment comme étant les meilleures le sont également dans le cas de la gêne de nuit, les graphiques relatifs à ces quatre méthodes ont été portés sur les figures 10 à 13.

CONCLUSION

Notre échantillon, tant du point de vue de la nature des planchers que de la classe socio-professionnelle des habitants n'était pas suffisamment fourni pour que nous puissions tirer de cette étude des conclusions définitives.

Il est toutefois apparu d'une manière incontestable que la manière de juger la qualité de la protection contre les impacts, dans les logements, telle que nous la pratiquons actuellement à l'aide de la machine à chocs normalisés, n'est pas satisfaisante, puisqu'un même indice de qualité peut correspondre à des pourcentages d'habitants gênés variant de 10 à 60 % et que des indices différant de 17 dB peuvent correspondre au même degré de gêne.

Pour la classe d'habitants interrogés (essentiellement des occupants d'H.L.M.), la sensibilité de la méthode aux composantes aiguës du bruit d'impact normalisé paraît exagérée eu égard à la réaction de ces habitants : un accroissement de sonorité aux fréquences aiguës résultant, dans le cas des planchers en maçonnerie, d'une moindre élasticité du revêtement de sol ne se traduit que par un accroissement relativement faible du pourcentage d'habitants gênés. Il est probable que, pour des appartements d'un standing plus élevé, où les réceptions seraient plus fréquentes et les modes de vie plus variés, la sensibilité aux fréquences aiguës serait plus accentuée.

Cette surestimation, dans le cas des logements sociaux, de l'importance de la sonorité aux fréquences aiguës fait que des planchers dont la sonorité excessive se manifeste essentiellement aux fréquences graves (planchers trop flexibles) sont moins pénalisés que des planchers peu sonores aux fréquences basses mais sonores aux fréquences aiguës (planchers en maçonnerie recouverts d'un revêtement peu efficace) alors que les habitants se plaignent davantage des premiers.

Pour remédier à cet état de choses, il existe plusieurs méthodes, entre lesquelles on ne pourra choisir qu'après les avoir testées sur un échantillon de planchers et d'occupants plus représentatif.

La méthode d'utilisation la plus pratique serait, soit la canadienne, soit celle basée sur la moyenne des fréquences graves et moyennes.

Les méthodes fondées sur l'emploi du décibel B ou du décibel C, apparemment plus simples, sont en fait plus compliquées car la durée de réverbération d'un local est rarement indépendante de la fréquence et, en tout état de cause, doit être préalablement mesurée. Il est donc indispensable de faire une analyse par 1/3 d'octave, comme précédemment, mais de plus il faut se livrer à un calcul assez laborieux pour en déduire le résultat cherché. Ces deux méthodes ne conviendraient donc réellement qu'à des locaux habités pour lesquels on voudrait évaluer la gêne dans les conditions d'ameublement données.

LISTE BIBLIOGRAPHIQUE

1. H. REIHER, « Uber den Schallschutz durch Baukonstruktionsteile », *Beih. Ges. ing.* 2 Nº 11, 2-28 (janvier 1932).

2. D. OLYNYK et T. D. NORTHWOOD, « Subjective Judgments of Footstep Noise-Transmission through Floors », *J. A. S. A.* 38, 1035-1039 (1965).

23

CONCERNING IMPACT NOISE

A. Coblentz and R. Josse

This English summary was prepared expressly for this Benchmark
volume by T. D. Northwood, National Research Council
of Canada, from "A propos des bruits d'impact,"
Centre Sci. Tech. Batiment Cahiers,
92(800):1–9 (June 1968).

INTRODUCTION

Acousticians have for many years been searching for an objective measure of the insulation provided by floors in multifamily dwellings against impact noises such as footsteps. Reiher, in 1932 [1], employed an impact machine consisting of a wooden hammer weighing 280 g falling 3 cm at intervals corresponding to a person walking. This was unsuccessful because of difficulty in measuring the discontinuous noise produced, and because the levels were in any case too low to measure. In 1938 another machine, described in the German Standard DIN 4110, was devised. This consisted of five brass hammers weighing 500 g each and falling 4 cm, so arranged as to produce 10 impacts per second. This solved the limitations of Reiher's model, and has since become an international standard (ISO Recommendation 140).

This study is a comparison of objective ratings obtained in multifamily dwellings, using the ISO impact machine, and subjective assessments by the occupants of the dwellings. The subjective data were obtained during two social surveys; one in 1963 involving 266 families in six different types of buildings; a second in 1965 involving 142 families in three additional types of buildings chosen specifically to study the impact-noise problem.

1. PROCEDURE

1.1 Acoustical Measurements

Insulation against impact noise has been rated objectively in France according to the method given in CSTB Technical Note of 1 December 1958. The procedure is to operate the standard impact machine on the floor in one apartment and measure the $1/3$-octave spectrum of noise transmitted to the room below.

The average results of $1/3$-octave band levels are compared to certain reference values for three frequency ranges: for a living room the reference values are 66 dB for the mean level at low frequencies (100–315 Hz), 62 dB for middle frequencies (900–1250 Hz), and 51 dB for high frequencies (1600–3200 Hz). The "Index of Quality" (IQ) for each frequency range is determined by the amount by which the mean levels lie below the reference values; for example, if the average levels for the three ranges are 64, 52, and 37 dB, respectively, the IQ's would be 2, 10, and 14. The overall performance, denoted the Index of Quality for Impacts (IQI) is the smallest of the three individual values, 2 dB in this instance.

The International Organization for Standardization (ISO) prepared a draft proposal No. 880 (eventually approved as R 717) embodying the same physical measurements but interpreting the results by comparison with a standard reference
curve shown in Figure 1. The insualtion meets the reference condition if the
measured values do not exceed the reference curve by
   —more than an average of 2 dB over the total frequency range (16 $\frac{1}{3}$-octave
   bands)
   —more than 8 dB in any one band

1.2 Subjective Reactions of the Occupants

The surveys included two questions designed to determine the degree of disturbance of the occupants of the buildings under study:

Question J: During the day or evening, do you hear people walking in the living
room or bedroom above?

Responses: Not heard; heard; heard and disturbed by.

Question N: During the night do people walking overhead—prevent you from
sleeping? ___ waken you? ___

Responses: yes; no.

The degree of disturbance from footstep noises in a given building can then be
represented by the percentages of people making each of the listed responses.

2. TEST STRUCTURES

The six building types in the first survey were similar in floor construction, consisting of 14-cm concrete floor slabs finished with hardwood flooring, linoleum, or
plastic floor covering. These were supplemented by a second study involving three
other floor systems. One (Creil) consisted of 11-cm wood joists with 2.3-cm oak
flooring nailed on top and 1-cm "Isorel" nailed on the underside; another (Strasbourg-Meinau) consisted of prefabricated concrete beams supporting hollow
masonry, a 4-cm concrete screed and linoleum above, and a plaster ceiling below;
the third (Strasbourg-Esplanade) consisted of a floating floor of 3-cm concrete
over a 6-mm mat of cork particles, over a 14-cm concrete slab, with 8 mm of oak
parquet on top.
   Results of the acoustical measurements are shown graphically in Figure 2.
   Subjective inquiries were made among occupants of 408 apartments distributed
among various building types. Particulars concerning the population were not
recorded during the first study, but in the second group it was found that generally
there were at least two children in each dwelling unit.

3. RESULTS

3.1 Results of the Social Survey

Responses to Question J, concerning the sounds of footsteps from overhead,
yielded the results shown in the first table, in which are shown for each location

the percentages of people who did not hear footstep noises (X), who heard them but were not disturbed (E), and who were disturbed by them (G).

Responses to Question N (Do footstep noises prevent you from sleeping or wake you up?) are similarly tabulated in the second table. The first column shows the percentage of occupants prevented from sleeping by footstep noises and column 2 shows the percentages wakened by footstep noise.

3.2 Correlation Between Objective Measures and Subjective Responses

3.21 Evaluation of the IQI

Values of the IQI were calculated for each building type. In Figure 3, these are plotted against the percentages of people who indicated that they were disturbed by footstep noises. Aside from the Creil results, all the others except one (Strasbourg-Meinau) are for solid concrete slab floors with various surface finishes. Although there is a degree of correlation between the subjective assessments and the IQI, the latter is clearly not a good measure of impact-noise isolation. For a large range of values of IQI, the corresponding variation in subjective assessment is very small. For example, comparing Thionville and Choisy, for which the degree of disturbance is the same, there is a difference of 17 dB in the IQI. The three points at the extreme right of the graph are characterized by the amount of high-frequency noise produced in these instances by the standard tapping machine. Apparently the high-frequency noise produced by the tapping machine in these buildings bears but little relation to the degree of disturbance.

With the inclusion of the Creil result, for a lightweight wood-frame floor, it is evident that the IQI is not a good general indicator of disturbance due to footstep noise. For the same IQI (Choisy and Creil) the percentage of persons disturbed differs by 50 percent. This difference results in part from the exaggerated importance given to high-frequency sounds relative to those of low and middle frequencies.

One can also approach this question by comparing the sounds of real footsteps for the floors at Creil and Choisy. Experiments done at CSTB and elsewhere indicate that the footstep noise produced by a man wearing plastic or rubber-heeled shoes on a concrete floor such as those at Choisy is almost inaudible compared to the daytime ambient level. By contrast, the same person walking on a joist floor such as that at Creil vigorously shakes the floor, because of its high flexibility and low mass. The vibration, which can be strong enough to shake the light fixtures below, produces a hollow sound the level of which can be estimated (from the low-frequency tapping-machine results) as about 20 dB higher than in the case of a solid concrete floor. In the case of a woman wearing high-heeled shoes, the noise produced on the concrete floor at Choisy becomes quite audible because of the high-frequency components associated with the sharper impact.

In summary: at Creil the footsteps of both men and women will be manifest as a relatively loud hollow sound; whereas at Choisy only the sounds of sharp impacts will be heard. The importance of the latter will depend on the mode of life of the inhabitants; for example, women who change from high-heeled shoes at home might produce negligible impact noise. These considerations seem sufficient to explain why the high-frequency noise transmitted by the tapping machine bears less relation to disturbance than does the low-frequency noise.

Although the IQI ceases to be reliable for assessing the response to impact noise when it becomes strongly negative, it appears from Figure 3 that a floor having a positive or slightly negaitve index (higher than –5 dB) satisfies at least 90 percent of the occupants, and can be considered to provide adequate comfort.

3.2.2 Other Methods of Rating Impact Insulation

a) ISO Draft Proposal 880 (ISO Recommendation 717)

As mentioned earlier, the International Standards Organization has developed a rating system based on the use of the limiting curve shown in Figure 1. This method is practically identical to the one used in Germany (DIN 4109). The ISO indices calculated from the $\frac{1}{3}$-octave data are shown in Figure 4. The correlation with subjective assessments appears no better than for the IQI.

b) Canadian Method

This appellation designates the method proposed by Olynyk and Northwood [2], which differs from the ISO method only in that the reference curve is a horizontal line corresponding to a $\frac{1}{3}$-octave level of 70 dB. The value zero corresponds approximately to the level for a bare concrete slab.

The results of interest in this study are shown in Figure 5. Considering all the test results, it is evident that this method yields a significantly better correlation with the percentage disturbance than either of the previous two. If, however, one considers only the results for which the index is greater than zero (omitting Creil and Strasbourg-Meinau) there remains practically no correlation. One therefore cannot assert that a floor having an index +10 is more satisfactory than one with an index of +5. What one can conclude is that for an index beyond +5 fewer than 15 percent of the occupants feel disturbed.

In spite of the lack of sensitivity for positive values of the index, this method seems, at least for these test samples, to provide a good indication of the quality of impact isolation.

c) Method of the Mean of Low and Middle Frequencies

Since the IQI gives too much importance to the high frequencies, the effect of dropping the high-frequency value and taking only the mean of the low and medium frequency values was considered. The results so obtained are shown in Figure 6. This yields a much better correlation than that given by the IQI. It is found that the degree of correlation is about the same as for the Canadian method. An index greater than –2 dB corresponds to disturbance of less than 15 percent of the occupants.

d) Methods Based on A-, B-, and C-Weighted Sound Levels

From the $\frac{1}{3}$-octave data, the overall sound levels weighted according to the A-, B-, and C-filter characteristics were calculated, on the assumption that the reverberation time is 0.5 s at all frequencies. The results, given in Figures 7, 8, and 9, show that the A-weighted level displays the same difficulties as the IQI and the ISO index, but the B- and C-weighted levels seem to correlate as well as the Canadian index and the mean of the low- and middle-frequency indices.

3.22 Nighttime Disturbance

The conclusions reached with respect to daytime disturbance are found to apply equally to nighttime disturbance: poor correlation was found relative to the IQI, but the four methods that yielded good correlations with daytime disturbance do so also for the nighttime disturbance, as demonstrated in Figures 10 to 13.

CONCLUSIONS

The test sample, both with respect to the nature of the floors and the social status of the occupants, was not sufficiently broad to permit drawing completely definitive conclusions. Nevertheless it seems conclusively demonstrated that the present method of rating the quality of impact isolation is not satisfactory, since on the one hand the same IQI corresponds to a range from 10 to 60 percent of occupants disturbed, and on the other hand IQI's differing by 17 dB can correspond to the same degree of disturbance.

At least for the tenants questioned in this study (occupants of publicly subsidized housing), the method appears to exaggerate the importance of the high-frequency components: an increase in high-frequency sound resulting in the case of concrete floors with hard floor coverings corresponds to a relatively slight increase in the percentage of occupants disturbed. This has the result that floors that transmit excessive low-frequency sound (floors that are too flexible) are less penalized than floors that transmit little low-frequency sound but some high-frequency sound (concrete floors with ineffective covering), even though the occupants complain more about the former.

To remedy this state of affairs, there are several methods, among which a choice can be made only after testing a more representative range of floors and of occupants. At the moment the most practical rating system would seem to be either the Canadian method or the one based on the mean of the low- and medium-frequency ranges.

24

Reprinted from *Int. Congr. Acoust., 4th, Copenhagen, 1962, Proc.*, pp. 31-54

Sound Insulation Requirements between Dwellings

by Ove Brandt

In a number of countries it has, during more than the past two decades, become necessary to introduce acoustic insulation specifications for flatted dwellings. The reasons for this are several. One is that modern flats get poor insulation if such directives are not enforced one way or another. In many countries flats are no longer built the traditional way with thick and heavy floors and walls but instead they are erected by modern prefab methods which usually imply reduced mass and thickness for the sound insulating barriers between the flats. Even then a good insulation may be obtained but only by a very careful planning of the buildings. However, many building designers have little or no acoustic training to solve this problem and it is simply ignored in most cases if no acoustic requirements exist.

It is not necessary to remind the readers that the number and power of acoustic sources in flats have grown tremendously also and thus stress the need for insulation between neighbours.

We do not expect this problem to be taken so seriously in countries where most people live in their own house. But in England where only 5% of dwellings were built as flats between the two great wars, acoustic recommendations were issued during the 1950-ies nevertheless and they seem to be developing into strict requirements in Scotland where a tradition for living in flats exists. Such is also the case in the colder climates of Scandinavia—it is not at all surprising that Sweden where 73% of the dwellings produced are flats (1961) was among the first countries to introduce insulation requirements.

If we do not want our cities to grow enormously we simply have to build flats in place of houses. But people will not want to remain in their flats if we do not solve the sound insulation problem.

For such reasons and others acoustic specifications have now been introduced in at least 13 countries. I shall try to review the international situation within this field.

Do the insulation requirements give us enough protection?

When the first proposals for acoustic requirements were made in Germany in 1938 [1] little was known as to how much insulation is required between two flats. Our theoretical and experimental knowledge was to a great extent limited to laboratory conditions for partitions and floors. It became necessary to estimate what was required.

As to *airborne sound* the choice fell on the insulation equivalent to that provided by a 25 cm plastered brickwall. Thus, the first requirements were expressed as minimum average figures

289

principally based on laboratory measurements on this brickwall. The frequency range chosen was nearly the same as we have today: 100–3000 Hz. In Scandinavia the same estimation was also made and the same expressions used when requirements were introduced here shortly after the war.

However, the brickwall was often replaced by other types of partitions, very often lightweight double walls in lighter prefabricated buildings. It was then easy to get a very high average figure, especially if it was measured in a laboratory with good craftmanship and no flanking transmission. But, the result in the field as experienced by the tenant was not judged to be equally good. It was thought necessary to express the required insulation not as an average figure for the whole frequency range but as a curve, based on octave or $1/_3$-octave intervals, a *grading curve*. Thus constructions with a high average insulation based on the insulation curve of the double wall as in fig. 1 would not be permitted. Also the realities of field conditions were taken care of in introducing requirements based on field results and intended for field control.

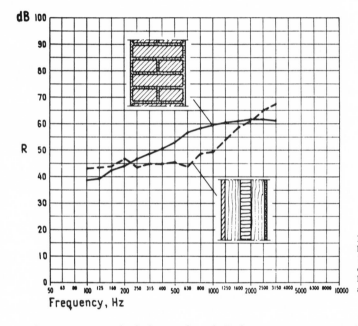

Fig. 1. Airborne sound insulation of a 25 cm plastered brickwall ($\overline{R}$ = 52 dB) and a double wall consisting of two leaves of 8 cm plastered stone and Rockwool ($\overline{R}$ = 50 dB).

In Germany, a new single figure, *the Schallschutzmass*, was proposed to replace the average arithmetical figure [2]. For airborne sound, the figure Luftschallschutzmass (LSM) was based on the proposed grading curves: it is the number of dB's that a measured curve has to be lifted or lowered in order to satisfy the required grading curve. LSM becomes 0 if the requirement is exactly satisfied, has positive and rising figures for accepted insulation curves but negative for insulation below the grading curve. Similar figures were proposed for the impact sound insulation, Trittschallschutzmass (TSM).

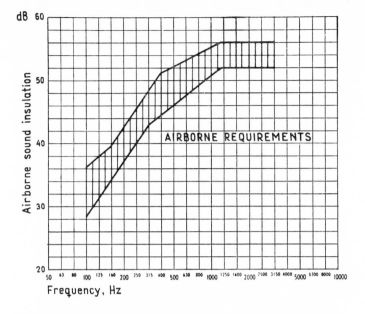

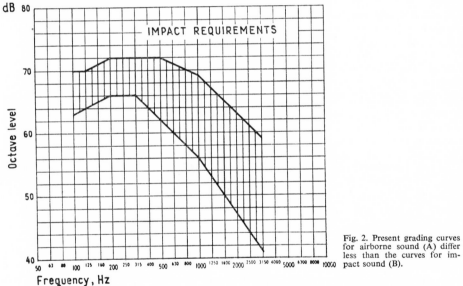

Fig. 2. Present grading curves for airborne sound (A) differ less than the curves for impact sound (B).

Even with these refinements, the background was still the same assumption that the 25 cm brickwall had sufficient insulation. The grading curve, first introduced in Germany after the war, was based on a number of laboratory and field measurements on this type of wall. However, with changing building technique towards prefabs in some countries one might ask why the insulation provided by this brickwall should be a divine answer to the need for acoustic protection as interpreted in the laboratory as well as in the actual buildings in the form as average figure and as minimum curve with the correct value at all frequency bands. We have had a similar development for the requirements on impact sound insulation. However, in this case different countries have apparently not had a common construction to suppose was adequate as with the brickwall for airborne insulation. It seems that in each country a choice has been made between current floor constructions and the better of them have become the standard and this has lead to a much greater spread in requirements for impact insulation compared with airborne insulation, fig. 2. So even more for impact insulation the question may be raised: "Which is the "right" answer for adequate protection against impact noises?"

The direct method to find out an answer to these questions is simply to ask people living in flats what they think about the acoustic insulation against the noise in the other flats and then make an objective measurement of the insulation in order to find out what the answer means in dB-requirements. It sounds very easy, but in fact it is not the easiest way to do it. The need for acoustic insulation may vary much from family to family. Some families produce a lot af sound with radios, TV:s, children and many more sources and do not care much about the noise they may hear from the neighbours in pauses between their own noises, and they may be honestly surprised if they get noise complaints from their neighbours. Some families may be at the other extreme: producing very little sound themselves and thus creating no masking to the neighbours' noises which may upset them very much and perhaps disturb rest and sleep thereby leading to strong complaints about the insulation. Of great importance is also the outside background noise level, with traffic as the main source: a high level leads to masking of the interior noises and thus an impression that the sound insulation is good.

For these and many more reasons it is of no use to make such a survey on a little scale if anything useful shall be concluded. The survey must comprise several hundreds of flats, carefully selected to give a typical picture of the numerous variations in the human reaction and activity and the objective sound insulation. In practice it is not really possible to get enough material to answer all the questions one might like to have answered.

Such social surveys have been carried out in England, Holland, Norway and Sweden [3, 4, 5, 6, 7]. The English surveys shall be briefly reviewed. In a flat survey the material was divided in 3 groups of flats with a difference in floor insulation of roughly 5 dB between each group, but having the same insulation in the horizontal direction. In a similar survey for row houses the material was divided in 2 groups, one having an average airborne insulation between neighbouring houses of 50 dB, the other with an insulation of 55 dB. These dwellings were all chosen amongst local authority houses or flats which, as I understand, means that they are built in an economic way in order that people with a low income can afford to live there. The results are therefore, as pointed out by the investigators, not necessarily valid for other sorts of dwellings with higher rent and standard.

In the *row houses* only the airborne sound insulation in the horizontal direction was measured. The two groups, comprising 250 pairs of houses, each had, as mentioned, an average insulation of 50 and 55 dB, for a single, plastered 25 cm brickwall and for a double wall of two leaves of 11 cm brick and an airspace of 5 cm, respectively. The insulation curves reported from field measurements on these two walls are given on fig. 3. It was found that there was no distinguishable difference in the disturbance in the two groups of houses. As the difference in insulation is found primarily at high frequencies it was concluded that

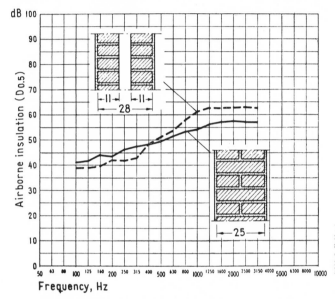

Fig. 3. Airborne insulation ($D_{0.5}$) for the English party walls in the social survey. Average of twenty-one 23 cm solid brick walls and five 27 cm cavity brick walls in houses.

better high-frequency insulation, obtained with a double wall, gives no appreciable advantage for the tenants. This is explained by the fact that it is the low and medium frequencies that are heard through walls as such frequency components dominate in the source which is verified by other investigations.

These results were ready at about the same time as the first grading curves, still based on the insulation of the 25 cm brickwall, were proposed in Germany. As the same type of wall was concluded to be sufficient for row houses in England, even here a grading curve was used based also on the brickwall. The two grading curves do not agree very well as seen from fig. 4.

The English social surveys in *flats* comprised 3 groups of about 1500 flats arranged according to different floor insulations for both airborne and impact sound. As mentioned before the average floor insulation differed 5 dB between each of the 3 groups while the horizontal airborne insulation was equivalent to a 25 cm plastered brickwall, i.e. roughly 50 dB in average. Group I had an average airborne insulation of 49 dB, Group II 44 and Group III 39 dB. Insulation curves for the Group I-III floors are given in fig. 5. The difference be-

3*

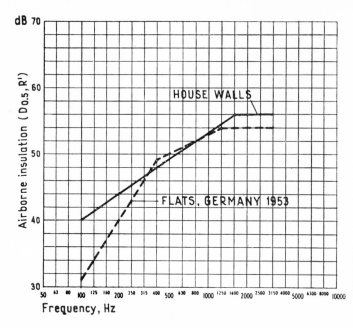

Fig. 4. The English row house recommendation and the original German requirement (1953), both based on the brickwall insulation do not coincide.

tween these insulation Groups is so big that one expects a clear indication of annoyance, least in Group I. The results of the survey did also verify this expectation for the Groups I and II: In the first Group 22% said they were disturbed by the noise, in the second Group the number of disturbed increased to 36%. In Group III this number surprisingly decreased to 21%. This unexpected relative satisfaction with acoustic insulation was explained by the fact that the tenants in Group III previously had had very bad dwellings and still seemed to compare the present improved conditions with their preceding living conditions.

In Group I noise from the neighbouring flats was no more annoying than so much else attached to living in a flat—as mentioned before England is not a country where it is considered a natural thing to live in a flat in place of a traditional house. In Group II flats noise was one of the biggest disturbances. Another measure for these Groups is that in Group I only 7% did not complain of anything, while this figure in Group II increased to 14%, and in the immune Group III these uncomplaining people were no less than 42%. This last Group was not used as a basis for recommendation as its tenants were uncritical in general. It was concluded from this survey that the insulation obtained with the floors in Group I flats should be used as a minimum recommendation for flatted dwellings, as these tenants apparently equally complained about noise as so much else in the flats. The average insulation curve was somewhat simplified, fig. 6, and was called Grade I.

A grade II was defined as a 6 dB lower curve at all frequencies. It was stated when employing this Grade that the tenants must be expected to find their neighbours noise the worst thing to endure in the flats.

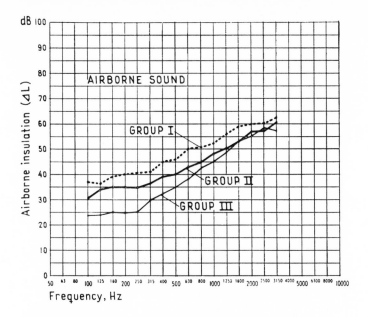

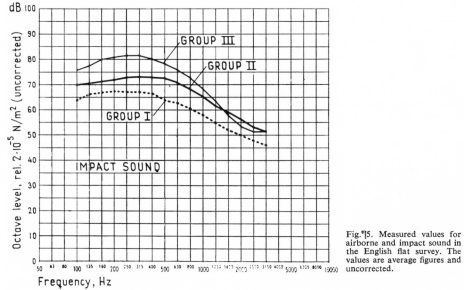

Fig. 5. Measured values for airborne and impact sound in the English flat survey. The values are average figures and uncorrected.

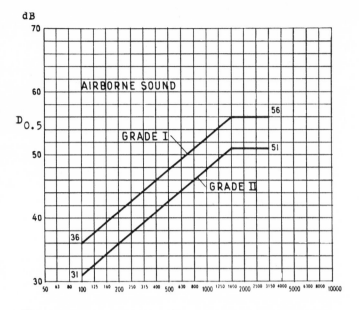

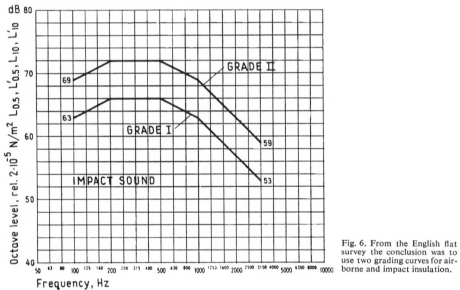

Fig. 6. From the English flat survey the conclusion was to use two grading curves for airborne and impact insulation.

It must be recalled when using Grade I for planning a block of flats that noise then is considered equally bad as draught, dampness, faults in the heating system etc. If we get rid of such shortcomings—which must be quite easy in a modern flat—one must expect that the complaints against the sound insulation increase. Also it should be remembered that this Group of flats was taken amongst local authority flats with, perhaps, relatively uncritical tenants. It must finally be remembered that flats are not the traditional type of dwellings for an Englishman and he may not complain so much because he considers his flat as only a provisional state before finding his definite dwelling in a house. Apparently, the Grade I recommendation cannot be expected to give a very good acoustic protection for the tenants. A few results from the Swedish survey complete this picture. It was carried out in about 500 flats at about the same time independantly of the British surveys. As a criterion for the airborne insulation the average figure in the range 100–3200 Hz was used, which is possible because very few of the walls or floors showed anomalies in the insulation curves as they were heavy, single leaf constructions. It was found that amongst people in flats with an average airborne insulation of about 45 dB 21% were disturbed by the neighbours airborne sounds. For flats with an insulation of 48–50 dB—roughly equivalent to the 25 cm brickwall—16% expressed dissatisfaction with the airborne insulation. At the highest insulation, 50–55 dB, only 7% were disturbed by these sources.

From these surveys we see that a decent protection is gained against airborne noise with the traditional brickwall, but we can hardly expect that this standard of protection is to be considered sufficient when the general standard of flats is raised. This is especially the case in countries where the flatted dwellings tend to dominate and people do not consider a flat as a provisional place to live. Also the noise sources seem to increase in number and power and this increases the need for airborne insulation.

Most specifications for noise protection are now expressed as a grading curve. As stated before a grading curve based on the measured insulation for a 25 cm plastered brickwall is not necessarily the correct answer at all frequencies, even if such an assumption may serve us well for a provisional standard. To find out what is the correct curve is, however, not easy. It can hardly be done with the same sort of social surveys as the ones mentioned, be-

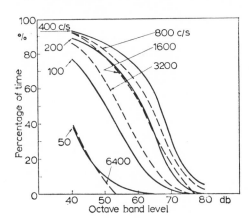

Fig. 7. Statistics of peak levels in radio-programmes. For each of eight octave bands the percentage of the time that certain octave band levels are surpassed is indicated. Based on about 55 hours of mixed radio-programmes (v. den Eijk).

cause we then need a very big material and we should have to ask people about frequency distribution etc. in terms that they are not familiar with. Other methods must be found. One method has been used by v. den Eijk in Holland. [8] He uses the fact that radio and TV-sets are the most annoying noise sources in flats and in order to find out how much insulation is needed he makes field studies on the time and frequency distribution of radio sounds in the source room in dwellings. He presents the results of such studies of 17 mornings and afternoons in fig. 7. Then he requires the level in the receiving room to be 0 phon

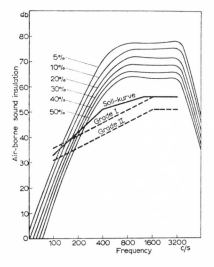

Fig. 8. Required airborne sound insulation based on a disturbing neighbour's radio level surpassing 0 phone during, in the mean, 5, 10, 20, 30, 40 or 50 per cent of the time. For comparison the German (Soll-Kurve) and the Brittish (Grade I and II) requirements for dwellings are added (v. den Eijk).

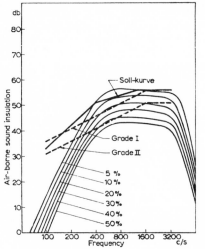

Fig. 9. Required airborne sound insulation based on a disturbing neighbour's radio level surpassing 20 phones during, in the mean, 5, 10, 20, 30, 40 or 50 per cent of the time. For comparison the German (Soll-Kurve) and the Brittish (Grade I and II) requirements for dwellings are added (v. den Eijk).

using the Fletcher-Munson 0-phon contures for pure tones. In this way he can get the shape of required level difference. As this requirement is very high he gets curves that lie very much higher than the present grading curves in Germany and Great Britain, fig. 8. He finds it more realistic to ask for a reduction to the 20 phon-contours. This leads to required level differences which by comparison with the German grading curve can be reached with the traditional brickwall, fig. 9. As normal insulation curves are less steep below 400 Hz and usually increase above this frequency he raises the question if there is any need to have requirements outside the important frequency range 400–800 Hz. Fasold, Germany, gets similar results. [9]

The correct shape of the grading curves have also been studied by Rademacher and Venzke, Germany. [10] They simulate the insulation curves of the walls with electric filters and arrange a receiving room similar to a normal dwelling room in volume and acoustics. The observers enter this room one by one and listen to different complex sounds from loudspeakers, filtered through the "wall" filters, and compare the loudness with a third-octave band of random noise centered around 1000 Hz. The selected source sounds are male and female speech, music and random noise of different band widths—all with little dynamics to make it easier for the observers to compare with the 1000 Hz random noise.

With this technique they demonstrate how different insulation curves influence the loudness of typical sounds in a receiving room. For each type of sound they ask the observers to compare the loudness of the sound filtered through different wall filters. The results of these subjective judgements are then compared with different objective figures such as the average arithmetical insulation and different German Luftschallschutzmass based on a number of grading curves, including the one in use and others proposed in Germany. They find that quite different grading curves can be used as a basis for the Schutzmass without appreciably changing this objective measure compared with the subjective one based on loudness. Even the average figure seems to follow the subjective measure surprisingly well, fig. 10. This fact is further studied and seems to be explained by the phenomena that two frequency ranges with good and bad insulation can compensate each other. This is further studied with the wallfilters as examplified in fig. 11. The higher insulation of K at medium

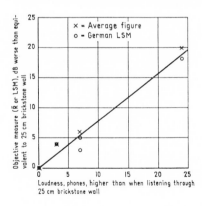

Fig. 10. Example from Rademacher and Ventze's work ([10]). For taped music, listeners have judged the equal loudness (phones) of this sound which they listened to "behind" different walls, evaluated by the average figure $\bar{R}$ or the Luftschallschutzmass LSM. All results are reduced to the case of 25 cm brickwall (0 dB and 0 phone).

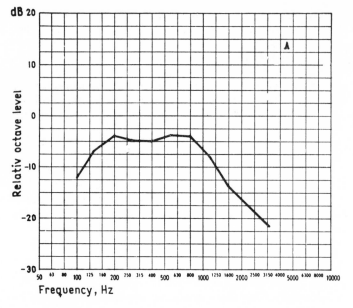

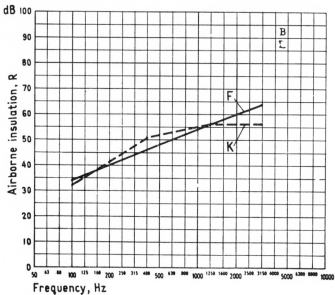

Fig. 11. When the observers listened to "coloured" noise, see A, through the wall filters with the responses F (6 dB/octave) and K, see B, it was judged to be equally loud. The average figure $\bar{R}$ and Luftschallschutzmass LSM are for F: $\bar{R}$ = 49 dB, LSM = 0 dB, for K: $\bar{R}$ = 50 dB and LMS = 2 dB.

frequencies seems to be compensated by the better insulation of F at frequencies above 1600 Hz so that the two loudnesses are alike. This result is most interesting as the main objections against the classical average figure have been its unrealisticly high values for steep insulation curves. It must be remembered that these results have been obtained according to loudness levels judged at 20–30 dB higher levels in the receiving room compared with what is usually experienced in a dwelling room. When one is exposed to the sound in a building some of the frequency range of the neighbours sound may be masked by the background noise and we do not know the distribution in time and frequency of this masking noise.

That the background noise must be very important for the judgement of the interior insulation is demonstrated for instance in the Swedish social survey mentioned above. Here the flats were put into 3 groups according to the exposure to outdoor noise—the noise was characterized as 1) high level, 2) normal town level and 3) low or very low noise level. The tenants who said they were annoyed by the outdoor noise were 19, 13 and 6% for the Groups 1)–3) respectively. When they were asked about the annoyance caused by noise from other flats the percentage disturbed were 26, 42 and 50% for the same 3 groups 1)–3), a very clear indication of the influence of the outdoor noise on the subjective experience of indoor insulation.

As to *impact sound insulation* our knowledge is so far quite limited. From the English surveys in flats we could draw some conclusions which lead to Grade I and II with similar remarks as for airborne sound insulation. It was also concluded that the light wooden floors had not sufficient impact insulation, even if Group III was not aware of insulations defects. As a matter of fact, in England it was recommended to use floating, concrete floors in order to satisfy Grade I, even if usually a floating floor well done should give more insulation than the required curve. From the Dutch survey we can conclude that the light floors and especially the light wooden floors are not usually sufficient for impact insulation. Finally the Swedish survey indicates that impact sounds do not seem to be a big problem if we use solid concrete floors. For tenants with floors without a separate screeding course only 7% were disturbed by impact sounds. This percentage fell to 2% for floors with a floating course on a mineral wool mat. Remembering that in the same survey the percentage of people who were annoyed by airborne sounds was 16—when airborne insulation requirements were just satisfied—one must conclude that impact sound insulation is not a big problem if the floors are not expecially light as e.g. wooden floors. This is perhaps also the explanation why grading curves for impact insulation in different countries vary so much. It thus seems that the present requirements give us a moderate protection against the neighbours' noise, at least for airborne noise: probably some more insulation is required, especially at the low and medium frequencies, but investigations made on the frequency response have used loudness and not annoyance as a subjective criterion for sound insulation. Further masking has not been considered. We have little evidence about how closely the present grading curves must be followed before the tenants are aware of such a change. Grading curves can hardly be taken as more than a rough indication as to what sort of insulation curves we want. It is probably too early to establish new single figures based on such grading curves as they may have to be changed as new research results appear.

How is „sound insulation" defined?

As mentioned before the first insulation specifications grew out of studies in traditional transmission laboratories where only the direct sound reduction factor for a test panel is measured. For such tests we have a very reliable measuring method which we have agreed upon in the International Standardization Organization. [15] We determine the airborne sound reduction factor R in measuring the level difference ΔL between two neighbouring rooms divided by the test panel of area S the absorption A in the receiving room and thus get R from the formula:

$$R = \Delta L - 10 \log A/S \text{ dB}$$

This formula is valid if all sound in the receiving room is transmitted through the test panel. Also, diffuse fields are required in the rooms. Such conditions are not difficult to satisfy in a stationary laboratory. But we want to make the specifications in building codes valid also for the field. If we could only test or check in the laboratories rules would be of little value and certainly not gain much respect in practice.

But can we expect to have enough diffuse sound fields in normal dwelling rooms, furnished or unfurnished to make sensitive measurements? Can we use the same relationship between level difference and the reduction factor as is used in the laboratory according to the formula above? Or do we have more practical relationships to base our requirements on?

As a matter of fact, it is easier to make reliable measurements in dwelling rooms than one might expect. Of course we do have some troubles at very low frequencies when the room dimensions are of about the same size as the wavelength. Usually not more in a furnished room than in a smaller laboratory as we get some diffusion from the furniture. At higher frequencies we expect to get difficulties as the porous damping of the higher frequencies tend to make the field look like a free field in place of a diffuse field. Gösele [12] in Germany has, however, shown, that we do measure one or two dB higher levels in the pressure field in the receiving room, but if we correct to a constant absorption we get too low values for the absorption determined from the Sabine formula and from short reverberation times, which compensates for the error in the level measurements. He showed that by changing the reverberation time in the receiving room from less than 0.5 seconds to more than 3 seconds the corrected impact sound level changed less than 2 dB at the individual frequencies for the same floor.

In one sense there is a great difference between the laboratory and field conditions: we cannot guarantee that the sound in the receiving room has arrived only through the partition or the floor in the building. Rather it is so, that a good deal is transmitted through flanking elements, *flanking transmission*. Of course, we can still use the same formula above, but then we must include the flanking transmitted sound in the reduction factor (which is then nominated R') if we still take S as the area of the common surface for source and receiving room. This method is used with success in e.g. the German requirements and its advantage lies in its simplicity for the building designers as we shall see.

In some other building codes the level difference is used as a measure for sound insulation in a dwelling, but this magnitude must be normalized in one way or another. If we only used the level difference in a requirement, sound insulation would depend on the acoustics of the receiving room. If we increase the amount of absorption we get an apparent increase of the sound insulation observed in the diffuse field of the receiving room. We then have the possibility to correct to a certain time of reverberation or to an absorption of the dwelling room. What is to be preferred?

In order to answer this question some reverberation measurements have been made in e.g. England and Denmark. [13] It might be expected that the reverberation time increases with the room volume as we know is the case for classical concert rooms. This is also the case for unfurnished rooms and for rooms with little furniture, but not for furnished rooms. For living-rooms Larris found that the reverberation time varied only between 0.35 and 0.7 seconds with an average value around 0.5 seconds when the room volume varied from 19 to 118 m³, fig. 12. For the same furnished rooms the absorption computed from Sabine's formula varied from 6.5 to 38 m². This is explained by the fact that the principal absorption

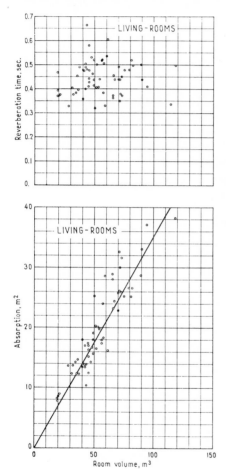

Fig. 12. Reverberation time and absorption in furnished living-rooms. Average values for 125—4000 Hz (Larris).

in living-rooms such as stuffed furniture and mats is connected with the floor. When the floor area increases with the volume the absorption must also increase and thus it is easy to show for a rather constant density of furniture the reverberation time must be quite constant. This is less true in bed-rooms where the total furniture is more constant, fig. 13. The frequency dependance has a peak in the mean frequency range, as the low frequency absorption is procured by panel absorbents and the high frequency absorption by porous absorbents.

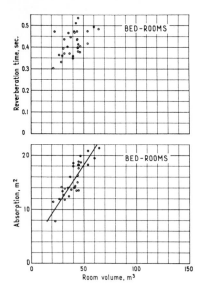

Fig. 13. Reverberation time and absorption in furnished bed-rooms. Average values for 125—4000 Hz (Larris).

If we state that a dwelling room has a reverberation time of 0.5 seconds we must have in mind that this is primarily so for living-rooms, less for bed-rooms—which in some countries tend to disappear in smaller flats—and it is not the case for rooms like kitchens, bath-rooms, halls and similar rooms with little or no furniture where we expect the reverberation time to increase with the volume.

The fact that the reverberation time in a furnished living-room is nearly constant independant of volume has lead some countries to use the level difference normalized to the reverberation time of 0.5 seconds as a basis for insulation specifications. Thus this required level difference $D_{0.5}$ is defined as:

$$D_{0.5} = \Delta L + 10 \log T/0.5 \text{ dB}$$

In this way the required level difference and also the measured one is a true picture of the observed level difference when having a living-room as a receiving room, a very important practical case in flats.

This normalized level difference is then a result of the reduction factor R' of the common surface S between two neighbouring rooms and the flanking transmitted sound from other surfaces. This is quite easy to understand for building planners without acoustic training, but it is in practice not always so easy to evaluate, not even when flanking transmission can be neglected. The fact is that $D_{0.5}$ also depends on the volume V of the receiving room which we see in expressing $D_{0.5}$ as a function of R':

$$D_{0.5} = R' + 10 \log \left(\frac{0.32 \cdot V}{S} \right) \text{ dB}$$

It will be noticed that this measure is not reciprocal if used between two rooms with different volumes; the building designer may suspect that sound insulation of a structure is not reciprocal. So the direction of the measured level difference must be stated in the reports. If we choose to normalize to a constant absorption we do not get this drawback. This measure D_{10} which has been standardized by ISO for field measurements is then defined as:

$$D_{10} = \Delta L - 10 \log A/10 \text{ dB}$$

thus normalizing the level difference to a reference absorption in the receiving room of 10 m^2. If we express this measure by R' and the common surface of two neighbouring rooms we get:

$$D_{10} = R' + 10 \log 10/S \text{ dB}$$

leaving to the building planner to make his calculations based upon the insulation R' with or without flanking transmission and the size S of the transmitting element.

Of course also this definition has its drawbacks. For instance, for big rooms separated by big surface areas this correction gives a false picture of the real insulation when the rooms are normally furnished. We correct then to a much smaller absorption and neglect that the real absorption is bigger. When we use the same value of D_{10} for all room sizes in dwellings —which we must for the sake of simplicity—the requirements tend to be too rigourous for big rooms and perhaps too mild for small rooms. The trend should of course be in the opposite direction.

Both $D_{0.5}$ and D_{10} are of course a little difficult to handle for the architect or builder with little acoustic training. To simplify this planning it may be better to specify permitted partitions, floors etc. in the building codes, completely omitting acoustic specifications. The drawback of this method is that it may put a brake on the development of building constructions and in many cases it is difficult to give information of all the permitted combinations of e.g. partitions and joining elements. What is usually preferred is *both* an acoustic requirement somehow in dB *and* a number of examples demonstrating how to satisfy the requirements.

Some countries have like Germany preferred to simplify the specifications and also the planning by using the same reduction factor as in the laboratory, here nominated by R'. The planner then need pay no attention to variations in wall surface or room volume, but can simply refer to measuring reports from the identical constructions, even combined with the right joining constructions. Then the requirements must be adjusted to cover even big surfaces. One of the only drawbacks of this principle is that it cannot be used for cases where a common surface S between two rooms are not defined, e.g. between a living-room and a staircase. It may also be a bit disturbing to attribute all defects of for instance a bad outer wall to the common surface S. In Germany laboratories have been built to measure R' for rooms with flanking walls but still referring to a constant area of the partition, here much better insulated than the flanking construction.

The three existing definitions on airborne sound insulation, R', D_{10} and $D_{0.5}$, may lead to quite different results when the same figures are required as some examples will show.

If we require D_{10} and R' to be equally big for the same case the wall surface must be 10 m². If we even demand D_{10} to equalize $D_{0.5}$ for horizontal insulation the volume of the receiving room must be 31.3 m³. For a room height of 2.5 m we thus get a standard receiving room with the dimensions $4 \times 3.1 \times 2.5$ m³, which is quite a normal room in a modern flat. But quite big deviations from these dimensions may occur.

If we look at quite a big room with the floor size of 8.0×3.1 m² and standard height of 2.5 m, we get the following differences (vertical insulation):

$$D_{0.5} - R' \ = 6 \text{ dB}$$
$$D_{10} - R' \ = 3 \text{ dB}$$
$$D_{10} - D_{0.5} = 9 \text{ dB}$$

If we turn to small rooms the differences are usually not quite as big. A minimum standard floor for a Scandinavian bed-room is about 2.1×3.3 m². With the room height of 2.5 m and vertical transmission we get:

$$D_{0.5} - R' \ = 0.25 \text{ dB}$$
$$D_{10} - R' \ = 2.8 \ \text{ dB}$$
$$D_{0.5} - D_{10} = 2.5 \ \text{ dB}$$

For *impact sound transmission* we have luckily only two alternatives for definitions. One of these is to refer the measured level in the receiving room to 0.5 seconds for the same reason as for airborne insulation. This leads to the definition:

$$L_{0.5} = L + 10 \log 0.5/T \text{ dB}$$

Unlike $D_{0.5}$ we have no such drawback as lack of reciprocity because the direction of transmission is given.

The other alternative which is recommended by ISO for field and laboratory measurements is to correct to 10 m² of absorption:

$$L_{10} = L + 10 \log A/10 \text{ dB}$$

Both these alternatives have the drawback that we get a higher figure for decreasing insulation, but this disadvantage does not seem to bother building planners so much as they apparently quickly get used to it.

Obviously, we get cases when these two definitions give different figures, even if the difference is not so big as for the measures for airborne sound. Still we get the same figure if the room volume is 31.3 m³ which means a floor size of 12.5 m² for the room height of 2.5 m. A great majority of modern flats have floor sizes of this order. If the floor increases to 25 m² the difference is 3 dB. A small room has the size of about 6 m² which still gives us a difference of about 3 dB.

It is easy to show that a correction to a constant absorption is the same thing as to assume a constant power from the ceiling independently of its size. Thus we should get the same results for the same floor construction even if we measure on different floor areas. This seems to be the case for floor sizes in the range from 6–25 m² according to German [14] (Gösele) and Swedish measurements. Thus L_{10} would seem to be a good physical magnitude, but with corrections not fitted for the normal acoustics in living-rooms as for D_{10}. We can also show that the correction to a constant time of reverberation as for $D_{0.5}$ is the same as to assume a radiated power from the ceiling growing with its surface. This measure then has the advantage to follow the variation in room volume as is done in furnished rooms but it has as mentioned its physical disadvantages.

Obviously, the three definitions for airborne sound insulations and the two for impact sound have its advantages and disadvantages and it is a matter of taste which is to be preferred. However, it should be a step forward if we could agree internationally on this subject in order to reduce confusion.

Insulation requirements or recommendations in different countries

In the preceding sections we have looked a little at the present background and terminology for insulation requirement. Let us now look at some of the principles used in different countries for acoustic specifications. A detailed report is being prepared by ISO.

In some countries such *specifications* are presented as *requirements*, in others as *recommendations*. There may be little difference in practice. The recommendations *may* have much stronger power than strict requirements which may be only writting table products completely ignored by building designers. The advantage with recommendations is that the real acoustic claim may be expressed without too much compromize with other factors from the very start. An example of this is the British Grade I recommendation for impact noise which is based on floating floors. In Austria a 5 dB higher Luftschallschutzmass (based on the German Sollkurve) is recommended. Germany gives us a good example with requirements which work well and many stationary and mobile labs are available to control the results in practice. In such a case the specifications must be somewhat milder and roughly be intended to cut off the extremely bad cases. The danger in this system is that the standards must be compromised and consequently are only partly sufficient in the majority of cases. Building planners may easily get the impression that all is well if they build just to satisfy the requirements. In fact, it might be better to have a minimum requirement combined with an uncompromised recommendation but this leads to complicated specifications without the simplicity which must characterize rules for building planners with little acoustic training.

Today at least 13 countries have insulation specifications for dwellings. In the great majority grading curves are used to express the minimum values. For airborne sound 10 countries use one of the curves presented in fig. 14 and 15.

To evaluate a measured curve in relation to a grading curve many countries follow the German system of computing the average negative deviations in the whole frequency range and setting positive deviations equal to 0. In Germany this average deviation must not exceed 2.0 dB, based on third-octave frequencies, while e.g. USSR, Bulgaria and

4

Czeckoslovakia base this average deviation on octave frequencies and add the rule that no single negative deviation may exceed 8 dB. In Great Britain and Scandinavia this procedure is somewhat simplified as only the sum of negative deviations is computed and not permitted to exceed 16 dB.

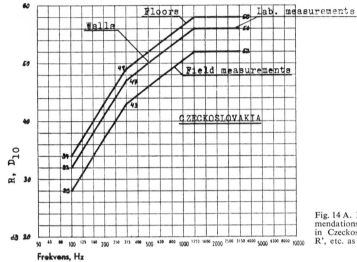

Fig. 14 A. Requirements and recommendations for airborne insulation in Czeckoslovakia. Symbols for R, R', etc. as in the preceeding section.

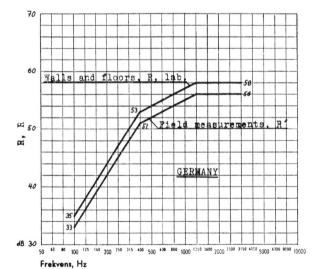

Fig. 14 B. Requirements and recommendations for airborne insulation in Germany. Symbols for R, R', etc. as in the preceeding section.

The present grading curves for impact insulation are presented in fig. 16 and the measured impact insulation should result in a curve *below* the grading curve. We have similar rules as for airborne insulation to decide on cases where part of the measured curve lies above the grading curve. The same 10 countries that have grading curves for airborne sound have such curves for impact sound transmission.

In Canada which was one of the first countries to introduce insulation specifications for airborne sound the average minimum figure of 45 dB has been recommended for the

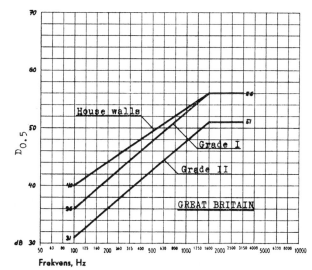

Fig. 14 C. Requirements and recommendations for airborne insulation in Great Britain. Symbols for R, R', etc. as in the preceeding section.

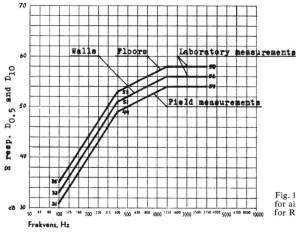

Fig. 14 D. Requirements and recommendations for airborne insulation in Scandinavia. Symbols for R, R', etc. as in the preceeding section.

4*

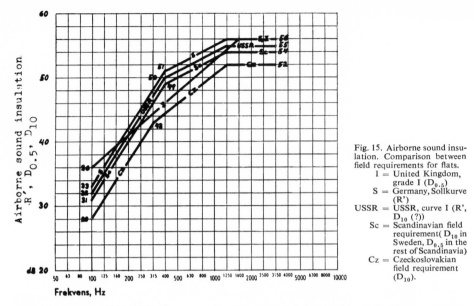

Fig. 15. Airborne sound insulation. Comparison between field requirements for flats.
I = United Kingdom, grade I ($D_{0.5}$)
S = Germany, Sollkurve (R')
USSR = USSR, curve I (R', D_{10} (?))
Sc = Scandinavian field requirement(D_{10} in Sweden, $D_{0.5}$ in the rest of Scandinavia)
Cz = Czeckoslovakian field requirement (D_{10}).

frequency range of 125–4000 Hz; now a grading curve is being prepared. In France, average figures for airborne and impact sounds are given for the frequency ranges 100–320, 400–1250 and 1600–3200 Hz. This is a very simple principle without troublesome evaluations. Nor does it pretend to more knowledge than we possess.

In some countries, e.g. Scandinavia and France, the specifications comprise both the reduction factor for the bounderies between flats as measured in a traditional laboratory and the normalized level difference in the completed building. This complication is made because one could reach a sufficiently high level difference even if a very small element in the partition has a very low reduction factor. However, if for instance a bed is placed close to such an element very low insulation is experienced.

Some effort has so far been made to get a quality figure for insulation to replace the traditional average insulation. A few countries have followed the German example to introduce a Schallschutzmass for airborne (LSM) and impact (TSM) sound. As mentioned before it is based on the grading curves and is defined as the number of dB's that a measured curve has to be lifted or lowered in order that it can be considered acceptable (average deviation 2.0 dB). The drawback for such a single figure is primarily that it is tied to a certain curve. If this is changed we get new quality figures which must be very confusing for building designers. This is the primary reason why some countries like Scandinavia and England have hesitated to introduce another single figure for sound insulation before we have got an international agreement on such requirements. In the meantime only the sum of deviations from the grading curve is used as a provisional quality figure but with the drawback that it gives the figure 0 for all cases that we get an insulation higher than required.

A grading curve may be difficult to change when finally it has become well established in a national building code. In place one can use two (like e.g. Great Britain) or more grading curves and require an appropriate curve to be satisfied in the specific case. But it is also possible to have only one curve (like e.g. Germany, for walls) and then require different Schallschutzmasses for different situations which is the same thing as choosing between a great number of parallel grading curves.

In view of these facts one might raise the question whether it is possible to establish some sort of international standardization on sound insulation requirements, a great advantage in the growing international exchange of knowledge and products. One might well be a little pessimistic as to the success of such a work considering the different grading curves already established. Further, we can hardly as acousticians expect to change building traditions in some countries which happen to accept for instance floors with low insulation and have no apparent tenants' reaction. Obviously, other countries with building technique which happens to favour sound insulation—or have strong public opinion on this subject— would not be ready to accept an international standard so compromised. Nevertheless I have some hope for such an attempt at international cooperation on this problem.

This feeling of optimism is supported by the success of a Scandinavian collaboration on this subject. We met five years ago to agree just on the measuring methods, but found it possible also to agree on requirements. These were then shaped as the grading curves shown in fig. 14 and 16. As to airborne sound our first proposal was a grading curve a little different from the British Grade I and the German Sollkurve. However, we found it wrong to introduce another curve and thus increase the international confusion. In place we accepted the German Sollkurve for airborne sound.

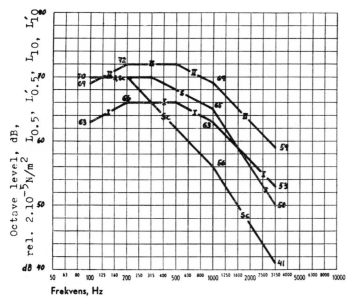

Fig. 16. Impact sound transmission. Comparison between requirements for flats.
I = United Kingdom, grade I ($L'_{0.5}$)
II = United Kingdom, grade II ($L'_{0.5}$)
S = Germany (Sollkurve) and USSR (curve IV) (L_{10}, L'_{10})
Sc = Scandinavia ($L_{0.5}$ and $L'_{0.5}$ except for Sweden where L_{10} and L'_{10} is used).

As we know the existing grading curves lead to very little change in tenants' reaction it should be possible to agree on an international grading curve, at least as a first step for airborne insulation. Also the present French method of having a number of average figures for part bands should be discussed because of its simplicity and leading to no new single figures. Also the appropriate definitions for sound insulation should be discussed and decided on.

While we discuss and perhaps accept such an international provisional recommendation we should organize more research on this subject to see how well the different systems function and also if it is possible to simplify—for instance in limiting the frequency range as suggested by v. den Eijk and others. Such an international discussion which already has started within ISO may also be a great help in countries where such specifications are not yet considered but probably needed.

REFERENCES

1. DIN 4110. Technische Bestimmungen für die Zulassung neuer Bauweisen. Abschnitt D 11, (Schallschütz) 1938.
2. DIN 52211 (Vornorm). Bauakustische Prüfungen. Schalldämmzahl und Norm-Trittschallpegel. Einheitliche Mitteilung und Bewertung von Messergebnissen (September 1953).
3. P. G. Gray, A. Cartwright and P. H. Parkin, Noise in three groups of flats with different floor insulations. National Building Studies. Research Paper no 27. HMSO, London, 1958.
4. P. H. Parkin, H. J. Purkis and W. E. Scholes, Field Measurements of Sound Insulation between Dwellings. National Building Studies. Research Paper no 33. HMSO, London, 1960.
5. C. Bitter and P. Van Weeren, Geluidhinder en Geluidisolatie in de Woningbouw I. TNO, Rapport no 24. September 1955.
6. Gunnar Ö. Jörgen, Stöy i Boliger (Noise in dwellings), Norges Byggforskningsinstitutt, Rapport nr. 16, Oslo 1955.
7. O. Brandt and I. Dalen, Är ljudisoleringen i våra bostadshus tillfredsställande? (Is the sound insulation in our dwellings sufficient?). Byggmästaren B6, 1952.
8. J. van den Eijk, Mijn Buurmans Radio (My neighbours' radio), TNO, De Ingenieur, no. 39, 1960, Gezondheidstechniek 5. see also: J. van den Eijk, My Neighbours' Radio, Proceedings of the 3rd International Congress on Acoustics, p. 1041, Elsevier Publishing Company, 1961.
9. W. Fasold, Untersuchungen zum zweckmässigsten Sollkurvenverlauf den Schallschutz im Wohnungsbau. Proceedings of the 3rd International Congress on Acoustics, p. 1038, Elsevier Publishing Company, 1961.
10. H. J. Rademacher and G. Venzke, Die subjektive und objektive Bewertung des Schallschutzes von Trennwänden und -Decken, Acustica, vol. 9 (1959), p. 409.
11. L. Cremer, Der Sinn der Sollkurven, Schallschutz von Bauteilen. Verlag von Wilhelm Ernst & Sohn, Berlin, 1960.
12. K. Gösele, Zur Durchführung von Trittschallmessungen in bewohnten Bauten, Gesundheits-Ingenieur, 80 (1959), Heft 2.
13. F. Larris, Undersøgelse over Efterklangstiden i Beboelsesrum (Reverberation time in dwellings), Teknologisk Instituts Forsøgsberetning 1946–48, Meddelse Nr. 5, Copenhagen 1948.
14. K. Gösele, Trittschall- Entstehung und Dämmung, VDI-Berichte Bd. 8 (1956), "Schall und Schwingungen in Festkörpern", p. 23.
15. ISO Recommendation R 140. Field and Laboratory Measurements of airborne and impact Sound Transmission. 1960 (E)

Part IV

SOUND INSULATION—TECHNIQUE AND THEORY

Editor's Comments
on Papers 25 Through 30

Sound insulation is concerned with the performance of a building structure in preventing the transmission of sound from one space to another. The usual measure of this performance is the "sound transmission loss" (sometimes called "sound reduction index"), defined as ten times the logarithm of the ratio of the sound power incident on the dividing structure to the sound power radiated on the receiving side. The total transmission process may thus involve room-acoustics aspects at the source and receiving ends as well as the interrelations with the structure.

The first theoretical formulation of the process appears to be that of Edgar Buckingham in 1925. Most of his paper is reproduced here (Paper 26). His Equation 24 forms the basis of the current standard method of measuring sound transmission loss. It is interesting to note that, whereas with today's technology this formula may easily be applied directly, Buckingham found it necessary to devise a method based on measuring the duration of a decaying sound field in each room; then he could utilize Sabine's apparatus, i.e., a stopwatch and a pair of ears. These were the most precise measuring instruments available in 1925.

Buckingham's treatment fits the classic case of a partition separating two reverberant rooms. In this case, the difference in average sound pressure levels in the two rooms is a measure of the sound transmission loss. For the basic theory to apply, the sound fields in the two rooms must be diffuse, and the sound field incident on the partition must comprise a random distribution of all possible angles of incidence. In ordinary rooms, however, the prescribed conditions may not be fully realized, and one must be cautious in generalizing from the results of one measurement of sound transmission loss. A recent report by R. F. Higginson (Paper 26) on the scatter among several measurements on the same wall illustrates some of the experimental difficulties in even a relatively simple measurement. If either of the rooms separated by a partition is very dead, or if transmission is into or out of free space, the reverberant-room approach becomes inapplicable and other concepts must be applied as discussed in an early paper by London[1] and more recently in an ASTM Recommended Practice.[2]

The interaction of the partition structure with the incident and radiated sound fields is usually separated from the room-acoustics part of the process. The major contributor to studies in this area has been Lothar Cremer of the University of Berlin, who over the years has published an extensive series of papers on the topic. One of the most well known, reproduced here as Paper 27, contains his development of the theory of sound transmission through a single homogeneous panel. Among other things, this paper included, for the first time in the context of architectural acoustics, an explanation of the effects of "coincidence" between flexural wavelengths in the panel and projected wavelengths of sound waves incident on the panel. Also considered was the effect of damping in the coincidence frequency range. An alternative treatment also appears in a recent book by Cremer, Heckl, and Ungar.[3]

Cremer treated the case of a single wall of infinite lateral dimensions, taking account of the effects of flexural stiffness and damping, but assuming mainly a diffuse incident sound field. Several writers have attacked the problems associated with finite dimensions and edge constraint and the limitations imposed by the sound fields existing in typical rooms. From a voluminous literature the papers by Schoch,[4] Gösele,[5]

315

Sewell,[6] de Bruijn,[7] and Kihlman and Nilsson,[8] exemplify theoretical and experimental studies of these effects.

It is well known that the sound insulation of a wall is improved if the total mass is divided into two or more relatively independent layers. Many attempts have been made to analyze this case and fit more-or-less plausible theories to more-or-less equivalent experimental evidence. Although no single definitive solution for the multileaf wall can be cited, a selection of typical studies might be mentioned: an early paper by Albert London deals with an idealized model consisting of two identical walls of infinite extent separated by an air space;[9] Cremer attacked the special problem of a relatively light flexible layer resiliently attached to a heavy stiff layer;[10] White and Powell approached the problem from energy considerations, developing an interesting thermal analogy;[11] Lyon and Maidanik used a linear coupled oscillator model;[12] and Price and Crocker used a statistical energy analysis,[13] which is the most fashionable current approach, especially useful in problems involving structureborne sound.

The second major sound-insulation problem is the transmission of footsteps and similar impact sounds that originate as disturbances in the structure itself. The main difficulty in analyzing this problem is that the structure is not merely part of the transmission system, but is also part of the excitation system as well. A second difficulty is that the sounds take the form of brief impulses; hence the usual reverberant-room method of monitoring the transmitted noise seems inapplicable. Cremer, in close association with K. Gösele of Stuttgart, made major contributions to the theoretical and experimental treatment of simple slab floors and floating floor systems.[14] In Paper 28, Gösele presents a brief status report on much of this work. More recently it was found that for certain floor structures it is possible to relate the airborne sound insulation to impact insulation. This was first done by Heckl and Rathe (Paper 29) using the reciprocity principle, and more recently by Vér (Paper 30) using a more direct approach. An experimental attack on the relation between impact forces and resultant sound radiation, by Ford, Hothersall, and Warnock,[15] further clarifies the impact transmission process.

In the final analysis, it is the sound insulation achieved in buildings that matters. Although a fair amount of information on specific elements is available, the best comprehensive study of buildings is by Parkin, Purkis, and Scholes,[16] of the British Building Research Station, who report airborne and impact insulation measurements on 464 building walls and floors, and analyze the results.

REFERENCES

1. A. London, "Methods for Determining Sound Transmission Loss in the Field" *J. Res. Nat. Bur. Stand.* (U.S.) **26**:419–453 (1941).

2. American Society for Testing and Materials, "Recommended Practice No. E 336," for measurement of airborne sound insulation in buildings (1971).

3. L. Cremer, M. Heckl and E. E. Ungar, *Structure-Borne Sound* (Springer-Verlag, New York, 1973), p. 506.

4. A. Schoch, [Asymptotic Relation of the Forced Vibrations of Plates at High Frequencies], *Akust. Z.* **2**:113–118 (1937).

5. K. Gösele, "Uber Prüfstande zur Messung der Luftschalldämmung von Wänden und Decken," *Acustica* **15**:317–324 (1965).

6. E. C. Sewell, "Errors in the measurement of the sound reduction index of masonry walls in the BRE laboratory facility," Building Research Establishment Current Paper 47/74 (1974).

7. A. de Bruijn, "Influence of Diffusivity on the Transmission Loss of a Single-Leaf Wall," *J. Acoust. Soc. Am.* **47**:667–675 (1970).

8. T. Kihlman and A. C. Nilsson, "Effects of Some Laboratory Designs and Mounting Conditions on Reduction Index Measurements," *J. Sound Vib.* **24**:349–364 (1972).

9. A. London, "Transmission of Reverberant Sound Through Double Walls," *J. Res. Nat. Bur. Stand.* (U.S.) **44**:77–88 (1950).

10. L. Cremer et al., op. cit., p. 501.

11. P. H. White and A. Powell, "Transmission of Random Sound and Vibration Through a Rectangular Double Wall," *J. Acoust. Soc. Am.* **40**:821–832 (1966).

12. R. H. Lyon and G. Maidanik, "Power Flow Between Linearly Coupled Oscillators," *J. Acoust. Soc. Am.* **34**:623–639 (1962).

13. A. J. Price and M. J. Crocker, "Sound Transmission Through Double Panels Using Statistical Energy Analysis," *J. Acoust. Soc. Am.* **47**:683–693 (1970).

14. L. Cremer, "Theorie des Klopfschalles bei Decken mit schwimmendem Estrich," *Acustica* **2**:167–178 (1952).

15. R. D. Ford, D. C. Hothersall, and A. C. C. Warnock, "The Impact Insulation Assessment of Covered Concrete Floors," *J. Sound Vib.* **33**:103–115 (1974).

16. P. H. Parkin, H. J. Purkis, and W. E. Scholes, *Field Measurements of Sound Insulation Between Dwellings*, (Her Majesty's Stationery Office, London, 1960).

25

Reprinted from pp. 194–214 of *Natl. Bur. Stand. Sci. Papers*
20(506):193–219 (1925)

THEORY AND INTERPRETATION OF EXPERIMENTS ON THE TRANSMISSION OF SOUND THROUGH PARTITION WALLS

By Edgar Buckingham

I. INTRODUCTION

Our scientific knowledge of the acoustic properties of closed rooms is due almost entirely to the work of W. C. Sabine, who was the first to apply trained common sense to the practical problems of architectural acoustics. In the course of 20 years of indefatigable labor, he developed methods of experimentation and the mathematical theory needed for interpreting the observations, beside obtaining the greater part of the quantitative data now available to architects for their guidance in securing acoustically satisfactory results in the design of theatres, lecture rooms, etc., or in correcting the defects of existing structures.

Unfortunately, the papers published up to Sabine's untimely death give only a very incomplete account of his work; and, as related by Professor Lyman in his preface to Sabine's Collected Papers on Acoustics (Harvard University Press, 1922), much that had already been accomplished was apparently left unrecorded or has been lost, so that while the fundamental work on reverberation can be followed in considerable detail, the later experiments on the measurement of sound transmission are barely outlined and the mathematical treatment of them is not given at all.

So far as can be judged from his published papers, Sabine developed his equations step by step, as he needed them for coordinating his experimental results and interpreting them to an audience not composed of professional physicists; and in the form in which he left it, the treatment seems somewhat cumbersome. But having the complete experimental investigation before us, it is possible to present the mathematical aspect of the subject of reverberation in more compact form by deduction from certain simple assumptions, the approximate truth of which may be regarded as having been established by the experiments.

The equations needed for interpreting experiments on the transmission of sound through partition walls depend upon and follow naturally from the theory of reverberation. Sabine doubtless developed this part of the theory, but he did not publish it and, so far as the writer has discovered, no one else has done so. The primary purpose of the present paper is therefore to discuss the principles of certain methods of measuring transmission, and to give the appropriate equations. But since frequent reference to the equations for reverberation is unavoidable, they will be given first, although the excellent paper by E. A. Eckhardt in the Journal of the Franklin Institute for June. 1923, makes it unnecessary to discuss them at length.

II. THE FUNDAMENTAL IDEA OF THE THEORY OF REVERBERATION

In the paper entitled "On the Absorbing Power of Wall Surfaces," Sabine states the following three general propositions regarding the dying out of sound in closed rooms after the source has ceased to emit sound:

(*a*) The duration of audibility of the residual sound is nearly the same in all parts of an auditorium.

(*b*) The duration of audibility is nearly independent of the position of the source.

(*c*) The efficiency of an absorbent in reducing the duration of the residual sound is, under ordinary circumstances, nearly independent of its position.

These three propositions contain the gist of the theory. In connection with the known fact that the loudness of a sound of given pitch and quality increases with the rate at which sound energy reaches the observer's ears, the first two show that, wherever the source may have been, the sound waves are very soon so distributed through the room by successive reflections from the walls—including. in this term the floor, ceiling, and all exposed surfaces of furniture, etc.—that the amount of sound energy falling on any small object in the room is nearly the same from all directions and for all locations. The third proposition corroborates this conclusion by reference to an independent receiver—the absorbent. Since the ear can not perceive very rapid variations of intensity, the conclusion of course refers only to time averages over a finite though short interval, and is not valid for each separate instant.

If the sound emitted by the source were instantaneously so effectively diffused and mixed up by reflection that the room at once became uniformly filled with sound energy and that the flow of energy at any point was the same in all directions, the three propositions quoted would be not "nearly" but exactly true. And they can not be "nearly" true, as experiment showed them to be, unless the actual state of affairs, so far as it affects the ear, is an approximation to the ideal just mentioned.

The assumption of complete uniformity and diffuseness is therefore a safe starting point for an approximate theory, no exact theory of such excessively complex phenomena being at all possible of attainment.

III. REMARKS ON THE FUNDAMENTAL ASSUMPTION

Before proceeding to mathematical developments, it is well to consider the physical meaning of the assumption and to inquire how reality differs from the ideal and what circumstances may be expected to make the approximation better or worse. Echoes and interference due to regular reflection will be discussed later, but at first we shall proceed as if sound were reflected diffusely, as light is from a perfectly matt surface.

Let us suppose, to begin with, that the whole internal surface of the room in question is perfectly nonabsorbent, so that if the source keeps on sounding, the total amount of sound energy in the air of the room continually increases. At any instant the volume density of energy will evidently be, on the whole, greater near the source than farther away, and all directions will not be quite equivalent.

After a certain amount of sound energy has thus been given out, let the emission of the source be cut off. Since the sound waves already started are not weakened by absorption, they continue indefinitely to be reflected back and forth, and become more and more mixed up and diffused. The initial influence of the positions of the source and the point of observation is gradually obliterated by the successive reflections, and the state of affairs approaches the ideal, perfectly uniform and diffuse distribution of energy.

Since all real surfaces absorb sound to some extent, the theoretically infinite number of reflections needed for perfect attainment of the ideal state can not take place. But smooth rigid surfaces reflect sound more completely than even the best mirrors reflect light; and in a room with such walls, a great many reflections do occur before a moderately loud sound dies down so far as to be inaudible, so that the approach to the ideal state is closer than might be expected at first sight. Moreover, the speed of sound is so high that in a room of moderate size a great many reflections occur in a short time, and the approach to the ideal state is rapid.

If a continuous source of sound is started in any real, and therefore absorbent, room, it gradually fills the room with sound of increasing intensity until the increasing absorption by the walls just balances the emission of the source and a steady state is established. The energy within any volume element of the room may then be regarded as the sum of a nearly uniform and diffuse part due to sound waves which were emitted some time ago and have already been reflected a great many times, and a nonuniform component due to waves which have been emitted recently and have suffered only a few reflections or none at all. This latter part will, on the whole, be directed away from the source.

The relative importance of these two parts depends on the absorbing power of the room. If the walls were perfectly absorbent and did not reflect at all, there would be no uniform part; all the sound received at any place in the room would come from the direction of the source, and its intensity would fall off with the inverse square of the distance, just as if there were no walls and the sound were being emitted in the open air. The higher the reflecting power of the interior surface of the room, the less important is the nonuniform component in relation to the whole, and the more nearly will the ideal state be approached.

If the emission of the source ceases after a steady state has been set up, the nonuniform component, due to recent emission and early reflection, automatically passes over into the other; and as the residual sound dies down, it becomes more and more nearly uniform and perfectly diffuse.

These elementary conclusions have next to be somewhat modified by the consideration of regular reflection. If the interior surface of a room were, in whole or in part, of highly polished silver, a source of light in the room would produce very different degrees of illumination on a white screen placed at various points and turned in various directions. Reflection from the walls would produce bright and dark regions, which would shift about if the source were moved.

Something analogous is often observed with sound; and in large empty rooms with highly reflecting walls, the loudness with which a given source of sound is heard may vary considerably when either the source or the observer changes position. The analogy of light suggests the examination of such cases by means of the conception of sound rays and a law of reflection like that for light rays, and this method may be very useful in some cases; but it is liable to be misleading and should be used with caution.

Completely regular reflection, whether of sound or of light, occurs only from surfaces which are large compared to the wave length, in both linear dimensions and radius of curvature. As the surface is made smaller and smaller, a larger and larger fraction of the incident energy is scattered, or reflected diffusely, until, when the dimensions of the surface are of the same order of magnitude as the wave length, most of the incident energy which is not absorbed is scattered, and very little is sent back according to the law of regular reflection.

Now, the note an octave above middle C, which is somewhere near the average of the pitches that are of most practical interest and importance, has a wave length of about 2 feet. Hence, although notes of this pitch may be reflected quite regularly from a large flat wall or ceiling, small isolated surfaces, such as the back of a chair, a column, or the narrow front of a balcony, will not give much regular reflection but will scatter the sound waves that strike them and tend to make the distribution of sound through the room more uniform. Decorative elements, such as pilasters, cornices, and coffered ceilings, tend to produce a similar scattering and reduce the amount of regular reflection, so that it is usually much less disturbing than might be expected offhand. Nevertheless, the effects of regular reflection are often quite appreciable, so that the loudness with which a constant source is heard and the duration of the residual sound, are not quite the same in all parts of the room but only "nearly" so.

The foregoing leads directly to the consideration of interference. With a constant source of fixed pitch and quality, regular reflection results in interferences and gives rise to maxima and minima of intensity distributed in a fixed pattern throughout the room. The sound impulses which produce the resultant effect at a particular point in the room will have arrived at that point after different numbers of reflections and some will have been more weakened by absorption than others, because the different parts of the interior surface of the room do not all reflect equally well. Hence, if the emission is stopped and the residual sound is left to die out, the interference pattern changes and shifts about in a very complicated and quite unpredictable manner, and the decay of sound intensity at any one point is not regular but rapidly fluctuating. The same sort of thing happens during the period when the source has been started but has not yet built up the steady state in which absorption just balances emission.

For this reason, a receiving instrument which responded instantaneously to variations of intensity within a very small region would give records so complicated as to be unintelligible. Observations on residual sound have, therefore, to be made by a slower instrument which will average a good many fluctuations without showing their details, or else by the ear which has this same power of averaging by reason of the persistence of sensation.

During the steady state, the difficulty of the fixed interference pattern remains, and observations at a fixed point by an instrument of small dimensions, even if equally sensitive in all directions, can not be relied on to tell anything about the average intensity of the sound throughout any large region. The advantages of ear observations, in which the observer gets an average impression from two points and can readily move his head about so as to get a space average, led Sabine very early in his investigations to abandon the use of artificial receiving instruments and plan the experiments so that direct measurements of intensity were unnecessary and time measurements of the duration of audibility were sufficient. Now that the initial obscurities of the subject have been cleared up, more attention may profitably be devoted to perfecting purely instrumental methods of observation which, if they can be made satisfactory and reliable, will immensely decrease the effort demanded of the experimenter and save a great deal of time.

It is evident from the foregoing discussion that the assumption that the sound energy in a closed room is always uniformly distributed in space and perfectly diffuse as regards direction of propagation, is very far from being true in instantaneous detail. But

Sabine's investigations showed that with sufficient skill and patience it is possible to obtain average results over small regions and short intervals of time, which differ but little from those that would be obtained if the assumption were strictly true; and his theoretical treatment of reverberation while different in appearance is the same in substance as the following deductive treatment which starts from the assumption of perfect uniformity and diffuseness.

IV. NOTATION, DEFINITIONS, AND ASSUMPTIONS

The following notation and nomenclature will be adopted:

C = the speed of sound at the temperature of the air in the room, which is assumed to be uniform.

E = the power of the source, or the rate at which it gives out energy in the form of sound waves.

V = the volume of the room.

S = the internal exposed area of the room and its contents.

α = the absorption coefficient or absorptivity of any element dS. It is defined as the fraction of the energy falling on dS which is not thrown back into the room but is either dissipated into heat in the substance of the wall or transmitted and given out elsewhere, outside the room.

a = the absorbing power or absorptance of the whole room, with its contents, defined by the equation

$$a = \int_0^s \alpha \, dS = \bar{\alpha}S \tag{1}$$

where

$\bar{\alpha}$ = the average absorptivity of the interior exposed surface.

ρ = the volume density of the sound energy at any time t; by the fundamental assumption discussed above, ρ is the same everywhere.

ρ_0 = the steady value of ρ when the source has already been emitting at the constant rate E for a long time.

ρ_m = the least density that produces any sensation in the ear or the density at the limit of audibility.

r = the reverberation time of the room. Following Sabine, it is defined as the time required for the residual sound, remaining after the emission of the source has been cut off, to decrease to one millionth of its initial intensity, or the time required for the energy density to decrease in the ratio 10^{-6}.

$b = \dfrac{Ca}{4V}$; Sabine calls this the "rate of decay of the sound."

$\epsilon = \dfrac{\rho C}{4}$; it is the rate at which sound energy strikes unit area of the walls.

41732°—25——2

The absorptivity of a solid surface depends on the pitch of the sound and is usually greater for high than for low notes, although resonance may cause marked exceptions to this general rule. In order, therefore, to give α, a, and $\bar{\alpha}$ definite meanings, it will be supposed that the source emits a pure musical note of fixed pitch.

It will also be assumed that absorptivity does not depend on the intensity of the incident sound, so that α is a constant for sounds of a given pitch. It is quite conceivable that this assumption may not be very accurately true, especially for soft, highly absorbent surfaces; but no variation with intensity has been noted and, to the degree of accuracy hitherto attained in the experiments which the theory undertakes to represent, any such effect may safely be neglected.

The air in the room has certain natural frequencies of free vibration, and the same is true of the walls, furniture, etc., though these latter vibrations will usually be strongly damped. The room and the air constitute an imperfectly elastic connected system, capable of being set into very complicated states of vibration by suitably timed sound impulses from the source. Such a system may respond strongly, by resonance, to notes near certain frequencies, and when this happens, the system may react on the source to change its frequency. Moreover, vibrations of other frequencies than that of the exciting impulses may be set up in the system, and energy transferred back and forth among the various modes of vibration. The result is that though the source may give out a pure note, the sound in the room may be a confused mixture of notes of various pitches, which are not simply related and are differently absorbed by the walls.

In theory, such a state of affairs will always exist if the frequency of the source is close to any one of the numerous natural frequencies of the system; but usually there are only a very few notes, within the practically important range of pitch, to which the system responds strongly enough to make these resonance effects seriously disturbing. It will be postulated that the notes to which the theory refers are not within the narrow critical regions of pitch where appreciable resonance occurs; and upon this condition, each element of a compound sound may be regarded as separately subject to the theory, with different values of E and α for each component, so that the total result is the summation of the results for the separate elements.

V. THE RATE AT WHICH SOUND ENERGY STRIKES THE WALLS OF THE ROOM

In Figure 1, let dS be an element of area of the wall of a room in which the energy density ρ is uniform and perfectly diffuse; and let dV be a volume element, at the distance r from dS, in a direction which makes the angle φ with the interior normal n.

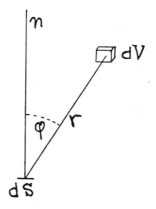

Fig. 1

Of the energy ρdV which is contained within $d\dot{V}$ at any instant, that portion will ultimately strike dS which is moving in directions included within the solid angle $d\omega = dS \cos \varphi / r^2$ subtended by dS at dV. This fraction is $d\omega/4\pi$, so that the amount of energy inside dV which will ultimately strike dS is

$$\frac{\rho \, dVdS \, \cos\varphi}{4\pi r^2} \tag{2}$$

Sound waves leaving dV toward dS will reach dS within one second, if the distance r to be traversed is not greater than the speed of propagation C. Hence the total amount of energy that falls on dS in one second from all possible directions is the sum of the values of the expression (2) for all the volume elements of a hemisphere of radius C described about dS as a center.

Setting

$$dV = 2\pi r^2 \sin\varphi \, drd\varphi \tag{3}$$

substituting in (2), and integrating, we have

$$\frac{\rho ds}{2} \int_0^C dr \int_0^{\pi/2} \sin \varphi \cos \varphi d\varphi = \frac{\rho CdS}{4} \tag{4}$$

and the rate at which energy of sound waves falls on unit area of the walls is

$$\frac{\rho C}{4} = \epsilon \tag{5}$$

For slightly different deductions of this result the reader may be referred to papers by G. Jaeger, Wiener Sitzungsberichte, May, 1911; and E. A. Eckhardt, Journal of the Franklin Institute, June, 1923.

VI. THE GROWTH AND DECAY OF SOUND IN A CLOSED ROOM

The whole amount of sound energy in the room at any instant is $V\rho$, and the rate at which it increases is the difference between the rate of supply from the source and the rate of absorption by the walls. By (1) and (5) the rate of absorption is

$$\int_0^S \epsilon\, \alpha dS = \epsilon a = \frac{C\rho}{4} a = bV\rho \tag{6}$$

so that we have the equation

$$V\frac{d\rho}{dt} = E - bV\rho \tag{7}$$

in which V and b are constants for a given room, pitch, and temperature. It is assumed here that the dissipation of sound energy into heat in the air, by viscosity, conduction, and radiation, is negligible, as Sabine showed it to be in practice.

If the emission E is constant, the result of integrating (7) from t_1, to t is

$$\log\frac{E - bV\rho_1}{E - bV\rho} = b(t - t_1) \tag{8}$$

where ρ_1 is the value of ρ at the time t_1.

The energy density t seconds after the source has started emitting in the previously quiet room, is found from the general equation (8) by setting $\rho_1 = t_1 = o$ and solving for ρ; and setting $t = \infty$ in the resulting expression gives the final steady value ρ_0. The residual density t seconds after the emission has ceased, is found by setting $E = o$, $\rho_1 = \rho_0$, $t_1 = o$ and solving for ρ. The results are as follows: During the initial period of growth

$$\frac{\rho}{\rho_0} = 1 - e^{-bt} \tag{9}$$

in the steady state

$$\rho_0 = \frac{E}{bV} = \frac{4E}{aC} \tag{10}$$

during the decay of the residual sound

$$\frac{\rho}{\rho_0} = e^{-bt} \tag{11}$$

Since $b = aC/4V$, equation (9) shows that the larger the room and the smaller its absorbing power, the slower it is in filling up with sound; and (11) shows a similar slowness in the decay of the residual sound after the emission has been stopped. Equation (10) shows that the steady intensity is proportional to E/a and independent of the volume, except in so far as an increase of volume usually increases the absorbing area.

VII. THE MEASUREMENT OF ABSORPTION

Substituting $b = aC/4V$ in (11) and solving for a gives the equation

$$a = \frac{4V}{tC} \log \frac{\rho_0}{\rho} \tag{12}$$

but this suggestion for determining a from simultaneous values of t and ρ/ρ_0 is worthless because of the lack of a satisfactory method of measuring ρ/ρ_0 directly. The suggestion contained in equation (10) or

$$a = \frac{4E}{C\rho_0} \tag{13}$$

suffers from the same defect; what is needed is a method which does not require a measurement of ρ/ρ_0 or E/ρ_0, and Sabine's ingenuity provided this method. It presupposes only that the power of the source E can be varied in a known ratio, and this requirement was met by using, as a source, different combinations of identical organ pipes, placed far enough apart in the room that it could be assumed that they emitted independently and that the total power of any combination was the sum of the powers of the separate pipes used in that combination.

Let t_1 be the duration of audibility of the residual sound when the initial steady intensity is $\rho_0 = \rho_1$; and let t_2 be the corresponding time when $\rho_0 = \rho_2$, the final intensity in both experiments being that of minimum audibility or ρ_m. Then by (11)

$$\frac{\rho_m}{\rho_1} = e^{-bt_1}; \qquad \frac{\rho_m}{\rho_2} = e^{-bt_2} \tag{14}$$

whence

$$\frac{\rho_1}{\rho_2} = e^{b(t_1 - t_2)} \tag{15}$$

or

$$b = \frac{1}{t_1 - t_2} \log \frac{\rho_1}{\rho_2} \tag{16}$$

But if, in each experiment, the source had been sounding long enough to establish a steady state, $\rho_1/\rho_2 = E_1/E_2$, by equation (10);

and the value of this is known, having been fixed beforehand. Hence, substituting $b = aC/4V$ and solving (16) for a, the result is

$$a = \frac{4V}{C} \log \frac{E_1}{E_2} \frac{1}{t_1 - t_2} \qquad (17)$$

in which C is a known constant; V is obtained simply from geometrical measurements of the room; E_1/E_2 is known from the arbitrarily chosen ratio of the powers of the two sources; and only the two durations of audibility t_1 and t_2 have to be determined by the experiments.

This method of determining the absorbing power of a room dispenses with measurements of power E or energy density ρ, and requires only time measurements by the ear and a chronograph. It assumes that the intensity for minimum audibility ρ_m is the same in both experiments, and Sabine's experiments showed that it was surprisingly constant for any one observer. An observer whose hearing is more acute hears the residual sound a little longer, but by the same amount in each case, so that $(t_1 - t_2)$ is unaffected and a is correctly determined without regard to the sensitiveness of the observer's ears, so long as the same observer makes both experiments.

Sabine's first attack on the problem of measuring absorption was by the much simpler and more obvious method of substitution, which can not be used for a whole room but is applicable to objects that can be brought in or carried out. If the substitution of one thing for another does not change the duration of the residual sound from a given fixed source, the two bodies are obviously equivalent as regards absorption, and the absorptivities of their surfaces are inversely proportional to their exposed areas. Comparisons of the absorptivities of different surfaces can thus be effected, and absolute values may be obtained by comparison with known areas of open window, which reflect nothing and so provide a standard of unit absorptivity.[1]

The details of this and other methods of measuring absorptivity can be studied in Sabine's papers far better than anywhere else, and there is no need of discussing them here. For the present purpose of outlining the mathematical side of the subject, it suffices to note that the results confirm the assumption that, aside from the rapid fluctuations due to changing interference patterns, the sound energy is distributed nearly uniformly and diffusely through the room during the period of decay to inaudibility.

---

[1] Dr. Eckhardt has pointed out to me that an assumption is implied here. It is very natural to assume that the amount of sound energy which escapes through an open window, from a room in which the energy density is uniform and diffuse, is proportional to the area of the opening. But when the dimensions of the opening are comparable with the wave length, the opening has to be considered as consisting of edges as well as area. In other words, diffraction may falsify the assumption. This point requires further study.

VIII. MEASUREMENT OF THE POWER OF A SOURCE IN TERMS OF THE MINIMUM AUDIBLE INTENSITY

If t_a is the duration of audibility of the residual sound which had the initial steady intensity ρ_0 due to the continued emission of a source of strength or power E, equation (11) gives us

$$o_o = \rho_m e^{\frac{aC}{4v}} t_a \tag{18}$$

and by (10)

$$E_1 = \frac{aC\rho_m}{4} e^{\frac{aC}{4v}} t_a \tag{19}$$

Hence, if the volume of the room is known and its absorbing power a has been determined, a measurement of t_a permits of computing the value of E in terms of ρ_m, though the result is very sensitive to errors in t_a. To reduce the effect of a given absolute error in the determination of the time t_a, it is evidently well to use a room of large volume and low absorbing power, since both these properties tend to increase the duration t_a and so diminish the error in E.

IX. THE REVERBERATION TIME OF A ROOM

This is defined as the duration of audibility of a sound which had initially the standard intensity 10^6 times the minimum audible intensity. Denoting this time by r and setting $\rho_0 = 10^6 \rho_m$ in equation (11) gives us

$$10^6 = e^{br}$$

whence

$$r = \frac{6 \log_e 10}{b} = \frac{13.82}{b} = \frac{55.3\,V}{aC} \tag{20}$$

The commonest acoustic defect of modern, hard-surfaced, lecture rooms and auditoriums is the overlapping and confusion of successive sounds, such as the syllables of a sentence, by too great reverberation. Equation (20) shows that the remedy is to increase the absorbing power of the room, and it permits of computing the reverberation time, in advance of construction, from the known absorptivities of the materials of which the exposed internal surface is to consist.

X. THE TESTING OF SOUND-INSULATING PARTITION WALLS

The obvious procedure for testing the power of a certain kind of wall to transmit sound from the air on one side to that on the other, is to set the wall up as a partition between two rooms, produce a sound in one, and compare the intensity of the sound transmitted to the second room with the intensity of the original sound. This idea leads to the type of installation adopted by W. C. Sabine for the Riverbank Laboratories and described by Paul E. Sabine in The

American Architect for July 30, 1919, in which two closed rooms are in acoustic communication through the wall under investigation but are otherwise as nearly as possible completely insulated and soundproof, both toward each other and toward their surroundings. The same general scheme has been followed in the sound laboratory of the Bureau of Standards, and it is illustrated by Figure 2.

The piece of wall upon which the experiments are to be made is set up as a panel *P* in the otherwise soundproof partition which separates the "sound chamber" *I* from the "test chamber" *II*. The source of sound *O* is in room *I*, while room *II* receives sound

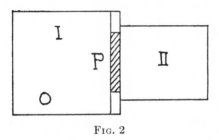

Fɪɢ. 2

only by transmission from the air in room *I* through the panel; and the purpose of the observations is to determine the ratio of the intensities in the two rooms when the source is emitting at a constant rate. This ratio, however, depends on the properties of the rooms as well as those of the panel, so that a certain amount of mathematical theory is needed in interpreting the observations so as to give a result which is characteristic of the panel alone and independent of the peculiarities of the laboratory where it is tested.

XI. SUPPLEMENTARY DEFINITIONS, ETC.

When sound waves fall on the panel, a part of their energy is reflected and a part is taken up by the panel and disposed of in several ways. In the first place, if the panel is porous, energy is absorbed and dissipated into heat by the heavily damped waves which run into the air contained in the pores. If there are cracks or holes of appreciable size, sound waves pass through the air in these passages from one room to the other, but we shall assume that the partition is tight, so that this direct leakage of sound through air passages is negligible.

A second part of the energy received by the panel goes to setting up true sound waves; that is, longitudinal elastic vibrations, in the solid material. These are propagated perpendicularly through the panel and affect the air on the farther side; but on account of the great difference of density of the air and the panel, the energy of these true sound waves in the panel must always be extremely small, and its influence on the transmitting power of the panel must be absolutely negligible under all ordinary circumstances.

The third thing to be considered is bodily motion of the central parts of the panel, which is forced, by the varying air pressure on one side, to bend back and forth and so sets up vibrations of the same frequency in the air on the other side. The panel acts as an imperfectly elastic plate, which is more or less rigidly supported at its edges and is set into forced vibration. Energy is dissipated in the plate and at the imperfectly supported edges, and some is conducted away laterally, but a certain residue remains and is transmitted and given out as sound to the air in the test chamber. It is this transmitted residue of the energy absorbed by the panel with which we are now concerned.

The fraction of the energy falling on the panel which is thus transmitted will be called the transmissivity of the panel and denoted by τ; and the quantity

$$T = \tau S \tag{21}$$

where S is the area of one face of the panel, will be called the transmitting power or transmittance of the panel.

Unless the panel is perfectly inelastic, it will have certain natural frequencies of free vibration and it will respond to sounds of these frequencies and transmit them more freely than sounds of other frequencies. The transmittance thus varies with the pitch of the sound, and to make T and τ definite it will be supposed, as in the consideration of reverberation, that the source employed emits a single pure note of fixed pitch.

It is quite conceivable that the transmittance of a panel may vary with the intensity of the sound incident upon it, faint sounds being more perfectly transmitted than loud ones of the same pitch, or vice versa; and it seems by no means safe to assume that τ and T will always be independent of the intensity of the sound; that is, of the amplitude of the motion of the panel. For the present, however, we shall assume that there is no such variation with intensity, and that τ and T are constant for any one pitch.

The two rooms, with their inclosed masses of air and the panel between them, form an even more complicated vibrating system than a single room. Two different and nearly independent kinds of resonance are possible in this system—resonance of the panel, which would be nearly the same in one laboratory as in another, and resonances of the two rooms, which are peculiar to the laboratory and are not greatly affected by the properties of the panel. It will be stipulated that the pitch of the source used shall not be such as to excite strong resonance in either of the two rooms; but no restriction is placed on the pitch, as regards the free periods of the panel.

The symbols V_1, S_1, a_1, b_1, etc., shall have, for room I, the meanings explained in section 4; and V_2, S_2, etc., shall denote the corresponding quantities for room II. The steady energy densities, after the source in room I has been emitting at the constant rate E for a long time shall be denoted by ρ_{01} and ρ_{02}.

The absorbing powers a_1 and a_2 and the reverberation times r_1 and r_2 are to be understood as referring to the properties of the two rooms with the panel in place, S_1 and S_2 including the area S of one face of the panel.

XII. STATIC DETERMINATION OF TRANSMITTANCE

If the source has been sounding long enough to establish a steady state in both rooms, the time rate at which sound energy strikes the face of the panel in the sound chamber is $S\rho_{01}C/4$ (see equation 5); and the rate at which the panel gives out energy to the air of the test chamber is

$$\tau \frac{S\rho_{01}C}{4} = \frac{\rho_0 CT}{4} \tag{22}$$

The rate of absorption by the walls of room II is

$$\frac{1}{4} S_2\rho_{02}C\overline{\alpha_2} = \frac{1}{4}\rho_{02}Ca_2 \tag{23}$$

and since transmission and absorption are equal, in the steady state, we have by (22) and (23)

$$\frac{T}{a_2} = \frac{\rho_{02}}{\rho_{01}} \tag{24}$$

Instrumental devices for measuring the ratio of the two steady intensities ρ_{01} and ρ_{02} need not be discussed here; but if such devices can be perfected, equation (24) offers the simplest imaginable means for computing the transmittance of the panel from the absorbing power of the test room, which may be assumed to be already known. Experiments by this method with different absolute values of the intensity ρ_{01} would answer the question whether the transmittance did or did not vary appreciably with the intensity of the sound.

If quantitative receiving instruments can not be employed in such a way as to give satisfactory mean values of ρ_{02}/ρ_{01} which are independent of interference patterns and loud spots due to regular reflection, the instrumental measurement has to be replaced by Sabine's procedure of measuring the duration of audibility of residual sound, and it becomes necessary to consider how the intensity of the sound in the two rooms dies down after the source in room I has ceased to emit.

XIII. THE DECAY OF THE RESIDUAL SOUNDS

If the emission of the source is cut off after a steady state has been established, the rate of decrease of the total energy in either room is equal to the rate of absorption by that room minus the rate at which energy is being received from the other room by transmission through the panel. We therefore have the simultaneous equations

$$-V_1 \frac{d\rho_1}{dt} = \frac{C\rho_1}{4} a_1 - \frac{C\rho_2}{4} T \tag{25}$$

$$-V_2 \frac{d\rho_2}{dt} = \frac{C\rho_2}{4} a_2 - \frac{C\rho_1}{4} T \tag{26}$$

or

$$\frac{d\rho_1}{dt} = \frac{b_1 T}{a_1} \rho_2 - b_1\rho_1 \tag{27}$$

$$\frac{d\rho_2}{dt} = \frac{b_2 T}{a_2} \rho_1 - b_2\rho_2 \tag{28}$$

which are to be solved subject to the condition that at $t=0$ (see equation 24),

$$\left(\frac{\rho_2}{\rho_1}\right)_{t=0} = \frac{\rho_{02}}{\rho_{01}} = \frac{T}{a_2} \tag{29}$$

In the practical testing of sound-insulating partitions, the transmittance is small and the amount of energy transmitted from the sound chamber to the test chamber and then back again will be too small to have any appreciable effect on the intensity in the sound chamber, or on the duration of audibility of the residual sound in it. This means that the term "$b_1 T\rho_2/a_1$" may be omitted from equation (27) and that, to a very close approximation, we have

$$\frac{d\rho_1}{dt} = -b_1\rho_1 \tag{30}$$

or, after integrating from 0 to t,

$$\rho_1 = \rho_{01}e^{-b_1 t} \tag{31}$$

Substituting this value of ρ_1 in (28) we have

$$\frac{d\rho_2}{dt} = me^{-b_1 t} - b_2\rho_2 \tag{32}$$

where

$$m = \frac{b_2 T}{a_2} \rho_{01} \tag{33}$$

and upon substituting the new variable

$$y = \frac{e^{-b_1 t}}{\rho_2} \tag{34}$$

equation (32) may be thrown into the form

$$\frac{dy}{y(b_1 - b_2 + my)} = -dt \tag{35}$$

Integrating from o to t, eliminating y by means of (34), solving for ρ_2, and setting $\rho_{02} = \rho_{01} T/a_2$ (equation 29) gives us

$$\rho_2 = \left(\rho_{01} \frac{T}{a_2} + \frac{m}{b_1 - b_2} \right) e^{-b_2 t} - \frac{m}{b_1 - b_2} e^{-b_1 t} \tag{36}$$

which, with (31), constitutes the solution of equations (27) and (28), on the supposition that the influence of the second room on the first is negligible.

It may be noted here that the T which appears in (28), (29), (33), and (36) is the transmittance T_{12} from room I to room II. The transmittance T_{21}, in the opposite direction disappeared when the term $(b_1 T \rho_2 / a_1)$ was dropped from equation (27), and it has therefore *not* been assumed that the transmittance is the same in both directions, but only that it is small.

XIV. THE RELATION OF TRANSMITTANCE TO DURATION OF AUDIBILITY

Let t_1 and t_2 be the durations of audibility of the residual sounds left in rooms I and II, respectively, after the emission of the source in room I has ceased, so that we have

$$\left. \begin{array}{l} \rho_1 = \rho_m \text{ at } t = t_1 \\ \rho_2 = \rho_m \text{ at } t = t_2 \end{array} \right\} \tag{37}$$

Substituting these values in (31) and (36) and equating the results gives us the equation

$$\rho_{01} e^{-b_1 t_1} = \left(\rho_{01} \frac{T}{a_2} + \frac{m}{b_1 - b_2} \right) e^{-b_2 t_2} - \frac{m}{b_1 - b_2} e^{-b_1 t_2} \tag{38}$$

and, after eliminating m by means of (33) and solving for T/a_2, we have

$$\frac{T}{a_2} = \frac{(b_1 - b_2) e^{-b_1 t_1}}{b_1 e^{-b_2 t_2} - b_2 e^{-b_1 t_2}} \tag{39}$$

If the two rooms have the same reverberation time, so that $b_1 = b_2$, equation (39) becomes indeterminate; but the difficulty may be avoided by returning to (35), which now reduces to the form

$$-\frac{dy}{y^2} = m\,dt \qquad (40)$$

Integrating from o to t and eliminating y, y_0, and m, we have

$$\rho_2 = \rho_{01}\frac{T}{a_2}(1 + b_2 t)e^{-b_1 t} \qquad (41)$$

in place of (36).

Setting $b_1 = b_2 = b$, we have by (31) and (41)

$$\rho_m = \rho_{01}e^{-bt_1} = \rho_{01}\frac{T}{a_2}(1 + bt_2)e^{-bt_2} \qquad (42)$$

whence

$$\frac{T}{a_2} = \frac{e^{-b(t_1 - t_2)}}{1 + bt_2} \qquad (43)$$

which may be compared with the more general equation (39) which is applicable when the reverberation times of the rooms are different.

In the case of a very weak or flexible panel, the transmittance may be so high that the influence of room *II* on room *I* is no longer negligible.   Equations (27) and (28) in the forms

$$\left.\begin{array}{l} \dfrac{d\rho_1}{dt_1} = \dfrac{b_1 T_{21}}{a_1}\,\rho_2 - b_1\rho_1 \\[3mm] \dfrac{d\rho_2}{dt_2} = \dfrac{b_2 T_{12}}{a_2}\,\rho_1 - b_2\rho_2 \end{array}\right\} \qquad (44)$$

must then be solved generally, subject to the condition

$$\frac{\rho_{02}}{\rho_{01}} = \frac{T_{12}}{a_2} \qquad (45)$$

and the result which corresponds to equation (38) is

$$e^{-b_1 t_1}(D_1 K_2 e^{D_2 B t_1} - D_2 K_1 e^{D_1 B t_1}) = e^{-b_1 t_2}(K_1 e^{D_1 B t_2} - K_2 e^{D_2 B t_2}) \qquad (46)$$

where

$$B = \frac{b_2 T_{12}}{a_2} = \frac{C T_{12}}{4 V_2} \qquad (47)$$

$$D_1,\ D_2 = \frac{b_1 - b_2}{2B} \pm \sqrt{\left(\frac{b_1 - b_2}{2B}\right)^2 + \frac{V_2 T_{21}}{V_1 T_{12}}} \qquad (48)$$

$$\left.\begin{array}{l} K_1 = \rho_{01}\left(1 + D_1\dfrac{T_{12}}{a_2}\right) \\[3mm] K_2 = \rho_{01}\left(1 + D_2\dfrac{T_{12}}{a_2}\right) \end{array}\right\} \qquad (49)$$

When the substitutions indicated by equations (47) to (49) are made in (46), the resulting equation is so complicated that it appears impossible to obtain a solution corresponding to (39) or (43). While it would be interesting to have this solution for the case of high transmittance, the matter is fortunately of no great importance, because in the practical testing of sound-insulating partitions, the solutions (39) and (43) are sufficiently approximate.

XV. THE DETERMINATION OF TRANSMITTANCE FROM MEASUREMENTS OF DURATION OF AUDIBILITY

The values of the quantities

$$\left.\begin{array}{l} b_1 = \dfrac{Ca_1}{4V_1} = \dfrac{13.82}{r_1} \\[2mm] b_2 = \dfrac{Ca_2}{4V_2} = \dfrac{13.82}{r_2} \end{array}\right\} \tag{50}$$

may be computed from the dimensions of the rooms, if the absorptivities of their surfaces are known; or they may be found from measurement of the reverberation times r_1 and r_2 by using a source of known power (see equations 10 and 16). Assuming the values of b_1 and b_2 to be known, equations (39) and (43) are available for computing the value of T/a_2 from the observed values of t_1 and t_2.

1. ROOMS WITH EQUAL REVERBERATION TIMES

If the reverberation times are equal, the results may be reduced by means of (43) which is much the simpler of the two, and which may also be written in the form

$$\frac{T}{a_2} = \frac{e^{-13.82\frac{t_1-t_2}{r}}}{1+13.82\frac{t_2}{r}} \tag{51}$$

The condition is satisfied if the two rooms are identical and the absorptivity of the panel is the same on both sides; but even if the rooms are somewhat different, the longer reverberation time may be brought down to equality with the shorter by introducing absorbent materials, and the condition for the validity of (43) or (51) may thus be fulfilled.

To minimize the effect of a given absolute error in the times, t_1, t_2, and r should evidently be made as long as possible, which means that the rooms should be large and have highly reflecting walls, and that the initial steady intensity ρ_{01} should be made large by using a powerful source.

There then remains the final question whether the panel should be large or small, and to this there is no definite answer, because there are conflicting requirements. In the steady state, and to some extent after the emission of the source has ceased, the panel acts as a source for room *II*; and in order to satisfy the condition assumed in the theory, viz, that the sound shall be uniform and diffuse throughout the room, the area of the panel should evidently be as small as practicable in relation to the whole wall area. But on the other hand, if *S*, and therefore $\tau S = T$, is very small, the intensity in the test chamber will always be small, the duration of audibility will be short, and the percentage error in t_2 will be large. Some sort of compromise is necessary but there does not seem to be any a priori method for selecting a best value of S/S_2.

There are, however, quite other considerations which make it apparent that measurements on very small panels are of no practical value, no matter how accurate they may be; but since they are equally applicable to all methods of measuring transmittance, and are, moreover, sufficiently obvious, they need not be discussed here.

2. ROOMS WITH UNEQUAL REVERBERATION TIMES

If the two rooms are so different that their reverberation times can not be made equal without both being short, it is necessary to revert to (39) which may also be put into the form

$$\frac{T}{a_2} = \frac{(r_1 - r_2)\, e^{-13.82 t_1/r_1}}{r_2 e^{-13.82 t_2/r_2} - r_1 e^{-13.82 t_2/r_1}} \tag{52}$$

And since this expression approaches o/o as r_1 and r_2 approach equality, the reverberation times should evidently be made very different in order to minimize the effects of errors in the times.

The question then arises whether the smaller or the larger room should be used as the test chamber, and a little consideration shows that the test chamber should be the one with the long reverberation time. For the initial steady intensity ρ_{02} will always be small compared to ρ_{01}, and unless r_2 is long, the duration of audibility of this initially weak sound in the test chamber will be so short that its value t_2 is liable to a very large percentage error, whereas the duration t_1 in room *I* will be of the same order of magnitude as r_1 and can be measured with the same accuracy.

Another line of reasoning also points to the desirability of putting the source in the smaller rather than in the larger room. In room *I*, the absorptivity of the panel will usually not be very different from that of the remainder of the walls and the presence of the panel will not cause any serious departure from the uniformity of energy distribution assumed in the theory. But in room *II*, the panel alone is acting as a source and, as remarked above, the uniformity of distribution will be improved if the area of the walls of the room is made large in comparison with the area of the panel.

On the whole, the conditions of dissimilarity which require the use of equation (39) or (52) do not appear favorable to accuracy of determinations by the present method. If the only two rooms available are of very different sizes, the larger should be used as the test chamber; but in building a new laboratory for testing transmission, the rooms should be designed for equal reverberation times and should be as large and nonabsorbent as practicable, under the imposed limitations of cost of construction.

[*Editor's Note:* Material has been omitted at this point.]

26

Reprinted with the permission of the Controller of Her Britannic Majesty's
Stationery Office from *J. Sound Vib.* **21**(4):405–429 (1972)

A STUDY OF MEASURING TECHNIQUES FOR AIRBORNE SOUND INSULATION IN BUILDINGS

R. F. HIGGINSON

Building Research Station, Garston, Watford WD2 7JR, England

(*Received* 15 *November* 1971, *and in revised form* 18 *February* 1972)

Some comparison measurements of airborne sound insulation have been made in a laboratory house, with 12 organizations taking part. Using their current procedures resulted in a large spread in the results obtained, due to differences in equipment and techniques. The various steps in the measurements have been studied in detail, to determine how discrepancies could arise, and to establish a rationalized technique by which they would be minimized. The measuring teams repeated the comparison exercise using such a technique, and a smaller spread in the results was found. The instrumentation was developed further and measurements were made between rooms of various sizes. The results were repeatable within small limits. The consequences for further standardization in the measuring procedure are discussed.

1. INTRODUCTION

Non-repeatability in measurements of sound transmission has been reported from a number of different places. Kihlman [1] and Scholes [2] were concerned with field measurements, while Schultz [3] referred to laboratory investigations. In all cases, measurements were made in accordance with the standard procedures [4, 5] which have changed little since they were first proposed [6].

The usual reason for the discrepancies is that the sound fields set up in the measuring rooms are not uniform, at least at low frequencies. Individual room modes are prominent and if these are not equally excited, or if the degree of excitation changes from one room to the other, the measured average sound pressure levels and level differences vary. In some cases coupling between the modes of vibration of the building structure and the acoustic modes of the rooms aggravates the difficulties. The whole problem is made worse if care and precision are not exercised in measuring the sound fields.

Possible shortcomings in the basic method of measurement were appreciated even in 1948, when it was put forward as a provisional code by an informal international working group. Since that time, measurements of sound transmission have assumed an increasing importance. Many national building codes now incorporate minimum requirements for sound insulation between dwellings. Moreover, in other types of building, specifications are being extended to include this subject. The measuring technique therefore needs to be reliable.

The Building Research Station has for many years been conducting field measurements of sound insulation [7]. The measuring difficulties in buildings are particularly severe because of the relatively small size of the rooms encountered (of the order of one acoustic wavelength or less at low frequencies) and because practical considerations rule out the use of sophisticated sound source arrangements and large diffusers. Hence a study has been made of field measurement techniques for airborne sound transmission in order to establish how they might be rationalized. From this study suggestions for revision of the standard are made.

The building used for the most part was a house of traditional brick construction. The investigations fell into four stages.

(i) Standard measurements of the airborne sound insulation between two neighbouring rooms in the house were obtained by several teams engaged in field measurements (consultants, university departments, etc.). They each used their own equipment and techniques. This showed the spread in results obtained by different teams measuring the same thing.

(ii) A detailed study was next made of the various constituent parts of the measurement method. The generation of the sound fields in the source and receiving rooms was investigated, and different methods of determining the mean sound pressure level differences between the rooms were tried. The method of measurement of reverberation time in the receiving room and the effect of variation of the total absorption in the room on normalization of the level difference were both studied. The influence of temperature differences between the measuring rooms was investigated.

(iii) Based on the findings from study (ii), a measurement procedure was devised to give repeatable results, and which still complied with the general method described in the standard. The same measuring teams that took part in study (i) then each did a repeat measurement using this procedure.

(iv) A modified version of the rationalized measuring technique was finally applied in buildings with rooms of different sizes.

It has been suggested that repeatability of measurements might be improved if they are obtained in octave bands rather than 1/3-octaves [8]. This is already done in the Netherlands [9]. However, in the studies reported here the terms of reference taken were those of the existing Building Regulations [10, 11] where 1/3-octaves are specified.

2. DESCRIPTION OF TEST BUILDING

The building used for the main part of the investigation was of traditional brick construction (see Plate 1). It had two storeys with four rooms on the upper floor and four on the lower. The room dimensions were 4·56 m long × 3·63 m wide × 2·52 m high. A floor plan of the

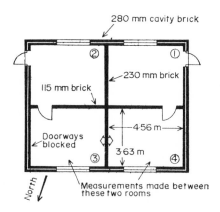

Figure 1. Floor plan of test building.

building is shown in Figure 1. The external walls were 280 mm cavity brickwork and the internal walls were 230 mm and 115 mm solid brick, all plastered. The ground floor was solid concrete; the first floor was of wood joist construction with tongue and grooved floorboards on top and skim-coated plasterboard ceiling underneath; the roof was timber-raftered and tiled.

[*Editor's Note:* Plate 1, a photograph of the test building; Plate 5, a photograph of a microphone on a turntable; and Plate 6, a photograph of a moving microphone, have been omitted due to limitations of space.]

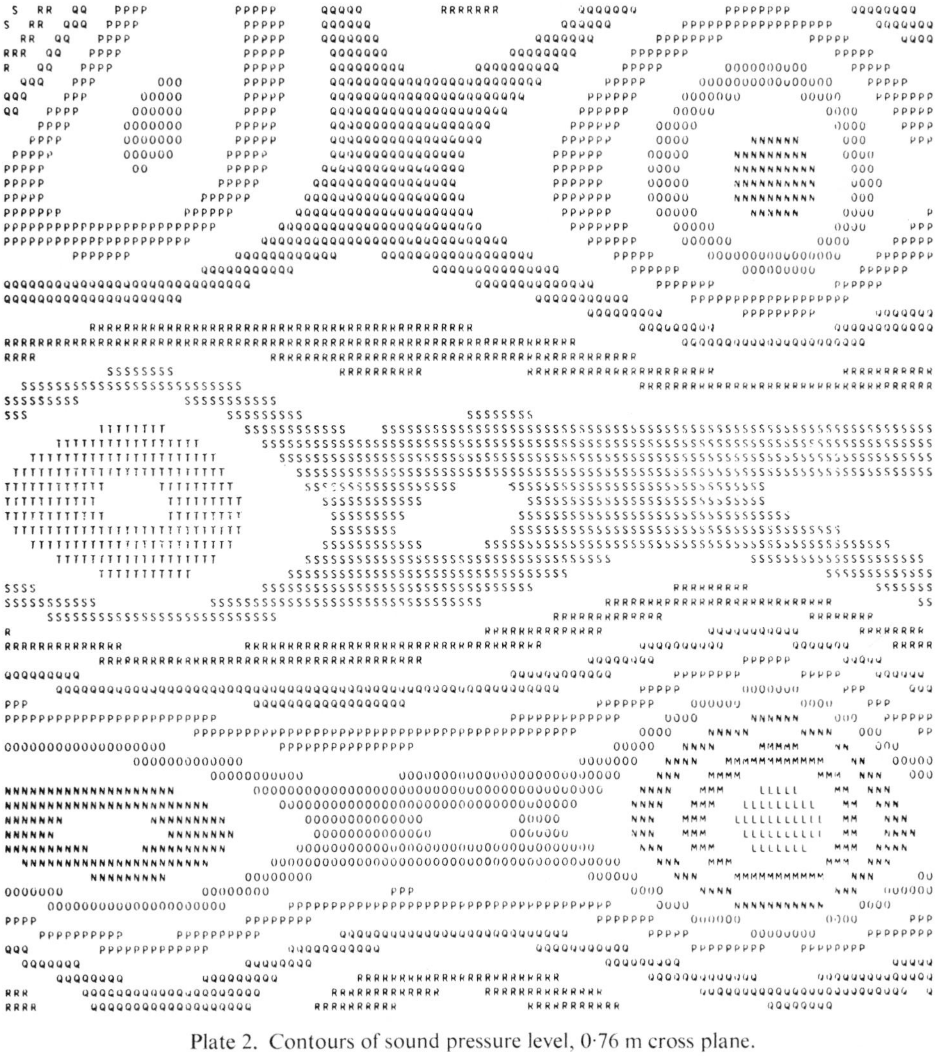

Plate 2. Contours of sound pressure level, 0·76 m cross plane.

The two rooms used for the measurements (numbered 3 and 4, see Figure 1) were on the ground floor, separated by the 230 mm solid brick wall. In order to minimize temperature fluctuations in the rooms a number of precautions were taken. First, the rooms were on the north side of the building. Second, the external access doorways were sealed with 115 mm brickwork having 12 mm plaster finish on the inside. Access was then made through the other ground floor rooms. Finally, the windows were double-glazed (with 4 mm glass and an air cavity of 190 mm) and covered on the inside with 12 mm plasterboard panels.

3. COMPARISON MEASUREMENTS OF SOUND INSULATION— CURRENT METHODS

3.1. INTRODUCTION

The standards [4, 5] describe a general procedure for the measurements. It consists, in the case of airborne sound insulation, of the setting up of a sound field "as diffuse as possible" in the transmitting room, and of the measurement, over a range of frequencies, of the average sound pressure level differences between the transmitting room and a neighbouring receiving room. The level differences are adjusted for the amount of sound absorption in the receiving room.

Twelve organizations took part in the comparison exercise (including BRS) all of which actively engage in investigations of sound insulation as part of their regular activities. They are listed in the Acknowledgments. From these, 17 measurements between the same pair of rooms in the test building were obtained. This came about by some organizations supplying two measuring teams and some teams doing measurements with and without absorbent materials in the receiving room. It is the normal practice of those teams that used these materials to employ them for actual field measurements. In all cases the quantity measured was the normalized level difference, corrected to a reverberation time in the receiving room of 0·5 s, at 1/3-octave intervals.

In addition, one team obtained four measurements, to give a measure of the typical spread of results by the same team.

3.2. DESCRIPTION OF EQUIPMENT AND PROCEDURES

Loudspeakers were used to generate the sound fields for measurements of both level difference and reverberation time (for absorption correction) in all cases but one, where an impulsive source (noise from bursting a rubber balloon) was used for the reverberation time. Three teams used warble tones as a signal source to the loudspeakers and the remainder used random noise. Amplifiers and loudspeakers varied widely in power output. Five teams used one loudspeaker and the remainder used two. They all positioned the loudspeakers near corners of the source room, but none were very precise in the positions and directions they chose.

For the measurements some teams used moving coil microphones and some the condenser type. Half measured in both rooms simultaneously, and half in one room at a time, the latter relying on the monitored signal input to the loudspeaker to set up the same sound field in the source room. Sound pressure levels were averaged by eye, either from a meter or from a calibrated level recorder chart, and in some cases tape recordings were made. The number of microphone positions varied from one to six, with fewer positions at the high frequencies. One team measured at 1/4- and 1/2-octave intervals.

A final variation was that four teams used absorbent materials in the otherwise empty receiving room. The materials used were fibre board, polyurethane foam, and panels of perforated hardboard with a fibreboard backing. One team employed an observer in the receiving room, reading sound pressure levels off a sound level meter.

The team that repeated its own measurements used two loudspeakers fed by random noise, took six microphone positions for the frequency range 100–400 Hz and three positions for 500–3150 Hz, and measured between empty rooms. In order to check the measuring teams' results, a measurement was obtained in empty rooms, with one loudspeaker arrangement and a large number of microphone positions. Further details of this are given in sections 4.2 and 4.3.

3.3. RESULTS

The results from all the measuring teams are plotted in Figure 2, grouped to show the range of values at each frequency. The total spread from minimum to maximum varied from 15 dB at 100 Hz down to 5 dB at 800 Hz, and then remained fairly constant up to 3150 Hz. The subsequent check measurement is also shown in Figure 2. Average values and standard deviations from 16 of the 17 measurements (omitting the ones at 1/4- and 1/2-octaves) are compared with those from the four measurements by one team in Figure 3.

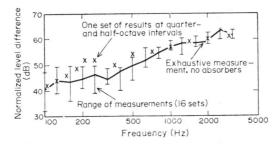

Figure 2. Comparison measurements of sound insulation—current methods.

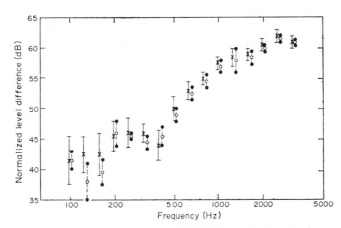

Figure 3. Comparison measurements of sound insulation—spread of results by current methods. 16 measurements by 12 teams: ×, mean; I, 2 standard deviations. 4 measurements by 1 team: ○, mean; ⦙, 2 standard deviations.

The results have been subdivided to show the effect of added absorption in the measuring rooms. In Figure 4 mean values and standard deviations from 11 measurements in bare rooms and from four measurements with absorbent materials in the receiving room are plotted. The measurement with an observer in the receiving room as well as that at 1/4- and 1/2-octaves are omitted here.

The results are discussed in section 6. Before this, the detailed studies of the measurement procedure are reported, and the results obtained by using the rationalized technique given.

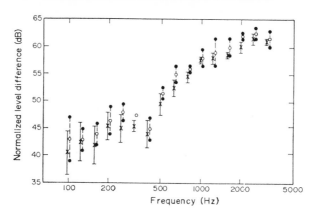

Figure 4. Comparison measurements of sound insulation—effect of absorption in receiving room on results from current methods. From 11 results with bare rooms: ×, mean; I, 2 standard deviations. From 4 results with added absorption in receiving room: ○, mean; ‡, 2 standard deviations.

4. STUDIES OF MEASUREMENT PROCEDURES

4.1. GENERATION OF SOUND FIELDS FOR MEASUREMENT OF LEVEL DIFFERENCE

The object was to compare the uniformity of the sound fields in source and receiving rooms with various loudspeaker arrangements.

4.1.1. *Method of comparing sound fields*

Sound pressure levels were plotted over a grid of measuring points (see Figure 5). The grid had rows of eight points along the length of the room and seven across the width, with columns of six vertically, a total of 336 points. The intervals between them were chosen for convenience to fit in with the room dimensions. On grid planes across the width of the room (each plane including 42 points) the levels were cubic spline interpolated between measuring

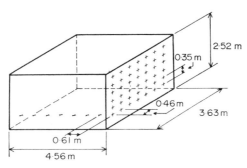

Figure 5. Measuring point grid.

points, a computer being used, and contours were printed out with letter symbols representing areas of constant level. Some typical source room contours, on grid planes 0·76 m, 1·98 m and 3·81 m from the separating wall between measuring rooms and for the 125 Hz 1/3-octave are shown in Plates 2, 3 and 4. The key to the levels corresponding to the letters used is given in Table 1.

Measures of two characteristics of the contours are necessary for a full description of the uniformity of sound pressure levels on a grid plane: the excursions of sound pressure level from the mean value, and the rate of change of sound pressure level with distance (in two dimensions).

TABLE 1

Key to letter symbols used to represent areas of constant sound pressure level in Plates 2, 3 and 4†

Symbol	Range of sound pressure levels (dB)
A	70– 71
B	72– 73
C	74– 75
D	76– 77
E	78– 79
F	80– 81
G	82– 83
H	84– 85
I	86– 87
J	88– 89
K	90– 91
L	92– 93
M	94– 95
N	96– 97
O	98– 99
P	100–101
Q	102–103
R	104–105
S	106–107
T	108–109
U	110–111
V	112–113
W	114–115
X	116–117
Y	118–119
Z	120–121

† Areas shown blank in Plates 2, 3 and 4 nave sound pressure levels varying between those of the ranges corresponding to the letters surrounding the blanks.

The deviation from the mean value across an individual measuring plane was measured by a quantity termed d_{plane}:

$$d_{\text{plane}} = \frac{\sum\limits_{r=1}^{42} (L_r - \bar{L}_p)^2}{\bar{L}_p},$$

where the quantities L_r are the measured levels at each of the 42 points, and $\bar{L}_p$ is the mean level over the plane.

For the rate of change of sound pressure level, a computation was made based on differences between measured levels at neighbouring points on a grid plane. A quantity termed r_{plane} was derived:

$$r_{\text{plane}} = \left[\tfrac{1}{71} \sum\limits_{m=1}^{6} \sum\limits_{n=1}^{7} (L_{mn} - L_{m(n-1)})^2 + (L_{mn} - L_{(m-1)n})^2 \right]^{1/2},$$

```
R     QQ     PPP      000000     PPPP     QQQQ     RRRRRRRRRRRRRRRRRRRRRRR     QQQQQQ        PPPPPPPP
R     QQ     PPP      0000000    PPPP     QQQQQ    RRRRRRRRRRRRRRRR           QQQQQQ         PPPPPPPPPPPPP
R     QQQ    PPP      0000000    PPPPP    QQQQQQ   RRRRRRRR                   QQQQQQ          PPPPPPPPPPPPPPPPPPP
      QQQ    PPPP     000000000  PPPPP    QQQQQQQ                             QQQQQQQQ        PPPPPPPPPPPPPPPPPPPPPPPP
      QQ     PPPP     0000000000 PPPPP    QQQQQQQ            QQQQQQQQ         PPPPPPP          PPPPPPPPPPPP
      QQQ    PPP      0000000000 PPPPP    QQQQQQQQQQQQQQQQQQQQQQQQQQQ         PPPPPP            PPPPPPPP
      QQQ    PPP      00000000000 PPPPPP  QQQQQQQQQQQQQQQQQQQQQQQ             PPPPPP            PPPPPPP
      QQQ    PPP      0000000000  PPPPPPP QQQQQ                              PPPPPPP
      QQQ    PPPP     000000000   PPPPPPPP                                   PPPPPPP
      QQQQ   PPPP     0000000     PPPPPPPPPP                                 PPPPPPPP
      QQQQ   PPPPP      0         PPPPPPPPPPPPPPPP                           PPPPPPPPP
      QQQQ   PPPP                 PPPPPPPPPPPPPPPPPPP                        PPPPPPPPP
      QQQQ   PPPPP               PPPPPPPPPPPPPPPPPPPPPPPP                    PPPPPPPPPPP
      QQQQ    PPPPP       PPPPPPPPPPPPPPPPPPPPPPPPPPPPPPPPPP                 PPPPPPPPPPPP
      QQQQ    PPPPPPPPPPPPPPPPPPPPPPPPPPPPPPPPPPPPPPPPPPPPPP                 PPPPPPPPPPPPPP             PPPPPPP
      QQQQQQ  PPPPPPPPPPPPPPPPPPPP                                          PPPPPPPPPPPPPPPPPPPPPP      PPPPPPPPPP
       QQQQQQQ                                                              PPPPPPPPPPPPPPPPPPPPPPPPPPPPPPPPPPPPP
         QQQQQQQ                                                            PPPPPPP
RR          QQQQQQQQQQQQ   QQQQQQQQQQQQQQQQQ
RRRR            QQQQQQQQQQQQQQQQQQQQQQQQQQQQQQQQQQ
RRRRRR                  QQQQQQQQQQQQQQQQQQ                                   QQQQQQQQQQQQQQQQQQQQQ
RRRRRRRRRRRRR                    QQQQQQQQQQQQQ              QQQQQQQQQQQQQQQQQQQQQQQQQQQQQQQQQQQQQQQQQQQQQ
RRRRRRRRRRRRRRRRRRRRRRRRRRRRRRRRRRRRRRRRRRRRRR        QQQQQQQQQQQQQQQQQQQQQQQQQQQQQQQQQQQQQQQQQQQQQQQQQQQQQQQQ
            RRRRRRRRRRRRRRRRRRRRRRRRRRRRRRRRRR       QQQQQQQQQQQQQQQQQQQQQQ                       QQQQQ
                        RRRRRRRRRRRR                QQQQQQQQQQQQQQQQQQQQ
                        RRRRRRRRRRR                 QQQQQQQQQQQQQ          RRRRRRRR
                        RRRRRRRRRR                                        RRRRRRRRRRRR
                        RRRRRRRRRR                                        RRRRRRRRRRRRRRRRRRRRRR
            SSSSSSSSSSS  RRRRRRRRRR                                       RRRRRRRRRRRRRRRRRRRRRRRRRRRRR
            SSSSSSSSSSSS RRRRRRRRRR                                       RRRRRRRRRRRRRRRRRRRRRRRRRRRRR
            SSSSSSSSSSS  RRRRRRRRRR                                       RRRRRRRRRRRRRRRRRRRRRRRRRRRRR
            SSSSSSSSSS   RRRRRRRRRR                                       RRRRRRRRRRRRRRRRRRRRRRRRRRRR
             SSSSSS      RRRRRRRRRR                                       RRRRRRRRRRRRRRRRRRRRRRRRRRRR
R                       RRRRRRRRRR                                        RRRRRRRRRRRRRRRRRRRRRRRRRRR
RRR                     RRRRRRRRRR                                        RRRRRRRRRRRRRRRRRRRRRRRRRR
RRRRR                   RRRRRRRRRR
RRRRRRRRRR    RRRRRRRRRRRRRR                    QQQQQQQQQQ
RRRRRRRRRRRRRRRRRRRRRRRRRRRRRRRR        QQQQQQQQQQQQQQQQQQQQ                  QQQQQQQQQQQQQQQQQQQQQQQQ
            QQQQQQQQQQQQQQQQQQQQQQQQQQQQQQQQQQQQQQQQQQQQQQQQQQQQ             QQQQQQQQQQQQQQQQQQQQQQQQQQQQ
         QQQQQQQQQQQQQQQQQQQQQQQQQQQQQ                QQQQQQQQQQQQQQQQQQQQQ    QQQQQQQQQ
QQQQQQQQQ                                            QQQQQQQQQQQQQQQQQQ
QQQQQQ        PPPPPPPPPPPPPP                         QQQQQQQQQQQQQ          PPPPPPPPPPPP
QQQQQ       PPPPPPPPPPPPPPPPPPPPPPPPPPPPPP                                 PPPPPPPPPPPPPPPPPPPPPPP  P
QQ      PPPPP               PPPPPPPPPPPPP             PPPPPP                              PPPPPP
Q       PPPP    00000                    PPPPPPPPPPP  PPPPPP      00000000000       PPPPP
    PPP          0000000000000000000     PPPPPPPPPPPP PPPPPP       000000          000000
    PPP    0000           000000         PPPPPPPPPPPP PPPPPP       0000             000000
    PPP    0000                 00000    PPPPPPPPPPP  PPPPPPP      0000    NNNNNN   00000000
    PPP    0000    NNNNNNNNN     00000    PPPPPPPPPPP  PPPPPPP      0000    NNNNNNNNN   0000000
    PP     000    NNNNNNNNNNNNN   00000   PPPPPPPPPP   PPPPPP      000     NNNNNNNNNN    000000
    PP     00     NNNNNNNNNNNN    0000    PPPPPPPPP    PPPPPP      0000    NNNNNNNNNNNNN   00000
    PP     000    NNNNNNNNNNNN    0000    PPPPPPPPP    PPPPPP      0000    NNNNNNNNNNNNN   00000
    PP     000    NNNNNNNNNNNN    0000    PPPPPPPP     PPPPPP      0000    NNNNNNNNNNNN    00000
    Q      PPP    000             00000   PPPPPPP      PPPPP       0000    NNNNNNN        00000000
QQ     PPP   000            00000    PPPPPP            QQQQQQ      PPPP     00000           0000000
    QQ     PPP    00000      0000000   PPPPPP       QQQQQQQQQQQQQQ         PPPPP    000000        0000000
    QQQ    PPPP   000000000000    PPPPPP    QQQQQQQQQQQQQQQQQQQQ           PPPPP     00000000000
R   QQQ    PPPP   000000000000    PPPPPP    QQQQQQQQQQQ         QQQQQQQQ   PPPPP
RR     QQQQ   PPPPPP       PPPPPPPP     QQQQQQQQQ                QQQQQQQQ      PPPPPPPPPPPPPPPPPPPPPPPPPP
RRR    QQQQ    PPPPPPPPPPPPPP      QQQQQQQQ                QQQQQQQQ       PPPPPPPPPPPPP     P
    RRRR   QQQQQ       QQQQQQQQ      RRRRRRRRRRRRR    QQQQQQQQ
     RRR    QQQQQQ     QQQQQQQQ      RRRRRRRRRRRRRRRRRRRRR     QQQQQQQQQQQ
S       RRRR   QQQQQQQQQQQQQQQQQQQQQQQ       RRRRRRRRRRRRRRRRRRRRRRRRR     QQQQQQQQQQQQQQQQQQQQQQ
SS      RRRR   QQQQQQQQQQQQQQQQQQQQQQ        RRRRRRRRRRRRRRRRRRRRRRRRRRR    QQQQQQQQQQQQQQQQQQQQQQ
```

Plate 3. Contours of sound pressure level, 1·98 m cross plane.

346

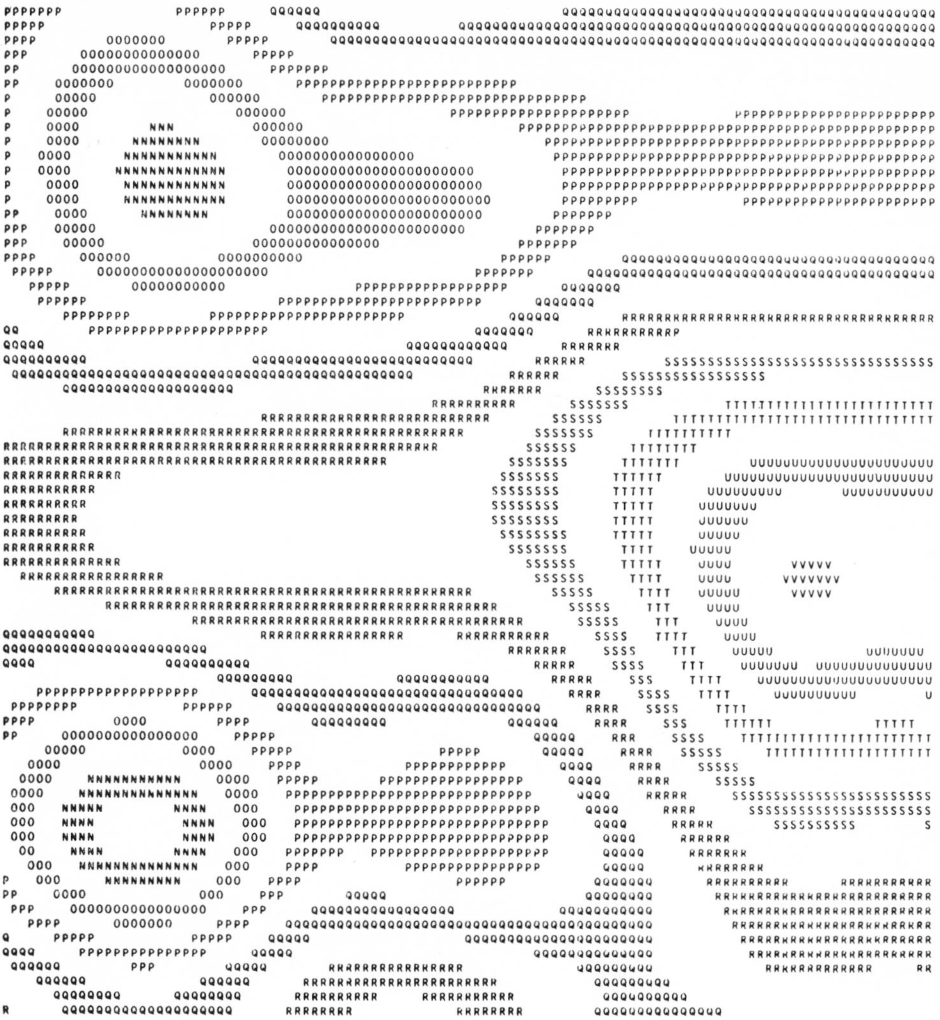

Plate 4. Contours of sound pressure level, 3·81 m cross plane.

where L_{mn} is the measured level on the mth row of measuring points, at the nth column of a grid plane, and 71 is the total number of level differences. d_{plane} and r_{plane} are of the same order, though they have to be considered separately. The smaller they are, the more uniform are the sound pressure levels.

The contours in Plates 2, 3 and 4 were obtained for a particular loudspeaker arrangement. The relevant values of d_{plane} and r_{plane} are as follows:

	d_{plane}	r_{plane}
0·76 m plane	6·0	4·6
1·98 m plane	2·4	3·1
3·81 m plane	5·5	3·4

It will be seen that the sound pressure levels on the 1·98 m plane are more uniform than those on either the 0·76 m or 3·81 m planes. However, while the deviations from the mean level on both 0·76 m and 3·81 m planes are similar in magnitude, the rates of change of level of the latter are lower, and the sound field is therefore less "peaky" than on the former.

TABLE 2

Uniformity of sound pressure levels in source room—one loudspeaker in corner of room at 45° to walls and floor. 125 Hz 1/3-octave band

Plane reference: distance from separating wall (m)	d_{plane}	r_{plane}
0·15	2·5	3·0
0·76	6·0	4·6
1·37	2·7	3·1
1·98	2·4	3·1
2·59	3·6	3·8
3·20	2·7	3·1
3·81	5·5	3·4
4·42	3·3	3·0

In order to make use of d_{plane} and r_{plane} to compare the uniformity of sound fields with different loudspeaker arrangements, it is convenient to use a single grid plane of measuring points to represent each case. The full set of figures from all eight source room cross planes, for the arrangement illustrated above, is given in Table 2. It will be noted that the highest values of d_{plane} and r_{plane} occur at 0·76 m and 3·81 m from the separating wall. The latter is at a distance of 0·76 m from the wall opposite the separating wall. These two cross planes consistently registered the greatest non-uniformity for other arrangements (see below). Therefore the plane 0·76 m from the separating wall was selected for general comparisons of sound fields. This was done in the source room only. From the arrangements tried, certain ones were selected for a full investigation over the whole grid of measuring points in both source and receiving rooms.

4.1.2. *Loudspeaker arrangements studied*

A number of arrangements of one and two loudspeakers were investigated. The speakers were high quality 250 mm diameter, dual cone, each mounted in a lined plywood cabinet.

26

Three basic speaker positions were tried: (i) in a corner standing on the floor, with axis horizontal and at 45° to the two walls; (ii) in a corner with axis inclined at 45° to the floor as well as the walls; (iii) in a corner, mid-way between floor and ceiling, with axis again horizontal and at 45° to the walls.

In all cases, the loudspeakers were set up at a distance of 0·75 m from the corner. Variations were with and without a back panel on the cabinets, and source signals of 1/3-octave random noise and warble tones.

Values obtained for the two uniformity terms, d_{plane} and r_{plane} on the 0·76 m cross plane in the source room were consistently between 4·0 and 5·0 in all but two loudspeaker arrangements. The two arrangements which gave notably increased uniformity of sound pressure levels were each of one loudspeaker fed by random noise, with the cabinet back removed

TABLE 3

Uniformity of sound pressure levels in source room—one loudspeaker in corner of room, mid-way between floor and ceiling. 125 Hz 1/3-octave band

Plane reference: distance from separating wall (m)	d_{plane}	r_{plane}
0·15	0·8	1·5
0·76	1·6	2·3
1·37	1·2	1·5
1·98	1·4	2·0
2·59	1·0	1·7
3·20	1·1	1·7
3·81	2·7	3·0
4·42	1·2	1·8

TABLE 4

Uniformity of sound pressure levels in receiving room— one loudspeaker in source room: arrangement A, in corner mid-way between floor and ceiling; arrangement B, in corner at 45° to walls and floor. 125 Hz 1/3-octave band

Plane reference: distance from separating wall (m)	Arrangement A		Arrangement B	
	d_{plane}	r_{plane}	d_{plane}	r_{plane}
0·15	2·3	2·0	4·0	2·2
0·76	6·2	3·6	4·4	2·9
1·37	3·5	1·9	5·7	2·5
1·98	3·7	2·1	6·3	2·7
2·59	4·1	2·5	4·8	2·5
3·20	2·4	1·8	2·6	1·8
3·81	3·5	2·7	3·4	2·7
4·42	4·7	3·0	5·4	2·8

and set up in positions (i) and (iii) above. Values of d_{plane} and r_{plane} for these two were, respectively, 2·0, 2·4 and 1·6, 2·3. The corresponding case with set-up (ii) gave a d_{plane} value of 6·0 and an r_{plane} value of 4·6. The arrangement with the loudspeaker mid-way between floor and ceiling was then selected for a full plot of the sound field in the source room, over

TABLE 5

Uniformity of sound pressure levels on 0·76 m cross plane in source room—two arrangements of loudspeakers: A, one speaker in corner, mid-way between floor and ceiling; B, one speaker in corner at 45° to walls and floor. 125 Hz 1/3-octave band except where otherwise stated

Condition	Arrangement A		Arrangement B	
	d_{plane}	r_{plane}	d_{plane}	r_{plane}
(a) Loudspeaker size (mm diameter)				
(i) 200	4·2	3·8		
(ii) 250	3·4	3·5		
(iii) 300	3·0	3·3		
(iv) 450	2·4	3·0		
(b) Cabinet/baffle design				
(i) No baffle	9·8	5·2		
(ii) Open-back cabinet	1·6	2·3		
(iii) Large baffle	4·2	3·8		
(c) Baffle orientation				
(i) 30°/60°	3·7	3·0		
(ii) 45°/45°	4·2	3·8		
(iii) 60°/30°	3·2	3·1		
(d) Signal level (dB)				
(i) 80	1·7	2·0		
(ii) 90	1·8	2·2		
(iii) 100	1·5	2·2		
(iv) 110	1·4	2·2		
(f) Frequency variation (Hz)				
(i) 125	1·6	2·3	6·0	4·6
(ii) 160	1·4	2·0	2·4	3·0
(iii) 200	1·2	2·1	1·8	2·7
(iv) 250	1·1	2·0	1·0	1·9
(v) 500	0·5	1·4	0·3	1·0
(vi) 1000	0·1	0·6	0·1	0·8
(vii) 2000	0·2	0·8	0·1	0·8
(viii) 4000	1·4	1·7	1·0	1·5

the whole measuring grid. The uniformity terms on the eight cross planes are listed in Table 3. These can be compared with the corresponding figures in Table 2, obtained with the speaker in the corner at 45° to the walls and floor. For reference, the uniformity terms on cross planes in the receiving room, for the same two cases, are evaluated in Table 4.

The arrangement with a loudspeaker mid-way between floor and ceiling was used for further investigation of the effects of detail variations in sound field generation. The conditions studied were as follows: (i) *Loudspeaker size*—speakers of diameter 200 mm, 250

mm, 300 mm and 450 mm in an unlined cabinet with open back; (ii) *cabinet/baffle design*—
the 300 mm speaker was mounted first in a 2 m square baffle board of 19 mm thick block-
board, then in an open-back cabinet, and finally, in a small baffle board (effectively the
condition of no cabinet at all); (iii) *baffle orientation*—the 300 mm speaker in the same 2 m
square baffle as before was arranged, still with axis horizontal, first at 30°/60° to the two side

TABLE 6

*Uniformity of sound pressure levels in receiving room (one
loudspeaker in source room, in corner at 45° to walls and
floor)—effect of absorption. 125 Hz 1/3-octave band*

Plane reference distance from separating wall (m)	Room empty		Absorbent panels in room	
	d_{plane}	r_{plane}	d_{plane}	r_{plane}
0·15	4·0	2·2	0·4	0·9
0·76	4·4	2·9	2·1	2·2
1·37	5·7	2·5	0·9	1·4
1·98	6·3	2·7	1·6	1·9
2·59	4·8	2·5	2·2	2·5
3·20	2·6	1·8	0·6	1·1
3·81	3·4	2·7	1·9	2·1
4·42	5·4	2·8	0·7	1·2

TABLE 7

*Uniformity of sound pressure levels on 0·76 m cross plane
in receiving room (one loudspeaker in source room, in corner
at 45° to walls and floor)—different frequencies*

1/3-octave band centre frequency (Hz)	Room empty		Absorbent panels in room	
	d_{plane}	r_{plane}	d_{plane}	r_{plane}
125	4·4	2·9	2·1	2·2
160	3·3	2·9	2·0	2·2
200	2·1	2·3	1·0	1·6
250	2·4	2·1	2·0	1·7
500	0·3	0·9	1·1	1·2
1000	0·2	0·7	1·4	0·8
2000	0·2	0·6	0·7	0·7

walls, then at 45°/45° and finally at 60°/30°; (iv) *signal level to loudspeaker*—the signal level
to the 250 mm diameter speaker mounted in the open-back cabinet was varied, such that the
sound pressure level monitored at a point 100 mm in front of the speaker face was, in turn,
80 dB, 90 dB, 100 dB and 110 dB.

In all cases measurements were made in the 125 Hz 1/3-octave band only. Further checks
were next made at different frequencies, and additionally with a loudspeaker in the corner at
45° to the wall and floor: (v) *frequency variation*—with the 250 mm diameter speaker in the
open-back cabinet as before, and the source room empty, measurements were made in
1/3-octaves centred on 125, 160, 200, 250, 500, 1000, 2000 and 4000 Hz.

In all cases sound pressure levels were measured on the 0·76 m cross plane in the source room. The uniformity terms obtained are summarized in Table 5.

Sound pressure levels were measured in the empty receiving room with two loudspeaker arrangements in the source room, for the single condition in each case of the 250 mm diameter speaker in an open-back cabinet, frequency 125 Hz 1/3-octave band. The results have already been referred to (see Table 4). Sound pressure levels in the receiving room with six absorbent panels were also plotted. The panels were 1·25 m square, of perforated hardboard with a fibreboard backing and a 25 mm cavity. Further measurements were made on the 0·76 m plane both with the receiving room empty and with the absorbent panels in place, at frequencies 160, 200, 250, 500, 1000 and 2000 Hz. The results showing the effect of added absorption are listed in Tables 6 and 7.

4.1.3. *Discussion*

The results from this part of the investigation can be summarized briefly.

(i) The arrangement of a single loudspeaker in an open-back cabinet, set up in a corner of the source room mid-way between floor and ceiling and relaying a band-noise signal, generated more uniform sound pressures in the source room at low frequencies, relative to those from other arrangements tried. At higher frequencies the sound fields generated by two selected loudspeaker arrangements became more uniform, and the advantage of a particular arrangement disappeared.

(ii) In the source room, sound field uniformity was notably affected by cabinet/baffle design. Loudspeaker size, baffle orientation and signal level to the loudspeaker did not appear to have any great influence on the sound fields.

(iii) In the receiving room, no source room loudspeaker arrangement gave notably increased uniformity of sound pressure levels. Extra absorption brought about a small increase in sound field uniformity at low frequencies. At higher frequencies the panels were more effective, and caused considerable distortion of the sound field. Their effect was then reversed, and they made the situation worse relative to the empty room condition.

4.2. MEASUREMENT OF SOUND FIELDS—LEVEL DIFFERENCE

4.2.1. *Introduction*

Sound fields in the two measuring rooms, with one loudspeaker in the source room in a corner at 45° to the walls and floor, were plotted extensively. Mean sound pressure level differences were then compared with figures obtained at a small number of microphone positions (typical of the positions used by the measuring teams for the comparison exercise) and with further measurements from a moving microphone.

4.2.2. *Extensive measurements and measurements with moving microphone*

With the loudspeaker relaying 1/3-octave random noise, sound pressure level sin the source room were measured at all microphone positions on the grid of Figure 5 and at 1/3-octave intervals over the frequency range 100–3150 Hz. Mean levels (on a pressure-squared basis) were determined at each frequency from measuring points within a restricted volume of the room, chosen so as to avoid the near field effects of the hard bounding surfaces. The volumes assumed for this purpose varied with frequency. Ideally, measurements closer to the walls than half a wavelength should not have been used. At the lowest frequency this meant a limiting distance of 1·7 m, which in these measuring rooms could not be achieved. Therefore, the distances from the bounding surfaces within which the measured levels were accepted are listed in Table 8. The limits were chosen to coincide with particular planes of measuring points in the grid. For illustration, Figure 6 shows some typical variations of sound pressure level along a single row of measuring points. The row chosen for the figure was near the mid-length and mid-height of the room.

Further estimates of the mean sound levels in the source room were made by using a moving microphone. The equipment for this consisted of a Brüel and Kjær turntable, Type 3921, with a condenser microphone clamped to an arm mounted in the chuck (see Plate 5). The microphone described a plane circular path, with a period of 80 s for a complete revolution.

TABLE 8

Distances from boundary surfaces of measuring rooms within which measured sound pressure levels were admitted for the purpose of determining mean values

Frequency (Hz)	Distances (m) in direction indicated		
	Length	Width	Height
100– 200	1·37	1·37	1·07
250– 500	0·76	0·91	0·71
630–1000	0·76	0·46	0·36
1250–3150	0·15	0·46	0·36

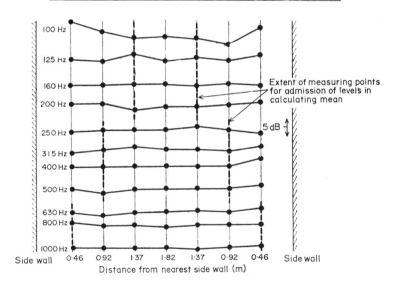

Figure 6. Typical variation of sound pressure level along a row of measuring points across width of source room.

The length of the arm was 0·8 m, so that the microphone cleared the two side walls in its traverse by 1·0 m. The turntable was tilted over at an angle to ensure that at its highest and lowest positions the microphone passed just 1·0 m from the ceiling and floor respectively. The microphone output was averaged by eye from a level recorder trace.

The receiving room was measured empty and with six absorbent panels distributed around the room. The panels were the same as used previously. The mean sound pressure level differences between source room and receiving room (empty and with absorbers) by the two methods of measurement are plotted in Figures 7 and 8. The level differences from the extensive measurements with the receiving room empty were the ones used in checking the measuring teams' results in the comparison exercise (see section 3.3).

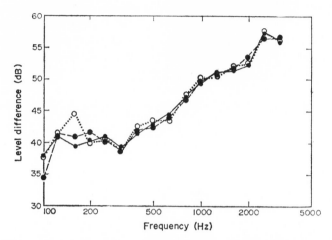

Figure 7. Level differences between source room and empty receiving room. ●—●, Large number of microphone positions; ○···○, small number of microphone positions; ●---●, moving microphone.

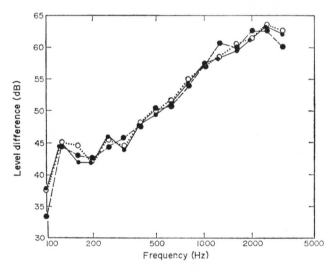

Figure 8. Level differences between source room and receiving room with absorbent panels. ●—●, Large number of microphone positions; ○···○, small number of microphone positions; ●---●, moving microphone.

4.2.3. *Measurements from a small number of microphone positions*

Six microphone positions were selected in source and receiving rooms. The six chosen were at selected points on the grid, coinciding with typical positions used by the measuring teams earlier. They were at two heights from the floor, one pair near a corner close to the separating wall, the second pair at approximately the mid-length of the room, and the third pair was near a corner distant from the separating wall. At frequencies above 500 Hz only three of these points were used.

The geometric mean sound pressure levels from these points were determined, and level differences obtained. The results are plotted for comparison in Figures 7 and 8.

4.2.4. *Discussion*

The mean levels from the small number of microphone positions and from the moving microphone are to be regarded as estimated values of the means from the large number of microphone positions. Figures 7 and 8 show that the differences between all three sets of measurements are small. The few microphone positions and the moving microphone both give deviations from the extensive measurements of several decibels at individual frequencies, but on average over the whole frequency range they are both accurate to within ± 0.25 dB. Such a close alignment is considered fortuitous, however. The spread in the extensive measurements of sound pressure level, even within the restricted volumes from which results were accepted, was such that the standard deviations were as follows.

Source room: 2·9 dB at 100 Hz, reducing to 0·8 dB at 3150 Hz.

Receiving room empty: 2 dB at 100 Hz, reducing to 0·6 dB at 1000 Hz, but then increasing again to 1 dB at 3150 Hz.

Receiving room with absorbent panels installed: 2 dB at 100 Hz, reducing to 1 dB at 3150 Hz.

Statistical analysis [12] shows that in order to determine mean values to within 1 dB (90% confidence) from a population with such divergencies, at least 15 observations would be necessary at low frequencies, and 5 or 6 at the high frequencies. These numbers would be trebled if the accuracy required was 0·5 dB for the same confidence limit.

Conversely, if the results from the few microphone positions are considered, standard deviations were as follows.

Source room: 4 dB at 100 Hz, reducing to 0·3 dB at 3150 Hz.

Receiving room empty: 2 dB at 100 Hz, reducing to 0·5 dB at 3150 Hz.

Receiving room with absorbent panels installed: 2·5 dB at 100 Hz, reducing to 0·5 dB at 3150 Hz.

By a similar analysis, the accuracies of the means from these results, to within 90% confidence, are 2–3 dB at low frequencies, reducing to 0·5–1 dB at high frequencies. This explains the divergencies which were found at individual frequencies.

The mean levels from the moving microphone at three turntable positions varied by ± 1 dB at low frequencies, and by ± 0.2 dB at high frequencies. However the plane circular path which the microphone traced out was not very satisfactory for sampling a sound field in which large variations of level occurred. A traverse in both vertical and horizontal directions should be more satisfactory, and results which bear this out have been reported [13].

The mean levels from the small number of microphone positions were deliberately calculated as the geometric means (averaging dB's), and those from the moving microphone were also such by their nature. This introduces the possibility of further error relative to the extensive measurements when the spread in levels from point to point in space is large, usually at low frequencies. The geometric mean level then tends to be low. However, measurement of airborne sound insulation relies on level differences between the rooms, and such errors are to some extent cancelled.

4.3. MEASUREMENT OF SOUND FIELDS—REVERBERATION TIME IN RECEIVING ROOM

4.3.1. *Measurements*

Three sets of reverberation times were obtained by measuring arrangements similar to those used in determining level differences. Reverberation times were determined from the decay of sound level following cut-off of the signal to the loudspeaker. With the moving microphone, the turntable was revolving continuously, and at the speed of rotation the microphone typically moved 50 mm during a decay.

Results were obtained with the room empty and with six absorbent panels installed, and are plotted in Figures 9 and 10. The reverberation times from the large number of microphone positions, with the receiving room empty, were used to normalize the corresponding level differences of section 4.2.2 for the check on the measuring teams' results in the comparison exercise.

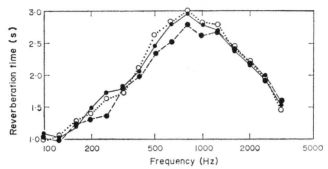

Figure 9. Mean reverberation times in empty receiving room. ●—●, Large number of microphone positions; ○···○, small number of microphone positions; ●---●, moving microphone.

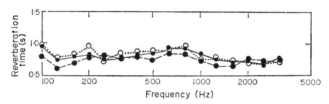

Figure 10. Mean reverberation times in receiving room with absorbent panels. ●—●, Large number of microphone positions; ○···○, small number of microphone positions; ●---●, moving microphone.

4.3.2. *Discussion*

Figures 9 and 10 show some divergencies between the results from the moving microphone and the means from both large and small numbers of static microphone positions. The mean deviations from the extensive measurements are −0·13 s with the room empty and −0·06 s with absorbers installed. The fact that the deviations are negative—that is, the moving microphone tends to indicate a shorter reverberation time—is significant. The cause is to be found in the limited region of the room sampled by the microphone in its traverse, namely a circle near the centre. The sound field decays detected in this region were generally regular and smooth. It was found in the extensive measurements that decays nearer the room boundaries were less regular, and pronounced double slopes were the rule rather than the exception. Reverberation times from such decays were estimated on the basis of the mean slope over the total fall in sound pressure level, and were longer than if they had been calculated from the initial decay.

The dominance of the results near the room boundaries is demonstrated by the fact that the discrepancies are largest, in the case of the empty room, at frequencies 250 Hz and 630 Hz. These were the frequencies at which the limits of the regions from which the means were determined were extended closer to the boundaries (see Table 8). At higher frequencies wavelengths were shorter and the near fields of the walls contracted. Reverberation times were then closer to those determined from the moving microphone.

The differences in reverberation time measurements were small and for the most part such as to give rise to errors in normalization of level difference of no more than 0·5 dB. Adequate sampling procedures are still necessary, however: with both room conditions studied the extensive measurements showed standard deviations of reverberation time at low frequencies of 0·10–0·15 s, falling to 0·05 s at the highest frequencies. Statistically [12], for an accuracy of 0·1 s (90% confidence) the minimum numbers of observations over the frequency range are 7, reducing to 4. For an accuracy of 0·05 s (90% confidence) the numbers are 20 reducing to 6.

4.4. EFFECT OF ABSORPTION IN THE RECEIVING ROOM

4.4.1. *Measurements*

The following conditions in the receiving room were studied: (i) room empty; (ii) three absorbent panels arranged around the room, one flat on the floor, one upright on the wall opposite the separating wall from the source room, and one upright on a side flanking wall; (iii) six panels arranged around the room, two, two and two where there were one, one and one in arrangement (ii). Normalized level differences between the rooms (corrected to a reverberation time in the receiving room of 0·5 s) were determined, the moving microphone being

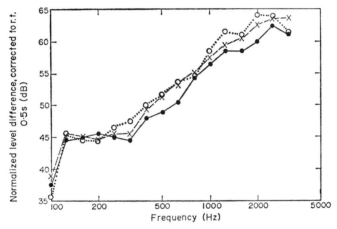

Figure 11. Effect of absorption in the receiving room on normalized level difference. ●——●, Receiving room empty; ×---×, three absorbent panels installed; ○···○, six absorbent panels installed.

used for the measurements. In fact, the relevant results for conditions (i) and (iii) were extracted from those of sections 4.2 and 4.3, and supplemented with further measurements for condition (ii), all other conditions being the same.

The measured values of total absorption in the receiving room varied as follows: (i) 6·7 m² at 100 Hz, down to 2·2 m² at 800 Hz, and then up to 4·4 m² at 3150 Hz; (ii) 7·4 m² at 100 Hz, fluctuating around 7·0 m² through the frequency range; (iii) 7·4 m² at 100 Hz, increasing to 9·5 m² at 3150 Hz. Normalized level differences for the three cases are plotted in Figure 11.

4.4.2. *Discussion*

Three absorbent panels in the receiving room increased the measured values of normalized level difference by 1·3 dB on average over the frequency range, and six panels added another 0·2 dB to make 1·5 dB average difference, relative to the empty room condition. The increases were greatest at high frequencies.

The effect of the absorbent panels is due in part to their influence on the uniformity of the sound fields in the room. It was noted earlier that at frequencies above 250 Hz their presence made the distribution of sound pressures more non-uniform than in the empty room. In such distorted sound fields, neither diffuse nor free-field, the elementary normalization procedure is not justified. Further, the panels induce some mis-match between the acoustic modes of source and receiving rooms. This results in an increase in the measured sound insulation.

4.5. EFFECT OF TEMPERATURE DIFFERENCE BETWEEN THE MEASURING ROOMS

4.5.1. *Measurements*

Normalized level differences between the rooms were determined, the moving microphone again being used. One of the rooms was heated by means of a bank of tubular heaters, arranged round the perimeter near floor level. They were thermostatically controlled, and it was ascertained that after a period of adjustment a constant temperature was achieved

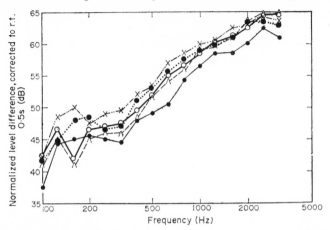

Figure 12. Effect of temperature difference between measuring rooms on normalized level difference—source room heated, receiving room empty. Temperature differences: ●—●, 0°C; ∧---∧, 5·8°C; ○—○, 9·0°C; ●···●, 15·0°C; ×----×, 20·6°C

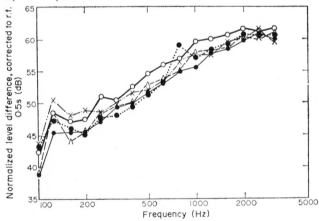

Figure 13. Effect of temperature difference between measuring rooms on normalized level difference—receiving room heated and empty. Temperature differences: ●—●, 0°C; ∧---∧, 5·6°C; ○—○, 8·8°C; ●···●, 15·5°C; ×----×, 20·8°C.

throughout the whole room. The temperature in this room could be varied up to 20 deg C above that in the other (unheated) measuring room. The same room was heated while loudspeakers were changed about, so that results were obtained in turn with the source room and the receiving room at the elevated temperature. This was done with the rooms empty. The measurements were repeated with the source room heated and with six absorbent panels in the receiving room. Nominal temperature differences for which results were obtained were 0, 5, 10, 15 and 20 deg C. Results are plotted in Figures 12, 13 and 14, and the mean deviations of the measurements at various temperature differences from those at zero temperature difference are listed in Table 9.

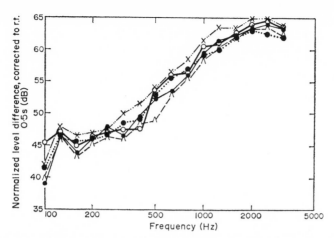

Figure 14. Effect of temperature difference between measuring rooms on normalized level difference—source room heated, six absorbent panels in receiving room. Temperature differences: ●—●, 0°C; ∧ ---∧, 5·4°C; ○—○, 8·3°C; ●···●, 16·0°C; ×----×, 19·0°C.

TABLE 9

Mean deviations of normalized level difference, for various temperature differences between measuring rooms, from values with rooms at the same temperature

Condition of measuring rooms	Temperature difference (°C)	Mean deviation (dB)
Source room heated, receiving room empty	5·8	+1·5
	9·0	+2
	15·0	+2·5
	20·6	+4
Receiving room heated and empty	5·6	+0·5
	8·8	+2·5
	15·5	+1
	20·8	+1·5
Source room heated, six absorbent panels in receiving room	5·4	−0·5
	8·3	+0·5
	16·0	0
	19·0	+2

4.5.2. *Discussion*

It appears that with the rooms empty, the measured sound insulation increases with temperature difference, although in the case of the receiving room heated the increases are erratic. Temperature differences of 20 deg C are unlikely to be encountered in practice in field measurements. Differences of 5 deg C are quite likely, however, due to solar gain through the windows, variations in heating conditions, etc., and a trend is indicated as predicted by Scholes [2]. The phenomenon is caused by mis-matching of acoustic modes of the rooms, due to the difference in temperatures, when the rooms are of the same geometry. Figure 12 shows that the largest increases occur at low frequencies, where mode densities are lowest.

While Figures 12 and 13 show the tendency for measurements to follow temperature difference, they indicate that it is erratic, with unexpected variations at individual frequencies.

In some cases the variations are hardly enough to exceed experimental error. This is confirmed by the mean deviations of Table 9, and it is therefore difficult to devise a correction. Moreover, Figure 14 shows that when extra absorption is introduced into the receiving room, the effect is largely nullified, except in the extreme case of 19 deg C difference. This again was predicted by Scholes, and results from the mis-matching of modes already brought about by the absorbent panels themselves. The same argument would apply in the case of dissimilar measuring rooms.

5. COMPARISON MEASUREMENTS OF SOUND INSULATION— RATIONALIZED TECHNIQUE

5.1. DESCRIPTION OF TECHNIQUE

The instructions to the measuring teams were as follows.

(i) A single loudspeaker was to be employed for the measurement of both level difference and reverberation time. The source signal was 1/3-octave limited random noise, and a brief specification for the minimum requirements of the equipment was given. The loudspeaker was to be 300 mm diameter, good quality, with 20 W minimum power rating, and mounted in an open-back cabinet. It was set up in a corner of the room at 45° to walls and floor.

(ii) The measurements were to be obtained using moving microphones (one in each room) as described in sections 4.2 and 4.3.

(iii) Absorbent panels were required in the receiving room. The panels used were the ones described in 4.1.2, and the instruction was to ensure that the reverberation time of the room at 125 Hz was less than 1 s.

This time the teams made one measurement each, of normalized level differences corrected to a reverberation time of 0·5 s. For the most part, since they were not equipped to carry out the measurement as instructed, they used equipment provided (the same throughout) by BRS. Further, there was practically no variation in the way the equipment was set up. One team did tape record the microphone output for laboratory analysis, and not everybody used the same number of absorbent panels in the receiving room—half used four panels, and the rest used six.

To check the results obtained, normalized level differences from the extensive measurements described in sections 4.2 and 4.3, obtained with the absorbent panels in the receiving room, were used.

5.2. RESULTS

The 12 sets of results are summarized in Figure 15. The total spread was 8·5 dB at 100 Hz, reducing to 2 dB at 1250 Hz, but then it increased again to 6 dB at 3150 Hz. The extensive measurement result (with the absorbent panels) is shown in Figure 15 for comparison.

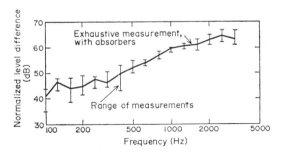

Figure 15. Comparison measurements of sound insulation—rationalized technique.

6. ASSESSMENT OF COMPARISON MEASUREMENTS

Mean values and standard deviations from the results of the current methods and the rationalized technique are plotted in Figure 16. It is interesting to compare this with Figure 3, from which it is seen that the spread of results from repeated measurements by one team was up to half that from all the teams together using their current methods, and comparable to the spread from all the teams using the rationalized technique. The implication from this and from the comparison of the effect of extra absorption (Figure 4) is that in order to reduce non-repeatability of measurements of sound insulation, three things are necessary.

(i) The sound source arrangement should be closely specified. Since the studies reported in section 4.1 did not suggest a loudspeaker arrangement which gives notably improved uniformity of sound pressure levels over the frequency range, in both source and receiving

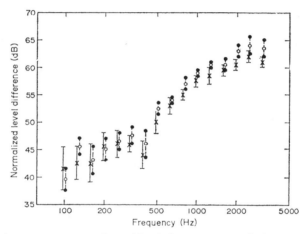

Figure 16. Comparison measurements of sound insulation—current methods and rationalized technique. From current methods: ×, mean; I, 2 standard deviations. From rationalized technique: ○, mean; 2, standard deviations.

rooms, selection of a particular arrangement can be made to suit practical convenience. Therefore, one loudspeaker should be satisfactory, provided it is sufficiently powerful. Further, a corner position for the loudspeaker, some way above floor level, would allow the same arrangement to be used for measurements on floors and walls.

(ii) The measurements should be obtained with sufficient precision. It would appear that with static microphones, six positions at the lowest frequencies is not a sufficient number. With moving microphones, a plane circular path for the microphone does not give a sufficiently representative measure of the sound fields. A requirement with moving microphones is a facility for averaging the square of the sound pressure.

(iii) The measurements should be obtained with a representative amount of sound absorbing material in the receiving room.

The object of the final phase of this investigation was to make some measurements according to the above specifications, in rooms of various sizes.

7. EFFECT OF ROOM SIZE—COMPARISON OF MEASURING TECHNIQUES

7.1. DESCRIPTION OF TEST ROOMS

Sound insulation was measured in test houses between three pairs of rooms of different sizes; in each case two different measuring arrangements were used. One pair of rooms was that in the brick-built house used for all the foregoing investigations. The room size is typical of average living rooms and double bedrooms in modern dwellings [14], and these will be referred to later as medium-sized. A second pair of rooms was on the upper floor of the same house. They were on either side of the 230 mm brick wall, but timber partitions were used to give rooms 2·29 m wide × 3·63 m long × 2·52 m high. This size is typical of single bedrooms in dwellings [14] and the rooms will be referred to later as "small".

The third pair of rooms was in another BRS test house. This also had two storeys, and the room layout was similar to that in the first house. The construction was steel frame, with cavity brick panel external walls, and concrete beams with hollow claypot infill making up the upper floor. The internal dividing wall across the width of the house was two leaves of 50 mm clinker block with a 200 mm cavity. The rooms used for the measurements were on the ground floor. One extended across the full width of the house. Its dimensions were 7·42 m long × 4·56 m across × 2·52 m high, nominally twice the size of the larger pair of rooms in the brick-built house. The other one was formed by a partition of 75 mm clinker block across the room on the other side of the internal cavity wall. Its size was 4·56 m × 3·63 m × 2·52 m. The large room is typically the size of the largest living rooms in dwellings [14], and the combination of room sizes is often encountered in real field measurements of sound insulation. In references below these rooms are termed medium and large.

7.2. MEASURING ARRANGEMENTS

In both arrangements the sound fields were generated by means of one loudspeaker. The loudspeaker was of 300 mm diameter, in an open-back cabinet, mounted in a corner with its axis horizontal 1 m above the floor and directed at 45° to the two side walls. Further, all measurements were obtained with absorbent panels (as described in section 4.1.3) in the receiving room.

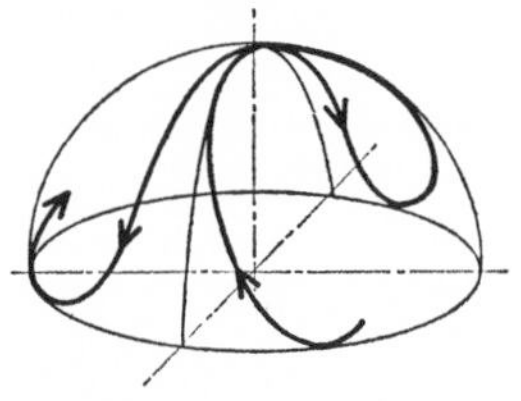

Figure 17. Motion of moving microphone.

The two set-ups differed in the equipment used to measure the sound fields. The first had static microphones at a sufficient number of positions for average level differences and reverberation times to be determined within narrow limits. There were two microphones clamped to a stand at heights of 0·75 m and 1·68 m, and the stand was sited in turn at randomly selected positions around the rooms. The total number of microphone measuring positions in the large room varied from 30 at the lowest frequencies down to 5 at the highest. In the medium-sized rooms they varied from 16 down to 3, and in the small rooms from 12 to 3. The positions were always well spaced out around the rooms, never closer to one another than 1 m, nor closer to any of the walls of the rooms than 0·5 m or to the loudspeaker source (in the transmitting room) or absorbent panel (in the receiving room) than 1·5 m. The transmitting and receiving rooms were measured separately, and in order to maintain constant sound levels in the former the voltage to the loudspeaker was monitored. The outputs from the microphones were fed *via* a two-way switch and a frequency spectrometer to a level recorder. The sound pressure levels at each microphone position were averaged by eye from the level recorder trace, and the level differences were derived from the average mean square sound pressures.

In the second set-up moving microphones were used, one in the transmitting room and one in the receiving room, measuring simultaneously. The microphone was clamped to the end of an arm which swung backwards and forwards through 180° vertically while at the same time it swept through 360° horizontally. The mechanism was designed and constructed at BRS (see Plate 6). The motion of the microphone is illustrated in Figure 17: it performed four reversals vertically for one horizontally, describing a complete cycle in 60 s. The length of the arm was adjustable. In the two larger pairs of rooms it was set at 0·9 m and the microphone swept between the same heights as those of the two fixed microphones on the stand in the first set-up. In order to retain a minimum clearance of 0·5 m from the side walls in the small rooms the length of arm was reduced to 0·65 m there. The outputs from the two moving microphones were taken separately to third-octave filters and then to logarithmic amplifiers. The latter were equipped with averaging networks, having variable integrating time constants, before the log converter, so that true average sound pressure levels could be measured over the microphone path. For level difference measurements, the outputs from the logarithmic amplifiers were taken to a differential amplifier, and the difference signal to a chart recorder. In the medium-large rooms the average level difference at each frequency was obtained from two pairs of moving microphone stand positions, a pair consisting of one position in the transmitting room and one in the receiving room. In the medium–medium and small rooms they were from one pair of stand positions. In the medium and small rooms the stand was sited arbitrarily near the middle of the room each time, while in the large room the two positions used were widely spaced apart along the length. Reverberation times were always measured while the microphone continued to revolve.

With static and moving microphone measurements, five sets of normalized level differences were obtained in each pair of rooms. Within the limits described the microphone positions were chosen at random in every case.

7.3. RESULTS AND DISCUSSION

Mean values and standard deviations of the normalized level differences in the three pairs of rooms are plotted in Figures 18, 19 and 20. At the frequencies where no standard deviation is indicated in the figures there was no scatter in the results. The mean deviations over the 16 frequencies of the mean values obtained by the moving microphones from those obtained by the static microphones are as follows:

medium-large rooms,	+0·25 dB;
medium sized rooms,	−0·5 dB;
small rooms,	−0·25 dB.

There are differences at one or two individual frequencies in all three pairs of rooms (at most 3·5 dB at 200 Hz in the small rooms) for which there is no obvious explanation. The standard deviations from the five sets of measurements by each measuring arrangement in the three pairs of rooms are generally similar. The practical point is that overall the moving microphone results are identical with those from large numbers of static microphone positions, and yet they were obtained in only a quarter of the time needed for the latter.

Moving microphone techniques lend themselves to automation of the instrumentation and of the measuring process. With or without automation, however, the necessary measuring instruments are likely to be more expensive than those for static microphone techniques, in view of the need to average mean square sound pressures and not sound pressure levels.

The spread in results shown in Figures 18–20 is probably due to the fact that it is impossible to find a sufficient number of randomly chosen measuring points spaced sufficiently far apart (at least half a wavelength). The moving microphone sampled over a limited volume, and while it measured continuously, the equivalent number of widely spaced positions, though difficult to estimate, was probably quite small, at least at low frequencies. Even with this, therefore, it was necessary to take two widely spaced stand positions in the large room for

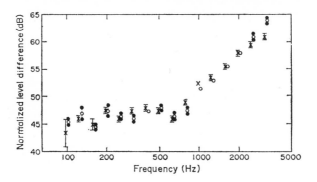

Figure 18. Effect of room size: measurements in medium-large rooms. Measurements at static microphone positions: ×, mean; Ι, 2 standard deviations. Measurements with moving microphones: ○, mean; ↕, 2 standard deviations.

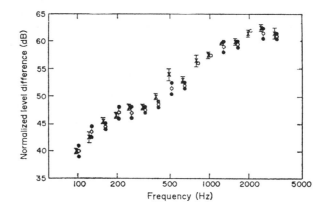

Figure 19. Effect of room size: measurements in medium sized rooms. Measurements at static microphone positions: ×, mean; Ι, 2 standard deviations. Measurements with moving microphones: ○, mean; ↕, 2 standard deviations.

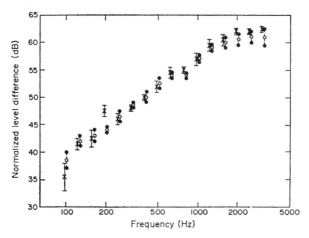

Figure 20. Effect of room size: measurements in small rooms. Measurements at static microphone positions: ×, mean; Ι, 2 standard deviations. Measurements with moving microphones: ○, mean; ↕, 2 standard deviations.

each set of measurements. The difficulty is reflected in that the standard deviations of the results are largest in the small rooms, where the number of microphone positions used was least, and the moving microphone arm length smallest.

8. CONCLUSIONS

The differences in results of sound insulation measurements from current standard measuring methods have been demonstrated. It has been shown that they are due in part to variations in the amount of sound absorption in the rooms and that the standard normalization of results does not work. In addition, variations in detail of the arrangements used to generate the sound fields and inadequate sampling to determine average level differences contribute to the non-repeatability.

Some improvement in repeatability was shown when a closely defined measuring technique was followed. Even so the results from the second comparison exercise still showed a spread, and a tendency to err on the low side. This was probably due to a measuring set-up (a microphone sweeping out a circular path, with visual averaging of sound pressure levels) that was inadequate. The comparisons in rooms of various sizes with more refined measuring arrangements showed the extent to which the scatter in results could be reduced.

The implication of this work is that the experimental errors of measurement of sound insulation can be reduced by describing equipment and techniques more explicitly. A Swedish recommendation [15] goes some way towards achieving this. A Standard should include the following: (i) precise instructions in the specification and arrangement of equipment to generate the sound fields; (ii) recommendations either to measure in furnished rooms or to introduce sufficient absorbent material to the rooms to represent the furnishing; (iii) guidance on the extent of measurements for the accurate determination of average sound level differences and reverberation times.

Moving microphones are useful for the measurements because of the reduced time taken to achieve a given accuracy. However they must be used with instrumentation capable of averaging sound pressures.

Finally, a comment is relevant on the use of measurements for comparison with performance requirements. The requirements sometimes incorporate a tolerance (e.g., in the U.K. a total adverse deviation over all frequencies of 23 dB against Party Wall Grade, Grade I, etc.) which is in part to allow for the unreliability of measurements. If the unreliability is reduced, and yet the tolerance is retained, building constructions can be designed to give a performance close to the lower end of the tolerance and standards of sound insulation achieved will fall. Performance specifications ought to allow for this, therefore, and should be re-defined statistically so as to ensure that the intended requirement is achieved in most cases.

ACKNOWLEDGMENTS

The experimental work and computation was mostly done by Mr R. S. Alphey, with assistance from Mrs A. E. Watts. The following organizations supplied measuring teams for the comparison exercise: Acoustical Investigation and Research Organisation Ltd; BPB Industries (Research and Development) Ltd; British Broadcasting Corporation Research Department; G. H. Buckle and Partners; Greater London Council Scientific Branch; Heriot–Watt University; H. R. Humphreys; University of Liverpool; McClaren Ward and Partners; James Moir and Associates; Sound Research Laboratories Ltd.

The work described has been carried out as part of the research programme of the Building Research Station of the Department of the Environment and this paper is published by permission of the Director.

REFERENCES

1. T. Kihlman 1965 *Proceedings of the Fifth International Congress on Acoustics, Liège*. On the precision and accuracy of measurements of airborne sound transmission.
2. W. E. Scholes 1969 *Journal of Sound and Vibration* **10**, 1–6. A note on the repeatability of field measurements of airborne sound insulation.
3. T. J. Schultz 1971 *Journal of Sound and Vibration* **16**, 17–28. Diffusion in reverberation rooms.
4. International Organization for Standardisation 1960 *ISO Recommendation R*140. Field and laboratory measurements of airborne and impact sound transmission.
5. British Standards Institution 1956 *British Standard* 2750. Recommendations for field and laboratory measurement of airborne and impact sound transmission in buildings.
6. P. H. Parkin 1949 *Report of the 1948 Summer symposium of the Acoustics Group, The Physical Society*, 36–44. Provisional code for field and laboratory measurements of airborne and impact sound insulation.
7. P. H. Parkin, H. J. Purkis and W. E. Scholes 1960 *National Building Studies Research Research Paper* 33. Field measurements of sound insulation between dwellings. London: HMSO.
8. W. A. Utley 1971 *Journal of Sound and Vibration* **16**, 643–644. The accuracy of laboratory measurements of transmission loss.
9. J. van den Eijk 1966 *Journal of Sound and Vibration* **3**, 7–19. The new Dutch code on noise control and sound insulation in dwellings and its background.
10. Statutory Instrument 1970 No 1137 *The Building Standards (Scotland) (Consolidation) Regulations* 1970 *Part H: Resistance to the transmission of sound*. London: HMSO.
11. Statutory Instrument 1971 No 1600 *The Building (Seventh Amendment) Regulations 1971 Part G: Sound insulation*. London: HMSO.
12. British Standards Institution 1935 *British Standard* 600, 61. The application of statistical methods to industrial standardisation and quality control.
13. K. Rohrberg 1967 *Institut für Technische Physik, Stuttgart. BS* 6/67. Erhöhung der Messgenauigkeit bei Schalldämm—Messungen durch integrierende Geräte.
14. Ministry of Housing and Local Government 1963 *Design Bulletin* 8. Dimensions and components for housing. London: HMSO.
15. Swedish National Institute for Materials Testing 1969 *Circular No* 38. Recommendations for field measurement of sound insulation. Stockholm: Statens Provningsanstalt.

27

Reprinted from pp. 81–102 of *Akust. Z.* 7:81–103 (May 1942)

Theorie der Schalldämmung dünner Wände bei schrägem Einfall*)

Von L. Cremer

(Mit 6 Textabbildungen)

I. Teil: Übersicht über den Forschungsstand

§ 1. Die Wand als träge Masse bei senkrechtem Einfall

In der vorliegenden Arbeit wollen wir uns mit der Schalldämmung einfacher, fester Trennwände befassen, deren Dicke klein gegenüber derjenigen Wellenlänge ist, die dem Schall im Wandmaterial zukommen würde, und welche ferner zwischen ein schubspannungsfreies Medium gesetzt sind, dessen Schallwellenwiderstand klein gegenüber den Wellenwiderständen des Wandmaterials ist. Aus diesen Voraus-

setzungen folgt, daß sich die Wandteilchen in Dickenrichtung als Ganzes verschieben. Derartige „Wände" pflegt man in der Mechanik als „Platten" zu bezeichnen. Doch sind hierzu auch noch einsteinstarke Ziegelwände im tiefen und mittleren Frequenzbereich zu rechnen.

Betrachten wir nun zunächst den senkrechten Einfall und nehmen wir dabei die Wand als beliebig breit an, so können wir alle Größen auf die Flächeneinheit beziehen und erhalten das einfachste Schema einer Trennwand, bei welcher die Druckdifferenz zwischen Vorder- und Rückseite eine Beschleunigung der Wandmasse bewirkt:

$$(1.1) \qquad p_1 - p_2 = m \frac{dv}{dt}$$

(p_1 Schalldruck vor, p_2 Schalldruck hinter der Wand, m Wandmasse je Flächeneinheit, v Wandschnelle, t Zeit).

Indem wir uns in Anbetracht der Linearität des Problems gleich auf reine Töne beschränken und uns der symbolischen komplexen Darstellung bedienen, nimmt Gleichung (1.1) die Form an:

$$(1.1a) \qquad \bar{p}_1 - \bar{p}_2 = i \, \omega \, m \, \bar{v}$$

$(i = \sqrt{-1}, \omega$ Kreisfrequenz).

Die Balken über p und v sollen andeuten, daß es sich um komplexe Größen handelt.

Da wir später zu der Massenwirkung noch andere Einflüsse hinzuzufügen haben, die auch der Wandschnelle proportional sind, wollen wir allgemein den Quotienten aus Druckdifferenz und Schnelle, der hier zunächst durch den Massenblindwiderstand $i \, \omega \, m$ ge-

---

*) Die Arbeit wurde im wesentlichen am 13. Januar 1942 im Seminar für Mechanik der T. H. Berlin vorgetragen.

[*Editor's Note:* An English summary of this article prepared by T. D. Northwood follows.]

geben ist, als Trennwiderstand $\overline{T}$ einführen:

$$(1.2) \qquad \bar{p}_1 - \bar{p}_2 = T\,\bar{v}\,.$$

Da jenseits der Wand eine ebene Welle abgestrahlt wird, für welche gilt:

$$(1.3) \qquad \bar{p}_2 = Z\,\bar{v}$$

(Z Wellenwiderstand des schubspannungsfreien Mediums), so unterscheidet sich der eingeführte Trennwiderstand von dem in der Raumakustik sonst geläufigen Wandwiderstand $\overline{W}$

$$(1.4) \qquad \overline{W} = \frac{\bar{p}_1}{v}$$

(W = Wandwiderstand)

nur um den Wellenwiderstand:

$$(1.5) \qquad \overline{W} = \overline{T} + Z\,.$$

Durch den Trennwiderstand wird allgemein das uns interessierende Verhältnis der je Flächeneinheit durchgelassenen Energie zur aufgefallenen, der sog. Durchlaßgrad $\varDelta$, bestimmt. Da wir auf Vorder- und Rückseite das gleiche Medium annehmen wollen, läßt sich der Durchlaßgrad auch durch die Quadrate der entsprechenden Schallschnellen ausdrücken:

$$(1.6) \qquad \varDelta = \frac{v^2}{v_a^2}$$

(v_a = Schnelle der auffallenden Welle).

Zerlegen wir nun den Druck vor der Wand p_1 in den auffallenden und reflektierten Anteil:

$$(1.7) \qquad \bar{p}_1 = \bar{p}_a + \bar{p}_r\,,$$

die wir ihrerseits durch die entsprechenden Anteile der Schnelle ersetzen können:

$$(1.7\,\text{a}) \qquad \bar{p}_1 = Z\,\bar{v}_a - Z\,\bar{v}_r$$

und addieren wir hierzu die entsprechende Gleichung für die resultierende Schnelle:

$$(1.8) \qquad Z\,\bar{v} = Z\,\bar{v}_a + Z\,\bar{v}_r\,,$$

so läßt sich die reflektierte Welle eliminieren und wir erhalten einerseits:

$$(1.9) \qquad \bar{p}_1 + Z\,\bar{v} = 2\,Z\,\bar{v}_a\,.$$

Andererseits gilt nach (1.2) und 1.3)

$$(1.10) \qquad \bar{p}_1 = (\overline{T} + Z)\,\bar{v}\,.$$

Die Elimination von $\bar{p}_1$ durch Zusammenfassung beider Beziehungen liefert schließlich für den Durchlaßgrad nach (1.6):

$$(1.11) \qquad \varDelta = \frac{v^2}{v_a^2} = \frac{1}{\left|1 + \dfrac{T}{2Z}\right|^2}\,.$$

Es leuchtet ein, daß mit verschwindendem Trennwiderstand der Durchlaßgrad auf 100% gehen muß.

Da Durchlaßgrade andererseits selbst dann noch interessieren, wenn sie nur 10^{-5} betragen, hat es sich in der technischen Praxis eingebürgert, stattdessen ein logarithmisches Maß in db, die sog. Schalldämmung D zur Kennzeichnung zu verwenden, die definiert ist durch:

$$(1.12) \qquad D = 10\log\frac{1}{\varDelta}\,.$$

Bei dem zunächst behandelten einfachsten Schema einer Trennwand als träge Masse:

$$(1.13) \qquad \overline{T} = i\,\omega\,m\,.$$

erhalten wir:

$$(1.14) \qquad D = 20\log\left|1 + \frac{i\,\omega\,m}{2Z}\right|\,,$$

oder da der reelle Summand 1 fast in allen praktischen Fällen bei der Bildung des Absolutwertes vernachlässigt werden kann:

$$(1.14\,\text{a}) \qquad D \approx 20\log\frac{\omega\,m}{2Z}\,.$$

Die hier zunächst gegebene unmittelbare Darstellung der Wand als eine träge Masse wurde zuerst von G. Jäger[1]) in seiner durch die Theorie des Nachhalls bekanntgewordenen Arbeit gegeben. Das Ergebnis findet sich schon bei Rayleigh[2]), bei dem es als Grenzfall einer endlich starken schubspannungsfreien Zwischenschicht von größerer Dichte und Steifigkeit auftritt. Da bei senkrechter Anregung in der unendlich breiten Wand keine Schubspannungen erzeugt werden, verhalten sich feste Körper dabei wie Flüssigkeiten und Gase. Die gleiche Ableitungsart wurde später von Berger[3]) und in neuerer Zeit von Wintergerst[4]) benutzt. Die Existenz dieses Grenzüberganges zeigt deutlich, daß der Schalldurchgang durch Mitschwingen der Wand bzw. durch Verdichtungswellen im Innern, nicht, wie oft irrtümlich angegeben, zwei verschiedene parallele Wege darstellt, sondern verschiedene Betrach-

[1]) G. Jäger, Wiener Akad. Ber. math.-nat. Kl. 120, Abt. IIa (1911), S. 613.

[2]) Lord Rayleigh, Theory of Sound II, § 271.

[3]) R. Berger, Diss. T. H. München 1911, sowie Gesundheits-Ing. 38, S. 49 und 67.

[4]) E. Wintergerst, Schalltechnik 4 (1931), S. 85.

tungsweisen ein und desselben Vorganges bedeutet.

§ 2. Vergleich mit statistisch gewonnenen Meßergebnissen

Obschon die bisher gegebene einfachste Darstellung zwei wesentliche Vereinfachungen zur
Voraussetzung hat, nämlich einmal die Beschränkung auf senkrechten Einfall und zwei-
tens die Vernachlässigung einer Einspannung
am Rande, so stimmt die so gefundene Formel (1.14) für die Schalldämmung doch in zwei
wichtigen Punkten mit der Erfahrung überein.
Einmal hat sich bei den verschiedensten Baustoffen und Konstruktionen gezeigt, daß die
Masse pro Flächeneinheit hauptsächlich die

die schrägen Einfallswinkel maßgebend be-
teiligt sind. Man muß daher eher zum Vergleich mit theoretischen Ergebnissen die Annahme einer idealen, statistischen Einfallswinkelverteilung machen, wonach jede Raumrichtung gleich häufig ist (siehe unten § 4).
Derartige Meßergebnisse sind u. a. in dem
Buch von Schoch[5]) zusammengestellt, aus
welchem wir die folgenden Abb. 1a und 1b,
sowie 2a und 2b entnehmen.

In Abb. 1a ist die in 6 verschiedenen Laboratorien gemessene Abhängigkeit der mittleren
Schalldämmung vom Gewicht je Flächeneinheit
aufgetragen; man kann feststellen, daß die Meßpunkte sich gut um eine Mittelwertkurve grup-

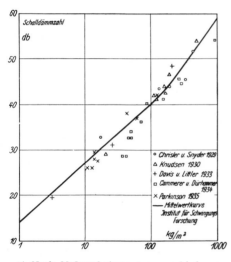
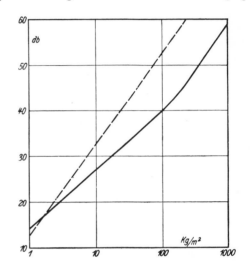

a) Nach Meßergebnissen aus verschiedenen Laboratorien

b) Vergleich der Mittelkurve aus a (ausgezogen) mit der einfachen Theorie (gestrichelt)

Abb. 1. Mittlere Schalldämmung in Abhängigkeit vom Wandgewicht. (Nach dem Buch von Schoch)

Schalldämmung bestimmt, und zwar in dem
Sinne, daß schwere Massen besser isolieren als
leichte, und ferner hat sich gezeigt, daß die
Dämmung mit der Frequenz wächst.

Dabei sind die Messungen, in denen diese
Gesetzmäßigkeiten zum Ausdruck kommen,
nicht bei senkrechtem Einfall gemacht, sondern
sie sind, in Anpassung an die Verhältnisse der
Praxis, aus dem Vergleich der mittleren Energie-
dichten in zwei, durch das Meßobjekt getrennten, hallenden Räumen gewonnen. Es ist
sicher, daß bei einer solchen Anordnung auch

pieren. In Abb. 1b ist diese mit der gestrichelt
eingezeichneten Dämmkurve verglichen, die
sich nach Formel (1.14) ergeben würde. Dieser
Vergleich lehrt, daß eine Übereinstimmung
im Absolutwert nur bei leichten Wänden vorliegt, daß dagegen die schweren Wände
alle weit schlechter dämmen, als nach
der einfachen Theorie zu erwarten
wäre.

[5]) A. Schoch, Die physikalischen und technischen Grundlagen der Schalldämmung im Bauwesen, Leipzig 1937.

Abb. 2a zeigt in gleicher Weise die Frequenz-abhängigkeit der Schalldämmung für 6 verschiedene, unter der Abbildung näher bezeichnete Baustoffe. Sie weisen wohl ein Anwachsen der Dämmung mit der Frequenz auf. Aber, wie die in Abb. 2b für die Fälle 1 bis 3 gestrichelt eingetragenen Dämmkurven nach

einfachen Theorie erwarten müßte, hat man zunächst die Randeinspannung herangezogen, welche es unmöglich macht, daß die Bewegungsamplitude an allen Stellen gleich groß ist. Durch sie mußte die Biegesteifigkeit von Einfluß sein. Da aber jede Steifigkeitswirkung, wie der Federungswiderstand beim einfachen

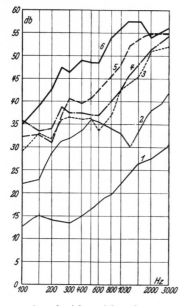

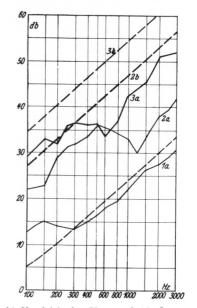

a) es beziehen sich auf:

b) Vergleich der Kurven 1 bis 3 aus a mit der einfachen Theorie

	Gewicht kg/m²
Kurve 1: Sperrholz, lackiert	2,05
„ 2: Spiegelglas	30
„ 3: Leichtbauplatte, beiderseits verputzt	70
„ 4: ¹/₄ Stein Vollziegelwand, beiderseits verputzt	153
„ 5: ¹/₂ Stein Vollziegelwand, beiderseits verputzt	223
„ 6: ¹/₁ Stein Vollziegelwand, beiderseits verputzt	457

Abb. 2. Frequenzgang der Schalldämmung für verschiedene Baustoffe. (Nach dem Buch von SCHOCH)

Formel (1.14) erkennen lassen, steigen die gemessenen Frequenzgänge im Mittel weniger an. Dabei zeigt dieser Anstieg deutlich in verschiedenen Frequenzgebieten verschiedene Tendenz. Im allgemeinen ist der Frequenzgang bei niedriger Frequenz wesentlich flacher, bei höherer oft steiler als der einfachen Theorie entspricht[6]).

Um nun zu erklären, warum die Schalldämmung nicht so gut ist, wie man nach der

Schwinger, zu den Trägheitswiderständen gegenphasig ist, so leuchtete ein, daß sich der Einfluß der Biegesteifigkeit in einer Verringerung des Trennwiderstandes und somit auch in einer Verringerung der Schalldämmung bemerkbar machen mußte.

Man versuchte dabei zunächst diesen Einfluß dadurch zu berücksichtigen, daß man die Platte als ein Schwingungssystem von einem (kinematischen) Freiheitsgrad betrachtete, nämlich wie eine in ihrer Grundschwingung erregte Telefonmembran[7]). Dementsprechend wurde

---

[6]) Auch von E. LÜBCKE wurden bei verputzten Ziegelwänden und bei tiefen Frequenzen sehr schwache Anstiege mit der Frequenz gefunden. Z. techn. Phys. 17 (1936), S. 54.

[7]) K. W. WAGNER, S.-B. preuß. Akad. Wiss., physik.-math. Kl. (1931).

der Trennwiderstand angesetzt als Reihenschaltung einer Masse m, einer Federung k und eines linearen Reibungswiderstandes r[8]) zur Berücksichtigung innerer Verluste:

$$(2.1) \qquad \overline{T} = i \left(\omega\, m - \frac{k}{\omega} \right) + r.$$

Da in diesem Falle die Schallschnelle über der Platte notwendigerweise verschiedene Werte hat, so ist der Trennwiderstand als Quotient des Schalldruckes vor der Wand zu einer „mittleren" Schnelle einzusetzen. Dabei kommen für die Trägheitswirkung und für die Abstrahlungswirkung verschiedene Mittelwertbildungen in Frage. Wir können aber darauf verzichten, auf diese Feinheiten näher einzugehen, weil diese Art der Darstellung bereits aus anderen Gründen nicht aufrechterhalten werden kann.

Dieselbe ist nämlich theoretisch widerspruchsvoll. Man muß einerseits annehmen, um im wesentlichen auf einen Trägheitswiderstand zu kommen, daß die Grundfrequenz der Membran

$$(2.2) \qquad \omega_0 = \sqrt{\frac{k}{m}}$$

sehr tief unter der Schallfrequenz ω liegt, so daß in dem entsprechend umgeformten Ausdruck für den Trennwiderstand

$$(2.3) \qquad \overline{T} = i\, \omega\, m \left(1 - \left(\frac{\omega_0}{\omega} \right)^2 - i\, \frac{r}{\omega\, m} \right)$$

die von Federung und Dämpfung herrührenden Glieder als kleine Korrekturen aufgefaßt werden können. Auf der anderen Seite kann aber nur in der Nähe der Grundschwingung und unterhalb derselben die Platte als ein System von einem Freiheitsgrad aufgefaßt werden. Da sie aber entsprechend ihrer zweidimensionalen, kontinuierlichen Verteilung von Massen und Steifigkeiten über eine zweifach unendliche Schar von Eigenfrequenzen verfügt, ist es ausgeschlossen, daß man weit oberhalb des Grundtones diesen Eigenfrequenzen höherer Ordnung entgeht und die Platte als einfachen Schwinger auffassen darf.

Dies zeigt sich auch in zwei wichtigen Versuchsergebnissen, die dieser sog. „Membran-Theorie" widersprechen. Der eine Versuch wurde von R. Berger[9]) unternommen. Er setzte auf eine Trennwand, welche zwei untereinander befindliche Räume als Decke trennte, ein Gewicht auf, das mehr als das Vierfache des gesamten Wandgewichtes betrug. Nach der Membran-Theorie hätte damit eine wesentliche Erhöhung der Schalldämmung verbunden sein müssen, während sich experimentell im Mittel kaum eine Abweichung bemerkbar machte. Dies spricht dafür, daß die Wand nur an der Stelle, an der das Gewicht aufgesetzt war, für die Schallübertragung ausfiel, an den übrigen Stellen aber nicht durch das Gewicht behindert wurde.

Der zweite Versuch bezieht sich auf schrägen Einfall. In dieser Hinsicht versagt jede Auffassung, die die Wand als Ganzes behandeln möchte von vornherein, weil die bei schrägem Einfall an den verschiedenen Stellen auftretenden gegenphasigen Anregungen bei Integration über die gesamte Wand sich gegenseitig kompensieren würden. Stattdessen hat man aber bei den wenigen Versuchen, bei welchen die zu untersuchende Wand nicht 2 Hallräume trennte, sondern im Gegenteil 2 schalltote Räume, und bei welchen der Schall in Form einer breiten ebenen Welle auf die Wand einwirkte, Schalldurchgang feststellen können[10])[11]). Ja, es ergab sich sogar, daß die Welle im sekundären Raum die gleiche Richtung beibehielt. Das letzte verlangt aber nach dem Huygensschen Prinzip, daß die einzelnen Massenelemente sich mit der gleichen Phasenverschiebung gegeneinander bewegen, wie sie der Schalldruck der einfallenden Welle vor der Trennwand aufweist. Man gelangt damit zu einem Wellenbild, wie es im Prinzip in Abb. 3 skizziert ist.

Wenn sich überdies auch bei schrägem Einfall die Wandmasse als ausschlaggebend erwies, so konnte man sich hiernach die Wand als zusammengesetzt aus lauter schweren Mas-

---

[8]) H. A. Davis, Phil. Mag., Ser. 7 XV (1933), S. 309.

[9]) R. Berger, Forsch. Ing.-Wes. 3 (1932), S. 193.

[10]) F. R. Watson, Univ. Illinois Eng. Exp. Stat. Bull. 127 (1922).

[11]) A. H. Davis und T. S. Littler, Phil. Mag. 7 (1929), S. 1050.

sen vorstellen, die in ihrer gegenseitigen Verschiebung seitlich kaum behindert sind. Für
ein solches gegenüber der komplizierten Wirklichkeit erstaunlich einfaches Modell, das wir
kurz als „Massenmodell" bezeichnen wollen,
fehlte aber zunächst jede theoretische Basis.

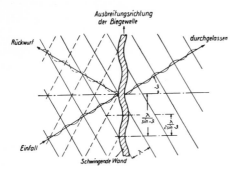

Abb. 3. Schematische Skizze zum Schalldurchgang
bei schrägem Einfall

§ 3. Das asymptotische Gesetz von SCHOCH

Es bedeutete daher einen wesentlichen Fort-
schritt für die Theorie der Schalldämmung,
daß es A. SCHOCH[12]) gelang, das „Massenmodell" als ein unter gewissen, noch näher zu
beschreibenden Voraussetzungen geltendes
asymptotisches Verhalten von Einfachwänden
bei hohen Frequenzen darzustellen. Wir wollen
ihrer Wichtigkeit halber die SCHOCHsche Ableitung unter Anpassung an die hier gewählte
Darstellung kurz wiederholen.

SCHOCH macht zunächst keinerlei Einschrän-
kungen in bezug auf Form und Einspannung
der Wand. Er geht allgemein davon aus, daß
man stets die Bewegung einer solchen Wand
als Summe von Eigenfunktionen $u_n(x, y)$ zu-
sammensetzen kann, d. h. aus solchen Ver-
teilungen, wie sie zu den einzelnen Eigenfrequenzen mit der Kreisfrequenz ω_n gehören:

$$(3.1) \qquad \bar{v}(x, y) = \Sigma \, \bar{v}_n \, \bar{u}_n(x, y).$$

Hierin bedeutet $\bar{v}_n$ die zur normierten Eigenfunktion $\bar{u}_n(x, y)$ gehörige Teilamplitude. Die
gleiche Entwicklung läßt sich auch für eine
beliebige Druckverteilung vor der Wand heran-

[12]) A. SCHOCH, Akust. Z. 2 (1937), S. 113.

ziehen:

$$(3.2) \qquad \bar{p}(x, y) = \Sigma \, \bar{p}_n \, \bar{u}_n(x, y).$$

Der Druck hinter der Wand kann näherungsweise vernachlässigt werden.

Für jeden einzelnen Summanden in (3.1)
und (3.2) läßt sich nun die Wand als ein
System von einem Freiheitsgrad auffassen,
für welchen zwischen der komplexen Amplitude
des Druckes und derjenigen der Schnelle, wie
oben eine Beziehung der Form:

$$(3.3) \qquad \frac{p_n}{\bar{v}_n} = \overline{T}_n = i\,\omega\,m\left(1 - \left(\frac{\omega_n}{\omega}\right)^2 (1 + i\,\eta)\right)$$

angesetzt werden kann. Dabei haben wir mit
SCHOCH die Dämpfung durch einen Korrekturfaktor $i\,\eta$ in der Klammer eingeführt.

Die Gesamtlösung setzt sich dann in der
Form zusammen:

$$(3.4) \qquad \bar{v}(x, y) = \frac{1}{i\,\omega\,m} \sum \frac{p_n u_n(x, y)}{1 - \left(\frac{\omega_n}{\omega}\right)^2 (1 + i\eta)}.$$

Tragen wir nun einmal das Spektrum der
Teiltonamplituden des Druckes auf, so wird
dasselbe in jedem Falle schließlich eine beliebig abnehmende Tendenz zeigen (siehe
Abb. 4).

Dabei ist der spezielle Fall konstanten
Druckes über der Platte, also senkrechten Einfalls, gewählt, und ferner ist die Platte der

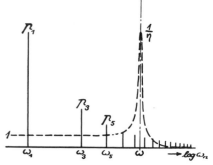

Abb. 4. Amplitudenspektrum für senkrechten Einfall auf eine gelenkig gelagerte Platte und Vergrößerungsfaktor (gestrichelt) für Dämpfung $\eta = 0,1$

Einfachheit halber in der einen Richtung beliebig breit, in der anderen Richtung an beiden
Rändern aufgestützt angenommen. Dadurch
werden die Eigenfunktionen zu Sinuskurven
und die Eigenfrequenzen verhalten sich wie

1 : 4 : 9 usw., wobei die geraden Teiltöne wegen der Symmetrie der Anregung wegfallen.

Zeichnen wir nun in dieselbe Skizze gestrichelt den Vergrößerungsfaktor

$$\left| \frac{1}{1 - \left(\dfrac{\omega_n}{\omega}\right)^2 (1 + i\,\eta)} \right|$$

ein, mit dem dieses Spektrum zu multiplizieren ist, so erkennt man, daß das Produkt aus beiden Größen, welches das Spektrum der Geschwindigkeitsverteilung kennzeichnet, unterhalb der Erregerfrequenz auf das Spektrum der Druckverteilung führt. Wenn wir nun unter Beibehaltung der gleichen Druckverteilung, also des gleichen Amplitudenspektrums, mit der Frequenz höher und höher gehen, so gilt dies schließlich für das ganze Spektrum, sofern der Resonanzgipfel dabei nicht beliebig wächst. Dies ist der Fall, wenn man den Korrekturfaktor — im Gegensatz zu der früheren Einführung der Dämpfung durch einen äußeren Reibungswiderstand — als konstant annimmt, was mit den Beobachtungen für innere Materialdämpfungen übereinstimmt.

Diese Eigenschaft der Materialdämpfung führt also zu dem allgemeinen Ergebnis, daß das Verteilungsbild der Bewegung der Platte das gleiche ist wie das des einwirkenden Druckes, sofern die Eigenfrequenzen aller in diesem Verteilungsbild wesentlichen Teilschwingungen tiefer liegen als die Erregerfrequenz:

$$\omega_n \ll \omega.$$

Ferner ist dabei die Bewegung nur durch die Massenwiderstände gehemmt, so daß man sich die Wand unter den genannten Umständen im Sinne des erwähnten „Massenmodells" als aus lauter einzelnen unabhängig voneinander beweglichen Massen zusammengesetzt vorstellen kann.

Wir können somit das SCHOCHsche asymptotische Gesetz mathematisch folgendermaßen formulieren:

$$(3.5) \qquad \boxed{\lim_{\substack{\omega \to \infty \\ p_n = \text{konst.}}} \bar{v}\,(x, y) = \frac{\bar{p}\,(x, y)}{i\,\omega\,m}}\,.$$

Die Einschränkung $p_n = $ konst. ist hier ausdrücklich hinzugefügt worden. An sich

würde bereits die allgemeinere und geringere Einschränkung genügen, daß die Teiltonamplituden p_n unterhalb dem jeweiligen Resonanzgipfel in Abb. 6 sehr klein geworden sein müssen; dabei dürften sie sich in diesem Bereich noch verändern. Das asymptotische Gesetz hat also einen etwas breiteren Gültigkeitsbereich, als ihn die Einschränkung $p_n = $ konst. abgrenzt. Dieser Zusatz soll auch nur deutlich machen, daß die p_n einer besonderen Bedingung genügen müssen, der mit $p_n = $ konst. jedenfalls genügt ist.

Wir werden nämlich sehen, daß, wenn wir bei schrägem Einfall die Frequenz beliebig wachsen lassen, die von SCHOCH gemachte Voraussetzung nicht mehr erfüllt ist und daß dann wesentliche Abweichungen von dem asymptotischen Gesetz und von dem einfachen „Massenmodell" auftreten.

Bei tieferen Frequenzen sind durch das Auftreten von Resonanzen Abweichungen im Sinne einer Erniedrigung der Schalldämmung zu erwarten. Der asymptotische Bereich wird nun bei um so niedrigeren Frequenzwerten bereits erreicht, je niedriger die ω_n sind, je niedriger also insbesondere die tiefste unter ihnen, die Grundfrequenz liegt. SCHOCH hat daher in

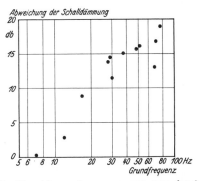

Abb. 5. Abweichung der gemessenen von der berechneten mittleren Schalldämmung in Abhängigkeit von der Grundfrequenz. (Nach SCHOCH)

seinem Buch[5]) sein asymptotisches Gesetz noch besonders dadurch experimentell gestützt, daß er die Abweichungen zwischen gemessener und theoretisch zu erwartender Dämmung in Abhängigkeit von der Grundfrequenz der Platten aufgetragen hat. Abb. 5 zeigt das Ergebnis. In der Tat wachsen die Abweichungen mit der Grundfrequenz an.

§ 4. Der Komponenteneffekt

Es erhebt sich nun die Frage, wieso bei Gültigkeit des asymptotischen Gesetzes sich die gemessenen Dämmkurven nicht an die bisherigen theoretischen bei hohen Frequenzen annähern, sondern auch dort erheblich abweichen.

Hierzu kann zunächst gezeigt werden, daß bei schrägem Einfall auch unter Beibehaltung des einfachen Massenmodells ein Effekt auftritt, der die Dämmung verkleinert. Wir wollen diesen Effekt, den auch SCHOCH bereits in seinem Buch[5]) behandelt, als „Komponenteneffekt" bezeichnen, weil er darauf beruht, daß bei schräger Einstrahlung und Abstrahlung, die Wandschnelle v zur Erfüllung der kinematischen Randbedingung nur gleich der wandnormalen Komponente der Schnelle v_2 der durchgelassenen Welle

$$(4.1) \qquad \bar{v} = \bar{v}_2 \cos \vartheta$$

(ϑ Einfallswinkel)

bzw. gleich der Summe der entsprechenden Komponenten der ankommenden und reflektierten Welle

$$(4.2) \qquad \bar{v} = \bar{v}_a \cos \vartheta + \bar{v}_r \cos \vartheta$$

sein muß. In den meisten früheren Beziehungen muß nun statt v ausdrücklich v_2 geschrieben werden, was dort gleichbedeutend war. So gilt für den Durchlaßgrad:

$$(4.3) \qquad \varDelta = \frac{v_2^2}{v_a^2},$$

ebenso kann nach (4.1) und (4.2) v_2 gesetzt werden:

$$(4.4) \qquad \bar{v}_2 = \bar{v}_a + \bar{v}_r;$$

ferner kann für die Drücke geschrieben werden:

$$(4.5) \qquad \bar{p}_1 = Z\,\bar{v}_a - Z\,\bar{v}_r,$$

$$(4.6) \qquad \bar{p}_2 = Z\,\bar{v}_2.$$

Nur wenn wir in der Trennwiderstandsformel unter Beibehaltung der früheren Definition v durch v_2 ersetzen, tritt abweichend der $\cos \vartheta$ als Faktor auf:

$$(4.7) \qquad \bar{p}_1 - \bar{p}_2 = (\overline{T} \cos \vartheta)\,\bar{v}_2.$$

Daher tritt auch in der aus (4.3) bis (4.7)

wie früher sich ergebenden Formel für den Durchlaßgrad der Trennwiderstand in Verbindung mit dem Faktor $\cos \vartheta$ auf:

$$(4.8) \qquad \varDelta = \frac{1}{\left|\, 1 + \dfrac{T \cos \vartheta}{2\,Z} \,\right|^2}.$$

Man kann auch sagen, statt des Wellenwiderstandes Z ist der Ausdruck $Z/\cos \vartheta$ einzusetzen. Dabei sei daran erinnert, daß derselbe Ersatz auch in die für senkrechten Einfall gültige Beziehung zwischen Schluckgrad und Wandwiderstand einzuführen ist, wenn man zum schrägen Einfall übergeht[12a]).

Auch hinsichtlich der Mittelwertbildung über die Durchlaßgrade bei statistischer Einfallswinkelverteilung gelten von der Schluckgradmessung im Nachhallraum her bekannte Gesichtspunkte. Es ist einerseits zu beachten, daß die schrägen Einfallswinkel entsprechend einem Faktor $\sin \vartheta$ häufiger sind, andererseits, daß eine gegebene Wandfläche F aus einer schräg einfallenden Welle nur eine Fläche $F \cos \vartheta$ herausschneidet. Somit ist für den mittleren Durchlaßgrad auszusetzen:

$$(4.9) \qquad \varDelta_m = \frac{\int\limits_0^{\frac{\pi}{2}} \varDelta \cos \vartheta \sin \vartheta\, d\vartheta}{\int\limits_0^{\frac{\pi}{2}} \cos \vartheta \sin \vartheta\, d\vartheta} = \int\limits_0^1 \varDelta\, d\,(\cos^2 \vartheta).$$

Wenden wir die Formeln (4.8) und (4.9) nun auf das „Massenmodell" mit dem Trennwiderstand $i\,\omega\,m$ an, so erhalten wir unter Einführung des Durchlaßgrades für senkrechten Einfall

$$(4.10) \qquad \varDelta_0 \approx \left(\frac{2\,Z}{\omega\,m}\right)^2,$$

die Beziehung:

$$(4.11) \qquad \varDelta_m \approx \varDelta_0 \ln \frac{1}{\varDelta_0}$$

bzw.:

$$4.12) \qquad D_m = D_0 - 10 \log 0{,}23\, D_0,$$

worin D_0 den nach (1.14a) zu errechnenden Schluckgrad für senkrechten Einfall bedeutet. Hiernach würde zu einem D_0 — Wert von 40 db eine mittlere Dämmung von nur etwa 30 db hören.

[12a]) Vgl. z. B. Elektr.-Nachr.-Techn. 10 (1933), S. 244.

Nun ist aber zu bedenken, daß der größte Teil dieses Abfalls auf das Gebiet streifenden Einfalls zurückzuführen ist, weil dort Δ sich theoretisch dem Wert 1, d. h. also dem Totaldurchgang nähert. Dieser Grenzfall dürfte hier aber ebenso ausscheiden, wie umgekehrt das Verschwinden des Schluckgrades bei streifendem Einfall praktisch bedeutungslos ist, weil die Voraussetzungen der unendlich breiten Welle, die auf eine unendlich breite Wand fällt, sich hier nicht mehr erfüllen lassen. Hinzu kommt, daß die zu messenden Trennwände — man denke etwa an Türrahmen — oft etwas tiefer liegen, so daß sie streifenden Einfällen gar nicht zugänglich sind. Es erscheint daher durchaus vernünftig, die Integration, wie auch SCHOCH es macht, nur bis zu einem bestimmten Winkel, etwa bis zu 75^0 oder zu 80^0 durchzuführen.

Das Unbefriedigende an diesem Verfahren liegt nur darin, daß die sich so ergebende Korrektur wesentlich von der Wahl der Grenzen abhängt. Man erhält nämlich dann

$$(4.14) \quad \Delta'_m = \int\limits_{\cos^2 \vartheta_1}^{1} \frac{d(\cos^2 \vartheta)}{1 + \dfrac{\cos^2 \vartheta}{\Delta_0}} = \Delta_0 \ln \frac{1}{\Delta_0 + \cos^2 \vartheta_1} .$$

Wählt man nun einmal $\vartheta_1 = 75^0$, also $\cos^2 \vartheta_1 = 0{,}067$ und andererseits $\vartheta_1 = 80^0$ also $\cos^2 \vartheta_1 = 0{,}0304$, so kann in beiden Fällen bei Dämmwerten über 30 db, also bei Durchlaßgraden unter 10^{-3}, Δ_0 gegen $\cos^2 \vartheta_1$ vernachlässigt werden, und die Dämmungskorrekturen werden unabhängig vom Dämmungspegel. Für $\vartheta_1 = 75^0$ ergibt sich

$$(4.15a) \qquad D_m - D_0 = 4{,}3 \text{ db}$$

für $\vartheta_1 = 80^0$ entsprechend

$$(4.15b) \qquad D_m - D_0 = 5{,}4 \text{ db} .$$

Diese grundsätzliche Unsicherheit haftet nicht nur der Auswertung der statistischen Messung an, sondern auch den Messungen selbst, da die ideale statistische Winkelverteilung praktisch gar nicht zu erreichen ist und es andererseits von großem Einfluß ist, ob zufällig etwa mehr oder weniger Schallwellen streifend auf die Wand fallen.

§ 5. Die allgemeinen Ansätze für Wände beliebiger Stärke

Wir dürfen aus der erwiesenen Gültigkeit des asymptotischen Gesetzes von SCHOCH für senkrechten Einfall jedenfalls schließen, daß bei hohen Frequenzen Wandform und Einspannungsart beliebig an Bedeutung verlieren und daß man hierbei jedes Wandelement als gleichberechtigt mit seinem Nachbarn ansehen kann. Nur unter dieser Voraussetzung kann von einer Schalldämmung je Flächeneinheit und ebenso von einem auf die Flächeneinheit bezogenen Trennwiderstand gesprochen werden, und nur dann kann aus den Isolations-Ergebnissen einer Wand auf die einer anderen mit anderer Gesamtfläche geschlossen werden.

Wir wollen uns daher im folgenden auf die Betrachtung von seitlich unendlich ausgedehnten Wänden beschränken. Wir wollen aber versuchen, das einfache Massenmodell dadurch zu ergänzen, daß wir die seitliche Kopplung zu erfassen suchen, die sich notwendigerweise der bei schrägem Einfall auftretenden gegenseitigen Verschiebung der einzelnen Massen widersetzen muß.

Auch in dieser Form ist das Problem des Schalldurchtritts durch eine feste Wand schon behandelt und gelöst worden, sogar in einer Allgemeinheit, wie wir sie hier gar nicht benötigen, nämlich ohne Beschränkung auf dünne Platten.

Es ist dabei die endliche Ausbreitungszeit für elastische Wellen im festen Körper berücksichtigt, wobei zwei verschiedene Wellenarten in Frage kommen, nämlich Verdichtungswellen (Longitudinalwellen) und Schubwellen (Transversalwellen).

Schon DRUDE[13] hatte den Übergang von Schallwellen aus einem festen Körper in einen zweiten behandelt und dabei gezeigt, daß eine reine Verdichtungswelle oder eine reine Schubwelle an einer Trennfläche im allgemeinen beide Wellenarten als reflektierte oder gebrochene Wellen auslöst.

BERGER[3] hat zuerst die DRUDEschen Ergebnisse für die Bauakustik zu verwerten ge-

---

[13] P. DRUDE, Wied. Ann. Phys. 41 (1890), S. 759.

sucht und dabei die Theorie sowohl dahin etwas ausgebaut, daß das primäre Medium schubspannungsfrei ist, als auch dahin, daß die Trennwand endliche Stärke besitzt. Es entstehen dann infolge der mehrfachen Reflexionen an Vorder- und Rückseite vier resultierende Wellenzüge in der Wand, nämlich je ein hineilender und ein rückeilender Verdichtungs- und Schubwellenzug.

Nun hatte RAYLEIGH[2]) schon für den Fall eines schubspannungsfreien Zwischenmediums mit der Verdichtungswellengeschwindigkeit c_d gezeigt, daß stets dann Totaldurchgänge zu erwarten sind, wenn die Projektion der Wandstärke d auf die Wellenausbreitungsrichtung im Material ϑ_d gerade ein Vielfaches der halben Wellenlänge λ_d im Material beträgt, also wenn gilt:

$$(5.1) \qquad d \cos \vartheta_d = n \frac{\lambda_d}{2}, \qquad n = 1, 2, 3, 4 \ldots$$

Wir wollen diese Erscheinung kurz als Dickenresonanz bezeichnen. Sie tritt bei wachsender Frequenz frühestens auf, wenn die Wandstärke gleich der halben Wellenlänge λ_d ist:

$$(5.2) \qquad d = \frac{\lambda_d}{2}.$$

Bei fester Zwischenwand waren solche Dickenresonanzen sowohl bei den Verdichtungs- wie bei den Schubwellen zu erwarten.

Sie wurden auch an Glasplatten in Wasser von BÄR und WALTI[14]) beobachtet, die sich bemühten, mit solchen Schalldurchtrittsversuchen auf neue Art die elastischen Materialkonstanten zu bestimmen.

Durch diese Forscher veranlaßt hat dann H. REISSNER[15]) — offenbar ohne Kenntnis der genannten Vorarbeiten — das Problem des Schalldurchganges durch feste Platten beliebiger Stärke nochmals in Angriff genommen und bis zur Angabe von Formeln für den Durchlaßgrad durchgeführt. Nach ihm haben nochmals BÄR[16]), sowie LEWI und NAGENDRA NATH[17]) versucht, seine Ergebnisse, wie auch

seine Ableitungen, etwas zu vereinfachen. Wie sich denken läßt, führt die vollständige Lösung des Problems auf sehr umfangreiche und undurchsichtige Ausdrücke.

Im Rahmen der Raum- und Bauakustik kann aber fast in allen Fällen die Wanddicke als so klein gegenüber den Wellenlängen im Wandmaterial angesehen werden, daß man, wie schon eingangs erwähnt, auf Vorder- und Rückseite einer Wandstelle die gleiche Bewegung ansetzen, die Wand also in Dickenrichtung als Ganzes betrachten kann. Das Spiel der Longitudinal- und Transversalwellen im Innern geht dann in die einfachen Biegeschwingungen über, wie dies TIMOSHENKO[18]) für die Eigenschwingungen eines Balkens gezeigt hat. Auch in den allgemeinen Formeln von REISSNER[15]) ist dieser Grenzfall enthalten, doch ist ihre Form gerade für diesen Fall sehr wenig durchsichtig, und es ist einfacher auch in der Ableitung von vornherein sich auf die Biegungs-Beziehungen zu stützen, dabei aber die seitliche Unbeschränktheit der Platte beizubehalten.

Merkwürdigerweise ist dieser Weg, der zwischen dem zu stark vereinfachten Schema des „Massenmodells" und den sehr allgemeinen, aber darum auch ziemlich verwickelten Darstellungen von SCHOCH einerseits und REISSNER andererseits liegt, noch nie beschritten worden.

Dies soll nun im zweiten Teil der vorliegenden Arbeit geschehen. Dabei werden wir nicht nur eine neue Ableitung für das „Massenmodell" erhalten, sondern wir werden gleichzeitig eine Übersicht gewinnen über den Einfluß der diese Massen seitlich koppelnden Biegesteifigkeit, und wir werden dabei zu einem wichtigen Effekt geführt werden, der bisher für die Schalldämmung unberücksichtigt geblieben ist und der erlaubt, manche der noch ungeklärten Abweichungen zwischen Theorie und Erfahrung, insbesondere die geringen Absolutwerte der Dämmung und die schwächer ansteigenden Frequenzgänge zu erklären.

Wie mir nach Abschluß der nachfolgenden Untersuchungen bekannt wurde, ist auch dieser

[14]) R. BÄR und A. WALTI, Helv. phys. Acta 7 (1934), S. 658.

[15]) H. REISSNER, Helv. phys. Acta 11 (1938), S. 140.

[16]) R. BÄR, Helv. phys. Acta 11 (1938), S. 397.

[17]) E. LEWI und NAGENDRA NATH, Helv. phys. Acta 11 (1938), S. 408.

[18]) S. P. TIMOSHENKO, Phil. Mag. Ser. 6, 43 (1922), S. 125.

Effekt mit Hilfe von Ultraschall von etwa $6 \cdot 10^6$ Hz an dünnen Metallfolien in Wasser von F. H. SANDERS beobachtet worden[19]). SANDERS erklärt ihn auch bereits durch Biegewellenbetrachtungen, wie es hier geschehen wird. Er befindet sich nur im Irrtum, wenn er meint, daß dieser Fall in den allgemeinen Formeln von REISSNER nicht enthalten sei. Er tritt in ihnen allerdings nicht deutlich zutage. Man kann aber mit einiger Rechenarbeit die in diesem Falle einfachere Biegewellendarstellung aus ihnen entwickeln. Dabei ergibt sich auch, daß die Biegewellendarstellung namentlich bei festen Wänden in Luft ein großes Frequenzgebiet umfaßt und somit selbständige Bedeutung hat, während bei festen Körpern im Wasser man sehr bald Korrekturen benötigt, weil dort in dem Frequenzgebiet, wo die Biegesteifigkeit Bedeutung gewinnt, auch schon die Dicken nicht mehr klein zur Wellenlänge sind.

II. Teil: Die Behandlung als erzwungene Biegewelle

§ 6. Der „Koinzidenz"-Effekt bei einer gespannten Membran

Als Voraufgabe wollen wir zunächst als dämmendes Gebilde eine gespannte Membran von unendlicher Ausdehnung betrachten. Auf diese möge unter dem Einfallswinkel ϑ eine ebene Welle laufen. Das Problem ist zweidimensional und spielt sich in der Einfallsebene ab. Die Membran tritt dabei nur als schwingende Saite in Erscheinung, für deren freie Bewegung bekanntlich die Differentialgleichung gilt:

$$(6.1) \qquad S \frac{\partial^2 w}{\partial x^2} = m \frac{\partial^2 w}{\partial t^2}.$$

Dabei bedeuten w den Ausschlag, S die Membranspannung und m ihre Masse je Flächeneinheit. Da diese Gleichung die Anwendung des dynamischen Grundgesetzes auf ein Massenelement der Länge und Breite 1 darstellt, wobei links die Resultierende aus den rück-

[19]) F. H. SANDERS, Canad. J. Res., Vol. 17 Sec. A. (1939), S. 179. Ich habe für diesen Literaturhinweis Herrn A. SCHOCH zu danken.

treibenden Komponenten der Spannkräfte und rechts das Produkt Masse mal Beschleunigung steht, so haben wir im Falle der Anregung durch eine auftreffende Schallwelle nur noch links die Drücke p_1 und p_2 vor und hinter der Membran hinzuzufügen:

$$(6.2) \qquad p_1 - p_2 + S \frac{\partial^2 w}{\partial x^2} = m \frac{\partial^2 w}{\partial t^2}.$$

Unter Beschränkung auf reine Töne und unter Einführung der Schnelle statt des Ausschlages

$$(6.3) \qquad \bar{v} = i \, \omega \, \bar{w}$$

nimmt diese Gleichung die Form an:

$$(6.4) \qquad \bar{p}_1 - \bar{p}_2 + \frac{S}{i \, \omega} \frac{\partial^2 \bar{v}}{\partial x^2} = i \, \omega \, m \, \bar{v}.$$

Nun wird bei der durch die auffallende Schallwelle erzwungenen Welle der Membran nicht nur die zeitliche, sondern auch die räumliche Periodizität durch die auffallende Welle bestimmt, d. h. alle Phasenänderungen müssen über die Membranoberfläche mit der gleichen Spurgeschwindigkeit $\frac{c}{\sin \vartheta}$ laufen. Die x-Abhängigkeit der Wandschnelle ist somit durch den Ansatz:

$$(6.5) \qquad \bar{v} \sim e^{\pm i \left(\frac{\omega}{c} \sin \vartheta \right) x}$$

gekennzeichnet. Damit kann aber Gleichung (6.4) auf eine der Beziehung (1.2) entsprechende Form gebracht werden:

$$(6.6) \qquad \bar{p}_1 - \bar{p}_2 = \left[i \, \omega \, m - i \, \omega \, \frac{S}{c^2} \sin^2 \vartheta \right] \bar{v}.$$

Der Trennwiderstand setzt sich also aus zwei Teilen zusammen, dem schon eingangs eingeführten Trägheitswiderstand und einem durch die seitliche elastische Kopplung bedingten Anteil, der seiner Phase entsprechend als eine Art Federungswiderstand aufgefaßt werden kann, der aber ebenfalls mit der Frequenz wächst. Ein Abfallen der Schnelle umgekehrt proportional der Frequenz bei konstanter Erregung braucht also kein Beweis dafür zu sein, daß eine Wand nur als träge Masse wirkt.

Das Wichtigste ist nun, daß der „Federungswiderstand" vom Einfallswinkel abhängt. Für senkrechten Einfall reduziert sich der Trennwiderstand, wie eingangs angesetzt auf den Trägheitswiderstand. Mit schräger werdendem Einfall wächst aber der Federungswiderstand, um schließlich bei streifendem Einfall seinen Höchstwert zu erreichen. Bei diesem Übergang wird der Trennwiderstand durch die gegenphasige Wirkung beider Anteile zunächst verringert, und es ergibt sich hieraus ein neuer Hinweis dafür, warum die im statistischen Mittel gewonnenen Dämmwerte viel kleiner sind, als die theoretischen Überlegungen für senkrechten Einfall ergeben würden.

Nun könnte aber auch für einen bestimmten Einfallswinkel der Fall eintreten, daß, ähnlich wie bei der Resonanz eines einfachen Schwingers für eine bestimmte Frequenz, sich Trägheits- und Federungswiderstand gerade gegenseitig aufheben; und zwar tritt das auch hier gerade dann ein, wenn die Bedingungen für eine „erzwungene" Bewegung mit denen einer „freien" Bewegung übereinstimmen.

Die Ausdrücke „erzwungen" und „frei" sind hierbei allerdings so zu verstehen, daß in beiden Fällen die Frequenz durch die ankommende Welle diktiert sein soll, daß aber bei der „erzwungenen" Bewegung auch noch die Phasengeschwindigkeit der Wellenausbreitung festgelegt ist. Eine „freie" Welle würde in diesem Sinne die Welle genannt werden müssen, die von einer punktförmigen Erregung mit der Kreisfrequenz ω ausgehen würde. Derartige „freie" Wellen sind zu überlagern, wenn man das Problem einer nach endlicher Länge eingespannten Membran behandeln will.

Bezeichnen wir die Ausbreitungsgeschwindigkeit einer solchen „freien" Welle der Membran mit c_1, so ist die x-Abhängigkeit bei einem Ton der Kreisfrequenz ω durch einen Ansatz der Form

$$(6.7) \qquad \bar{v} \sim e^{\pm i \frac{\omega}{c_1} x}$$

gekennzeichnet. Für c_1 gilt aber, wie sich auch durch Einsetzen von (6.7) in (6.1) ergibt, die bekannte Formel:

$$(6.8) \qquad c_1 = \sqrt{\frac{S}{m}}.$$

Unter Einführung dieses Wertes in (6.7) kann man den Trennwiderstand auch darstellen:

$$(6.9) \qquad T = i \omega m \left(1 - \left(\frac{c_1 \sin \vartheta}{c} \right)^2 \right).$$

Der Trennwiderstand verschwindet, wenn die anschauliche Beziehung:

$$(6.10) \qquad \boxed{c_1 = \frac{c}{\sin \vartheta}}$$

gilt, d. h. wenn die Spurgeschwindigkeit der einfallenden Welle gleich der Ausbreitungsgeschwindigkeit einer freien Membranwelle ist.

Dieser Effekt, für welchen die räumliche Periodizität dieselbe Bedeutung hat, wie die zeitliche Periodizität im Falle der Resonanz, wird zum Unterschied von dieser zweckmäßig mit einem besonderen Namen gekennzeichnet. Ich möchte hierfür die mir von Herrn H. GOETZ[20]) vorgeschlagene Bezeichnung „Koinzidenz" wählen, weil in ihr sowohl der Begriff des Zusammenpassens, wie der des Einfallswinkels anklingt, und derselbe jedenfalls in der Akustik nicht für andere Erscheinungen mit Beschlag belegt ist. Ich hoffe, daß sie auch auf anderen Gebieten der Physik nicht zu Verwechslungen führen wird; denn der so bezeichnete Effekt ist auch bei allen anderen Wellenarten möglich, vielleicht auch schon beobachtet und erklärt[21]).

Um eine Koinzidenz erhalten zu können, muß jedenfalls $c_1 > c$ sein oder, indem wir rechts und links durch die Frequenz dividieren, muß gelten $\lambda_1 > \lambda$, wenn λ_1 die Wellenlänge einer freien Welle gleicher Tonhöhe auf der Membran bedeutet. Für die e-Saite einer Geige gilt, daß ihre Länge, die $\frac{\lambda_1}{2}$ entspricht,

---

[20]) Herr Dipl.-Ing. H. GOETZ hat ferner im AEG-Forschungsinstitut eine Apparatur entwickelt, die es gestattet, diesen Effekt mit Hilfe von Ultraschall an Eisenplatten in Wasser quantitativ zu untersuchen. Diese Apparatur wurde von ihm am 13. Januar 1942 im Seminar für Mechanik der Technischen Hochschule Berlin im Anschluß an meinen Vortrag vorgeführt. Herr GOETZ wird hierüber im Rahmen einer Dissertation an gleicher Stelle berichten.

[21]) Es sei darauf hingewiesen, daß die „Koinzidenz-Beziehung" (6.10) in die Gleichung für den MACHschen Winkel übergeht, wenn c_1 die Geschoßgeschwindigkeit darstellt. Auch für die von OSWALD v. SCHMIDT, Ann. Phys. V. 19 (1934), S. 891 beobachteten Grenzwellen ergibt sich formal die gleiche Beziehung, wobei c_1 die Geschwindigkeit im zweiten Medium bedeutet.

etwa 1,2mal so lang ist, wie die Länge einer beiderseits offenen Pfeife für den gleichen Ton, die ihrerseits $\frac{\lambda}{2}$ entspricht. Vergleichen wir damit die tiefe Abstimmung einer kleinen Trommel, so leuchtet ein, daß derartige Spannungen, wie sie für $c_1 > c$ nötig sind, bei Membranen praktisch kaum auftreten werden.

Wenn aber umgekehrt gilt, daß $c_1 \ll c$ ist, so folgt daraus, daß die seitliche Kopplung nur eine vernachlässigbare Korrektur am Trennwiderstand in (6.8) bewirkt, und wir haben damit eine neue Ableitung für das „Massenmodell" gewonnen.

Da ferner bei gegebenen Abmessungen die Grundschwingung um so tiefer liegt, je kleiner c_1 ist, läßt auch diese Ableitung im Sinne der SCHOCHschen Beobachtungen eine um so bessere Übereinstimmung mit dem Massenmodell erwarten, je tiefer die Membran abgestimmt ist.

Zusammenfassend ist also zu sagen:

1. Die Betrachtung einer Trennwand als gespannte Membran führt bei tiefer Grundschwingung zu dem Modell aus ungekoppelten trägen Massen (Massenmodell).
2. Sie würde das vergleichsweise Absinken der Dämmwerte bei höheren Grundschwingungen erklären.
3. Sie würde aber keine Abweichung im Frequenzgang gegenüber der reinen Trägheitswirkung ergeben.

§ 7. Der Koinzidenz-Effekt bei der Platte

Wenn auch die Behandlung einer Trennwand als eine gespannte Membran der Wirklichkeit infolge des kontinuierlichen Charakters und der Berücksichtigung seitlicher Kopplungen bereits näher kommt, so befriedigt sie doch physikalisch nicht, weil ja die Rückstellkräfte bei einer Platte durch Biegespannungen hervorgerufen werden. Es bedeutet zudem keine besondere Erschwerung, wenn wir die im letzten Paragraphen angeschnittene Behandlungsweise auf eine biegesteife Platte übertragen, die dann in der Einfallsebene als Balken in Erscheinung tritt, nur mit dem Unterschied, daß infolge der verhinderten Querkontraktion ein etwas erhöhter Elastizitätsmodul nämlich (unter Be-

nutzung der POISSONschen Zahl μ) $E/1 - \mu^2$ statt des aus dem Zugversuch gewonnenen Elastizitätsmodul E einzusetzen ist.

Für das Massenelement der Länge 1 eines Balkens (der Breite 1) ergibt das dynamische Grundgesetz bekanntlich eine Differentialgleichung 4. Ordnung[22])

$$(7.1) \qquad - B \frac{\partial^4 w}{\partial x^4} = m \frac{\partial^2 w}{\partial t^2},$$

d. h. die durch die relative Verschiebung gegenüber der Nachbarschaft entstehenden Rückstellkräfte sind hier der vierten Ableitung nach dem Ort proportional.

Als Proportionalitäts-Konstante tritt die Biegesteifigkeit B auf, die ihrerseits gleich dem Produkt aus der erhöhten Dehnsteifigkeit $E/1 - \mu^2$ und dem Trägheitsmoment J je Breiteneinheit ist:

$$(7.2) \qquad B = \frac{E J}{1 - \mu^2}.$$

Unter Beschränkung auf reine Töne der Kreisfrequenz ω und unter Einführung der Schnelle v nimmt Gleichung (7.1) die Form an:

$$(7.3) \qquad i \frac{B}{\omega} \frac{\partial^4 \bar{v}}{\partial x^4} = i \omega m \bar{v}.$$

Die an einer Stelle mit der Kreisfrequenz ω eingeleitete „freie" Bewegung setzt sich aus 4 Lösungsanteilen zusammen, entsprechend den vier Anfangsbedingungen für Ausschlag, Neigung, Moment und Querkraft. Davon werden zwei durch reelle Exponentialfunktionen dargestellt, die mit der Entfernung von der Störung abklingen. Für die unendlich breite Platte interessieren nur die beiden anderen Anteile, die wieder Wellenausbreitungen mit der Geschwindigkeit c_1 entsprechen, die wir hier als Biegewellengeschwindigkeit bezeichnen können. Durch Einsetzen des Ansatzes

$$(7.4) \qquad \bar{v} \sim e^{\pm i \frac{\omega}{c_1} x}$$

erhält man aus (7.3) für die Biegewellen-

---

[22]) Hinsichtlich einer ausführlichen Ableitung der Biegewellengleichung in einer der modernen Akustik eigens angepaßten Darstellung sei u. a. auf die Arbeit des Verfassers in den Sitzungsberichten der Berliner Akad., math.-phys. Kl. (1934), Heft **1** verwiesen.

geschwindigkeit die ebenfalls bekannte Formel:

$$(7.5) \qquad c_1 = \sqrt[4]{\frac{B}{m}} \sqrt[2]{\omega} \,,$$

welche besagt, daß die Biegewellenausbreitung dispergierenden Charakter hat, und zwar daß die Phasengeschwindigkeit c_1 mit der Wurzel aus der Frequenz wächst. In dieser Frequenzabhängigkeit liegt der Hauptunterschied gegenüber den Verhältnissen bei der gespannten Membran.

Bei der Behandlung der durch eine schräg auffallende Welle erzwungenen Bewegung tritt in (7.3) wieder links unter den Kräften an der Flächeneinheit noch die Differenz der Drücke vor und hinter der Wand auf:

$$(7.6) \qquad \bar{p}_1 - \bar{p}_2 + i\,\frac{B}{\omega}\,\frac{\partial^4 \bar{v}}{\partial x^4} = i\,\omega\,m\,\bar{v}\,.$$

Ferner ist für die x-Abhängigkeit wieder die Spurgeschwindigkeit der einfallenden Welle maßgebend, so daß (7.6) durch Einsetzen des Ansatzes.

$$(7.7) \qquad v \sim e^{\pm i\,\frac{\omega}{c}\sin\vartheta\,x}$$

umgeformt werden kann zu:

$$(7.8) \qquad \bar{p}_1 - \bar{p}_2 = \left(i\,\omega\,m - i\,\frac{B\,\omega^3\sin^4\vartheta}{c^4}\right)\bar{v}\,,$$

d. h. der Trennwiderstand nimmt hier die Form an:

$$(7.9) \qquad \overline{T} = i\,\omega\,m - i\,\frac{B\,\omega^3\sin^4\vartheta}{c^4}\,.$$

Der „Federungswiderstand" wächst hier sogar mit der dritten Potenz der Frequenz. Ein Überwiegen desselben würde also ein noch wesentlich stärkeres Wachsen der Dämmung mit der Frequenz bedeuten, als es im allgemeinen beobachtet wird.

Unter Einführung der Phasengeschwindigkeit der Biegewellen c_1 läßt sich der Trennwiderstand auch darstellen in der Form:

$$(7.9a) \qquad \overline{T} = i\,\omega\,m\left(1 - \left(\frac{c_1\sin\vartheta}{c}\right)^4\right).$$

Wir erhalten also, abgesehen vom Auftreten der vierten Potenz statt der zweiten, dieselbe Form, wie in (6.9); der Trennwiderstand verschwindet wieder bei der Koinzidenz-Bedingung, daß die Ausbreitungsgeschwindigkeit der „freien" Welle längs der Wand gleich der Spur-

geschwindigkeit der einfallenden Welle ist:

$$(7.10) \qquad \boxed{c_1 = \frac{c}{\sin\vartheta}}\,.$$

Der wesentliche Unterschied gegenüber dem vorigen Paragraphen besteht nun darin, daß diese Bedingung frequenzabhängig geworden ist. Es gibt also nicht nur bestimmte Koinzidenzwinkel ϑ_k, sondern auch bestimmte Koinzidenzfrequenzen f_k. Für diese erhält man durch Einsetzen von (7.5) in (7.10):

$$(7.11) \qquad f_k = \frac{c^2}{2\,\pi\sin^2\vartheta_k}\sqrt{\frac{m}{B}}\,.$$

Es gibt daher grundsätzlich zu jedem gegebenen Einfallswinkel eine Koinzidenzfrequenz, unterhalb welcher die Massenwirkung, oberhalb welcher die Steifigkeitswirkung überwiegt. Diese Frequenz hat ihren niedrigsten Grenzwert f_g bei streifendem Einfall $\vartheta = 90^0$, nämlich:

$$(7.12) \qquad f_g = \frac{c^2}{2\,\pi}\sqrt{\frac{m}{B}}\,.$$

Unterhalb dieser Frequenz ist überhaupt keine Koinzidenz möglich, oberhalb derselben ist dagegen stets eine möglich, wobei der Koinzidenzwinkel mit wachsender Frequenz immer kleiner wird. So gehört z. B. zur doppelten Grenzfrequenz ein Koinzidenzwinkel von 45^0, zur vierfachen Grenzfrequenz ein Koinzidenzwinkel von 30^0, zur achtfachen ein solcher von etwa 20^0 usf. Weiter hinauf läßt sich die Beziehung (7.11) bei festen Wänden in Luft meist nicht mehr anwenden, weil dann die Voraussetzungen dünner Wände im Sinne der einfachen Biegungswelle nicht mehr statthaft sind.

Wenn wir nun auch auf diesem Wege zu dem asymptotischen Gesetz von SCHOCH und damit zu dem Wandmodell aus unabhängigen trägen Massen kommen wollen, müssen wir annehmen, daß die Grenzfrequenz im allgemeinen hoch über den Beobachtungsfrequenzen liegt. Dies wäre um so eher erfüllt, je kleiner $\sqrt{\frac{B}{m}}$ ist und das besagt, in Übereinstimmung mit der von SCHOCH gestellten Bedingung, daß die Grundfrequenz einer Platte von gegebenen Abmessungen möglichst tief sein soll.

Um nun diese Verhältnisse quantitativ zu kontrollieren, wollen wir (7.12) auf den praktisch hauptsächlich interessierenden Fall einer homogenen Platte der Stärke d und Dichte ϱ anwenden. Wir haben dann für B einzusetzen:

$$(7.13) \qquad B = \frac{E}{1-\mu^2} \cdot \frac{d^3}{12}$$

und für m:

$$(7.14) \qquad m = \varrho\, d\,.$$

Damit geht die Formel für die Grenzfrequenz über in:

$$(7.15) \qquad f_g = \frac{c^2}{2\pi d} \sqrt{\frac{12\,\varrho\,(1-\mu^2)}{E}}\,.$$

Die Materialeigenschaften lassen sich unter Vernachlässigung der geringen Querkontraktionseinflüsse ($\mu^2 \ll 1$) durch die meist aus Tabellen[23]) bekannte Longitudinalwellengeschwindigkeit c_l in Stäben:

$$(7.16) \qquad c_l = \sqrt{\frac{E}{\varrho}}$$

kennzeichnen, so daß wir aus (7.15) die einfache Näherungsformel:

$$(7.17) \qquad f_g \approx \frac{c^2}{2\,c_l\,d}$$

gewinnen, welche sich unter Einführung der zu f_g gehörigen Wellenlänge in Luft λ_g auf die anschauliche Proportion bringen läßt:

$$(7.18) \qquad \boxed{\frac{d}{\lambda_g} = \frac{c}{2\,c_l}}\,.$$

In Abb. 6 ist nun über der Grenzfrequenz als Abszisse die zugehörige Plattendicke als Ordinate aufgetragen. Da beide Skalen logarithmisch gestuft sind, tritt die Bedingung (7.17) als fallende Gerade in Erscheinung. Als Baustoffe sind Glas, Ziegelstein und Eichenholz längs und quer zur Faser gewählt. Da die Unterschiede in den Longitudinalgeschwindigkeiten nur ein Verhältnis $1:2$ umfassen, liegen die zugehörigen Geraden gegenüber den um 3 Zehnerpotenzen variierenden Werten der Frequenz und Dicke verhältnismäßig dicht beieinander. Es ist nun beachtlich, daß die erhaltenen Geraden, rechts oberhalb welcher

Koinzidenzen zu erwarten sind, nicht, wie zunächst zu vermuten wäre, nur ein kleines Ausnahmegebiet abtrennen, sondern daß sie diagonal durch den ganzen interessierenden Bereich laufen, dessen Abszisse die Hörfrequenzen von 50 bis 10000 Hz und dessen Ordinate sämtliche bautechnisch interessierenden Stärken von 1 mm bis zu einer einsteinstarken Wand von 25 cm umfaßt.

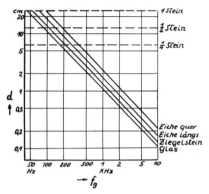

Abb 6. Zusammenhang zwischen Wanddicke d und Grenzfrequenz f_g für verschiedene Baustoffe

Hiernach ist nur für ganze dünne Wände, vorzugsweise aus Holz wie sie als „mitschwingende Schallschlucker" verwendet werden, die Gültigkeit des Massenmodells im ganzen Hörbereich zu erwarten. In diesem Sinne zeigt auch die in Abb. 2 unter Nr. 1 angeführte Sperrholzplatte als einzige kaum eine Abweichung vom einfachen Massendämmungsgesetz (1.14). Dagegen fällt schon bei einer 1,2 cm dicken Glasplatte, wie sie der Kurve Nr. 2 in Abb. 2 zugrunde liegt, die Grenzfrequenz etwas unter 1000 Hz. In der Tat zeigt die gemessene Kurve in diesem Gebiet einen deutlichen Abfall, worauf sie etwa bei 1200 Hz ein Minimum erreicht. Danach folgt ein stärkerer Anstieg, als er dem Massengesetz entspricht. Bei einem Frequenzgang für einen bestimmten Einfallswinkel hätte man dies auch bei Überwiegen des Biegesteifigkeitsgliedes im Trennwiderstand zu erwarten. Da bei statistischer Winkelverteilung aber oberhalb der Grenzfrequenz stets Koinzidenzen auftreten, liegen hierbei die Verhältnisse ver-

[23]) Z. B. E. Lübcke, Schallabwehr, Berlin (1940), S. 151.

wickelter. Wir werden darauf im nächsten Paragraphen eingehen [23a]).

Schließlich wollen wir noch die Ziegelwände Nr. 4 bis Nr. 6 in Abb. 2 betrachten. Sie zeigen zwar keine so deutlichen Einbrüche wie die Glasplatte, aber auch bei Nr. 4 und Nr. 5 ist der Verlauf anfangs flacher und wird dann steiler als dem Massengesetz entspricht. Dabei hat sich die Übergangsstelle bei der dickeren Wand nach links verschoben. Der zu der dicksten Wand gehörige flache Teil liegt bereits links außerhalb der Abb. 2. Die Grenzfrequenzen entnehmen wir — wenn wir von den Putzschichten absehen — der Abb. 6 zu 200 Hz für die $^1/_4$ Stein starke (Nr. 4), zu 100 Hz für die $^1/_2$ Stein starke (Nr. 5) und zu 50 Hz für die 1 Stein starke Ziegelwand (Nr. 6).

Es muß hier allerdings darauf hingewiesen werden, daß keine einheitlich anerkannten Angaben über die Longitudinalwellengeschwindigkeiten von Ziegelwänden vorliegen. In Abb. 6 wurde der bei LÜBCKE angegebene Wert von 4300 m/sec zugrunde gelegt. Dagegen hat R. SCHMIDT[23b]) den fast nur halb so großen Wert von 2325 m/sec gemessen. Unter Benutzung dieses Wertes würden die obigen Grenzfrequenzen 370, 185 und 92,5 Hz lauten.

Daß die Kurven Nr. 5 und 6 bei hohen Frequenzen wieder flacher werden, dürfte — sofern es sich nicht nur um Meßfehler infolge akustischer Nebenwege bei diesen hohen Dämmwerten handelt — auf Abweichungen von der hier benutzten einfachen Biegewellendarstellung zurückzuführen sein, die durch die im Verhältnis zur Verdichtungswellenlänge λ_d immer größer werdende Wanddicke bedingt sind. Dabei würde freilich die nach (5.2) zu

erwartende Dickenresonanz erst in der äußersten Ecke der Abb. 6 in Erscheinung treten. Aber schon 4 Oktaven tiefer macht sich eine Abweichung von der einfachen Biegewellentheorie bemerkbar, die sich darin äußert, daß man die Drehträgheit, was RAYLEIGH[24]) schon tat, und außerdem nach TIMOSHENKO[25]) die endliche Schubdeformation bei der Ableitung der Biegewellengleichungen berücksichtigen muß. Mit diesen Korrekturen käme man schätzungsweise etwas über 2 Oktaven höher, so daß man praktisch den ganzen interessierenden Bereich damit erfassen könnte.

Wir wollen nun noch untersuchen, wie sich der Koinzidenzeffekt zu dem asymptotischen Gesetz von SCHOCH verhält. Zunächst können wir nach (7.8) und (7.9a) die Ortsabhängigkeit der Schnelle bei einer unter ϑ einfallenden Welle von gegebenem Druck darstellen in der Form:

$$(7.19) \qquad \bar{v}(x) = \frac{\bar{p}\, e^{\pm i\frac{\omega}{c}\sin\vartheta\, x}}{i\,\omega\, m\left(1 - \left(\frac{c_1\sin\vartheta}{c}\right)^4 (1 - i\,\eta)\right)}.$$

Dabei haben wir im Sinne von SCHOCH den Druck auf der Rückseite und somit die Strahlungsdämpfung vernachlässigt, dafür aber eine innere Dämpfung in der Form des Korrekturgliedes $i\,\eta$ eingeführt. Formel (7.19) hat dadurch die Form der Beziehung (3.4) erhalten.

Da wir eine unendlich breite Wand voraussetzen, kann jede sinusförmige Druckverteilung von beliebiger Periodizität als Eigenfunktion $\bar{u}_n(x)$ aufgefaßt werden:

$$(7.20) \qquad \bar{u}_n(x) = e^{\pm i\frac{\omega}{c}\sin\vartheta\, x}.$$

Das Spektrum der Abb. 4 geht somit in eine einzige Linie über. Bei einer endlichen, aber sehr breiten Wand würden stattdessen mehrere Spektrallinien sich zu einem spitzen Gipfel vereinigen. Der Ort dieser Linie bzw. dieses Gipfels ist bestimmt durch die Eigenfrequenz ω_n, welche einer Biegewelle mit gleicher räumlicher Periodizität entsprechen würde:

$$(7.21) \qquad \frac{\omega_n}{c_1\, n} = \frac{\omega}{c}\sin\vartheta.$$

[23a]) Ähnliche Verhältnisse sind zu erwarten bei Holzwänden von 2 cm Stärke, wie sie namentlich bei Türen Verwendung finden. Herr W. WEBER (Reichs-Rundfunk-Ges.) teilte am 13. Januar 1942 in der Diskussion mit, daß er bei solchen Türen ein deutliches Nachlassen der Dämmung über 1000 Hz festgestellt habe. Er wies mich ferner auf ähnliche Beobachtungen von BERG und HOLTSMARK hin. (Kongel. norske Vidensk. Selsk. Forh. Bd. VIII, Nr. 35 [1935], S. 123).

[23b]) R. SCHMIDT, Ing. Arch. V (1934), S. 360.

[24]) LORD RAYLEIGH, Theory of Sound I, § 186.

[25]) S. P. TIMOSHENKO, Phil. Mag. Ser. 6, 41 (1921), S. 744.

Dabei ist zu beachten, daß die Biegewellengeschwindigkeit c_{1n} bei der Eigenfrequenz ω_n zu derjenigen c_1, die zur Erregerfrequenz ω gehört, entsprechend dem Dispersionsgesetz (7.5) im Verhältnis steht:

$$(7.22) \qquad \frac{c_{1n}}{c_1} = \sqrt{\frac{\omega_n}{\omega}}.$$

Aus der Zusammenfassung beider Beziehungen ergibt sich, daß wir in (7.19) einsetzen können

$$(7.23) \qquad \left(\frac{c_1 \sin \vartheta}{c}\right)^4 = \left(\frac{\omega_n}{\omega}\right)^2,$$

wodurch die Identität mit der Formel (3.4) völlig hergestellt ist. Schließlich ergibt sich unter Benutzung von (7.5) für die Eigenfrequenz:

$$(7.24) \qquad \omega_n = \sqrt{\frac{B}{m} \frac{\sin^2 \vartheta}{c^2}} \, \omega^2.$$

Nehmen wir nun an, daß die Spektrallinie der anregenden Druckverteilung in einer Abb. 4 entsprechenden Darstellung zunächst links von dem Gipfel des Übertragungsfaktors liegen möge. Lassen wir dann im Sinne des Grenzüberganges von SCHOCH die Frequenz wachsen, so rückt der Gipfel des Übertragungsfaktors proportional mit ω, derjenige des Druckverteilungs-Spektrums dagegen mit ω^2 nach rechts. Der letzte muß also den ersten überholen und damit aus dem Gültigkeitsbereich des asymptotischen Gesetzes herausrücken. Die Koinzidenz aber entspricht in dieser Darstellung dem Zusammenfallen der beiden Gipfel.

§ 8. Analogie zwischen Resonanz und Koinzidenz

Wir hatten schon auf die Ähnlichkeit der Koinzidenz mit der Resonanz eines einfachen Schwingers hingewiesen, weil hier wie dort die Massen- und Federungswiderstände sich gegenseitig aufheben. Diese Analogie ist bei genügend kleiner Dämpfung sogar mathematisch so vollkommen, daß wir viele Gesetzmäßigkeiten, die uns von der Resonanz her geläufig sind, auf die Koinzidenz übertragen können.

Betrachten wir zunächst eine elastisch gelagerte Kolbenmembran, die quer in ein unendlich langes Rohr gespannt ist, in welchem eine ebene Welle entlanglaufe, dann gilt für

den Trennwiderstand wie in (2.3)

$$(8.1) \qquad \overline{T} = i \, \omega \, m \left(1 - \left(\frac{\omega_0}{\omega}\right)^2\right)$$

und für den Durchlaßgrad

$$(8.2) \qquad \Delta = \frac{1}{1 + \left(\frac{\omega m}{2Z}\right)^2 \left(1 - \left(\frac{\omega_0}{\omega}\right)^2\right)^2}.$$

In der Nähe des Resonanzgipfels, wo allein große Durchlaßgrade erzielt werden, läßt sich durch die Einführung der Verstimmung

$$(8.3) \qquad \varepsilon = \left(\frac{\omega}{\omega_0} - 1\right)$$

der Durchlaßgrad auf die übliche Resonanzform bringen:

$$(8.4) \qquad \Delta = \frac{1}{1 + \left(\frac{\omega_0 m}{Z}\right)^2 \varepsilon^2}.$$

Diese Kurve ist um so schmaler, je größer $\omega_0 m$ ist. Als Maß hierfür pflegt man die Halbwertsbreite, d. h. die Breite der Resonanzkurven, bei der die Energie auf den halben Gipfelwert gesunken ist, anzugeben, welche gleich ist:

$$(8.5) \qquad 2\varepsilon' = \frac{2Z}{\omega_0 m}.$$

Diese Größe ist andererseits gleich dem Quotienten aus 2δ und ω_0, wobei δ die Abklingkonstante einer freien Schwingung der Kolbenmembran ist:

$$(8.6) \qquad \delta = \frac{Z}{m}.$$

Diese Größe kann auch unmittelbar daraus gefunden werden, daß die in der Membran schwingende zeitlich abnehmende Energie E

$$(8.7) \qquad E = \frac{1}{2} m \, v^2 = \frac{1}{2} m \, v_0^2 \, e^{-2\delta t}$$

(hierbei bedeutet v den Scheitelwert) durch Abgabe von Strahlungsleistung nach beiden Seiten verringert wird:

$$(8.8) \qquad -\frac{dE}{dt} = \delta \, m \, v^2 = 2 \, Z \, \frac{v^2}{2}.$$

Bei der Gegenüberstellung der Koinzidenz wollen wir nur ihre Winkelabhängigkeit betrachten, welche sowohl bei der gespannten Membran, als auch bei der Platte auftritt, dagegen wollen wir die Frequenz konstant lassen. Damit übernimmt die Ortsabhängigkeit parallel zur Platte, die sich in dem Faktor

$$e^{\pm i \left(\frac{\omega}{c} \sin \vartheta\right) x}$$

kundtut, die Rolle, die früher der

Zeitabhängigkeit zukam. Führen wir nun z. B. bei der Platte in gleicher Weise in die Formel für den Durchlaßgrad:

$$(8.9) \qquad \varDelta = \cfrac{1}{1 + \left[\cfrac{\omega \, m \cos \vartheta}{2\,Z} \left(1 - \left(\cfrac{c_1 \sin \vartheta}{c} \right)^4 \right) \right]^2}$$

die „Verstimmung" des x-Faktors gegenüber dem Koinzidenzwert ein, indem wir setzen,

$$(8.10) \qquad \sin \vartheta = \sin \vartheta_k \, (1 + \varepsilon) \, ,$$

so ergibt sich für kleine Winkelabweichungen ebenfalls eine „Resonanzkurve", wenn wir allgemein die in (8.4) gegebene Abhängigkeit einer Größe von der Verstimmung so bezeichnen, nämlich:

$$(8.11) \qquad \varDelta = \cfrac{1}{1 + \left(\cfrac{2\,\omega \, m \cos \vartheta_k}{Z} \right)^2 \varepsilon^2} \cdot$$

Die Halbwertsbreite beträgt dabei:

$$(8.12) \qquad 2\,\varepsilon' = \cfrac{Z}{\omega \, m \cos \vartheta_k}$$

und hieraus kann nach Multiplikation mit dem für die räumliche Periodizität maßgebenden Faktor $\frac{\omega}{c_1}$ geschlossen werden, daß das räumliche Abnehmen einer „freien", d. h. einer nur an einer Störstelle mit der Frequenz ω erregten Welle, mit einer Dämpfungskonstanten

$$(8.13) \qquad \beta = \frac{\varepsilon' \, \omega}{c_1} = \frac{Z}{2 \, c_1 \, m \cos \vartheta_k}$$

erfolgen muß.

Auch dies Ergebnis kann andererseits unmittelbar aus der Leistungsbilanz einer fortschreitenden Biegewelle erschlossen werden; eine genauere Betrachtung zeigt nämlich, daß einer solchen bei der maximalen Schnelle v eine Intensität J

$$J = m \, c_1 \, v^2 = m \, c_1 \, v_0^2 \, e^{-2\beta x}$$

zukommt. Da nun dieser Welle auf der Wegeinheit die Leistung

$$2 \, \frac{Z}{\cos \vartheta_k} \, \frac{v^2}{2}$$

entzogen wird, so gilt die Bilanz:

$$(8.14) \qquad - \frac{d\,J}{d\,x} = 2 \, \beta \, m \, c_1 \, v^2 = \frac{Z}{\cos \vartheta_k} \, v^2 \, ,$$

aus welcher ebenfalls Formel (8.13) folgt.

Es ist nun bekannt, daß, wenn ein zeitlich begrenzter Wellenzug mit einer der Resonanzfrequenz entsprechenden Trägerfrequenz auf

eine solche federnd gelagerte Kolbenmembran trifft, dieselbe eine gewisse Zeit benötigt, um auf ihre volle Schwingungsweite aufgeschaukelt zu werden, und dafür nach Beendigung des Signals noch nachschwingt.

Auf Grund der gezeigten Analogie können wir folgern, daß ebenso bei Auftreffen eines räumlich begrenzten Wellenzuges, also eines „Strahls", der unter dem Koinzidenzwinkel einfallen möge, die Wand nicht auf der ganzen Strahlbreite gleich erregt wird, sondern daß sie vielmehr, wenn man ihre Schwingung in Richtung der Spurgeschwindigkeit abtastet, zunächst eine gewisse Breite zum Aufbau der vollen Schwingung braucht, und daß sie ebenso jenseits des primären Strahles erst ein allmähliches „Abklingen" aufweist. Denselben Charakter weist die allein von den Wandschwingungen bestimmte durchgelassene Welle auf [25a].

Der hier erwähnte Effekt ist vor allem für die Beurteilung von Messungen wichtig, weil bei Benutzung eines Strahles endlicher Breite oder auch einer Kugelwelle, die nur in einem schmalen Bereich der Koinzidenzbedingung genügt, die Energieverbreitung dazu führt, daß der an einer Stelle jenseits der Wand gemessene Schalldruck bei Koinzidenz nicht so stark ansteigt, wie das bei Benutzung einer unendlich breiten einfallenden Welle der Fall sein müßte.

Wir können dieses Ergebnis unter Benutzung der Fourier-Analyse auch so erklären, daß bei einem zeitlich begrenzten Wellenzug auch Nachbarfrequenzen mit merklicher Energie auftreten, und daß ebenso ein „Strahl" endlicher Breite aus vielen unendlich breiten Wellenzügen zusammengefaßt werden kann, deren Einfallswinkel zu dem Strahlwinkel benachbart liegen.

Die Frage, welche Energie bei statistischer Winkelverteilung durch die Wand tritt, welcher

---

[25a] Die dabei auftretende räumliche Verbreiterung der aufgefallenen Energie weist ebenfalls auf die Verwandtschaft mit den von O. v. Schmidt betrachteten „Grenzwellen" hin, die bei einem Strahl auftreten können, der unter dem Grenzwinkel der Totalreflexion auf die Trennfläche zwischen zwei verschiedene Medien fällt.

wir uns nunmehr zuwenden wollen, ist daher unter diesem Gesichtspunkt verwandt. Auch hier hilft uns die Analogie mit dem einfachen Resonanzsystem. Wir fragen zunächst, welcher mittlere Durchlaßgrad für die federnd gelagerte Kolbenmembran anzusetzen ist, wenn sich die Schallenergie der auftreffenden Welle auf ein kontinuierliches Spektrum mit konstantem Amplitudenbelag zwischen den Frequenzen ω_1 und ω_2, welche die Resonanzfrequenz ω_0 einschließen mögen, verteilt. Wir haben dann anzusetzen:

$$(8.15) \qquad \Delta_m = \frac{1}{\omega_2 - \omega_1} \int\limits_{\omega_1}^{\omega_2} \Delta \, d\omega \, .$$

Da der Durchlaßgrad nur in unmittelbarer Nähe der Resonanz große Werte annimmt und wir dort für Δ den Ausdruck (8.4) einsetzen können, formen wir das Integral unter Einführung der Verstimmung zweckmäßig um zu:

$$(8.16) \qquad \Delta_m = \frac{\omega_0}{\omega_2 - \omega_1} \int\limits_{-\varepsilon_1}^{+\varepsilon_1} \frac{d\varepsilon}{1 + \left(\frac{\omega_0 m}{Z}\right)^2 \varepsilon^2} \, .$$

Die hierbei symmetrisch zur Resonanzstelle gewählten Grenzen haben nichts mit den früheren Grenzen ω_2 und ω_1 zu tun, denn so weit würde die Darstellung (8.4) ohnehin nicht gelten. Sie sind vielmehr beliebig. Sie müssen nur so groß sein, daß sie praktisch das ganze Resonanzgebiet einschließen und eine Vergrößerung derselben zur Integralfläche nichts Nennenswertes mehr beträgt. Dies ergibt sich mathematisch daraus, daß

$$(8.17) \quad \int\limits_{-\varepsilon_1}^{+\varepsilon_1} \frac{d\varepsilon}{1 + \left(\frac{\omega_0 m}{Z}\right)^2 \varepsilon^2} = \frac{2Z}{\omega_0 m} \text{arc tg} \frac{\omega_0 m}{Z} \varepsilon_1 \approx \frac{\pi Z}{\omega_0 m} \, ,$$

also unabhängig von ε_1 ist, so lange nur die durch $Z/\omega_0 m$ zum Ausdruck kommende Dämpfung der Resonanzkurven genügend klein ist. Wir haben also erhalten

$$(8.18) \qquad \Delta_m \approx \frac{\pi Z}{(\omega_2 - \omega_1) m} \, .$$

Genau so können wir nun vorgehen, um den statistischen Mittelwert des Durchlaßgrades zu bilden, der nach § 4 anzusetzen ist als:

$$(8.19) \quad \Delta_m = \int\limits_0^1 \Delta \, d(\cos^2 \vartheta) = 2 \int\limits_0^1 \Delta \sin \vartheta \, d(\sin \vartheta) \, .$$

Auch hierbei ist, wenn wir wieder das Gebiet streifenden Einfalls ausschließen und der Koinzidenzgipfel bereits aus dem streifenden Gebiet herausgerückt ist, der Energiedurchtritt auf die unmittelbare Nachbarschaft des Koinzidenzwinkels beschränkt. Wir können daher auch hier unter Benutzung von (8.10) und (8.11) das Integral umformen zu:

$$(8.20) \qquad \Delta_m = 2 \sin^2 \vartheta_k \int\limits_{-\varepsilon_1}^{+\varepsilon_1} \frac{d\varepsilon}{1 + \left(\frac{2 \omega m \cos \vartheta_k}{Z}\right)^2 \varepsilon^2}$$

und erhalten hieraus:

$$(8.21) \qquad \Delta_m \approx \frac{\pi Z \sin^2 \vartheta_k}{\omega m \cos \vartheta_k} = \frac{\pi}{2} \sqrt{\Delta_0} \frac{\sin^2 \vartheta_k}{\cos \vartheta_k} \, .$$

In dieser Formel kommt die Frequenzabhängigkeit sowohl in Δ_0 als auch im Koinzidenzwinkel zum Ausdruck. Führen wir noch das Verhältnis ξ der jeweiligen Frequenz zu der in (7.11a) eingeführten Grenzfrequenz ein, so läßt sich einerseits ausdrücken:

$$(8.22) \qquad \Delta_0 = \frac{\Delta_{0g}}{\xi^2} \, ,$$

dabei bedeutet Δ_{0g} den Durchlaßgrad für senkrechten Einfall bei der Grenzfrequenz, andererseits

$$(8.23) \qquad \sin^2 \vartheta_k = \frac{1}{\xi} \, .$$

Damit erhalten wir schließlich für den mittleren Durchlaßgrad im Koinzidenzgebiet, d. h. etwa für $\xi \gtrsim 2$:

$$(8.24) \qquad \Delta_m \approx \frac{\pi}{2} \frac{\sqrt{\Delta_{0g}}}{\xi^2 \sqrt{1 - \frac{1}{\xi}}} \, .$$

Für die Dämmung ergäbe sich hieraus unter Vernachlässigung unwesentlicher Glieder die Formel:

$$(8.25) \qquad D_m \approx \frac{1}{2} D_{0g} + 20 \log \xi \, .$$

Darin bedeutet D_{0g} die Dämmung für senkrechten Einfall bei der Grenzfrequenz.

Dieser Wert hängt übrigens, wie man sich leicht durch Einsetzen von (7.18) in (1.14a) überzeugt, nur vom Wandmaterial ab:

$$D_{0g} = 20 \log \frac{\pi \varrho c^2}{Z c_l} \, .$$

Wir entnehmen dieser Formel vor allem das wichtige Ergebnis, daß die Dämmung im

Koinzidenzbereich wieder mit der Frequenz ansteigen muß, und zwar würde der Anstieg hiernach wie unterhalb der Koinzidenz mit dem 20fachen Logarithmus derselben erfolgen. Dies ist zum Teil darauf zurückzuführen, daß die beiden gegenphasigen Anteile des Trennwiderstandes mit der Frequenz wachsen, die Koinzidenzspitze also immer schärfer wird, zum Teil darauf, daß der Gipfel in statistisch immer unwichtigere Gebiete rückt.

§ 9. Berücksichtigung der Materialdämpfung

Nach Formel (8.25) würde die Dämmung oberhalb der doppelten Grenzfrequenz sich einer Geraden nähern müssen, die der früheren theoretischen Dämmung D_0 parallel ansteigt, aber die um $D_{0g}/2$, also um den halben Wert bei der Grenzfrequenz tiefer liegt.

Vergleicht man dieses Ergebnis nun mit den praktisch beobachteten Werten, wie sie etwa in Abb. 2 gezeigt sind, bei welchen der D_{0g}-Wert stets über 40 db beträgt, so ergibt sich, daß die so errechnete Schalldämmung oberhalb der Grenzfrequenz viel zu niedrige Werte liefert, denn die beobachteten Einbußen gegenüber der früheren Theorie betragen bei weitem keine 20 db. Wir sind also jetzt im Gegensatz zu dem früheren Vergleich zwischen Theorie und Experiment vor die Frage gestellt, warum die Dämmungen noch so gut sind.

Einen Hinweis hierfür erhält man, wenn man den Durchlaßgrad im Koinzidenzbereich nach Formel (8.9) über dem Einfallswinkel bzw. im Sinne der statistischen Mittelwertbildung über dem Quadrat seines Cosinus aufträgt. Dies ist in Abb. 7 für den in Abb. 2 als Nr. 2 gemessenen Fall einer Glasplatte von 1,2 cm Stärke geschehen. Die ganze Kurve besteht hierbei nur aus 2 Nadelspitzen, von denen die eine breitere über dem Koinzidenzwinkel, die andere bei streifendem Einfall auftritt, welche, wie oben auseinandergesetzt wurde, praktisch bedeutungslos ist. Hiernach dürften bei statistischer Verteilung der Einfallswinkel im primären Raum jenseits der Wand nur Wellen eines schmalen Richtungsbereiches abgestrahlt werden. Diese müßten

notwendigerweise zu großen Täuschungen über den Ort der Schallquelle jenseits der Wand führen. Es ist nun außerordentlich unwahrscheinlich, daß ein derartig krasser Effekt bisher der Beobachtung entgangen sein sollte.

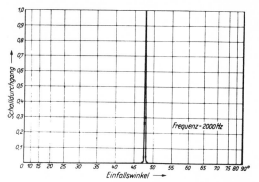

Abb. 7. Schalldurchgang durch eine 1,2 cm starke Glasplatte als Funktion des Einfallswinkels (gerechnete Werte, ohne innere Dämpfung)

Die starke Selektivität des Koinzidenzeffektes beruht offensichtlich darauf, daß wir als einzige Dämpfung bisher nur die (bei festen Platten in Luft geringe) Strahlungsdämpfung berücksichtigt haben. Die zusätzliche Beachtung einer, wenn auch sehr kleinen, Materialdämpfung führt sofort zu anderen Ergebnissen. Dabei darf es sich aber nicht um Dämpfungseinflüsse am Rande handeln; denn der Koinzidenzeffekt würde, wie die Ableitung erkennen läßt, ungeändert bleiben, wenn es etwa gelänge, den Rand in gedämpfter Weise so einzuspannen, daß jede Biegewellenreflexion unterbunden würde. Eine solche Einspannung würde dagegen sämtliche Eigenschwingungen der Platte wegdämpfen.

Wir führen daher gemäß den experimentellen Erfahrungen an festen Stoffen die Dämpfung durch Annahme eines komplexen Elastizitätsmoduls ein:

$$(9.1) \qquad \overline{E} = E\,(1 + i\,\eta)$$

worin η denselben Korrekturfaktor wie bei Schoch bedeutet und von der Frequenz unabhängig anzunehmen ist.

Mit dieser Einführung erhält der Trennwiderstand der Wand einen Realteil:

$$(9.2) \qquad \overline{T} = i\,\omega\,m\left(1 - \frac{c_1^4 \sin^4 \vartheta}{c^4}\right) + \eta\,\omega\,m\,\frac{c_1^4 \sin^4 \vartheta}{c^4}$$

und damit ändert sich die Formel für den Durchlaßgrad in:

$$(9.3) \quad \varDelta = \frac{1}{(1+\eta\,\xi^2\cos\vartheta\sin^4\vartheta/\sqrt{\varDelta_0})^2+\cos^2\vartheta\,(1-\xi^2\sin^4\vartheta)^2/\varDelta_0}.$$

Diese Abhängigkeit ist wieder unter Zugrundelegung der Daten der Glasplatte mit dem Wert von $\eta = 0{,}1$ in Abb. 8 eingezeichnet. Man erkennt daraus, daß auch noch bei so verhältnismäßig hoher Dämpfung die durchgelassene Energie im wesentlichen durch das Gebiet des Koinzidenzwinkels bestimmt wird.

Gleichzeitig ist in Abb. 8 die frühere Kurve aus Abb. 7 eingetragen. Ferner ist in Abb. 8 diejenige Kurve des Durchlaßgrades eingetragen, die sich bei gleichem $\varDelta_0$-Wert infolge des Komponenteneffektes ergeben würde, welche unterhalb der Koinzidenz eine Verringerung der Dämmung um etwa 5 db bewirkt. Der Vergleich zeigt, daß bei Auftreten der Koinzidenz die durch den Komponenteneffekt erzeugte Fläche auf einen ganz schmalen Bereich bei streifendem Einfall zusammenschrumpft. Daraus ist zu ersehen, daß der Komponenteneffekt im Koinzidenzgebiet keine Rolle mehr spielt.

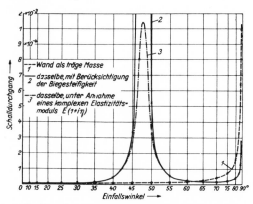

Abb. 8. Gerechnete Werte des Schalldurchgangs durch eine 1,2 cm starke Glasplatte in Abhängigkeit vom Einfallswinkel. Frequenz = 2000 Hz

Es genügt daher zur Ermittlung des mittleren Durchlaßgrades, die Integration wie früher über den Koinzidenzgipfel zu erstrecken, für welchen diesmal anzusetzen ist:

$$(9.4) \quad \varDelta = \frac{1}{(1+\eta\cos\vartheta_k/\sqrt{\varDelta_0})^2+(4\cos\vartheta_k/\sqrt{\varDelta_0})^2\,\varepsilon^2}$$

und hieraus ergibt sich wieder der statistische Mittelwert durch Bildung des Integrals

$$(9.5) \quad \varDelta_m = 2\sin^2\vartheta_k \int\limits_{-\varepsilon_1}^{+\varepsilon_1} \varDelta\, d\,\varepsilon$$

zwischen genügend weiten Grenzen $\pm\,\varepsilon_1$, wodurch man erhält:

$$(9.6) \quad \varDelta_m = \frac{\pi}{2}\,\frac{\sqrt{\varDelta_0}\,\sin^2\vartheta_k}{\cos\vartheta_k\,(1+\eta\cos\vartheta_k/\sqrt{\varDelta_0})}.$$

Drücken wir hierin wieder die Frequenz ausschließlich durch das Frequenzverhältnis ξ aus, so erhalten wir die Formel:

$$(9.7) \quad \varDelta_m = \frac{\pi}{2}\,\frac{\varDelta_{0g}}{\xi^2\sqrt{1-\dfrac{1}{\xi}}\left(\sqrt{\varDelta_{0g}}+\eta\,\xi\sqrt{1-\dfrac{1}{\xi}}\right)}.$$

Dieselbe geht für verschwindendes η in die frühere Beziehung (8.24) über.

Da der $\varDelta_{0g}$-Wert wie schon oben bemerkt, bei allen praktisch interessierenden Baustoffen kleiner als 10^{-4} ist und ξ nach den Voraussetzungen der Ableitung wenigstens 2 sein muß, so können wir diese Gleichung für nicht zu kleine Dämpfungen, d. h. etwa für $\eta > 0{,}04$ umformen zu:

$$(9.8) \quad \varDelta_m = \frac{\pi}{2}\,\frac{\varDelta_{0g}}{\eta\,(\xi^3-\xi^2)}$$

und hieraus würde sich für die mittlere Dämmung ergeben:

$$(9.9) \quad D_m = D_{0g} - 10\log\frac{1}{\eta} + 10\log\,(\xi^3-\xi^2)-2.$$

In Anbetracht der Unsicherheit in Meßergebnissen und Rechengrundlagen dürfte es genügen, sich die weiter vereinfachte (oberhalb $\xi = 2$ auf $\pm\,2$ db genaue) Formel

$$(9.10) \quad D_m = D_{0g} - 10\log\frac{1}{\eta} + 30\log\xi - 3$$

zu merken.

In Übereinstimmung mit den Kurven in Abb. 2 erhalten wir hiernach jenseits der Koinzidenz einen steileren Anstieg mit der Frequenz als unterhalb derselben, nämlich mit dem 30fachen statt mit dem 20fachen Logarithmus. Ferner ist diese Kurve unter Zugrundelegung eines geeigneten η-Wertes leicht den gemessenen Kurven anzupassen.

So ließ sich z. B. hieraus folgern, daß für die Glasplatte etwa der oben gewählte Dämpfungswert von $\eta = 0{,}1$ einzusetzen ist.

Wir haben nun für die Gebiete, die unterhalb oder wenigstens eine Oktave oberhalb der Grenzfrequenz liegen, einfache Formeln gefunden. Leider läßt sich für den dazwischenliegenden Bereich kein so übersichtlicher Ausdruck angeben. Man ist hier vielmehr darauf angewiesen, die Mittelwertbildung im einzelnen Falle graphisch oder numerisch durchzuführen. Herr H. GOETZ hat sich in dankenswerter

zwischen Theorie und Experiment nur als befriedigend bezeichnet werden.

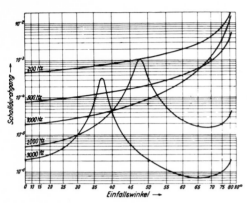

Abb. 9. Schalldurchgang durch eine 1,2 cm starke Glasplatte in Abhängigkeit vom Einfallswinkel, gerechnete Werte mit $E = E (1 + i \eta)$

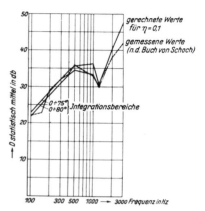

Abb. 10. Frequenzgang der statistischen Schalldämmung für eine 1,2 cm starke Glasplatte

Weise der großen Mühe unterzogen, diese Berechnung für die Glasplatte durchzuführen. In Abb. 9 sind die Abhängigkeiten $\Delta (\vartheta)$ unter Benutzung eines logarithmischen Ordinatenmaßstabes eingetragen. In Abb. 10 sind die aus ihnen sich errechnenden statistischen Mittelwerte in Vergleich mit der gestrichelt eingetragenen gemessenen Kurve nach Abb. 2 aufgezeichnet. Dabei sind die Integrationsbereiche einmal bis 75⁰ und einmal bis 80⁰ erstreckt. Oberhalb der Grenzfrequenz fallen beide Mittelwerte zusammen. Bei tiefen Frequenzen liegt die 80⁰-Kurve um etwa 1 db tiefer. Die größten Abweichungen zwischen beiden Kurven treten in der Nachbarschaft der Grenzfrequenz auf, weil hier die Gipfel des Komponenteneffektes und des Koinzidenzeffektes beide auf das Gebiet streifenden Einfalls fallen und daher das Ergebnis gegenüber einer Beschneidung des Integrationsbereiches in diesem Gebiet besonders empfindlich ist. Die gleiche Unsicherheit muß, wie schon erwähnt, auch bei Messungen in diesem Gebiet auftreten.

Trotz dieser Unsicherheit kann die in Abb. 10 zum Ausdruck kommende Übereinstimmung

[*Editor's Note:* Material has been omitted at this point.]

27

THEORY OF THE SOUND ATTENUATION OF THIN WALLS WITH OBLIQUE INCIDENCE

L. Cremer

This English summary was prepared expressly for this Benchmark volume by T. D. Northwood, National Research Council of Canada, from "Theorie der Schalldämmung dünner Wände bei schrägem Einfall," in Akust. Z. 7: 81–102 (1942).

PART I: SURVEY OF THE PRESENT STATE OF RESEARCH

1. The Wall as an Inert Mass Under Normally Incident Sound

This section deals with the simplest possible model of a wall, considered to be of infinite width, of thickness small compared to a wavelength in the wall material and immersed in a fluid medium such as air, the wave impedance of which is small compared to the wave impedance of the wall material.

The problem reduces to a one-dimensional one that can be expressed in terms of the pressure difference between the faces of the wall and the acceleration of a wall mass.

$$p_1 - p_2 = m\frac{dv}{dt} \tag{1.1}$$

Expressing the incident sound wave as a sinusoidal function, Equation (1.1) takes the form:

$$\bar{p}_1 - \bar{p}_2 = i\omega m\bar{v} \tag{1.1a}$$

where ω is the angular velocity and the bars over p and v signify complex quantities. We introduce the transmission impedance, $\bar{T}$, which is defined as the quotient of the pressure difference and the velocity of the partition surface. Then Equation 1.1 may be written

$$\bar{p}_1 - \bar{p}_2 = \bar{T}\bar{v} \tag{1.2}$$

On the second side of the wall, the radiated plane wave is described by

$$\bar{p}_2 = Z\bar{v} \tag{1.3}$$

where Z is the wave impedance of the medium into which the wall is radiating. The usual expression for wall impedance,

$$\bar{W} = \frac{\bar{p}_1}{v} \tag{1.4}$$

differs from the transmission impedance by the wave impedance of the medium:

$$\bar{W} = \bar{T} + Z \tag{1.5}$$

In this most simplified model, the sound transmission coefficient is given by

$$\Delta = \frac{v^2}{v_a^2}$$

where v_a is the velocity of the incident wave, and the sound transmission loss is given by

$$D = 10 \log \frac{1}{\Delta} \tag{1.12}$$

$$D = 20 \log \left| 1 + \frac{i\omega m}{2Z} \right| \tag{1.14}$$

2. Comparison with Measurements in Statistical Sound Fields

Figure 1(a) shows measured values of average sound transmission loss, as a function of the mass per unit area of the partitions. In Figure 1(b) the mean curve for the measured values is compared with the value calculated from the simple theory (Equation 1.14). It is evident that the agreement is poor except for very light partitions. Figure 2a shows that in general the variation of transmission loss with frequency does not follow the shape predicted by simple theory. In Figure 2(b), three of the measured curves are compared with calculated values, and it is seen that only the lightest wall corresponds reasonably with prediction. The lack of agreement may be attributed to the fact that, in contrast to the theoretical model, the measurements are with diffuse sound fields, containing waves incident at oblique angles.

3. Asymptotic Law of Schoch

Schoch treats the special case of constant pressure at normal incidence on a plate of arbitrary width in one direction and simply supported in the other.

Expressing the wall velocity and pressure in terms of the eigenfunctions $\bar{u}_n(x,y)$,

$$\bar{v}(x,y) = \Sigma \bar{v}_n \bar{u}_n(x,y) \tag{3.1}$$

$$\bar{p}(x,y) = \Sigma p_n \bar{u}_n(x,y) \tag{3.2}$$

For each term in (3.1) and (3.2), there is a single-degree-of-freedom relation of the form

$$\frac{\bar{p}_n}{\bar{v}_n} = \bar{T}_n = i\omega m \left(1 - \left(\frac{\omega_n}{\omega}\right)^2 (1 + i\eta)\right) \tag{3.3}$$

where η is a correction for internal damping in the plate material. Finally the complete solution may be written

$$\bar{v}(x,y) = \frac{1}{i\omega m} \sum \frac{\bar{p}_n \bar{u}_n(x,y)}{1 - \left(\frac{\omega_n}{\omega}\right)^2 (1 + i\eta)} \tag{3.4}$$

Two factors forming this solution are depicted in Figure 4. The vertical lines correspond to the eigenfunctions, which have frequencies in the ratio 1:4:9, etc., the even harmonics dropping out because of symmetry of the assumed uniform excitation. The denominator in (3.4), corresponding to the broken line in Figure 4, is an amplification factor with a peak value dependent on the damping. It is evident that as long as the fundamental frequency ω_n is much less than ω the wall velocity approaches the simple mass law relation, i.e.

$$\lim_{\substack{\omega \to \infty \\ p_n = \text{const.}}} \bar{v}(x,y) = \frac{\bar{p}(x,y)}{i\omega m} \tag{3.5}$$

This is consistent with experimental evidence shown in Figure 5, at least in that deviations from mass law increase with the frequency of the first resonance.

4. The Component Effect

Under oblique incidence, even the simple mass model requires modification to take account of the "component" effect, i.e., the fact that the wall velocity must equal the component normal to the wall. Thus Equation (1.2) becomes

$$\bar{p}_1 - \bar{p}_2 = (\bar{T} \cos \vartheta)\bar{v}_2 \tag{4.7}$$

and so we get for Δ

$$\Delta = \frac{1}{\left|1 + \dfrac{T \cos \vartheta}{2Z}\right|^2} \tag{4.8}$$

The resulting expression for random incidence depends significantly on the maximum angle ϑ_1.

$$\Delta'_m = \int_{\cos^2 \vartheta_1}^{1} \frac{d(\cos^2 \vartheta)}{1 + \dfrac{\cos^2 \vartheta}{\Delta_0}} = \Delta_0 \ln \frac{1}{\Delta_0 + \cos^2 \vartheta_1} \tag{4.14}$$

391

where

$$\Delta_0 \approx \left(\frac{2Z}{\omega m}\right)^2 \tag{4.10}$$

For a limiting angle of 75° and for transmission loss above 30 dB, the correction to the normal-incidence value is 4.3 dB, and for a limiting value of 80° the correction is 5.4 dB.

5. The General Derivations for a Wall of Arbitrary Stiffness

From Schoch's asymptotic law for normal incidence, one may infer that, for high frequencies at least, the shape and boundary conditions of the wall are of secondary importance. Only with this assumption is it valid to think of transmission loss as a property per unit area of a wall. In the following consideration this assumption is retained and the development will be for laterally infinite walls; but the previous model will be improved upon by considering the lateral coupling between adjacent mass elements under oblique incidence.

The problem has already been analyzed, in a degree of generality not required here, by Reissner. In building acoustics, it is almost always adequate to assume a wall thickness small compared to wavelengths in the wall material; then the wave motions in the wall reduce to simple flexural waves. This relatively simple model leads eventually to an explanation of the high-frequency performance found for real walls. The same effect was first discovered by F. H. Sanders, working with thin plates in water at ultrasonic frequencies.

PART II: TREATMENT AS A FORCED BENDING WAVE

6. The "Coincidence" Effect in the Case of a Stretched Diaphragm

As a preliminary exercise, the attenuation provided by a stretched diaphragm of infinite extent is considered. The problem is two dimensional, analogous to that of the stretched string, and may be represented by

$$p_1 - p_2 + S \frac{\delta^2 w}{\delta x^2} = m \frac{\delta^2 w}{\delta t^2} \tag{6.2}$$

where $p_1 - p_2$ is the net force on the diaphragm, w is the transverse deflection, S the diaphragm tension, and m its mass per unit area.

For a sinusoidal wave incident at angle ϑ, the relation becomes

$$\bar{p}_1 - \bar{p}_2 = \left(i\omega m - i\omega \frac{S}{c^2} \sin^2\vartheta\right)\bar{v} \tag{6.6}$$

The transmission impedance is thus composed of two terms of opposite phase. In principle, for a particular angle of incidence the two terms might be expected to cancel each other, leading to a region of low transmission loss. If the propagation velocity of a free wave in the diaphragm

$$c_1 = \sqrt{\frac{S}{m}} \qquad (6.8)$$

is introduced, the condition for a surface impedance to reduce to zero may be expressed as

$$c_1 = \frac{c}{\sin \vartheta} \qquad (6.10)$$

a condition described as "coincidence" between the free diaphragm wave velocity and the "trace velocity" of the incident wave.

7. Coincidence Effect in a Plate

A procedure analogous to that used in Section 6 can also be applied to the case of a rigid plate. With the same assumptions as before, the differential equation becomes similar to that of a simple beam, except that for a plate the modulus of elasticity becomes $E(1 - \mu^2)$, where μ is Poisson's ratio. Thus

$$-B\frac{\delta^4 w}{\delta x^4} = m\frac{\delta^2 w}{\delta t^2} \qquad (7.1)$$

where $B - EJ/(1 - \mu^2)$, and J is the moment of inertia per unit width. This leads to the expression

$$\bar{p}_1 - \bar{p}_2 = \left(i\omega m - i\frac{B\omega^3 \sin^4 \vartheta}{c^4} \right) \bar{v} \qquad (7.8)$$

Then the transmission impedance assumes the form

$$T = i\omega m - i\frac{B\omega^3 \sin^4 \vartheta}{c^4} \qquad (7.9)$$

This differs from the diaphragm case in that the reactance term varies as the cube of the frequency. As before, the coincidence condition occurs when

$$c_1 = \frac{c}{\sin \vartheta} \qquad (7.10)$$

but now c_1 is a function of frequency. In terms of frequency, coincidence occurs when

$$f_k = \frac{c^2}{2\pi \sin^2 \vartheta}\sqrt{\frac{m}{B}} \qquad (7.11)$$

393

It is evident that the lowest coincidence frequency corresponds to grazing incidence, for which

$$f_g = \frac{c^2}{2\pi} \sqrt{\frac{m}{B}} \qquad (7.12)$$

If this limiting frequency, the "critical" frequency, is well above the observation frequencies, i.e., if $\sqrt{(B/m)}$ is small, then the transmission impedance again reduces to Schoch's asymptotic law.

The formula for the critical frequency can be applied to a simple homogeneous plate of thickness d, for which

$$B = \frac{E}{1 - \mu^2} \cdot \frac{d^2}{12} \qquad (7.13)$$

Then Equation (7.12) becomes

$$f_g \approx \frac{c^2}{2c_l d} \qquad (7.17)$$

where c_l is the velocity of propagation of longitudinal waves along the plate which can be evaluated for typical building materials as in Figure 6. Here, for example, the predicted critical frequency for 1.2 cm glass is slightly below 1000 Hz, whereas the measurements shown in Figure 3 indicate a minimum at 1200 Hz. Generally, for plates ranging in thickness from a few millimetres up to a brick wall, the critical frequency falls within the audible frequency range, and there is a substantial deviation from the mass law above this frequency. In Figure 3, for example, only the very thin plywood (Curve 1) follows the mass law without significant deviation.

8. Analogy Between Resonance and Coincidence

The similarity has already been noted between coincidence and the resonance of a simple oscillator in that under certain conditions the resistance (mass) and compliance terms cancel each other. In this section, the analogy is developed in detail.

Considered first is an elastically supported piston diaphragm in an infinitely long tube in which a continuous wave passes. The transmission impedance is

$$\overline{T} = i\omega m \left[1 - \left(\frac{\omega_0}{\omega} \right)^2 \right] \qquad (8.1)$$

and the transmission coefficient is

$$\Delta = \frac{1}{1 + \left(\dfrac{\omega m}{2Z} \right)^2 \left[1 - \left(\dfrac{\omega_0}{\omega} \right)^2 \right]^2} \qquad (8.2)$$

In the neighborhood of the resonance peak, introduce

$$\epsilon = \left(\frac{\omega}{\omega_0} - 1\right) \tag{8.3}$$

thus obtaining for the transmission coefficient

$$\Delta = \frac{1}{1 + \left(\dfrac{\omega_0 m}{Z}\right)^2 \epsilon^2} \tag{8.4}$$

The width of the resonance curve, to the half-power points, is given by

$$2\epsilon' = \frac{2Z}{\omega_0 m} \tag{8.5}$$

For the coincidence case, if frequency is held constant, then the dependence on positions in the plate, which manifests itself in the factor

$$\exp\left[\pm i\left(\frac{\omega}{c}\sin\vartheta\right)x\right]$$

plays the part formerly played by time dependence. If, for the case of a plate, the transmittance is expressed by

$$\Delta = \frac{1}{1 + \left[\dfrac{\omega m \cos\vartheta}{2Z}\left(1 - \left(\dfrac{c_1 \sin\vartheta}{c}\right)^4\right)\right]^2} \tag{8.9}$$

then for small deviations from the coincidence angle ϑ_k a "resonance curve" is obtained by introducing

$$\sin\vartheta = \sin\vartheta_k(1 + \epsilon) \tag{8.10}$$

The half-width is

$$2\epsilon' = \frac{Z}{\omega m \cos\vartheta_k} \tag{8.12}$$

The question of what energy passes through the wall under random incidence conditions may be answered by first considering the familiar resonance case. If the elastically mounted piston is exposed to a uniform continuous spectrum of sound waves, the average transmission coefficient will be

$$\Delta_m = \frac{1}{\omega_2 - \omega_1}\int_{\omega_1}^{\omega_2}\Delta\,d\omega \tag{8.15}$$

395

and since the transmission coefficient will be low except in the vicinity of the resonance, the value given by (8.4) may be substituted in (8.15):

$$\Delta_m = \frac{\omega_0}{\omega_2 - \omega_1} \int_{-\epsilon_1}^{+\epsilon_1} \frac{d\epsilon}{1 + \left(\dfrac{\omega_0 m}{Z}\right)^2 \epsilon^2} \qquad (8.16)$$

Here the interval defined by the limits need only be broad enough to include the resonance region, as can be shown mathematically:

$$\int_{-\epsilon_1}^{+\epsilon_1} \frac{d\epsilon}{1 + \left(\dfrac{\omega_0 m}{Z}\right)^2 \epsilon^2} = \frac{2Z}{\omega_0 m} \text{ arc tg } \frac{\omega_0 m}{Z} \epsilon_1 \approx \frac{\pi Z}{\omega_0 m} \qquad (8.17)$$

Thus

$$\Delta_m \approx \frac{\pi Z}{(\omega_2 - \omega_1)m} \qquad (8.18)$$

In a similar way, the transmission coefficient of the plate exposed to a random incidence sound field may be found. Following Section 4, the average transmission coefficient may be written

$$\Delta_m = \int_0^1 \Delta \, d(\cos^2 \vartheta) = 2 \int_0^1 \Delta \sin \vartheta \, d(\sin \vartheta) \qquad (8.19)$$

Only the angular region around the coincidence angle will have an important contribution to the average transmission coefficient, and again the integral can be written in terms of this small region

$$\Delta_m = 2 \sin^2 \vartheta_k \int_{-\epsilon_1}^{+\epsilon_1} \frac{d\epsilon}{1 + \left(\dfrac{2\omega m \cos \vartheta_k}{Z}\right)^2 \epsilon^2} \qquad (8.20)$$

whence

$$\Delta_m \approx \frac{\pi Z \sin^2 \vartheta_k}{\omega m \cos \vartheta_k} = \frac{\pi}{2} \sqrt{\Delta_0} \frac{\sin^2 \vartheta_k}{\cos \vartheta_k} \qquad (8.21)$$

Now, introducing ξ, the ratio of frequency to the critical frequency given by (7.11) is

$$\Delta_0 = \frac{\Delta_{0g}}{\xi^2} \qquad (8.22)$$

where Δ_{0g} is the transmission coefficient at normal incidence at the coincidence frequency and

$$\sin^2 \vartheta_k = \frac{1}{\xi} \qquad (8.23)$$

Finally, for $\xi \geqslant 2$,

$$\Delta_m \approx \frac{\pi}{2} \frac{\sqrt{\Delta_{0g}}}{\xi^2 \sqrt{1 - \frac{1}{\xi}}} \qquad (8.24)$$

and the formula for random incidence transmission loss becomes (neglecting unimportant terms)

$$D_m \approx \frac{1}{2} D_{0g} + 20 \log \xi \qquad (8.25)$$

D_{0g} is a function only of the wall material, viz

$$D_{0g} = 20 \log \frac{\pi \rho c^2}{Z c_l}$$

Hence the sound transmission loss varies above the critical frequency as 20 log (frequency) just as it does below.

9. Consideration of the Material Damping

Equation (8.25) for transmission loss suggests that above about double the critical frequency it should approach a straight-line parallel to the low-frequency curve, but lower by $\frac{1}{2} D_{0g}$, i.e., by half the value at the critical frequency. This result is much lower than that observed experimentally, for example in Figure 2. The problem is illustrated by considering the passage of sound through a glass plate as a function of angle of incidence. Figure 7 shows the calculated value when material damping is neglected: here, apart from the grazing-incidence contribution which is of no practical importance, the only significant transmission is at a very sharply defined angle corresponding to coincidence. If such a phenomenon existed, it is most improbable that it would not have been noted subjectively.

The discrepancy is resolved by considering the effect of material damping, introduced as an imaginary component in the expression for modulus of elasticity:

$$E = E(1 + i\eta) \qquad (9.1)$$

This leads to a new expression for transmission coefficient

$$\Delta = \cfrac{1}{\left(1 + \cfrac{\eta \xi^2 \cos \vartheta \sin^4 \vartheta}{\sqrt{\Delta_0}}\right)^2 + \cos^2 \vartheta \; \cfrac{(1 - \xi^2 \sin^4 \vartheta)^2}{\Delta_0}} \tag{9.3}$$

Introducing the value $\eta = 0.1$ in the calculation for the glass plate yields curve 3 of Figure 8, whose peak value is much lower than for the undamped case.

On evaluating (9.3) for the coincidence region as before, the average transmission loss becomes

$$D_m = D_{0g} - 10 \, \log \frac{1}{\eta} + 10 \, \log(\xi^3 - \xi^2) - 2 \tag{9.9}$$

For ξ greater than 2 this can further be approximated (to ± 2 dB) by

$$D_m = D_{0g} - 10 \, \log \frac{1}{\eta} + 30 \, \log \xi - 3 \tag{9.10}$$

From this it follows that the transmission loss above twice the critical frequency should vary by 9 dB per octave.

In the octave above the critical frequency, the simple formula (9.10) is not applicable. Figure 9, however, shows the result of detailed calculations by H. Goetz from the more exact expression, for a range of angles of incidence. From these is calculated the average transmission loss over the interval 0 to 75° or 80°. Grazing incidence introduces an uncertainty: it cannot occur experimentally; and on the other hand the theoretical result is very sensitive to the limiting angle chosen. Nevertheless the agreement between calculated and measured values for the glass plate, shown in Figure 10, is very satisfactory.

FIGURE CAPTIONS

Figure 1   Average sound attenuation as a function of wall weight, according to Schoch's book. (Ordinate: sound transmission loss). (a) According to average measurement results from various laboratories (solid line). (b) Comparison of the average curve from (a) with the simple theory (dashed line).

Figure 2   Frequency dependence of sound attenuation for various building materials (according to Schoch).
(a) Experimental data:

Curve		kg/m^2
1	Plywood, lacquered	2.05
2	Plate glass	30
3	Lightweight wall, both sides plastered	70
4	¼-brick solid brick wall, both sides plastered	153
5	½-brick solid brick wall, both sides plastered	224
6	full-brick solid brick wall, both sides plastered	457

(b) Comparison of Curves 1 to 3 from (a) with simple theory.

Figure 3   Schematic diagram of sound transmission under oblique incidence, showing incident, reflected and transmitted waves, and consequent flexural wave propagating along the wall.

Figure 4 Pressure–amplitude spectrum (vertical lines) of a hinged supported plate for normal incidence, together with amplification factor (broken line) for damping ratio $\eta = 0.1$.

Figure 5 Discrepancy between measured and calculated mean sound transmission loss as a function of the fundamental frequency (from Schoch).

Figure 6 Relation between wall thickness d and limiting frequency f_g for various building materials. Key: Eiche quer = oak crosswise. Eiche langs = oak lengthwise. Ziegelstein = brick.

Figure 7 Sound transmission through a glass plate 1.2 cm thick as a function of the angle of incidence (calculated values, no internal damping). (Ordinate: transmission coefficient; abscissa: angle of incidence.) Broken lines refer to walls of one brick, ½-brick and ¼-brick thicknesses.

Figure 8 Calculated values of sound transmission through a glass plate 1.2 cm thick as a function of the angle of incidence. Frequency = 2000 Hz. (Ordinate: transmission coefficient; abscissa: angle of incidence). (1) Mass law. (2) Bending stiffness included. (3) Including stiffness and damping (complex elastic modulus).

Figure 9 Sound transmission at various frequencies through a glass plate 1.2 cm thick as a function of the angle of incidence, calculated values with $E = E(1 + i\eta)$. (Ordinate: transmission coefficient; abscissa: angles of incidence.)

Figure 10 Random incidence sound transmission loss for a glass plate 1.2 cm thick. (Ordinate: transmission loss, dB; abscissa frequency, Hz). Solid curves: calculated for $\eta = 0.1$ and for angles of incidence 0–$75°$ and 0–$80°$. Broken curves: measured values according to Schoch's book.

28

Reprinted from *VDI Ber.* **8**:23–28 (1956)

Trittschall — Entstehung und Dämmung

(Zusammenfassender Vortrag)

Von **K. Gösele**, Stuttgart

Institut für technische Physik. Stuttgart (Prof. Dr.-Ing. H. Reiher)

Zusammenfassung

Verfahren zur Kennzeichnung des Trittschallverhaltens; Verhalten verschiedener Deckentypen (homogene Massivdecken, Hohlkörperdecken, zweischalige Decken) unter Hinweis auf die physikalischen Grundlagen.
Trittschalldämmung durch schwimmende Estriche, wobei die verschiedenen Übertragungswege und die Einflußgrößen (Steifigkeit der Dämmschicht und der eingeschlossenen Luft, Strömungswiderstand der Dämmschicht) besprochen werden.
Trittschalldämmung durch weichfedernde Gehbeläge und die meßtechnisch sich ergebenden Einflußgrößen (Hammermasse und Form der Schlagflächen).

Summary

Methods for the characterisation of impact sound. Behaviour of different types of ceilings (homogeneous massive ceilings, hollow ceilings, double-layer ceilings) with reference to the underlying physical principles.
Isolation by floating floors, discussing the different possibilities of transmission and the significant parameters (stiffness of the isolating layer and of the enclosed air, flow resistance of the layer).
Reduction of impact sound by elastic floor coverings and discussion of the corresponding measuring parameters (mass of the hammer and shape of the striking surface).

Sommaire

Procédé pour caractériser la réponse au bruit d'impact. Comportement de différents types de planchers (massifs homogènes, à cavités, à double couche) au point de vue physique. Amortissement du bruit d'impact au moyen de revêtements flottants; examen des différents trajects de propagation et des paramètres ayant de l'influence (rigidité de la couche isolante et de l'air enfermé, résistance au passage de l'air dans cette couche). Amortissement du bruit d'impact par des enduits de grande élasticité; examen des paramètres ayant de l'influence dans les mesures (masse du marteau, forme des faces d'impact).

Für die schalltechnische Beurteilung von Decken reicht die Bestimmung der Luftschalldämmung nicht aus. Zwei Decken können dieselbe Luftschalldämmung aufweisen und sich trotzdem wesentlich verschieden gegenüber irgendwelchen punktförmig auf die Decke einwirkenden Wechselkräften verhalten, die im Wohnbetrieb sehr häufig auftreten und sich äußerst störend auswirken können. Zur vollständigen Kennzeichnung ist deshalb noch eine weitere Größe nötig, die das Verhalten gegenüber den genannten Wechselkräften kennzeichnet. Unter „Trittschall" versteht man im engeren Sinn des Wortes diejenigen Geräusche, die beim Begehen einer Decke im darunterliegenden Raum entstehen. Im weiteren Sinne versteht man darunter ganz allgemein das akustische Verhalten von Decken gegenüber örtlich begrenzt einwirkenden Wechselkräften, wobei es sich sowohl um Stoßanregungen (z. B. Stühlerücken, auffallende Gegenstände) als auch um stationäre Anregungen (z. B. Betrieb von Nähmaschinen, Kühlschränken) handeln kann. Im folgenden soll ein Überblick über die physikalischen Gesetzmäßigkeiten gegeben werden, die bezüglich der Entstehung des Trittschalls und seiner Dämmung bekannt sind. Die dabei auftretenden Gesetzmäßigkeiten sind nicht nur für Decken von Bedeutung. Sie lassen sich auch in manchen Fällen auf Wände übertragen (z. B. Anregung von Wänden durch Betätigen von Lichtschaltern, Übertragung von Installationsgeräuschen auf Wände).

Kennzeichnung des Trittschallverhaltens

In Analogie zu der Entstehung von Gehgeräuschen hat man bereits bei den ersten Versuchen [1] zur Charakterisierung des Trittschallverhaltens ein Hammerwerk benützt, bei dem eine Hammermasse aus einer bestimmten Höhe frei auf die Decke auffiel. Das in dem Raum unter der Decke sich ergebende Geräusch wurde bezüglich seiner Lautstärke als Maß für das Trittschallverhalten herangezogen (z. B. sog. Normtrittlautstärke nach DIN 4110). Diese Art der Anregung ist auch in neueren Normvorschriften (DIN 52210) beibehalten worden, wobei allerdings nicht mehr die gesamte Lautstärke, sondern der sich je Oktave ergebende Trittschallpegel des Geräusches gemessen und nach einer Umrechnung auf bestimmte Schallschluckverhältnisse des Empfangsraumes (10 m² Schallschluckfläche bzw. 0,5 s Nachhallzeit des Raumes) in Abhängigkeit der Tonhöhe angegeben wird. Wenn der Schallpegel auf 10 m² Schallschluckfläche bezogen wird, wird er nach DIN 52211 als „Norm-Trittschallpegel" bezeichnet.

Diese Kennzeichnung ist vom physikalischen Standpunkt aus insofern unbefriedigend, als das Ergebnis von gewissen Willkürlichkeiten bezüglich der Wahl des Hammerwerks abhängt. Alle Versuche — in Analogie zur Definition der Luftschalldämmung — ein Verhältnismaß zu finden, das unabhängig von der zufällig benutzten Anregungsart ist, sind jedoch bisher ergebnislos verlaufen.

An Stelle eines Hammerwerks kann zur Trittschallmessung nach *Th. Lange* [9] auch eine quasistationäre Anregung durch sinusförmige Wechselkräfte mit Hilfe eines elektrodynamischen Anregesystems benützt werden. Die Meßwerte, die nach den beiden Anregungsarten gewonnen werden, können bei bekanntem Kraftspektrum des Hammerwerks und bekannter Größe der sinusförmigen Wechselkraft ineinander übergeführt werden. Bild 1 enthält den rechnerisch zu erwartenden Unterschied in den Schallpegelwerten zwischen der Anregung mit dem Tritt-

[*Editor's Note:* An English summary of this article prepared by T. D. Northwood follows.]

schallhammerwerk und mit einer sinusförmigen Wechselkraft von 10⁶ dyn. Voraussetzung dafür ist u. a., daß die sinusförmige Wechselkraft über eine Masse m die Decke anregt, die der Hammermasse des Hammerwerks entspricht. Das Spektrum des Norm-Hammerwerks besteht aus einem Linienspektrum, dessen einzelne Komponenten gleich groß sind [3; 9]. Bezogen auf die Bandbreite einer Oktave ergibt sich ein mit der Wurzel aus der Frequenz ansteigendes Spektrum für die wirkende Kraft. Auf diesen Anstieg des Anregespektrums mit der Frequenz ist zu einem Teil auch das Ansteigen des Trittschallpegels von Rohdecken mit der Frequenz zurückzuführen.

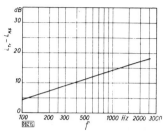

Bild 1. Rechnerisch nach *L. Cremer* zu erwartender Unterschied zwischen den Trittschallpegelwerten bei Anregung mit Trittschallhammerwerk (L_{Ti}) und bei sinusförmiger Anregung (L_{RS}) durch eine Wechselkraft von 10⁶ dyn.

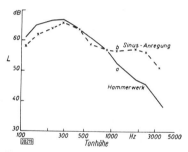

Bild 2. Vergleich des mit Hammerwerk und mit sinusförmiger Anregung erhaltenen Trittschallpegels für eine Decke nach *Th. Lange* [9].
Die mit sinusförmiger Anregung erhaltenen Werte sind auf gleiche, mit $\sqrt{f}$ ansteigende Kräfte umgerechnet, wie sie beim Trittschallhammerwerk auftreten und experimentell bestimmt worden sind.

Bild 2 zeigt nach *Th. Lange* [9] ein Beispiel für die Bestimmung des Trittschallpegels mit einem Hammerwerk (Kurve a) und einer sinusförmigen Anregung (Kurve b), wobei die Werte für die sinusförmige Anregung auf gleiche Kräfte wie bei der Hammeranregung umgerechnet sind. Das Spektrum des Hammerwerks enthält in bevorzugtem Maß hohe Frequenzen, wie sie z. B. beim Fall von kleinen Gegenständen auch vorkommen. Gehgeräusche sind dagegen wesentlich tieffrequenter [4]. Eine Betrachtung der Eigenheiten der beiden oben angeführten Anregungsarten ergibt keine Gesichtspunkte, die in jedem Fall die eine oder andere Anregungsart als unbedingt zweckmäßiger erscheinen lassen.

Die überwältigende Anzahl von verschiedenen möglichen Deckenausführungen[1]) läßt sich bezüglich ihres Trittschallverhaltens dadurch übersichtlich machen, daß man zwischen dem Verhalten der Decken o h n e Fußboden (sog. Rohdecken) und m i t Fußboden (sog. wohnfertige Decken) unterscheidet. Zur Kennzeichnung des Fußbodens wird eine sog. V e r b e s s e r u n g ΔL des Trittschallschutzes eingeführt [2; 4], wobei dieses ΔL die Differenz der Trittschallpegelwerte einer Decke m i t und o h n e den beobachteten Fußboden bedeutet.

─────────
[1]) Hierbei sollen nur Massivdecken und keine Holzbalkendecken betrachtet werden, bei denen eine Trennung in Decken ohne und mit Fußboden nicht immer möglich ist.

Dabei wird angenommen, daß dieser, in Abhängigkeit von der Frequenz angegebene Differenzwert ΔL sich für jeden Fußboden gleich groß ergibt, unabhängig von der Rohdecke, bei der die Messung vorgenommen worden ist [6]. Diese Annahme kann als streng richtig für weichfedernde Gehbeläge angenommen werden [6]. Auch für andere Fußbodenbeläge (Riemenböden, Estriche auf einer Dämmschicht) ist dies in erster Näherung zutreffend, wobei jedoch genaue Untersuchungen über die Gültigkeitsgrenzen noch ausstehen. Die Gültigkeit dieser angenommenen Gesetzmäßigkeit würde es ermöglichen, das Trittschallverhalten von wohnfertigen Decken aus den Eigenschaften der Rohdecken und der jeweils verwendeten Fußbodenelemente zu berechnen.

Im folgenden soll das Trittschallverhalten von Rohdecken und einigen typischen Fußbodenaufbauten besprochen werden.

Trittschallverhalten von Rohdecken

Von den vielen verschiedenen Arten von Massivdecken sollen hier nur drei wichtige Typen behandelt werden, womit allerdings die akustischen Eigenschaften der wesentlichsten Decken erfaßt sind. Bei der punktweisen Anregung einer Decke durch eine Wechselkraft F ist zunächst die an der Anregungsstelle entstehende Schnelle-Amplitude v von Bedeutung für das Trittschallverhalten der Decke. Diese Amplituden werden durch den Eingangswiderstand Z der Decke charakterisiert

$$Z = \frac{F}{v_0}.$$

Für homogene Platten haben *H.* und *L. Cremer* [3] diesen Eingangswiderstand errechnet. Er ist für schwere und genügend biegesteife Platten frequenzunabhängig und hängt von dem Produkt Masse je Flächeneinheit · Biegesteifigkeit der Platten ab.

An der Anregungsstelle breiten sich die Schwingungen, von einem kleinen Umkreis um die Anregungsstelle abgesehen, nach einem $1/\sqrt{r}$-Gesetz aus. Bei den endlich ausgedehnten Decken ergeben sich Reflexionen an den Einspann- bzw. Auflagerstellen der Decken, wodurch dem fortschreitenden Wellenfeld ein diffuses Wellenfeld überlagert wird. Dies mag ein Beispiel in Bild 3 erläutern.

Die verschieden starke Ausbildung des diffusen Feldes bei verschieden großen Decken macht es auch einigermaßen verständlich, daß der Trittschallpegel einer Decke bei nicht zu großer Variation der Deckenfläche (versuchsmäßig z. B. zwischen 7 und 20 m² vorgenommen) unabhängig von der Größe der Decke ist[2]).

Homogene Plattendecken

Die Körperschall-Amplituden von unendlich ausgedehnten homogenen Platten sind von *H.* und *L. Cremer* [3] berechnet worden. Die an sich wünschenswerte, genaue Berechnung des interessierenden Trittschallpegels für eine Plattendecke von endlichen Abmessungen aus den Körperschall-Amplituden stößt dabei auf gewisse Schwierigkeiten, da Annahmen über das Abstrahlverhalten (Umsetzung des Körperschalls in Luftschall) der Decken erforderlich sind und eine Berechnung des diffusen Körperschallfeldes der Decke noch nicht möglich ist. Die erstgenannte Schwierigkeit spielt bei genügend dicken Decken, deren Grenzfrequenz unterhalb des interessierenden Frequenzbereiches liegt, keine Rolle, weil dort ein nahezu frequenzunabhängiges, quantitativ bekanntes Abstrahlverhalten vorliegt. Der zweitgenannte Einfluß läßt sich an Hand von Meßergebnissen, wie sie Bild 3 zeigt, abschätzen.

In Bild 4 ist eine Abschätzung des Trittschallpegels unter Zugrundelegung der genannten Theorie für eine übliche Plattendecke gemacht und mit Meßwerten verglichen. Die Absolutwerte weichen in gewissem Umfang

─────────
[2]) Offensichtlich ist bei einer kleinen Deckenfläche infolge der geringeren Länge der Auflagerumrandung das diffuse Feld stärker ausgebildet als bei einer großen Deckenfläche, so daß der Unterschied in den Abstrahlflächen ausgeglichen wird.

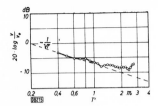

Bild 3. Abnahme der Körperschall-Amplitude v bei einer 14 cm dicken punktweise angeregten Vollbetondecke mit der Entfernung r von der Klopfstelle.

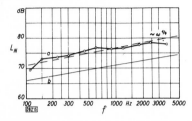

Bild 4. Norm-Trittschallpegel einer 12 cm dicken Massivplattendecke.

a gemessen

b auf Grund der Theorie von *L. Cremer* gerechnet (ohne Berücksichtigung des diffusen Biegewellenfeldes der Decke; für eine Deckenfläche von 16 m²)

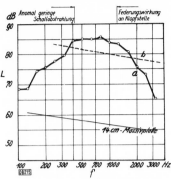

Bild 5. Trittschallpegel einer 7,5 cm dicken Porenbetonplatte, gemessen bei sinusförmiger Anregung mit 10⁶ dyn.

a Meßwerte

b auf Grund der Theorie von *L. Cremer* gerechnet (ohne Berücksichtigung des diffusen Wellenfeldes und einer anomalen Schallabstrahlung).

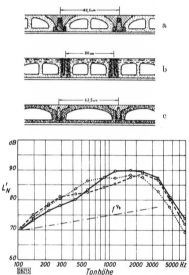

Bild 6. Norm-Trittschallpegel L'_N von verschiedenen Hohlkörperdecken. Der Trittschallpegel steigt mit der Frequenz wesentlich steiler an als bei Massivplatten ($f^{1/4}$).

——— a; ——— b; ····· c

bei Messung und Rechnung voneinander ab, wobei die Ursachen noch nicht bekannt sind. Die Übereinstimmung bezüglich des Frequenzverlaufes ist jedoch gut, so daß eine Reihe von Folgerungen der Theorie, obwohl sie im einzelnen noch nicht experimentell nachgewiesen worden sind, wohl als richtig angenommen werden dürfen. Danach wird der Trittschallpegel L_N durch folgende Eigenschaften der Decke bestimmt:

$$L_N = 20 \log \frac{f^{1/4}}{E^{3/8} \varrho_D^{5/8} h^{1,75}} + \text{const.}$$

Dabei bedeuten:

f die Frequenz, E den Elastizitätsmodul, ϱ_D die Dichte des Deckenbaustoffes und h die Dicke der Decke.

Die Masse je Flächeneinheit einer Decke spielt danach für das Trittschallverhalten keine entscheidende Rolle, während eine Änderung der Dicke der Deckenplatte, vor allem wegen der dadurch bedingten Erhöhung der Biegesteifigkeit von Bedeutung ist. Eine Erhöhung der Dichte des Deckenbaustoffes und damit der Masse der Decke, beispielsweise auf das doppelte, verringert den Trittschallpegel nur um 3,8 dB, eine Verdoppelung der Deckendicke dagegen um 10,5 dB.

Die vorliegenden Überlegungen gelten in vollem Umfang nur für Deckenplatten, bei denen im interessierenden Frequenzgebiet eine normale Abstrahlung vorliegt, bei denen somit die Grenzfrequenz genügend tief liegt. Bei dünnen Platten ergeben sich Abweichungen in dem Sinne, daß der Trittschallpegel bei tiefen Frequenzen durch eine unterhalb der Grenzfrequenz der Decke auftretende verringerte Abstrahlung verkleinert wird. Dies mag ein Beispiel in Bild 5 darlegen, wo bei einer 7,5 cm dicken Porenbetonplatte der Trittschallpegel unterhalb 400 Hz stark absinkt. Schließlich treten noch Abweichungen von dem theoretisch zu erwartenden Verlauf bei höheren Frequenzen auf, indem oberhalb einer bestimmten Frequenz der Trittschallpegel erheblich abfällt. Die Ursache ist entweder auf eine elastische Zusammendrückung der Deckenplatte an der Schlagstelle (nach *L. Cremer* [7]) oder auf eine Federungswirkung von Oberflächen-Rauhigkeiten zurückzuführen.

Hohlkörperdecken

Aus wärmetechnischen Gründen und zur Verringerung des Transportgewichtes werden zahlreiche Decken aus Hohlkörpern aufgebaut, die Lufthohlräume aufweisen. Von diesen Decken wäre zunächst ein relativ günstiges Trittschallverhalten zu erwarten, da die Decken infolge ihrer großen Bauhöhe ziemlich biegesteif ausgebildet sind. Das trifft jedoch nicht zu. Eine theoretische Behandlung dieser kompliziert aufgebauten Decken liegt nicht vor. Eine Betrachtung des Trittschallpegels von derartigen Decken [5] in Bild 6 zeigt, daß der Trittschallpegel steiler als mit $f^{1/4}$, nämlich etwa mit $f^{3/4}$, ansteigt. Eine wesentliche Ursache für diesen starken Anstieg bilden Resonanzerscheinungen der dünnen Hohlkörperschalen. Dies verdeutlicht Bild 7 an einem Versuch, bei dem Hohlkörper auf eine schwere Massivplattendecke mit Mörtel befestigt waren. Bei der Körperschallanregung der Massivplatte zeigten die Hohlkörperschalen wesentlich höhere Schwingungsamplituden als die Massivplatte. Auch bei Rippendecken mit dünner oberseitiger Druckplatte sind ähnliche Erscheinungen zu beobachten, die auf den inhomogenen Aufbau der Decken

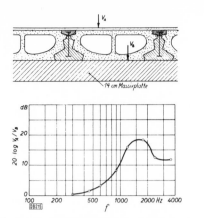

Bild 7. Nachweis der Resonanz-Eigenschaften der Hohlkörperschalen von Hohlkörperdecken.
Auf eine schwere, mit einem Hammerwerk zu Schwingungen angeregte Massivplattendecke waren Hohlkörper aufbetoniert. Die gemessenen Amplituden V_k der Hohlkörperschalen ergaben sich oberhalb etwa 700 Hz wesentlich höher als bei der unmittelbar angeregten Massivplatte.

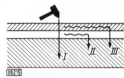

Bild 8. Zur Trittschall-Übertragung bei schwimmenden Estrichen.

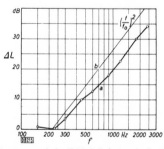

Bild 9. Verbesserung ΔL des Trittschallschutzes durch einen Zement-Estrich auf 0,8 cm dicken Korkschrotmatten ($S_F = 13$ Kp/cm³).
a Meßwerte
b Rechnung nach *L. Cremer*, unter alleiniger Berücksichtigung des Weges I in Bild 8.

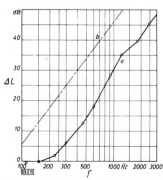

Bild 10. Verbesserung ΔL des Trittschallschutzes durch eine frei aufgehängte Betonplatte (2 m × 2 m) bei einem Luftabstand von rd. 1 cm (Kurve a).
b nach der Theorie von *L. Cremer* zu erwarten bei verhinderter seitlicher Kopplung im Lufthohlraum.

zurückzuführen sind. Dadurch kommt die oft sehr große Steifigkeit dieser Decken bei mittleren und hohen Frequenzen nicht zur Wirkung.

Zweischalige Decken

Decken, bei denen an der Unterseite — unter Belassung eines gewissen Luftabstandes — eine Deckenverkleidung durch Platten oder in Form einer Putzschale angebracht ist, zeigen eine verringerte Trittschallübertragung in den unter der Decke liegenden Raum [5]. Die Verbesserung beruht auf der verringerten Schallabstrahlung der auf die Verkleidung übertragenen Körperschall-Amplituden. Die Wirkung wird begrenzt durch die Übertragung des Körperschalls von der Deckenplatte auf die seitlichen Wände[3]. Der von diesen Wänden abgestrahlte Trittschall hängt in erster Linie von der Art der Wände ab, so daß sich in Bauten mit leichten Wänden höhere Trittschallpegelwerte ergeben als bei schweren Wänden. Näheres mag aus [14] entnommen werden.

Zusammenfassend ergibt sich auf Grund der bisherigen Erkenntnisse folgende Reihenfolge für das Trittschallverhalten von Decken, wobei die günstigeren Decken zuerst genannt sind:

1. zweischalige Massivdecken (bei lockerer Befestigung der Putzschale),
2. Massivplattendecken (bei genügender Dicke),
3. Hohlkörperdecken und
4. Rippendecken (ohne unterseitige Verkleidung).

Das Verhalten von Massivplattendecken ist weitgehend rechnerisch zu erfassen, während bei anderen Decken die Einflußgrößen nur zum Teil und dazu nur qualitativ bekannt sind, so daß eine Beurteilung lediglich auf Grund von Meßergebnissen möglich ist. Bei sämtlichen Massivdecken, auch bei zweischaligen Decken, muß der Trittschallschutz durch einen geeignet gewählten Fußboden zusätzlich verbessert werden, damit die Mindestanforderungen von DIN 52211 für Wohnungstrenndecken erfüllt werden.

Maßnahmen zur Trittschalldämmung

Zur Verringerung der Trittschallübertragung stehen im wesentlichen zwei Anordnungen zur Verfügung, deren physikalische Wirkungsweise hier besprochen werden soll.

Schwimmende Estriche

Sie bestehen aus einer lastverteilenden Estrichplatte, die auf einer federnden Dämmschicht verlegt ist. Auch von den umgebenden Wänden ist die Estrichplatte durch Dämmschichten getrennt. Die Übertragung des Trittschalls von dem Estrich auf die Rohdecke erfolgt nach unseren derzeitigen Kenntnissen auf drei verschiedenen Wegen, die in Bild 8 schematisch dargestellt sind. Die Übertragung auf dem Weg I ist von *L. Cremer* theoretisch behandelt worden unter der Voraussetzung, daß die einzelnen Dämmschichtelemente keine gegenseitige Kopplung besitzen. Die Verbesserung ΔL des Trittschallschutzes durch den schwimmenden Estrich ergibt sich danach zu

$$\Delta L = 40 \log f/f_0 \text{ für } f \gg f_0 \text{ mit}$$

$$f_0 = \frac{1}{2\pi} \sqrt{\frac{s}{m}}.$$

Dabei bedeuten f_0 Resonanzfrequenz des Estrichs, s auf die Flächeneinheit bezogene, dynamisch wirksame Steifigkeit der Dämmschicht und m Masse des Estrichs, bezogen auf die Flächeneinheit.

Die Art des verwendeten Estrichs geht danach in die Dämmwirkung nur über die Masse m (je Flächeneinheit) des Estrichs ein. Bei vorgegebenem Estrich ist für die Dämmung ausschließlich die dynamisch wirksame Steifigkeit der Dämmschicht maßgebend. Als Dämmschichten

---

[3]) Siehe *K. Gösele*: Abstrahlverhalten von Wänden. Dieses Heft, Bild 8, S. 100.

werden vor allem Matten oder locker gebundene Platten aus Mineralfasern, Kokosfasern, Holzfasern u. a. verwendet. Bei weichfedernden Dämm-Matten ist neben der Federung des Dämmstoffgefüges auch die Steifigkeit des in dem Dämmstoff eingeschlossenen Luftvolumens zu berücksichtigen [5; 7].

Die dynamisch wirksame Fasersteifigkeit kann für die interessierenden Dämmschichten an kleinen Dämmschichtproben gemessen werden; die Steifigkeit der Luftschicht wird näherungsweise rechnerisch berücksichtigt. Bei weichfedernden Fasermatten überragt die Luftschichtsteifigkeit die Fasersteifigkeit um ein Vielfaches, so daß die spezielle Art der verwendeten Fasern die Trittschalldämmung kaum mehr beeinflußt [5; 7]. Für die sich ergebende Dämmwirkung ist in diesem Fall in erster Linie die Dicke der Dämmschicht im eingebauten Zustand maßgeblich. Auf dieser Erkenntnis beruhen Versuche, die Dämmwirkung von Dämm-Matten dadurch zu steigern, daß zusätzliche Luftschichten zugeschaltet werden [10]. Vor allem bei verhältnismäßig steifen Dämmschichten gibt die Formel von *Cremer* das experimentell ermittelte Trittschallverhalten von schwimmenden Estrichen mit guter Näherung wieder. Bild 9 enthält dafür ein Beispiel.

Der Weg II, nämlich die Ausbreitung des unter der Stoßstelle erzeugten Luftschalls im Lufthohlraum zwischen Estrich und Rohdecke ist von Bedeutung bei besonders weichfedernden Dämmschichten, die keinen nennenswerten Strömungswiderstand aufweisen. Dadurch treten erhebliche Abweichungen von der oben angeführten Beziehung auf. Die bei der theoretischen Ableitung gemachte Voraussetzung einer verhinderten seitlichen Kopplung zwischen den einzelnen Dämmschicht-Elementen ist nicht mehr erfüllt. Bild 10 gibt ein Beispiel für eine reine Luftschicht, die infolge der seitlichen Kopplung innerhalb der Luftschicht eine wesentlich kleinere Dämmwirkung besitzt als eine gleich dicke Fasermatte [5]. Allerdings ist nur ein Teil der Abweichungen — und zwar derjenige bei tiefen Frequenzen — auf die unmittelbare Ausbreitung des Luftschalls auf dem Wege II zurückzuführen. Bild 11 zeigt diesen Einfluß nach einer Messung von *L. Cremer* [12]. Verhindert man die seitliche Ausbreitung in der Luftschicht durch einen genügend großen Strömungswiderstand der Dämmschicht, dann bleiben noch nach Bild 12 gewisse Abweichungen übrig, die auf die Übertragung auf dem Weg III zurückzuführen sind. Die an der Stoßstelle angeregten Biegeschwingungen der Estrichplatte breiten sich auf dieser aus, wobei sich ein statistisches Wellenfeld im Estrich ergibt, dessen Amplituden — mit Ausnahme der unmittelbaren Umgebung der Schlagstelle — überall gleich groß sind [13]. Die Bedeutung der Übertragung auf dem Wege III ist aus den Ergebnissen eines Versuches ersichtlich, der in Bild 13 angegeben ist. Dabei ist das Trittschallhammerwerk einmal auf einem schwimmenden Estrich unmittelbar über der Versuchsdecke (Kurve a), bei einem zweiten Versuch auf einem über die Versuchsdecke hinausragenden Estrichstück (Kurve b) betrieben worden. Bei hohen Frequenzen unterscheiden sich die Verbesserungswerte nur noch wenig, bedingt durch die Übertragung auf dem Weg III (nach Bild 8). In diesem Zusammenhang ist auch die ungünstige Wirkung von einzelnen, festen Verbindungen zwischen Estrich und Rohdecke, den sog. „Körperschallbrücken", zu erwähnen. Sie sind von großer Bedeutung, da sie bei praktisch ausgeführten Estrichen häufig einen erheblichen Teil der an sich möglichen Dämmwirkung aufheben. Theoretische und experimentelle Untersuchungen über die grundsätzliche Wirkung von Schallbrücken bei Estrichen liegen von *L. Cremer* [11] vor. Dabei wurde insbesondere auf die große Bedeutung einer hohen inneren Dämpfung der Estriche für die Unterdrückung der schädlichen Wirkung von Körperschallbrücken hingewiesen.

Zusammenfassend kann aus den bisherigen Erkenntnissen geschlossen werden, daß die Verbesserung des Trittschallschutzes in erster Linie durch die Steifigkeit der

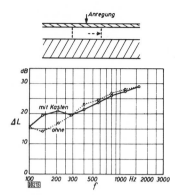

Bild 11. Einfluß des Weges II auf die Schallübertragung bei schwimmenden Estrichen nach Messungen von *L. Cremer* [12]. Durch einen eingebrachten Kasten unter der Anregestelle wird die Ausbreitung im Lufthohlraum vermindert und die Verbesserung von ΔL erhöht.

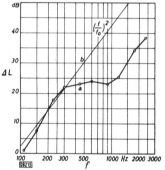

Bild 12. Verbesserung ΔL des Trittschallschutzes für einen Zement-Estrich auf 1 cm-Steinwollplatten.
a gemessen
b gerechnet unter alleiniger Berücksichtigung des Weges I in Bild 8.
Die Abweichungen zwischen a und b sind auf eine zusätzliche Übertragung auf dem Weg III zurückzuführen.

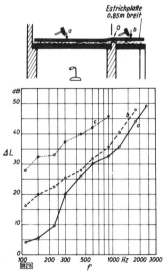

Bild 13. Nachweis der Trittschall-Übertragung über das statistische Körperschallfeld eines schwimmenden Estrichs (Weg III in Bild 8).
a Klopfstelle a, über der Decke
b Klopfstelle b, außerhalb der Decke
c wie bei b, jedoch Estrichplatte bei Stelle 0 unterbrochen.

Dämmschicht bestimmt wird, wobei bei ausgesprochen weiche lernden Schichten noch die Steifigkeit der Luftschicht berücksichtigt werden muß. Ein zu geringer Strömungswiderstand ist bei den letztgenannten Schichten schädlich. Die zunächst nur einen der drei in Bild 8 aufgeführten Übertragungswege erfassende Theorie von L. Cremer gibt die Verhältnisse in den meisten Fällen genügend gut wieder. Die physikalischen Zusammenhänge sind in einer für die Praxis ausreichend erscheinenden Weise geklärt.

Gehbeläge

Gehbeläge zeigen dann eine trittschalldämmende Wirkung, wenn sie eine genügend weiche Federung aufweisen. Nach H. und L. Cremer [3] bildet die Masse des Trittschallhammers zusammen mit der Federung des Gehbelages ein Schwingungssystem, wobei oberhalb der Resonanzfrequenz (f_R) eine Verbesserung ΔL sich ergibt, die einen gleichartigen Verlauf wie bei schwimmenden Estrichen aufweist:

$$\Delta L = 40 \log (f/f_R)\ f > f_R \text{ mit}$$

$$f_R = \frac{1}{2\pi} \sqrt{\frac{D}{M}}.$$

Dabei bedeuten D die Direktionskraft des am Stoßvorgang beteiligten Teils des Gehbelages und M die Masse des stoßenden Hammerwerks.

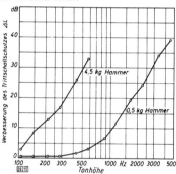

Bild 14. Verbesserung des Trittschallschutzes durch einen Gehbelag (Korkparkett mit Gummischicht) bei zwei verschieden großen Massen des Trittschallhammers.

Aus der Theorie ist zunächst zu folgern, daß die Verbesserung ΔL von der Masse M des stoßenden Hammers abhängt, und um so größer wird, je größer M ist. Dies konnte auch experimentell [8] bestätigt werden, wie dies Bild 14 zeigt. Die Direktionskraft D des Gehbelages hängt von der Steifigkeit des Gehbelages und der beim Stoß wirksamen Berührungsfläche zwischen Hammer und Belag ab. D ist um so kleiner, je kleiner diese Fläche ist. W. Bach hat dies durch Messungen mit verschiedenen gewölbten Schlagflächen der Trittschallhämmer nachgewiesen [15]. In Bild 15 sind einige Meßergebnisse wiedergegeben.

Die Dämmwirkung von weichfedernden Gehbelägen ist somit — im Gegensatz zu den Verhältnissen der schwimmenden Estrichen — in starkem Maße von der Art des stoßenden Körpers (Masse, Ausbildung der Stoßfläche) abhängig[4]). Meßwerte, die mit irgendeinem bestimmten Hammerwerk gewonnen werden, haben etwas Willkürliches an sich. Heute läßt sich noch nicht entscheiden, ob die derzeitigen, in DIN 52210 niedergelegten Daten des Trittschall-Hammerwerks eine den praktischen Wohnverhältnissen entsprechende Prüfung von Bodenbelägen ermöglichen. Die praktisch erzielbare Dämmwirkung von Gehbelägen ist, von Ausnahmen abgesehen (z. B. Bodenteppiche) infolge der zu fordernden Eindruckfestigkeit beschränkt und nicht mit den Wirkungen vergleichbar, welche schwimmende Estriche aufweisen können.

Übergangsformen

Zwischen den besprochenen beiden Grundtypen des schwimmenden Estrichs und des weichfedernden Gehbelages gibt es noch Übergangsformen, deren Dämmwirkung weder der einen, noch der anderen Gesetzmäßigkeit streng genügt. Dabei handelt es sich um Dämmschichten, welche mit einer harten und verhältnismäßig dünnen, lastverteilenden Schicht (in Form von Platten) abgedeckt werden. Ein praktisch bedeutsames Beispiel dafür ist z. B. ein Fußboden, der aus Hartfaserplatten besteht, die auf Weichfaserdämmplatten aufliegen oder auch fest mit ihnen verbunden sind. Die quantitative Wirkung derartiger Anordnungen ist bekannt. Theoretisch ist die Wirkung noch nicht erfaßt.

Der Eingangswiderstand der Platten ist in diesen Fällen gegenüber dem Masse-Widerstand des Hammerwerks vernachlässigbar klein. Infolgedessen ist die Verbesserung ΔL stark von der Masse M des Hammerwerks abhängig. Für die Übertragung ist nur die Schlagstelle von Bedeutung, so daß derartige Fußböden unempfindlich gegen etwaige Körperschallbrücken sind.

Grenzen des Trittschallschutzes

Im Gegensatz zur Luftschalldämmung, die infolge der Schallübertragung entlang der flankierenden Wände in praktisch sehr störender Weise begrenzt ist, kann die Trittschalldämmung so hoch getrieben werden, daß bei normaler Beanspruchung keine hörbaren Störung mehr auftreten. Unter Umständen erfordert dies die Verwendung schwerer Decken und dicker Dämmschichten. Eine Grenze für den Trittschallschutz, die im allgemeinen jedoch nicht stört, ergibt sich aus der Übertragung des beim Stoßvorgang auf der Deckenoberseite entstehenden Luftschalls über die seitlichen Wände.

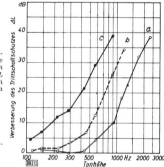

Bild 15. Einfluß der Wölbung der Schlagfläche von Trittschallhämmern auf die Verbesserung ΔL bei Gehbelägen nach Untersuchungen von W. Bach.
a 50 cm Krümmungsradius
b   5 cm Krümmungsradius
c   1 cm Krümmungsradius
Belag 0,6 cm Korkparkett

Schrifttum

[1] Reiher, H.: Über den Schallschutz durch Baukonstruktionsteile. Beihefte z. Ges. Ing. Reihe II, Heft 11 (1932).
[2] Ingerslev, F., A. K. Nielsen u. S. F. Larsen: The Measuring of Impact Sound Transmission through Floors. I. Acoust. Soc. Am. 19 (1947) S. 981.
[3] Cremer, H. u. L.: Theorie der Entstehung des Klopfschalls. Frequenz 2 (1948) S. 61.
[4] Gösele, K.: Zur Meßmethodik der Trittschalldämmung. Ges. Ing. 70 (1949) S. 66.
[5] Gösele, K.: Der Trittschallschutz von Decken, Dämmstoffen und Gehbelägen. Heft 11 (1951) der Veröffentlichungsreihe der Forschungsgemeinschaft Bauen und Wohnen, Stuttgart.
[6] Gösele, K.: Zur Berechnung der Trittschalldämmung von Massivdecken. Ges. Ing. 72 (1951) S. 224.
[7] Cremer, L.: Näherungsweise Berechnung der von einem schwimmenden Estrich zu erwartenden Verbesserung. Forschungen und Fortschritte im Bauwesen. Reihe D, Heft 2 (1952) S. 123 u. Acustica 2 (1952) S. 167.
[8] Gösele, K.: Der Schallschutz von Decken und Wänden. Forschungen und Fortschritte im Bauwesen. Reihe D, Heft 2 (1952) S. 50.
[9] Lange, Th.: Die Messung der Trittschalldämmung von Decken mit sinusförmiger Erregung. Acustica 3 (1953) S. 161.
[10] Gösele, K.: Neue Wege zur Entwicklung von Trittschall-Dämmstoffen. Ges. Ing. 75 (1954) S. 20.
[11] Cremer, L.: Berechnung der Wirkung von Schallbrücken. Acustica 4 (1954) S. 273.
[12] Cremer, L.: Ergänzungen zur Theorie des schwimmenden Estrichs. (Noch unveröffentlicht.)
[13] Gösele, K.: Experimentelle Untersuchungen über die Wirkungsweise von schwimmenden Estrichen. Forschungen und Fortschritte im Bauwesen. (Im Druck.)
[14] Gösele, K.: Das schalltechnische Verhalten von zweischaligen Decken. Forschungen und Fortschritte im Bauwesen. (Im Druck.)
[15] Bach, W.: Das Trittschallverhalten von Gehbelägen. (Noch nicht veröffentlicht.)
[16] Deutsche Normen DIN 4110; DIN 52210; DIN 52211.

---

[4]) Von der dritten möglichen Variablen, der Art des Materials für die Hammerschlagfläche, sei hier abgesehen (siehe [2]).

28

FOOTSTEP NOISE—ORIGIN AND REDUCTION

K. Gösele

*This English summary was prepared expressly for this Benchmark
volume by T. D. Northwood, National Research Council of
Canada, from "Trittschall—Entstehung und Dämmung,"
VDI Ber., 8:23–28 (1956).*

The airborne sound transmission loss of a floor slab does not sufficiently describe
its sound insulation performance. One must consider also its response to point
excitation by dynamic forces such as footsteps. In this note, a survey is made of the
known physical laws relating to the impact transmission process and to its attenuation. Although exemplified here by footsteps, the same considerations apply to
other point excitations, and to the excitation of walls, for example, by the operation of light switches or by attached plumbing appliances.

CHARACTERIZATION OF FOOTSTEP NOISE BEHAVIOR

In analogy with footstep excitation, use is made of a hammer machine consisting
of a series of hammers that drop on the floor from a specified height. This has
become standardized in DIN 4110 and DIN 52210 (and also in ISO R140). Transmitted noise levels are measured in a series of octave bands and normalized to a
room absorption of 10 m^2 (DIN 52211). The noise levels are of course dependent
on the particular source of excitation.

In place of the hammer machine one could use a sinusoidal force produced by an
electromagnetic exciter as devised by Lange [9]. This can be related to the hammer
excitation as shown in Figure 1, on the assumption that the force is applied via a
mass corresponding to the mass of a hammer. The line spectrum produced by the
hammer machine results in octave band levels increasing as $\sqrt{f}$ [3, 9]. Figure 2
shows experimental results for the two types of excitation.

The hammer-machine spectrum contains predominantly high frequencies, as
might be produced by the dropping of small objects, in contrast to which walking
noise contains substantially more low-frequency sound.

In characterizing the performance of various floor constructions with various
coverings, it is found feasible to treat the covering separately, in terms of ΔL,
the improvement realized by the addition of the covering to the bare floor.

PERFORMANCE OF BARE FLOOR SLABS

When an alternating force F is applied to a slab, producing a resultant velocity
v_0 underneath, the input impedance of the slab is given by $Z = F/v_0$. For stiff,
heavy, homogeneous plates this impedance is independent of frequency and depends on the product of the mass per unit area and the flexural stiffness of the
slabs [3].

From the point of excitation the disturbance propagates according to a $1/\sqrt{r}$

law, as illustrated by Figure 3. For finite plates, reflections arise at the edges and at supports, thus adding a diffuse field to the directly propagated field. The total impact noise is essentially independent of the floor boundaries for a wide range of floor areas (experimentally observed over 7 to 20 m^2).

Homogeneous Slabs

Calculations of wave amplitudes for infinite homogeneous plates, made by H. and L. Cremer [3], give values that are approximately correct also for finite plates large enough that their fundamental frequency is below the frequency range of interest. Results of such calculations are given in Figure 4: although absolute values differ substantially the variation with frequency seems correct.

Accordingly the footstep noise level is given by

$$L_N = 20 \log \frac{f^{1/4}}{E^{3/8} Q_D^{5/8} h^{1.75}} + \text{constant}$$

where f is frequency, E the modulus of elasticity, Q_D the density of the slab material, and h the slab thickness. Note that the density of the slab is less important than its thickness. For example doubling the density changes the level by only 3.8 dB, whereas doubling the thickness changes it by 10.5 dB.

Slabs with Cavities

When a floor slab is made up of hollow sections, one might expect that the performance would be better than for the same weight of solid slab, because the hollow one would be thicker and stiffer. Instead, the transmitted noise is substantially higher, increasing at least as $f^{3/4}$ instead of $f^{1/4}$ (see Figure 6). This appears to be a resonance effect associated with the thin shells. A further demonstration of the resonance effect is shown in Figure 7. Similar effects are observed in the case of ribbed slabs.

Double-Leaf Floors

Floors consisting of a main slab with a suspended ceiling below exhibit good resistance to footstep noise [5], because of the reduced radiation efficiency of the ceiling layer. The effect is limited only by flanking via the walls, so that this type of floor performs best when the walls are relatively heavy [14].

In sum, the floors considered perform with respect to footstep noise in the following order:

1. Double-leaf floors (solid slab with resiliently suspended plaster ceiling)
2. Solid slabs (of sufficient weight and thickness)
3. Hollow slabs
4. Ribbed slabs (exposed underneath)

MEASURES FOR ATTENUATION OF FOOTSTEP NOISE

To decrease footstep noise through a floor there are two main approaches, both involving additions above the structural floor.

Floating Floors

Floating floors consist of an upper plate supported from the slab (and from the perimeter walls) by a resilient damping layer. Transmission through such a system may take place by any of the three mechanisms depicted in Figure 8. Transmission by Path I has been calculated by Cremer, on the assumption of no lateral coupling in the soft layer. The improvement due to the floating floor is given by

$$\Delta L = 40 \, \log(f/f_0) \qquad f \gg f_0$$

with

$$f_0 = \frac{1}{2\pi} \sqrt{\frac{S}{m}}$$

where S is the stiffness of the damping layer and m is the mass per unit area of the floating slab. Typical damping layers are composed of mats of soft fibrous materials. For relatively stiff layers, Cremer's formula describes the performance fairly well (Figure 9). For very soft layers, the limiting stiffness may be that provided by the entrapped air [5, 7]. Then the thickness of the layer becomes the principal variable [10].

Path II consists of airborne propagation through the supporting layer, and is of importance for very softly sprung damping systems. The assumption of no lateral propagation no longer holds. Figure 10 compares a pure air layer with a fiber mat of the same thickness, showing the increased transmission when lateral propagation takes place [5]. However, only the low-frequency region can be explained in this way. This point is seen more clearly in measurements of Cremer [12], shown in Figure 11.

If lateral propagation is minimized by use of a damping layer of sufficiently high impedance, there remain certain residual effects attributable to Path III. These result from the diffuse field set up in the floating panel [13]. An experiment to demonstrate the effect is described in Figure 13.

In sum, the improvement due to a floating floor depends on the stiffness of the damping layer (limited by the stiffness of the air cavity). Too small a propagation impedance, leading to lateral propagation, is undesirable. For typical damping layers, Cremer's calculation, which considers only Path I of Figure 8, is sufficiently accurate for practical purposes.

Flexible Coverings

Floor coverings reduce the transmission of hammer machine noise, relative to a bare slab, according to the formula

$$\Delta L = 40 \, \log(f/f_R) \qquad f \gg f_R$$

with

$$f_R = \frac{1}{2\pi} \sqrt{\frac{D}{M}}$$

where f_R is the resonance frequency corresponding to the mass, M, of the impacting hammer, and D is the force transmitted through the covering, which in turn depends on the stiffness of the covering and the contact area. The effect of varying the impacting mass [8] is shown in Figure 14, and the effect of varying the contact area [15] is indicated in Figure 15.

In contrast to the floating floor, the attenuating effect of a soft floor covering is thus strongly dependent on the type of impacting body (its mass and the area of contact) [14]. Because of this there is uncertainty as to the value of the hammer machine in predicting the performance of such coverings under real footsteps.

Intermediate Types

Between the two basic types described above, there are many constructions that have some of the properties of each. An important example consists of a thin hard surface layer over a softer layer. Here the input impedance of the top surface is negligible compared to impedance of the hammer mass; again the improvement ΔL is strongly dependent on the hammer mass.

FIGURE CAPTIONS

Figure 1   Calculations according to L. Cremer of differences to be expected between footstep pressure level values under excitation with a footstep hammer machine (L_{Tr}) and under sinusoidal excitation (L_{KS}) due to a varying force of 10^6 dynes.

Figure 2   Comparison of the footstep noise level achieved with the hammer and sinusoidal excitation for a slab according to Th. Lange [9]. The values shown for sinusoidal excitation have been converted to equal forces rising as $\sqrt{f}$, as observed for the hammer-machine spectrum.

Figure 3   Decrease of the vibration amplitude v of a 14-cm thick solid concrete slab excited at a point, as a function of the distance r from the point of impact.

Figure 4   Normalized footstep noise level of a 12-cm thick solid slab: (a) Measured (for a slab area of 16 m^2). (b) Calculated on the basis of the theory of L. Cremer (without consideration of the diffuse field in the slab).

Figure 5   Footstep noise pressure level of 7.5-cm thick light-weight concrete slab, measured under sinusoidal excitation of 10^6 dynes. (a) Measured values. (b) Calculated on the basis of the theory of L. Cremer, without considering the diffuse wave field and anomalous sound radiation. (Below 400 Hz: anomalously small sound radiation. Above 1200 Hz: spring action at the point of impact. Solid straight line: 14-cm solid slab.)

Figure 6   Standardized footstep sound pressure level L'_N of various hollow slabs. The footstep sound-pressure level rises with frequency substantially more rapidly than with solid slabs.

Figure 7   Demonstration of the resonance properties of the cavities of hollow slabs. Hollow blocks were cemented to a heavy solid slab which was excited by means of a hammer machine. The measured amplitudes V_A at the top of the hollow blocks were substantially higher above 700 Hz than at the directly excited solid slab.

29

Copyright © 1963 by the Acoustical Society of America

Reprinted from *J. Acoust. Soc. Am.* 35(11):1825–1830 (1963)

Relationship between the Transmission Loss and the Impact-Noise Isolation of Floor Structures

M. Heckl* and E. J. Rathe†

Bolt Beranek and Newman Inc., Cambridge 38, Massachusetts
(Received 7 January 1963)

Transmission loss and impact noise are two important aspects of the properties of floors. With the application of the principle of reciprocity, a relation is derived for floors with hard surfaces that makes it possible to estimate the impact noise from transmission loss or vice versa. If both are known, some insight can be gained on the existence of leakage paths and flanking by side walls, or the efficiency of floor-surfacing materials.

INTRODUCTION

IN many practical applications of building acoustics, it is not only the transmission loss for airborne sound that defines the conditions that the occupants of a building have to live with, but also the impact-noise isolation of the structure. Since the mechanism for impact-noise transmission is based largely on the same properties of a floor that define the transmission loss, it is not surprising that a relation should exist between the two. This relation can be derived using the principle of reciprocity.

I. THEORY

Reciprocity Theorem Applied to Point-Driven Structures

We derive a simple formula that relates the sound radiation from a point-driven structure to the vibrations of the same structure when it is excited by a reverberant sound field. Smith[1] and Lyon and Maidanik[2] have shown that such a relation exists, and they use it to investigate the vibrations of ribbed panels.[3] Similar ideas are also included in a paper by Lyamshev.[4] The formula given by these authors was derived by expanding the vibrations of a structure in normal modes and using the modal density of the structure. Since the modal density is hard to obtain for complicated floor structures, we propose a derivation that does not depend on a modal expansion.

We assume a closed reverberant space of arbitrary shape (see Fig. 1). One part of its surface consists of the floor surface that we are interested in, together with any supporting walls that may contribute to the sound radiation. We call this arrangement structure A. A second part of the surface is represented by a limp wall B. We need not define any of the remaining boundary surfaces except by the total sound absorption of all surfaces, which gives us a room constant R.

If we apply a force F_i to the structure A, it will impart a certain vibration to this structure. The amplitude of these vibrations may vary markedly at different positions on the structure. All parts of the structure will radiate sound into the reverberant space. We take this system to be linear, and therefore the radiated power P must be proportional to the square of the applied force. By defining a factor of proportionality α, we can write

$$P = \alpha F_i^2. \tag{1}$$

* Present address: Müller-BBN GmbH, Munich, Germany.
† Present address: Swiss Federal Laboratory for Testing Materials (EMPA) in Zürich, Switzerland.
[1] P. W. Smith, Jr., J. Acoust. Soc. Am. 34, 640 (1962).
[2] R. H. Lyon and G. Maidanik, J. Acoust. Soc. Am. 34, 623 (1962).
[3] G. Maidanik, J. Acoust. Soc. Am. 34, 809 (1962).
[4] L. M. Lyamshev, Soviet Phys.– Doklady 6, 410 (1961).

Fig. 1. Model for reciprocity argument.

The radiated power P will in turn produce a reverberant sound pressure $\langle p^2 \rangle$ in the space, and its frequency average is given by

$$\langle p^2 \rangle = 4\rho c P/R, \qquad (2)$$

where ρc is the radiation impedance of the medium and R is the room constant of the space.

We take the effect of pressure doubling at the solid boundary into account and calculate the velocity of the limp wall B, which behaves according to the mass law:

$$v_B^2 = 2\langle p^2 \rangle/\omega^2 m^2 = 8\rho c \alpha F_i^2/\omega^2 m^2 R, \qquad (3)$$

where m is the surface mass of wall B.

To find out how much the structure A is excited by a reverberant sound field, we use the same procedure in reverse. Since structure B is assumed to be limp, the power radiated for point force excitation is[5]

$$P' = \rho c k^2 F_i'^2/2\pi\omega^2 m^2. \qquad (4)$$

Here, $k = 2\pi/\lambda$ is the wavenumber in air, and the primes denote excitation of structure B.[6] Since we are considering the same space, we again have

$$\langle p'^2 \rangle = 4\rho c P'/R. \qquad (5)$$

The velocity of the structure A due to this sound pressure cannot easily be calculated directly. However, we can define a factor of proportionality β:

$$v_A'^2 = \beta\langle p'^2 \rangle. \qquad (6)$$

On the basis of the reciprocity theorem, we can interchange the point of application of a force and the point of measurement of a velocity without changing the relationship between them. This gives us

$$v_B^2/F_i^2 = v_A'^2/F_i'^2 = 8\rho c\alpha/\omega^2 m^2 R = 2\rho^2 c^2 k^2 \beta/\pi\omega^2 m^2 R, \qquad (7)$$

from which we obtain

$$\beta/\alpha = 4\pi/\rho c k^2. \qquad (8)$$

This relation shows that the response of a structure to a reverberant sound field as measured by β can be found if its radiation for point excitation as measured by α is known. Equation (8) is independent of size, shape, or damping of the structure. It also is not restricted to resonating modes. Thus, Eq. (8) can be applied to a great number of problems. The relation between impact-noise level and sound-transmission loss is only one out of many possible applications.

[5] M. Heckl, Acustica 9, 371 (1959).
[6] It can be shown that Eq. (4) holds just for those walls for which "mass law" holds. Thus, when a wall is so limp that Eq. (3) is correct, Eq. (4) is correct, too. Actually, instead of a "limp" structure B, we would have used any other structure provided its response to a reverberant sound field and the power radiated by a point force are known. The limp structure was used for the sake of convenience only.

Impact-Noise Level and Sound-Transmission Loss

We now relate the quantity α to the impact-noise level and β to the sound-transmission loss.

The standard tapping machine described by ISO (International Standards Organization) can be regarded generally as a source of constant force F_T. (The magnitude and the spectral distribution of this force are considered in the next section.) The sound pressure generated by excitation of a floor with the machine is, therefore,

$$\langle p_T^2 \rangle = 4\rho c\alpha F_T^2/R. \qquad (9)$$

The standard normalized impact-noise level L_N is defined by ISO as

$$L_N = 10 \log(\langle p_T^2 \rangle R/p_0^2 R_0), \qquad (10)$$

where R_0 is the normalized room absorption of $10m^2$, p_0 is the reference sound-pressure level of 0.0002 μb, and $\langle p_T^2 \rangle$ is to be measured in octave bands. Therefore, we obtain the following relation between L_N and α:

$$L_N = 10 \log(4\rho c\alpha F_T^2/R_0 p_0^2). \qquad (11)$$

In order to relate β to the sound-transmission loss (TL), we start with the transmission coefficient τ defined by $TL = -10 \log\tau$. We assume that the surface of the floor that is radiating away from the reverberant space (the side that was subjected to the tapping machine) has a low coincidence frequency. Then, the radiation impedance equals ρc, and we can give the transmission coefficient as

$$1/\tau = \langle p_s^2 \rangle/4v_s^2\rho^2 c^2, \qquad (12)$$

where p_s is sound-pressure level in the source room and v_s is the velocity of the partition due to sound excitation. (The factor 4 is due to averaging over all angles.)

Using Eq. (6), we get

$$\beta = v_s^2/\langle p_s^2 \rangle = \tau/4\rho^2 c^2. \qquad (13)$$

We now combine Eqs. (8), (11), and (13), and we get the desired relation between impact-noise level and the transmission loss as

$$L_N + TL = 10 \log k^2 F_T^2/4\pi R_0 p_0^2. \qquad (14)$$

The sum of the two standard measures L_N and TL is thus independent of the properties of the floor.

Calculation of the Spectral Distribution of the Force Generated by the Standard Tapping Machine

In order to apply Eq. (14), we have to know the force generated by the standard tapping machine. Fortunately, this force can easily be calculated, provided the floor surface is hard and its input impedance is large as compared to the mass impedance of the hammer.

The standard ISO tapping machine uses a series of 5 hammers, each of which has a mass M of 500 g. They are dropped at a rate of 10 impacts/sec ($f_s = 10$ cps) from a height of $h = 4$ cm. To obtain the spectral distribution of this series of equally spaced force pulses, we use a Fourier series expansion. If we place the origin at the time of the impact impulse, the spectral density is given by

$$F_n = 2/T \int_0^T F(t) \cos n\Omega t\, dt, \qquad (15)$$

where Ω is $2\pi f_s$. Since each impulse is very short, we can approximate $\cos n\Omega$ by unity and obtain

$$F_n = 2/T \int_0^T F(t)\, dt. \qquad (16)$$

This equation in turn is related to the momentum J:

$$F_n = 2J/T = 2Mv_0/T = 2f_s M (2gh)^{\frac{1}{2}}. \qquad (17)$$

The height of the individual lines of the impulse power spectrum is thus given by[7]

$$F_n{}^2 = 8 f_s{}^2 M^2 g h. \qquad (18)$$

Within an octave band of center frequency f_m, we have $f_m / f_s (2)^{\frac{1}{2}}$ spectrum lines, and the expression for the spectral density of the force in one octave, expressed as

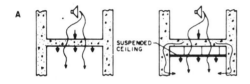

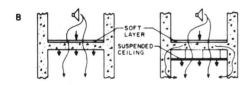

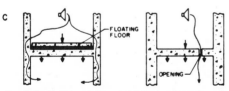

FIG. 2. Applicability of theoretical behavior. Examples under A: Both Eq. (14) and Eq. (21) hold. Examples under B: Eq. (14) [and Eq. (24)] holds. Examples under C: neither Eq. (21) nor Eq. (14) holds.

[7] The fact that the spectrum of the tapping machine consists of lines of equal height (up to 3 kc/sec) was verified by Th. Lange, Acustica 3, 161 (1953).

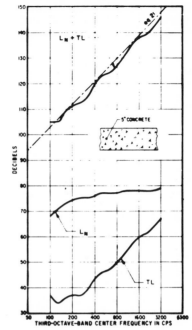

FIG. 3. Transmission loss and impact-noise levels of a 5-in. concrete floor. [After Heckl[9] and Gösele.[10]]

rms value, finally becomes

$$F_{rms}{}^2 = 4 f_m f_s M^2 g h / (2)^{\frac{1}{2}}. \qquad (19)$$

For the standard tapping machine, this turns out to be

$$F_{rms}{}^2 \approx 2\sqrt{2}\,10^{10} f_m (\text{dyn}^2). \qquad (20)$$

If we now introduce standard conditions (e.g., room absorption $R_0 = 10 m^2$), Eq. (14) takes the simple form

$$L_N + \text{TL} = 43 + 30 \log f_m (\text{dB}) \qquad (21)$$

(f_m is the octave-band center frequency in cps).

II. COMPARISON WITH EXPERIMENTAL RESULTS

Equation (21) was derived under the assumption that the coincidence frequency is low and the surface is hard and has a high input impedance. If these requirements are not met, especially if the floor surface is soft or has a low impedance, the force that excites the floor is smaller than the value given by Eq. (20). Thus, the right side of Eq. (21) would give values that are too high. This is not a deficiency of the formalism proposed in this paper, because Eq. (14) is still valid if the correct force F_T is introduced. However, there is one case where even Eq. (14) does not hold. This happens when the impact sound and the airborne sound travel along different paths. For example, if there is a hole in the floor so that the airborne-sound transmission is

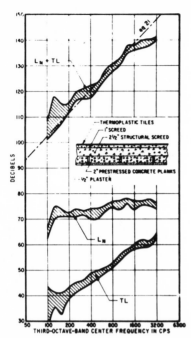

FIG. 4. Range of transmission loss and impact-noise levels of four 5½-in. concrete floors using precast concrete planks. [After Parkin, Purkis, and Scholes.[11]]

governed by the TL of the hole, Eq. (14) could not be expected to be correct. Another example is flanking transmission. If the sound transmission through the side walls is much greater than through the floor, Eq. (14) does not hold. Figure 2 shows some examples for which either Eq. (21) or Eq. (14) can be applied, or neither can be applied.

Fortunately, flanking transmission (through holes or side walls) is the only case where Eq. (14) does not hold, or, if the floor surface is hard, Eq. (21) does not hold. Therefore, if one finds in a practical situation that the sum of impact-noise levels and the TL values does not follow Eq. (21), even though the floor surface is hard,[8] one can conclude that the airborne sound is transmitted predominantly through flanking paths.

Floors with Hard Surfaces

Figures 3[9,10] and 4[11] show results obtained on floors with hard surfaces. As one can see, the agreement with

---

[8] As "hard" materials, one would consider concrete, gypsum, etc. Wood can be considered hard only up to approximately 600 cps. Thermoplastic tiles of less than $\frac{1}{8}$ in. generally behave like hard surfaces up to 1200 cps.

[9] M. Heckl, Acustica 10, 106 (1960).

[10] K. Gösele, Acustica 6, 69 (1956).

[11] P. H. Parkin, H. J. Purkis, and W. E. Scholes, *Field Measurements of Sound Insulation between Dwellings* (Her Majesty's Stationery Office, London, 1960), p. S153.

Eq. (21) is good. Figure 5[12] shows the behavior of four hollow-pot structural floors. In this case, the data spread is larger, but the general trend still follows Eq. (21).

Floors with Resilient Layers

Figures 6[13] and 7[14] give examples of floors that do not have a hard surface. For this reason, Eq. (21) is in

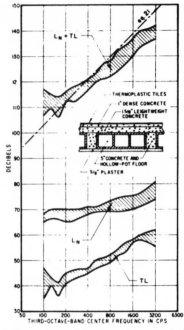

FIG. 5. Range of transmission loss and impact-noise levels of four concrete and hollow-pot floors. [After Parkin, Purkis, and Scholes.[12]]

agreement with the measured values only up to a certain frequency. At higher frequencies, the local elasticity of the surface comes into play, and, therefore, the force produced by the trapping machine is not transmitted fully to the floor.

It was shown by H. and L. Cremer[15] that the elasticity of the floor surface affects the transmitted force and, therefore, also the impact-noise levels only above the resonance frequency f_1 of the hammer mass-layer elasticity system. They also found that the reduction of the impact-noise level due to the resilient layer is

---

[12] See Ref. 11, p. S89.

[13] H. A. Müller, Schalltechnisches Beratungsbüro, Prüfbericht No. 374 (Munich, Aug. 1962) (unpublished).

[14] See Ref. 11, p. S5.

[15] H. Cremer and L. Cremer, Frequenz 2, 61 (1948).

given by

$$\Delta L_N = 40 \log f/f_1, \quad (22)$$

with

$$f_1 = (1/2\pi)(D/M)^{\frac{1}{2}}. \quad (23)$$

(D is the force reaction by the affected area of the resilient layer; M is the mass of a hammer.) The decrease in impact-noise levels can be seen very clearly from Figs. 6 and 7, where from a certain frequency f_1 the L_N curve changes its slope by 12 dB/octave.

Since the reduction of the impact-noise level is due to a proportional reduction of force, floors with resilient layers follow Eq. (24) instead of Eq. (21):

$$TL + L_N = 43 + 30 \log f_m - 10 \log(1 + f^4/f_1^4). \quad (24)$$

The theoretical curve according to Eq. (24) is also

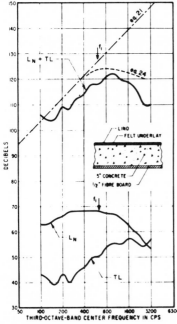

FIG. 7. Transmission loss and impact-noise levels of a concrete floor with soft surface material. [After Parkin, Purkis, and Scholes.[14]]

shown in Figs. 6 and 7; it is in reasonable agreement with the measured values.

In addition to this resonant behavior, the direct impedance mismatch between the input impedance of the floor and the mass impedance of the hammer leads to an impact-noise-level decrease of 6 dB/oct above a limiting frequency f_2. Based on this concept, L. Cremer[16]

[16] L. Cremer, "Näherungsweise Berechnung der von einem schwimmenden Estrich zu erwartenden Verbesserung," Forsch. Fortschr. Bauwesen D2, 123 (1952).

formulated the improvement in impact-noise levels as

$$\Delta L_N = 20 \log f/f_2; \quad (25)$$

$$f_2 = 2d^2(E\rho)^{\frac{1}{2}}/(3)^{\frac{1}{2}}\pi M. \quad (26)$$

E is Young's modulus, d is the thickness of the surface layer, ρ is the density of the surface layer, and M is the mass of the hammer.

Unfortunately, we were not able to find measurements on a floor that had a top layer such that the coincidence frequency and the input impedance were low (these two effects are somewhat contradictory). Therefore, no example for the influence of the low impedance can be given.

Floors with Flanking Paths

It is a well-known fact[17] that in most floating-floor constructions the transmission loss for airborne sound is mainly defined by flanking transmission through the side walls. Therefore, one would expect that for floating floors the sum $L_N + TL$ lies below the curve given by Eq. (21). Figure 8[18] gives a typical example of this behavior.

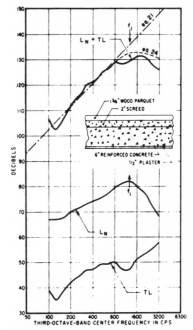

FIG. 6. Transmission loss and impact-noise levels of a concrete floor with wooden parquet surface. [After Müller.[17]]

[17] K. Gösele, "Experimentelle Untersuchungen über die Wirkungsweise von schwimmenden Estrichen," Forsch. Fortschr. Bauwesen D23 (1956).
[18] See Ref. 11, p. S33-1.

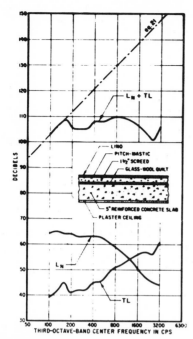

FIG. 8. Transmission loss and impact-noise levels of a concrete floated floor. [After Parkin, Purkis, and Scholes.[18]]

If we examine the L_N curve for this floor, we find that it behaves in a reasonable way. A slope of -9 dB/oct is to be expected above the resonant frequency for the

floating slab on its resilient support. But the transmission-loss curve shows almost no improvement over the values expected for a simple 5-in.-thick concrete floor. This is due to substantial flanking transmission over sidewalls, which in this case were 2 in. thick. Whereas the impact-noise levels were improved considerably by the floated floor, the transmission loss was probably not affected at all. The difference between the actual values of L_N+TL and the theoretical behavior according to Eq. (21) can be used here to judge the severity of the flanking transmission. This need not imply the inadequacy of a structure as such, but it can be very helpful in showing up the areas where improvement can be obtained if required. In the particular case of Fig. 8, no change in floor construction is likely to improve the transmission loss, whereas it can be improved considerably by changes in the wall structures.

CONCLUSIONS

The experimental results show good agreement with the relation for L_N+TL. The discrepancies that exist can be explained and, in many cases, predicted with adequate accuracy. We hope, therefore, that this approach will be of help in the general understanding of the behavior of partitions. In particular, this can be in the evaluation of systems where not all the data are available, where unfavorable situations in buildings require a careful evaluation to provide improvements and a checking method for many research projects in floor components. It may also be possible to substitute impact-noise measurements for the determination of the transmission loss in panels with high transmission loss where flanking paths can cause serious difficulties.

Reprinted from *J. Acoust. Soc. Am.* **50**:1414–1417 (1971)

Relation between the Normalized Impact Sound Level and Sound Transmission Loss *

Istaván L. Vér

Bolt Beranek and Newman Incorporated, Cambridge, Massachusetts 02138

An expression relating the normalized impact sound level (L_n) and the sound transmission loss (TL) of a floor is derived. It takes into account both the resonant and the forced response of the floor, thus extending the utility of the formulas derived previously by Heckl and Rathe into the frequency range below the critical frequency. This relation between normalized impact sound level and sound transmission loss permits an experimental evaluation of the potential TL of a floor partition in the presence of certain types of flanking.

INTRODUCTION

Whether a floor is excited by the hammers of a tapping machine, or by an airborne sound field in the source room, the sound transmitted into the receiver room below depends on the sound radiation properties of the floor. Accordingly, there must be a close relation between the airborne sound transmission loss (TL) and the normalized impact sound level (L_n) for a given floor. Heckl and Rathe,[1] using a reciprocity argument, derived an expression giving the connection between TL and L_n in the frequency range above the critical frequency of the floor. In this paper we derive a more general expression that takes into account the resonant, as well as the forced, response of the floor, without making use of the reciprocity theory.

I. SIMPLE FLOORS

The relationship between sound transmission loss and normalized impact sound level can be most easily explored by assuming that the source-room sound pressure p_1 is so adjusted that the sound power transmitted by the floor into the receiving room is the same as when the floor is excited by the standard tapping machine. Figure 1 shows schematically the acoustical and impact excitation of the floor. The variable resistance symbols labeled "f" and "a" on the right-hand side of the figure indicate the adjustment of the spectrum shape and amplitude of the source-room pressure to yield identical sound pressure in the receiving room for both acoustical and impact excitation ($p_2 = p_2'$).

In the case of acoustical excitation of the floor, the sound power transmitted to the receiving room is made

up of the contribution of the forced waves, $W_{ac}{}^f$, and that of the resonant waves, $W_{ac}{}^r$. The forced waves usually control the sound transmission loss below the critical frequency of the slab, while the resonant waves are dominant at and above the critical frequency.

The sound power transmitted by the forced waves into the receiving room can be expressed in terms of the field-incidence mass-law transmission loss of the homogeneous isotropic ceiling slab and the space-average mean-square sound-pressure level in the source room

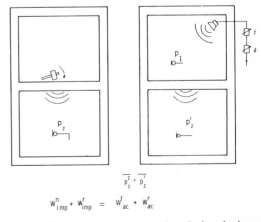

Fig. 1. Tapping machine and acoustic excitation of a heavy concrete floor. The acoustic pressure in the source room, p_1, is adjusted to yield the same pressure in the receiving room for impact and sound excitation ($p_2 = p_2'$).

$\langle p_1^2 \rangle$; namely,[2]

$$W^f_{ac} \doteq \frac{\langle p_1^2 \rangle A}{4\rho c} \left[1 + \frac{1}{3.1}\left(\frac{\rho_s \omega}{2\rho c}\right)^2 \right]^{-1} \doteq \langle p_1^2 \rangle A \frac{3.1\rho c}{\rho_s^2 \omega^2}, \quad (1)$$

where ρ is the density of air in kilograms/m³, c is the speed of sound in air in meters/sec, $\omega = 2\pi f$ is the angular frequency in radians/sec, ρ_s is the mass per unit area of the floor slab in kilograms/m², and A is the area of the floor slab (one side) in meters squared.

In the practical case where the radiation losses are small compared with the edge *and* dissipation losses, the acoustical power transmitted to the receiving room by the resonant modes of the floor can be calculated from statistical energy analysis[3] as

$$W^r_{ac} = \langle p_1^2 \rangle \frac{\pi}{2} \frac{(12)^{\frac{1}{2}}c^2 A \sigma_{rad}}{\rho_s c_L h \omega^2} \frac{\rho c \sigma_{rad}}{2\rho c \sigma_{rad} + \rho_s \omega \eta}$$

$$\doteq \langle p_1^2 \rangle \frac{\pi}{2} \frac{(12)^{\frac{1}{2}}c^2 A}{\rho_s c_L h \omega} \frac{\rho c \sigma^2_{rad}}{\rho_s \omega \eta}, \quad (2)$$

where σ_{rad} is the radiation efficiency of the floor slab, c_L is the propagation speed of longitudinal waves in the slab material in meters/sec, h is thickness in meters, and η is the composite loss factor of the slab (accounting for dissipation and conduction of power to adjacent structures).

When a rigid[4] bare concrete slab is excited by a standard tapping machine,[5] the effect of the impacting hammers on the slab can be well approximated[6] by a steady-state point force of spectral density

$$S_0 = 4 \quad N^2/Hz. \quad (3)$$

The mean-square force amplitude in an octave band of center frequency, f, and effective bandwidth of $\Delta f = f/\sqrt{2}$ is then

$$F^2_{rms} = (4/\sqrt{2})f \quad N^2. \quad (4)$$

The sound power transmitted to the receiving room for impact excitation of the floor is made up of the contributions of the nearfield, W_{imp}^n, and that of the reverberant resonant vibration field of the slab. The octave-band sound power radiated from the nearfield at the excitation point is given by[6]

$$W^n_{imp} = \frac{F^2_{rms}}{2\pi\rho_s^2 c} = \frac{S_0 f}{\sqrt{2}} \frac{\rho}{2\pi\rho_s^2 c}, \quad (5)$$

which is independent of the loss factor and radiation efficiency of the slab.

The mechanical power introduced to the slab is

$$W_{mech} = F^2_{rms} Y, \quad (6)$$

where Y is the point input admittance of the equivalent[7] infinite slab in meters/N-sec. The energy lost by the slab in unit time is given by

$$W_L = \langle v^2 \rangle \rho_s A \omega \eta. \quad (7)$$

Equating the input power and the lost power ($W_{mech} = W_L$) yields the mean-square slab velocity

$$\langle v^2 \rangle = F^2_{rms} Y / A\rho_s \omega \eta. \quad (8)$$

The octave-band sound power radiated into the receiving room by the resonant modes is

$$W^r_{imp} = \langle v^2 \rangle \rho c \sigma_{rad} A = \frac{S_0 f}{\sqrt{2}} \frac{Y\rho c\sigma_{rad}}{\rho_s \omega \eta}. \quad (9)$$

Expressing the point input admittance in terms of the physical parameters of the slab in Eq. 9, we get

$$W^r_{imp} = (S_0 f/\sqrt{2})(\rho c\sigma_{rad}/2.3\rho_s^2 c_L h\omega\eta). \quad (10)$$

Because of the special choice of the sound-pressure spectrum in the source room, we can equate the acoustical power transmitted to the receiving room for acoustical excitation with that transmitted for impact excitation of the floor, yielding

$$W^f_{ac} + W^r_{ac} = W^n_{imp} + W^r_{imp}. \quad (11)$$

The combination of Eqs. 1, 2, 5, 10, and 11 gives

$$\langle p_1^2 \rangle A \left[\frac{3.1\rho c}{\omega^2\rho_s^2} + \frac{\pi}{2} \frac{(12)^{\frac{1}{2}}c^2}{\rho_s c_L h\omega^2} \frac{\rho c\sigma^2_{rad}}{\rho_s \omega \eta} \right]$$

$$= \frac{S_0 f}{\sqrt{2}}\left(\frac{\rho}{2\pi\rho_s^2 c} + \frac{1}{2.3\rho_s c_L h} \frac{\rho c\sigma_{rad}}{\rho_s \omega \eta} \right). \quad (12)$$

If we express the mean-square sound pressure in the source room, $\langle p_1^2 \rangle$, in terms of the sound pressure in the receiving room, $\langle p_2^2 \rangle$, the airborne sound transmission loss of the floor, TL, and the absorption in the receiving room, $S\bar{\alpha}$, we get

$$\langle p_1^2 \rangle = \langle p_2^2 \rangle (S\bar{\alpha}/A)10^{(TL/10)} \quad N^2/m^4. \quad (13)$$

The definition of the normalized impact sound level[8] is

$$L_n \equiv 10\log\frac{\langle p_2^2 \rangle}{p^2_{ref}} - 10\log\frac{A_0}{S\bar{\alpha}} \quad dB \ re \ 2\times10^{-5}\frac{N}{m^2}, \quad (14)$$

which yields

$$\langle p_2^2 \rangle = p^2_{ref}10^{(L_n/10)}(A_0/S\bar{\alpha}) \quad N^2/m^4, \quad (15)$$

where $p_{ref} = 2\times10^{-5}$ N/m² and $A_0 = 10$ m².

Combining Eqs. 12, 13, and 15 yields the desired relationship between sound transmission loss, TL, and normalized impact sound level, L_n; namely,

$$L_n + TL = 88.5$$

$$+ 10\log\left(\frac{S_0 f}{\sqrt{2}}\left\{ \left(\frac{\rho}{2\pi\rho_s^2 c} + \frac{1}{2.3\rho_s c_L h} \frac{\rho c\sigma_{rad}}{\rho_s \omega \eta} \right) \Big/ \right.\right.$$

$$\left.\left. \left[\frac{3.1\rho c}{\omega^2\rho_s^2} + \frac{\pi}{2} \frac{(12)^{\frac{1}{2}}c^2}{\rho_s c_L h\omega^2} \frac{\rho c\sigma^2_{rad}}{\rho_s \omega \eta} \right] \right\} \right), \quad (16)$$

where L_n is measured in octave bands.

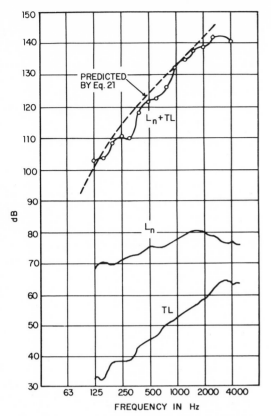

FIG. 2. Measured sound transmission loss (TL) and normalized impact sound level (L_n) and their sum (TL+L_n) of a bare high-stress concrete slab of 57-lb/ft^2 surface weight (Ref. 7). Dotted curves: TL+L_n as predicted by Eq. 17.

In the special case of a thick lightly damped slab, where $W^f_{ac} \ll W^r_{ac}$ and $W^n_{imp} \ll W^r_{imp}$, Eq. 16 yields

$$\mathrm{TL} + L_n \doteq 43 + 30 \log f - 10 \log \sigma_{rad}, \quad (17)$$

which, except for the last term, is the same as derived previously by Heckl and Rathe[1] using a reciprocity argument. Equation 17 states that, above the critical frequency (where $\sigma_{rad} \doteq 1$), *the sum* TL+L_n *is independent of the physical properties of the rigid floor slab.*

Figure 2 shows the measured[9] sound transmission loss, TL, and the normalized impact sound level, L_n, as well as their sum, for a bare high-stress concrete floor slab of 57-lb/ft^2 surface weight. The measured and predicted values for the sum are in good agreement.

Below the coincidence frequency of the floor slab, where the forced waves control the airborne sound transmission loss ($W^f_{ac} > W^r_{ac}$), but where the impact noise isolation is still controlled by the resonant vibrations of the impacted slab ($W^r_{imp} > W^n_{imp}$), Eq. 16

yields

$$L_n + \mathrm{TL} \doteq 88 + 10 \log \frac{f^2 \sigma_{rad}}{c_L h \eta}$$

$$= 40 + 20 \log f - 10 \log \frac{f_c \eta}{\sigma_{rad}}, \quad (18)$$

where f_c is the critical frequency of the floor. In this case the sum TL+L_n depends on the physical properties of the floor slab.

Equations 17 and 18 present powerful tools to determine the extent of flanking transmission in situations where both TL and L_n have been measured. Any difference between the calculated and measured values of the sum TL+L_n is a direct indication of flanking.

II. COMPOSITE FLOOR STRUCTURES

In order to improve the impact sound isolation, most finished floors usually have either a resilient surface layer or a second hard slab floating resiliently supported on the structural slab.

For an added resilient layer the sound transmission loss usually remains unchanged (the resilient layer does not add significant mass, stiffness, or damping to the structural floor), but the spectral density of the impact force decreases above a certain frequency f_1 which is given by the mass of the hammer, m, and the dynamic stiffness of the resilient layer under the striking surface of the hammer, k:

$$f_1 = (1/2\pi)(k/m)^{\frac{1}{2}}. \quad (19)$$

The force spectral density S becomes[10]

$$S \doteq \frac{S_0}{1 + (f_1/f)^4}. \quad (20)$$

By using the above value of S instead of S_0 in Eq. 16, it becomes immediately clear that at the frequency f_0 the slope of the sum L_n+TL changes from a positive 30 dB/decade to a negative 10 dB/decade slope as shown in Figs. 6 and 7 of Ref. 1. Adding a resiliently supported floating slab on the top of the structural slab increases both the sound transmission loss and decreases the normalized impact sound level of the floor by the same amount so that the sum TL+L_n for the floating floor should be the same as for the bare structural floor given in Eq. 17.

In practical installations where flanking paths through walls common to the source and receiving rooms are not controlled, the measured sound transmission loss of a floating floor is usually not much higher than that of the bare structural floor alone. However, the normalized impact sound level of the composite floor, which is not subject to direct flanking, decreases with increasing frequency with an approximate slope of 30 dB/decade[11] for resonantly reacting floating floors and with a slope of 40 dB/decade[6] for a locally reacting floating floor above the basic resonance frequency of the floating slab.

In the frequency range where the resonant motion of the structural slab controls both the airborne and impact noise transmission, the potential sound transmission loss, TL_{pot}, of a floating floor (i.e., which would be measured after complete elimination of the flanking paths) can be predicted by measuring the normalized impact sound level of the composite floor and calculating the TL of the composite floor from Eq. 17 as

$$TL_{pot} = 43 - L_n + 30 \log f - 10 \log \sigma_{rad}. \qquad (21)$$

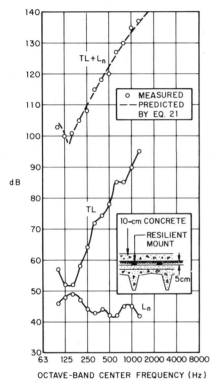

dB

FIG. 3. Measured (Ref. 12) sound transmission loss (TL) and normalized impact sound level (L_n) and their sum (TL+L_n) of a resonantly reacting floating floor assembly. Dotted curve: TL+L_n predicted by Eq. 21.

Figure 3 shows the measured sound transmission loss, TL, and normalized impact sound level, L_n, as well as their sum for a typical floating floor of the Consolidated Kinetics Corporation.[12] The structural floor consists of 14-in.-deep **T** sections with a 2-in. topping of concrete poured over the assembly and finished. The 4-in.-thick concrete floating slab is supported on type-L isolation pads, spaced 12 in. o.c. (on center), and the airspace is filled with 1.3-lb/ft³ density non-load-bearing glass fiber. The measured and predicted values for the sum are in good agreement, indicating that the precautionary measures taken to eliminate flanking have been successful.

In checking out the performance of floating floors in the field, it is advisable to measure both TL and L_n. The discrepancy between the measured TL and that calculated from Eq. 21 is a direct indication of flanking. By measuring the acceleration level on the wall surfaces in the source and receiving rooms during acoustical and impact excitation, the flanking paths can be immediately identified.

* Paper presented at the 79th Meeting of the Acoustical Society of America, Atlantic City, N. J., 21 April 1970.

[1] M. Heckl and E. J. Rathe, "Relationship between the Transmission Loss and Impact-Noise Isolation of Floor Structures," J. Acoust. Soc. Amer. **35**, 1825–1830 (1963).

[2] L. L. Beranek, Ed., *Noise Reduction* (McGraw–Hill, New York, 1960), p. 297, Fig. 13.7.

[3] P. W. Smith, Jr., "Response and Radiation of Structural Modes Excited by Sound," J. Acoust. Soc. Amer. **34**, 640–647 (1962).

[4] For less rigid structures, like a wooden floor where the effective length of the force pulse cannot be considered small compared with $1/f$, Eqs. 3 and 4 are not valid.

[5] Hammer mass 0.5 kg, free-falling height 4 cm, and repetition rate 10 times/sec.

[6] L. Cremer and M. Heckl, *Körperschall* (Springer, Berlin, 1967).

[7] R. H. Lyon, "Statistical Analysis of Power Injection and Response in Structures and Rooms," J. Acoust. Soc. Amer. **45**, 545–565 (1969).

[8] ISO Recommendation R-140, "Field and Laboratory Measurements of Airborne and Impact Sound Transmission," Ref. No. ISO/R 140-1960 (E) (1960).

[9] "Impact Sound Transmission and Airborne Sound Transmission Loss Test on Hi-Stress Flexicore Slabs," Test Rep. Cedar Knolls Acoust. Lab., Test No. 6612-12 (1966).

[10] H. Cremer and L. Cremer, Frequenz **2**, 61 (1948).

[11] I. L. Vér, "Acoustical and Vibrational Performance of Floating Floors," Bolt Beranek and Newman Inc., TIR No. 72 (Oct. 1969).

[12] Riverbank Acoust. Labs. Test Reps. TL-71-211 (3 May 1971); IN-71-15 (17 Apr. 1971).

AUTHOR CITATION INDEX

SUBJECT INDEX

About the Editor

THOMAS D. NORTHWOOD is head of the Noise and Vibration Section, Division of Building Research, National Research Council of Canada. He has been engaged since 1940 in various aspects of acoustics from underwater sound to seismology, in which he did his postgraduate work in 1950. He has been active in the affairs of the Acoustical Society of America for many years, and is currently an Associate Editor of the Journal of the Acoustical Society of America. He is a long-time member of ASTM Committee E-33 on Environmental Acoustics, and was recently awarded the ASTM Award of Merit. He is the author of about fifty papers, mainly in architectural acoustics.